Natural Fiber Composites for a Sustainable Environment

This book covers all the relevant details of natural fiber composites (NFCs) commencing from the reinforcement side to their fabrication routes, mechanical, thermal, and metallurgical characterizations, machining and joining aspects (conventional, advanced, and hybrid modes), surface analysis (defects and wettability characterization), and so on. It details the reuse, remanufacturing, and recycling of NFCs.

- Gives a complete overview of the mechanical, metallurgical, and tribological aspects of natural fiber composites
- Includes automotive applications and potential of natural fiber composites
- Covers tribo, thermal, and mechanical aspects apart from joining techniques
- Elaborate details of joining, reuse, remanufacturing, and surface modification
- Discusses all aspects of natural fiber composites toward sustainable manufacturing

This book is aimed at graduate students and researchers in sustainable composites, materials science, and polymers.

Advances in Design, Materials and Manufacturing for Sustainability

Series Editors: Jitendra Kumar Katiyar and T V V L N Rao

Conventional and Powder Mixed Electro-Discharge Machining: Biomedical Applications
Edited by Ahmad Majdi Abdul-Rani, Masdi Muhammad, T V V L N Rao, Saeed Rubaiee, Anas Ahmed and Mohd Danish

Micro- and Biofluidics: Structure, Properties, Technology and Spotlight into the Future
Avinash Kumar and Jitendra Kumar Katiyar

Natural Fiber Composites for a Sustainable Environment
Edited by Jibin T. Philip, Jerin Jose and Soney C. George

For more information about this series, please visit: www.routledge.com/Advances-in-Design-Materials-and-Manufacturing-for-Sustainability/book-series/CRCADMMS

Natural Fiber Composites for a Sustainable Environment

Edited by
Jibin T. Philip
Jerin Jose
Soney C. George

CRC Press
Taylor & Francis Group
Boca Raton London New York

CRC Press is an imprint of the
Taylor & Francis Group, an **informa** business

First edition published 2026
by CRC Press
2385 NW Executive Center Drive, Suite 320, Boca Raton FL 33431

and by CRC Press
4 Park Square, Milton Park, Abingdon, Oxon, OX14 4RN

CRC Press is an imprint of Taylor & Francis Group, LLC

ISBN: 978-1-032-52787-1 (hbk)
ISBN: 978-1-032-66532-0 (pbk)
ISBN: 978-1-032-66533-7 (ebk)

DOI: 10.1201/9781032665337

Typeset in Times
by SPi Technologies India Pvt Ltd (Straive)

Contents

List of Figures

List of Tables

About the Editors

Jibin T. Philip is currently serving as a National Post-doctoral Fellow (N-PDF) funded by the Science and Engineering Research Board (SERB), Department of Science and Technology (DST), Govt. of India, in the Department of Mechanical Engineering at the Indian Institute of Science (IISc), Bangalore, Karnataka, India. He is also an Assistant Professor (on lien) in the Department of Mechanical Engineering (Automobile) at Amal Jyothi College of Engineering, Kottayam, Kerala, India. Prior to this, he contributed as a Senior Research Fellow (SRF) for a project funded by SERB-DST, Govt. of India, and completed his Doctor of Philosophy (Ph.D.) in Mechanical Engineering from National Institute of Technology Mizoram, Aizawl, Mizoram, India, in collaboration with National Institute of Technology Calicut, Kerala, India. He acquired the M. Tech and B. Tech degrees from A P J Abdul Kalam Technological University, Thiruvananthapuram, Kerala, India, and Mahatma Gandhi University, Kottayam, Kerala, India, respectively. He has published 20+ international/national journal papers (SCI/Scopus), 28+ international/national conference papers, 4 book chapters, and is the recipient of several awards and accolades at international/national levels, inclusive of those from Institute of Physics Publishing (IOP, UK), The Institution of Engineers India (IEI), Indian National Academy of Engineering (INAE), Indian Society for Technical Education (ISTE), and International Travel Grant (SERB-DST India) for his excellence in research and academics. Philip is the Lead Editor for a book series and three book projects with the International Publisher: Taylor and Francis, USA, in the areas of tribology, solid state processing, and natural fiber composites. He is frequently called upon as a reviewer for many international peer-reviewed journals of publishers, invited speaker for seminars/webinars/FDPs/workshops, and is a member of the Board of Studies (BoS) at Saintgits College of Engineering (Autonomous), Kottayam, Kerala, India. His primary research interests are in the fields of tribology, solid state processing, process-structure-property correlation, advanced manufacturing/machining, surface engineering, and analysis/testing of materials (mechanical, metallurgical, morphological, corrosion, and bio-compatibility properties).

Jerin Jose currently works as a postdoctoral researcher in the Department of Civil Engineering and Architecture, University of Catania, Italy. The area of research is in the "recyclability of polymer materials used in additive manufacturing." He has also completed another postdoctoral fellowship from the University of Salento, Italy. Jerin has completed Ph.D., master's, and bachelor's from Karunya Institute of Technology and Sciences, Tamil Nadu, India. His research interests include biocomposites, machine learning for aerospace structural health monitoring, metal matrix composites, syntactic foams, and agro waste-based reinforcements. He has also been an Assistant Professor at Amal Jyothi College of Engineering (Autonomous), Kerala, India, for the past 2.5 years.

Soney C. George is working as Dean, Research, Amal Jyothi College of Engineering, Kanjirappaly, and Director, Center for Nanoscience Technology. George had completed M.Sc. in Polymer Chemistry with first rank from Mahatma Gandhi University in the year 1992. He took Ph.D. in Polymer Chemistry from Mahatma Gandhi University in 1999. He had successfully completed postdoctoral studies at the University of Blasé Pascal, France, and Inha University, South Korea, in 1999 and 2006, respectively. He had received the Young Scientist award from the Indian Science Congress Association in 2001 and was selected as SERC Visiting Fellow at IIT Chennai in 2001 by the Department of Science and Technology, New Delhi. He also received the Fast Track award for Young Researchers in 2005 from the Department of Science and Technology, New Delhi. He has to his credit more than 110 international publications and 45 chapters for reference books published by International Publishers. He has published 4 textbooks on Engineering Chemistry and edited 12 reference books on various fields such as membranes, nanocomposites, and supercapacitors, published by CRC Press and other reputed journals. His h-index is 34 with a total citation of 4772. Eight PhDs were awarded under his guidance, and he is currently supervising 5 PhD students. He has also guided 112 student projects at B.Tech, M.Tech, and M.Sc. levels.

He is ranked as the top-performing faculty in the professor category based on the feedback from students and the evaluation of the Faculty Appraisal Committee in 2008–2009, 2011–2012, 2013–2014, and 2015–2016. Based on his teaching and research contributions, he received the National Technical Teachers Award 2022, the best teacher's award from the Indian Society for Technical Education 2019, and the Best Researcher Award of the Year 2018 from Kerala Technological University.

List of Contributors

E. Bayraktar
Supmeca, Mechanical and Manufacturing Engineering School
Paris, France

Joseph Selvi Binoj
Department of Mechanical Engineering
College of Engineering, Alasala Colleges
P.O.Box 12666, Dammam-31483, Eastern Province
Kingdom of Saudi Arabia

Brailson Mansingh Bright
Department of Mechanical Engineering
Sri Ramakrishna Engineering College
Coimbatore, India

M. Brito
State University of Campinas
FEM-Dept. of Materials
Campinas, Brazil

R. H. L. da Silva
UTFPR – Federal Technological University of Paraná
Department of Mechanical Engineering
Cornélio Procópio, Brazil

Merin K. Eapen
Research Scholar
School of Energy Materials
Mahatma Gandhi University
Kottayam, India

Bin Fei
School of Clothing and Textile
The Hong Kong Polytechnic University, Kowloon
Hong Kong, SAR, China

J. S. G. Fracaro
Federal Technological University of Parana
Londrina-Brazil

F. Gatamorta
Mechanical Engineering Department
Rio de Janeiro State University (UERJ)
Rio de Janeiro, Brazil

K. Georgy
Postdoctoral Research Scholar
IIT Madras

Soney C. George
Centre for Nanoscience and Technology
Amal Jyothi College of Engineering
Kottayam, India

P. C. Gonçalves
Federal Technological University of Parana
Londrina-Brazil

Indira V.
Department of Chemical Engineering
Amal Jyothi College of Engineering
Kottayam, India

M. J. John
Principal Researcher
Centre for Nanostructures and Advanced Materials
The Council for Scientific and Industrial Research
Pretoria, South Africa

Rosemary Johnson
School of Energy Materials
Mahatma Gandhi University
Kottayam, India

Jerin Jose
Postdoctoral Researcher
University of Salento
Italy

and

Department of Mechanical Engineering
Amal Jyothi College of Engineering
Kottayam, India

Jomon Joy
Institute for Integrated Programmes and Research in Basic Sciences
M.G University
Kottayam, India

V. Kavimani
Karpagam Academy of Higher Education
Coimbatore, India

T. Satish Kumar
Amrita School of Engineering
Coimbatore, India

Rittin Abraham Kurien
Sustainable Materials Laboratory
Department of Materials Science and Engineering
Faculty of Engineering and Industrial Technology
Silpakorn University
Thailand

and

Department of Industrial Pharmacy
Faculty of Pharmacy
Silpakorn University
Thailand

and

Department of Mechanical Engineering
Saintgits College of Engineering (Autonomous)
Kottayam, India

Hanna J. Maria
School of Energy Materials
Mahatma Gandhi University
Kottayam, India

Vishal John Mathai
Department of Mechanical Engineering
Amal Jyothi College of Engineering
Kottayam, India

Mariya Mathew
School of Energy Materials
Mahatma Gandhi University
Kottayam, India

Arya Muraleedharan
TOMS College of Engineering and Polytechnic
Kottayam, India

N. Nomadolo
Centre for Nanostructures and Advanced Materials
The Council for Scientific and Industrial Research
Pretoria, South Africa

Aneesa Padinjakkara
School of Chemical Sciences
Mahatma Gandhi University
Kottayam, India

J. P. Paschoal
FEM-Dept. of Materials
State University of Campinas
Campinas, Brazil

Dr. Jibin T. Philip
National Post-doctoral Fellow (N-PDF) Indian Institute of Science (IISc)
Bangalore, Karnataka, India & Assistant Professor Amal Jyothi College of Engineering (AJCE) Kanjirappally
Kerala, India

Narayanan Rajesh Prabha
Department of Mechanical Engineering
Mohandas College of Engineering & Technology
Thiruvananthapuram, India

Vishnu Prasad
School of Mechanical and Materials Engineering
University College Dublin
Dublin, Ireland

and

Department of Mechanical Engineering Automobile
Amal Jyothi College of Engineering
Kottayam, India

Arun Raphel
Department of Mechanical Engineering
SRM Institute of Science and Technology
Chennai, India

Bismark Sarkodie
Department of Industrial Art
Kwame Nkrumah University of Science and Technology, Private Mail Bag
Kumasi, Ghana

Akash Paul Savio
Department of Mechanical Engineering
Viswajyothi College of Engineering and Technology
Vazhakulam, India

Sunil S. Suresh
Independent Researcher

Surej Rajan C.
Department of Chemical Engineering
Amal Jyothi College of Engineering
Kottayam, India

Shana Shaji
Caritas Hospital, Kottayam
Kerala, India

Benjamin Tawiah
Department of Industrial Art (Fashion and Textiles)
Kwame Nkrumah University of Science and Technology
PMB, Kumasi, Ghana

Karen Ter-Zakaryan
TEPOFOL Ltd.
Moscow, Russia

Titus Thankachan
Karpagam College of Engineering
Coimbatore, India

Manuel Thomas
Department of Mechanical Engineering
Amal Jyothi College of Engineering
Kanjirapally, India

Sabu Thomas
International and Interuniversity Centre for Nanoscience and Nanotechnology
Mahatma Gandhi University
Kottayam, India

Teena Varghese
Department of Basic Science
Indian Excellent School
Sharjah

Amal Alliyankal Vijayakumar
Department of Engineering for Innovation
University of Salento
Italy

Preface

In the pursuit of advanced sustainable development, the role of materials that balance the necessary performances with ecological responsibility has become gradually demanding. Natural fiber composites (NFCs), which incorporate fibers from different renewable resources with either synthetic or natural matrices, signify a noteworthy step to attaining this stability. The present book, *Natural Fiber Composites for Sustainable Development*, is a comprehensive collection of scientific writings that cover several facets of these innovative materials.

NFCs are promising alternatives to traditional composites, effectively addressing the negative impacts that hinder global sustainability efforts, such as resource depletion, environmental degradation, and waste management. The adaptability of natural fibers, such as sisal, hemp, flax, and jute, provides a diverse range of mechanical properties and applications, making them appropriate for numerous industries, including automotive, aerospace, consumer goods, and packaging. The inspiration behind this book was nurtured from the rising need to motivate, instruct, and inform students, engineers, researchers, and industry experts about the advantages and challenges associated with NFCs. By offering a detailed overview of current research related to the mechanical, tribological, and thermal advancements of NFCs, this book aims at connecting the gap between academic research and industrial applications.

This book is arranged to direct the reader through a logical progression of topics, starting with the fundamental principles of composite materials and the specific characteristics of natural fibers. Subsequent chapters describe the processing techniques and mechanical, tribological, and thermal characteristics of NFCs. Superior consideration is given to the environmental impacts and recyclability, underscoring the sustainability aspect of these materials. We explore a few innovative applications, showcasing how natural fiber composites are being integrated into modern engineering solutions. This showcases the everyday benefits, such as cost efficiency, weight reduction, and decreased environmental footprints, that are decisive for industries targeting to improve their sustainability credentials. A substantial portion of the book is devoted to debating the challenges that must be addressed to fully realize the potential of NFCs. Problems such as compatibility with matrices and changeability in fiber properties are examined. By understanding these trials, we hope to instigate additional development efforts and research intended to overcome these hurdles.

Sharing of knowledge and collaborations are critical in advancing the field of NFCs. This book tries to compile the contributions from leading specialists and researchers in the subject specialism from different parts of the world. Their collective proficiency offers rich content for readers looking to deepen their knowledge of this continuously expanding field. As the universal scientific community continues to search for sustainable solutions, the role of material science cannot be overstated. NFCs are not just a methodological innovation; they represent a pivotal shift to more sustainable and responsible practices in engineering and manufacturing. The contents in this book are intended to train students and budding researchers with a motivation to contribute to this change.

Finally, this scholarly work, *Natural Fiber Composites for Sustainable Development*, is a lot more than just an assembly of recent information. It is an invitation for action toward adopting and progressing in sustainable materials. We have currently reached a critical juncture where the selections we make at the moment will mold the future of our mother planet, Earth. Upon the application and progress of natural fiber composites, we hope we can take noteworthy steps in the ladder toward a resilient world. We anticipate that this book will assist as a treasured reference for researchers, PG and UG students, and industrial veterans alike. Whether you are a beginner or an experienced scientist, the wide-ranging coverage of topics included in this scientific text on NFCs will offer a rock-hard foundation and inspire further explorations and innovations toward better sustainability.

We, the three editors of the book, would like to wholeheartedly express our appreciation and gratitude to all the authors, the reviewers, and the direct and indirect supporters who have made this book project a reality. Their sincerity, timely response, and hard work have culminated in a resource that will make a helpful impact in the field of sustainable materials. A special word of gratitude to the editorial team members of CRC Press for their timely support and guidance in this venture. The editors are also obliged to the support and motivation extended by the Amal Jyothi management and family members. We request you to explore the seventeen chapters that trail and join us in the journey, aiming at a sustainable future.

Jibin T. Philip
Jerin Jose
Soney C. George

1 Introduction to Natural Fiber Composites (NFCs)

Jibin T. Philip, Jerin Jose, and Soney C. George

1.1 BACKGROUND

Natural fibers, both animal- and plant-based, play significant roles in various industrial applications, including textiles and composite materials. Among animal fibers, silk, wool, angora, and camel hair are highly valued for their unique properties, while cotton, flax, hemp, and jute dominate in production and use in textiles. In the realm of composite materials, natural fibers are increasingly being utilized, forming what are known as bio-composites. These composites typically consist of natural fiber embedded in a matrix composed of synthetic polymers. Natural fibers offer several advantages over synthetic reinforcing fibers, including biodegradability, renewability, low density, and lower processing costs. However, challenges such as poor strength compared to synthetic fibers and difficulty in achieving strong adhesion between fibers and the matrix exist. Nanocomposites, which involve fillers at the nanometer length scale, are desirable for their mechanical properties. Biological nanocomposites, such as bone and abalone shell, serve as examples of the remarkable mechanical properties achievable with nanosized elements. These structural proteins are processed into nanosized fibers and incorporated into synthetic matrices. The interface between the reinforcement and matrix is crucial for achieving favorable mechanical properties in nanocomposites. Fiber dispersity, aggregation tendencies, and additional processing of collagen sources to reach adequate purity are challenges in natural fiber nanocomposites. Natural fibers also hold promise as biomaterials, with chitin being particularly notable. Medical applications for chitin-based materials include medication delivery, tissue regeneration, and water pollution cleanup. The use of naturally synthesized proteins in medical implants has the potential to promote tissue integration or degradation, depending on the body's response to the material. Overall, natural fibers offer diverse opportunities in industrial and biomedical applications, showcasing their versatility and potential for sustainable innovation.

Traditionally, the matrices utilized in natural fiber composites (NFCs) encompass a variety of polymeric materials, often derived from renewable origins such as bio-resins or thermoplastics. These matrices serve as the binding medium that encapsulates and reinforces the natural fibers, providing structural integrity and imparting specific functionalities to the resulting composite materials. Bio-resins, derived from renewable biomass origins like vegetable oils or lignocellulosic materials, represent a prominent category of matrix materials in NFCs. These resins, including epoxies, polyesters, and polyurethanes, offer significant advantages such as biodegradability, reduced carbon footprint, and decreased dependence on petrochemical derivatives. Similarly, thermoplastic polymers are another common choice as matrix materials in NFCs due to their ease of processing, recyclability, and compatibility with various fiber types. Polylactic acid (PLA), polyethylene, polypropylene, and polyethylene terephthalate are among the thermoplastics frequently employed as matrices in NFC. The manufacturing process of NFCs involves impregnating natural fibers with the chosen polymeric matrix, followed by consolidation through processes such as injection molding, compression molding, or filament winding. In the process of construction, the matrix material interacts with the fibers in addition to binding them together, promoting reinforcement and load transfer inside the composite structure. Chemically speaking, the main components of natural

DOI: 10.1201/9781032665337-1

fibers are cellulose, hemicellulose, and lignin, in different amounts depending on the source of the fiber. The way these components interact with the matrix material determines how well NFCs perform overall, impacting things like adhesion, moisture absorption, and thermal stability [1].

In recent years, utilization of NFCs has gained traction across diverse sectors, including automotive, aerospace, construction, and consumer goods. In the automotive industry, NFCs are used to manufacture interior parts, such as door panels, dashboard trims, and seat backs, offering lightweight alternatives to traditional materials while reducing vehicle weight and improving fuel efficiency. Additionally, NFCs are increasingly being explored for exterior applications, such as body panels and structural components, due to their elevated strength-to-weight ratios and impact resistance. NFCs are used in the construction industry for a variety of purposes, including as wall panels, flooring, insulation, and roofing materials. These composites have superior acoustic and thermal insulation qualities, enhancing indoor comfort and energy efficiency in both commercial and residential structures. Moreover, NFCs are being investigated for use in sustainable building practices, such as green roofs and eco-friendly construction materials, to minimize environmental impact and promote sustainable development. In aerospace applications, NFCs are being developed for lightweight structural components, such as aircraft interiors, cabin fittings, and non-critical aircraft parts. When compared to conventional materials like aluminum and fiberglass, these composites offer significant weight savings, which lowers greenhouse gas emissions and saves fuel. Furthermore, NFCs can be tailored to meet specific performance requirements, including stiffness, strength, and fire resistance, making them suitable for aerospace applications where stringent performance standards must be met.

1.2 HISTORY OF NFCs

The history of NFC traces back to ancient civilizations, where humans utilized various plant and animal fibers for many uses. From the earliest recorded history, natural fibers have played a crucial role in meeting humanity's material needs, providing clothing, shelter, tools, and even artwork. The advent of agriculture marked a significant turning point, as humans began to cultivate plants like flax, cotton, and hemp for their fibers. These fibers were then processed and woven into textiles, laying the foundation for the development of NFCs. One of the earliest examples of NFCs can be found in ancient Egypt, where flax fibers were used as reinforcement to mud bricks in construction. The addition of flax fibers improved the strength and durability of the bricks, allowing for the construction of larger and more stable structures. Similarly, in ancient China, silk fibers were combined with other materials to create composite materials used in armor, weaponry, and even musical instruments.

Throughout history, NFCs have been used in a wide range of applications, from everyday items like baskets and furniture to more specialized uses in transportation, construction, and medicine. In medieval Europe, for example, wooden ships were reinforced with natural fibers such as hemp to increase their strength and resilience against the elements. Likewise, in ancient India, coconut fibers were used to reinforce concrete structures, providing added strength and durability. The Industrial Revolution marked a noteworthy milestone in the development of NFCs, as advances in manufacturing and technology enabled the mass production of textiles and other composite materials. During this time, natural fibers like cotton and wool became integral components of the burgeoning textile industry, fueling economic growth and industrialization in Europe and North America. Additionally, innovations in agriculture and processing techniques led to the widespread adoption of NFCs in various industrial sectors. In the 20th century, the advent of synthetic fibers like nylon and polyester posed a significant challenge to the dominance of natural fibers in the market. However, concerns over environmental sustainability and resource depletion have renewed interest in NFCs in recent years. Natural fibers offer several benefits over synthetic alternatives, including renewability, biodegradability, and lower environmental impact. These qualities make them increasingly attractive for use in sustainable manufacturing and green technologies. Today, NFCs are used in a wide range of applications, from building materials and automotive components to consumer goods and medical devices. In the automotive sector, for example, NFCs are being used as substitutes for traditional materials like fiberglass and carbon fiber in interior panels, seat backs, and door panels. These composite materials are more affordable, lighter, and ecologically friendly while maintaining comparable strength and durability qualities.

1.3 CLASSIFICATION OF NATURAL FIBERS USED AS REINFORCEMENT

Natural fibers are commonly reinforced and incorporated into various polymer-based matrices to produce NFCs, which offer a blend of renewable sourcing, enhanced mechanical properties, and reduced environmental impact. Among the majorly used matrix materials, bio-based resins and thermoplastics stand out as primary choices due to their compatibility with natural fibers and advantageous properties. A sustainable matrix option is represented by bio-resins, which are produced from renewable biomass feedstocks like vegetable oils or lignocellulosic materials. These resins, including epoxies, polyesters, and polyurethanes, not only provide excellent adhesion to natural fibers but also offer biodegradability and a diminished reliance on fossil fuels. Thermoplastic polymers, such as polypropylene, polyethylene, PLA, and polyethylene terephthalate, are widely used as matrix materials in NFCs. These thermoplastics offer ease of processing, recyclability, and good mechanical properties, making them perfect candidates for various applications. The incorporation of natural fibers into polymer-based matrices enhances the overall performance of composites by providing reinforcement, stiffness, and improved sustainability. Through meticulous selection and optimization of NF reinforcement.

The categorization of natural fibers stands as a pivotal element in comprehending the diverse array of materials sourced from plants and animals, which have played crucial roles in human civilization. These fibers form the backbone of textile production, offering indispensable attributes like durability, pliability, and breathability. Through a systematic classification, these fibers are organized according to their origin, structure, and inherent properties, enabling a thorough examination of their applicability across various industries. Figure 1.1 provides a visual representation of this classification, illustrating the distinct categories that encompass plant-based, animal-based, and mineral-based fibers. By segmenting natural fibers into distinct categories, researchers and practitioners can gain deeper insights into their unique characteristics, strengths, and limitations, facilitating informed decisions in material selection for specific manufacturing purposes. This introductory exploration into the classification of natural fibers embarks on a journey to uncover the complexities of plant-based and animal-based fibers, shedding light on the abundant resources that have sustained human endeavors for generations [2, 3].

1.4 ADVANTAGES OF NFCs

NFCs offer numerous merits, making them a compelling candidate in industries. One major advantage is its environmental sustainability. They are biodegradable and extracted from renewable sources, thus reducing dependence on fossil fuels and thereby curtailing the carbon footprints. These materials regularly exhibit superior mechanical properties such as good strength-to-weight ratios and stiffness. These characteristics are analogous to the traditional synthetic fibers like carbon fibers. The lightweight nature of these composites underlines the improved fuel efficiency in aerospace and automotive applications, and subsequently lesser greenhouse gas emissions. Another advantage of these materials is their lower abrasiveness, which encompasses the lifespan of processing equipment and condenses the maintenance costs. Additionally, NFCs deliver cost aids due to the fairly low price of natural fibers compared to their counterparts. These composites also provide better thermal and acoustic insulation properties, making them appropriate for construction and building industries. The adaptability of natural fibers, including hemp, sisal, jute, and flax, allows for a wide variety of applications and easy processing with traditional manufacturing procedures. NFCs are less toxic and

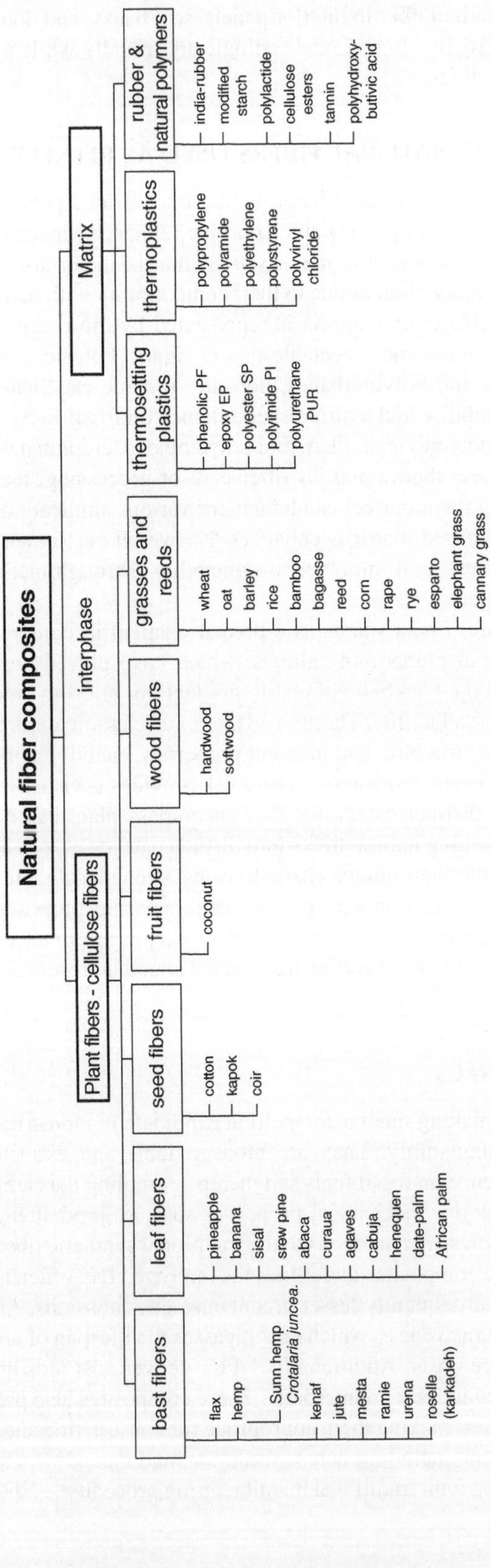

FIGURE 1.1 Classification of natural fibers and matrix materials [2]. Reproduced with permission from [2]. Copyright Taylor and Francis (2004).

safe to handle, thereby contributing to a healthier working atmosphere. They have better resistance to corrosion and fatigue, which improves the longevity of the fabricated parts. Also, NFCs support agricultural economies by giving a better market for natural fiber crops, encouraging rural development. The aesthetic appearance of natural fibers, along with their exceptional colors and textures, is highly cherished in consumer products, contributing an eco-friendly and distinctive substitute. NFCs also help in reducing waste, as these agricultural by-products can be successfully employed. Developments in surface modification and treatment methods have amended the bonding between polymer matrices and natural fibers, additionally enhancing the composite performance. NFCs' capability for reusability and recycling aligns with the circular economic ideologies, supporting justifiable development goals. These composites also show low density, which is beneficial for weight-sensitive opportunities. They also provide better energy absorption competencies, which are advantageous in impact-resistant applications. The custom of NFCs can also open ways to enhanced air quality as they release fewer volatile organic compounds (VOCs) than the synthetic composites. As the industries try to implement better eco-friendly materials, the merits of NFCs spot them as an integral candidate in the future of sustainable manufacturing.

1.5 CHALLENGES OF NFCs

NFCs, although offering abundant advantages, also face numerous challenges that hinder their wider application and adoption. One important problem is the inherent randomness in the properties of natural fibers, which can affect the uniformity and predictability of composite performance. Factors such as soil conditions, climate, and gathering techniques impact the fiber quality, leading to uneven mechanical properties. Moisture absorption is the other main challenge, as natural fibers can absorb water, leading to swelling followed by degradation, and compromising the mechanical properties over time. This makes NFCs less appropriate for applications in wet or humid environments without proper treatment. The compatibility between polymer matrices and natural fibers can be challenging, as it often necessitates chemical treatments or surface adjustments to advance adhesion. These added processing phases can raise the production complexity and costs. Furthermore, these natural fibers usually have minor thermal stability compared to the synthetic fibers, restraining their usage in high-temperature applications. The processing of natural fibers can also open doors to problems like poor dispersion and fiber breakage within the polymer matrix, which can deteriorate the properties of the composite material.

Reusability and recycling of NFCs, even though possible, are not as straightforward as with the synthetic composites. It is because the degradation of natural fibers during the lifespan of the composite can thwart its recycling efforts. Furthermore, the manufacture of NFCs frequently comprises lower manufacturing capacities compared to the traditional composites, leading to less competitive pricing and higher costs. The agricultural nature of natural fibers means that their accessibility may be seasonal and dependent on the agricultural yields, familiarizing supply chain uncertainties.

Biodegradability, though an advantage, can also pose challenges in applications necessitating long-term stability and durability. They may degrade faster than their synthetic substitutes, which can edge their lifespan in some products. Likewise, natural fibers can be prone to microbial/fungal attack, demanding the use of additives or preservative treatments that can contradict some of the environmental benefits. Standardization is a hurdle, as the natural fiber industry lacks the equal level of recognized quality control procedures found in synthetic fiber production. This can make it problematic for producers to safeguard the performance and reliability of NFC products. Moreover, the environmental aids of NFCs can be offset by the usage of synthetic resins and the energy-intensive fabrication methods essential to produce them. To realize the eco-friendly potential of NFCs completely, the advancement of bio-based or biodegradable resins is crucial. The transition to NFCs may demand substantial initial investments in novel equipment and training, which can be an obstruction for some manufacturers.

1.6 OVERVIEW OF THE CHAPTERS

This book, titled *Natural Fiber Composites for a Sustainable Environment*, is a scientific compilation of different aspects related to NFCs. The contents of this scholarly work, in the form of chapters, are a strategic culmination of efforts and contributions from various authors who are experts in the subject field. It is the valuable outcome of a year-long effort from 40+ authors, 3 editors, and the publisher (i.e., CRC Press, Taylor and Francis) working simultaneously. The book is organized into 17 distinct chapters, which individually have specific emphasis on various dimensions related to the production, analysis, testing, characterization, and applications of NFCs.

The list appended below provides a gist of the focus areas covered by each of the chapters in combination, forming a unified whole, that is, this book.

- **Chapter 1** Introduction
- **Chapter 2** Materials and reinforcements
- **Chapter 3** Procedures and techniques for the fabrication
- **Chapter 4** Surface analysis and characterizations of chemically modified NFCs
- **Chapter 5** Mechanical and metallurgical characterization
- **Chapter 6** Tribological aspects
- **Chapter 7** Thermal analysis and characterizations
- **Chapter 8** Conventional machining techniques for processing
- **Chapter 9** Advanced and hybrid route of machining procedures
- **Chapter 10** Joining of fiber reinforced composite
- **Chapter 11** Recycling and reuse possibilities
- **Chapter 12** Hybrid NFCs, their characterization and properties
- **Chapter 13** Surface-modified NFCs; their characterization and properties
- **Chapter 14** Natural reinforcements for composites
- **Chapter 15** Automotive engineering applications
- **Chapter 16** Synthetic fibers
- **Chapter 17** Future scope and recent trends

REFERENCES

1. Saxena, M.; Pappu, A.; Sharma, A.; Haque, R.; Wankhede, S. Composite Materials from Natural Resources: Recent Trends and Future Potentials. In *Advances in Composite Materials - Analysis of Natural and Man-Made Materials*; Tesinova, P., Ed.; InTech, 2011. https://doi.org/10.5772/18264
2. Kozlowski, R.; Wladyka-Przybylak, M.; Helwig, M.; Kurzydłoski, K. Composites Based on Lignocellulosic Raw Materials. *Molecular Crystals and Liquid Crystals*, 2004, *418* (1), 131–151. _https://doi.org/10.1080/15421400490479217
3. Thapliyal, D.; Verma, S.; Sen, P.; Kumar, R.; Thakur, A.; Tiwari, A. K.; Singh, D.; Verros, G. D.; Arya, R. K. Natural Fibers Composites: Origin, Importance, Consumption Pattern, and Challenges. *Journal of Composites Science*, 2023, 7 (12), 506. https://doi.org/10.3390/jcs7120506

2 Materials and Reinforcements for Sustainable Manufacturing of NFCs

J. P. Paschoal, M. Brito, F. Gatamorta, E. Bayraktar, and R. H. L. da Silva

2.1 INTRODUCTION

Materials research and development must align with current needs, and the manufacturing processes also need to observe the resources and the impacts of their use. Among these materials, composites are a class of important materials that combine different properties in the same product. New composites have natural fibers as the main focus of study, and natural fiber composites (NFCs) receive significant investment support for research and production. The investment interest is justified because these composites align with the modern needs and requirements of products, observing the sustainability characteristics needed today. Natural fibers significantly reduce the cost of processing and transport, reducing CO_2 emissions. Besides this, natural fiber is an important source of income in regions where the only possible economic activity is exploiting this type of raw material. The processing of natural fibers reaches around 60 million families directly involved in the production, and another 200 million in the production chain, providing income to almost 4% of the world's population (Townsend, 2020).

Climate change, water depletion, air pollution, and plastic pollution have shown the need for change in production processes. However, synthetic fibers are still the most popular choice as reinforcement to composites due to their superior physical and mechanical properties. Polymers are used as matrices because of their higher adhesive properties and chemical, mechanical, and physical characteristics. Despite their qualities, these materials are not biodegradable and can stay in the environment for hundreds of years (Karim et al., 2021).

Thinking on this issue, researchers and innovators work to reduce the environmental impacts of such compounds. The composites industry tends to follow the global requirement and move toward manufacturing more sustainable composites, thus trying to use recyclable, biodegradable, and sustainable materials. Besides, the environmental impacts of these natural fibers are significantly lower than synthetic fibers (Faris et al., 2015).

The total fiber produced in 2018 was 110 million tons, and only 30% comprised natural fibers. Despite the growth shown in production, the percentage values have decreased over the years. Natural fibers represented about 80% of total fiber production in the 1960s, and until the year 1900, natural fibers were the only type of fiber produced. Considering 2018, of all the natural fibers produced, cotton accounted for approximately 26 million tons. In contrast, non-cellulosic (fibers made from petroleum molecules) represent almost 80 million tons, and cellulosic and other natural fibers are less than 10 million tons. The cotton value represents just over 23% of the total and, like the other natural fibers, its percentage use has decreased over the years. The decline happens even though natural fiber has shown remarkable growth in the last 60 years, quadrupling its production. The other natural fibers with high-production examples are jute, with 3 million tons, and wool and

DOI: 10.1201/9781032665337-2

coconut fiber, with 1 million tons each. The remaining fibers summed together another 1.5 million tons. Concerning production regions, natural fibers are cultivated in about 100 countries, occupying 3% of arable areas worldwide. Finally, world trade in natural fibers is worth 60 billion dollars (FAOSTAT Statistics).

Natural fibers have the most diverse origins and can be obtained from animals, vegetables, or minerals. Natural fibers of animal origin are obtained from the animal's fur or its secretion, such as silk, and those of vegetable origin is obtained from parts of the plant, such as the seed, stem, leaves, and flowers. Those of mineral origin are extracted in fibers from the mineral stone. For its use to show growth, further development in its production is necessary. In terms of technology, the development may involve improving how fibers are obtained or genetic engineering in their source or application. Furthermore, mechanization applied to getting natural fibers began in the 19th century, initiating innovations in the planting, harvesting, and processing sector. Currently, the development of machines and equipment is presented at fairs and involves high technology, including the use of GPS monitoring and traceability drains. In bioengineering, genetic evolution is a reality with matrix modification to improve the yield of the plant and the fiber itself.

2.1.1 Considerations about Natural Fiber Composites Sustainable Manufacturing

The knowledge growth about the damage caused by synthetic materials has led to the development of eco-friendly materials (Thyavihalli et al., 2019). Currently, using natural fibers in engineering applications is typically recognized as a sustainable practice. Such green composites receive attention due to their potential to reduce environmental impact, especially in transportation, where lighter materials imply fuel savings (Baillie & Feinblatt, 2013). Thus, natural fibers have proven to be better substitutes for synthetic fibers, generally applied in low-loading and energy-conserving circumstances. Despite this, for example, there are studies on natural fibers reinforced with ceramic filaments, synthetic fibers, and polymers for high-loading applications (Singh et al., 2023).

Synthetic fiber composites' complete replacement with natural fiber composites (NFCs) is not possible due to the superior properties of synthetic fibers in several aspects (Arockiam et al., 2018). Environmental factors are among the most significant concerns in manufacturing composites today (Singh et al., 2023). Indeed, there is great potential and prospects for natural fiber composites, with several material- and processing-related challenges to be addressed, aiming to achieve long-term stability and performance (Balla et al., 2019).

NFCs have been highlighted not only in the research field, and the demand for commercial use has increased over the past few years. In this regard, the automotive sector is of great interest among the various industries that seek the replacement of synthetic fiber composites (Thyavihalli et al., 2019). It is due to NFCs' characteristics such as weight reduction (Jaafar et al., 2019), which implies reducing emissions in the vehicle use phase, natural fibers' low cost (Mansor et al., 2018), and reducing the embodied energy of these vehicles. In the aeronautical industry, the use of ecological composite materials, replacing aluminum structures, has already demonstrated its potential through life cycle assessments (LCAs) and the perspective is that these composites will contribute to the reduction of 15–20% of greenhouse gas emissions by 2050 (Khalid et al., 2021).

Such bio-based composites offer advantages to users and the environment. They are renewable and can be biodegradable and recyclable, reduce energy consumption, and pose less risk to human and animal health (Li et al., 2020; Mansor et al., 2018). In addition to these advantages, several other benefits related to NFCs can be pointed out and achieved, such as (Balla et al., 2019):

- The neutrality of greenhouse gas emissions.
- Abundant availability of natural fibers.
- Less harmful resulting waste after energy recovery by incineration.
- Reduction of the total production cost concerning polymeric composites.
- Safer manufacturing processes and greater occupational safety.

- Disposal routes being less aggressive to the environment.
- Due to their intrinsic characteristics, natural fibers bringing less damage to the equipment during manufacturing.
- Agriculture waste generating income and employment opportunities.

However, despite the advantages of NFCs, the planting/extraction, manufacturing, use, and subsequent disposal processes must consider the precepts of sustainability. Therefore, being better does not necessarily mean being the best, and there is potential for research and technology development to improve this type of product's sustainability.

When considering the environmental aspect, it is necessary to point out that renewable or natural-origin products do not necessarily imply a smaller environmental or sustainability impact (Le Duigou et al., 2012). The impacts can be even more significant than the synthetical equivalent material for specific categories (Ita-Nagy et al., 2020). In addition, the use of natural fibers cannot be restricted to species already identified, and new sources of bio-renewable materials must be sought, preventing the extinction of certain species, and disfavoring a phenomenon similar to what happened with some species of plants cultivated by humans (Vinod et al., 2020).

Meanwhile, regarding sustainability, NFCs' assessments should consider the environmental, economic, and social impacts. However, research on green composites tends to focus on sustainability's environmental and economic aspects, leaving aside the social aspect. Therefore, the production and use of social impacts must be considered. NFCs offer great potential as a waste management or poverty reduction tool, especially in developing communities that depend on exploiting or planting natural fibers (Baillie & Feinblatt, 2013).

Concerns with the communities involved in planting and producing different fiber types, the use of fertilizers, the chemical treatments necessary to improve properties, the recyclability and biodegradability of NFCs, the reduction of greenhouse gas emissions and eutrophication, for example, must be precept since there are some risks involved in the production of NFCs. Therefore, in the environmental aspect, the entire process of obtaining NFCs, from planting, extraction and fiber treatment, manufacturing, supply chain, use, and disposal, must be considered in a LCA from cradle to cradle.

For example, the emission of nitrogen and phosphorus through the use of fertilizers, the need for large arable areas, and the possible deterioration of ecosystems (Duflou et al., 2012) with the use of pesticides (Mansor et al., 2018) are examples of risks to the environment due to NFCs. There is also the issue of chemical treatments necessary to properly obtain natural fibers and their composites, which use elements such as sodium hydroxide and acetic acid (Thyavihalli et al., 2019). These chemical elements are known to pose risks to the environment and human health, and their impacts are often left aside in a life cycle assessment. Vigneshwaran et al. (2020) state that it was not possible to find literature information on disposal techniques for chemical products used in the treatment of fibers and their impact on the environment, which demonstrates a gap in the analysis of this type of material. Nevertheless, Mansor et al. (2018) consider it to be a fact that NFCs contain smaller amounts of hazardous materials when compared to synthetic fibers.

Consequently, the simple natural fibers used to obtain composites in engineering are insufficient for the material to become sustainable. Attention is needed throughout the product's life cycle, from planting and/or extracting the fiber to taking care of the sustainable characteristics of the composite matrix, passing through the manufacture, use, and disposal of the component produced. Kenaf (Mansor et al., 2018) is a good example, which is considered a fully biodegradable fiber without the need for chemicals and is also capable of absorbing nitrogen and phosphorus from the soil, which tends to reduce eutrophication. Therefore, an alternative may be to encourage the use of fibers of this type or research that develops such characteristics in other kinds of natural fibers.

Another issue is that natural fiber composites may not escape greenwashing. Thus, the created materials receive the green label without the necessary evidence, considering the product's life cycle (Baillie & Feinblatt, 2013). Elseify et al. (2021) point out that greenwashing and false sustainability are being used extensively for products, services, and companies that benefit from the growth of the eco-conscious consumer. Therefore, although convenient, the term "green" may not be adequate and much less sustainable. Many natural fibers depend on virgin plastics in their composition for reinforcement, fertilizers, and chemical products. They are not recyclable after production, etc. Indeed, there are good examples, and natural fiber composites have less impact in several categories when compared to synthetic fibers. Thereby, there is a promising study field for NFCs in situations with considerable impact.

There is no one-size-fits-all solution, which requires attention and sustainable concerns throughout the product's life cycle. Other relevant issues Elseify et al. (2021) addressed were that LCA prepared for natural fibers and their products require allocation since they usually generate by-products, in addition to dust and shivers. Furthermore, the highest emissions are linked to the use of fertilizers for natural fibers, in which organic fertilizers produce smaller consequences in the different impact categories (acidification, ecotoxicity, human toxicity, eutrophication, climate change, etc.). However, mineral fertilizers are necessary to produce certain types of natural fibers.

NFCs' recycling is one of the main alternatives at the end of the life cycle (Choudhary et al., 2023). Depending on the matrix type, it can be performed mechanically, chemically, or biologically. However, not all NFCs can be recycled, considering the elements that compose them and their manufacturing process. The recycling process of reinforced natural fiber composites needs research, and it is not well established, despite the European Union and some Asian countries starting to demand that a significant percentage of vehicle components must be recycled, including the parts manufactured in NFCs (Elseify et al., 2021).

According to Li et al. (2020), another possibility at the NFCs' end of the life cycle is sending biomass to landfills (landfilling of biomass), which sequesters carbon and reduces greenhouse gas emissions. The option of biomass incineration could recover energy and partially replace the energy produced by fossil fuels as an alternative route for greenhouse gas reduction. For example, Joshi et al. (2004) state that theoretically, the incineration of biomass from natural fiber composites can even reduce the carbon released into the atmosphere since the plant growth process sequesters carbon from the environment.

However, studies (Duflou et al., 2012; Ita-Nagy et al., 2020) indicate that concern should not be focused only on greenhouse gas emissions. It is also necessary to consider that recycling and recovery strategies are at the lowest levels of circularity, consuming more resources and producing more significant environmental pressure (Potting et al., 2017). Therefore, strategies for using and manufacturing smarter products or those extending their useful life are more appropriate when considering sustainability. Furthermore, improving the NFCs' recyclability and biodegradability are important features and should be seen as a purpose.

Even though biodegradability is a significant characteristic of NFCs, the degradation mechanisms of these composites have been little studied, and there is a long way to go for the full knowledge of these materials' biodegradation characteristics; an attempt is focused on modifying the matrix and its reinforcements (Vigneshwaran et al., 2020).

Finally, perspectives indicate that natural fibers will prevail over synthetic fibers (glass, carbon, etc.) in the future because of their low cost and environmental safety. However, the use of polymers and synthetic composites is growing at alarming rates, and only a small percentage of everything produced is recycled. Developing biocomposite production techniques that improve their physical, mechanical, and thermal properties, increasing the possibility of NFCs being substitutes for synthetic fiber composites in engineering applications, is also essential (Vinod et al., 2020).

2.2 MATERIALS AND REINFORCEMENTS

2.2.1 Natural Fibers from Vegetal Sources as NFC Reinforcements

The use of natural fibers as composite reinforcement emerges as a solution for replacing synthetic materials with undetermined end-of-life, contributing to environmental issues. The aim of disseminating their application is to replace traditional industrial materials, although there must be a comprehensive understanding of how its features can reinforce composites.

It is well known that intrinsic properties like the morphology, length, defects, and orientation of fibers modify the performance of composites (Bellil et al., 2023). Nevertheless, these factors are not enough to completely comprehend the entire scenario. A concern when choosing natural fibers as matrix reinforcements from vegetal origin lies in their high hydrophilicity, which causes dimensional instability and low interfacial properties between fiber and polymer matrixes (Bledzki et al., 2010). Besides that, stitched fibers also improve the mechanical performance of fiber-reinforced composites as their stitching patterns (Gukendran et al., 2022).

These structural problems happen because of the high polarity guaranteed by a large number of hydroxyl and carboxyl groups in the structure of vegetal fibers, making its surface present decreased wettability, resulting in a poor interaction between the polymer's matrix and fibers. Besides that, swelling and shrinkage caused by water absorption and desorption, consequently, change their morphology and result in the appearance of voids in their interfaces, decreasing the NFC's mechanical properties (Téraube et al., 2023). Thus, in most cases, fiber treatment/modification must be applied when producing NFC to increase load transfer between the polymer matrix and fiber reinforcement.

While synthetic fibers possess a homogeneous structure, vegetal fibers are composed of a multi-scale and multi-layer structure, depicted as a composite with elementary fibers glued by pectin to form a single plant fiber, also called technical fiber (Charlet et al., 2010). For example, Figure 2.1, based on the explanation of Li et al. (2017) diagram, describes a sisal fiber, which presents larger diameters than other plant fibers, making it possible to identify transition regions between cell wall layers in the nano-scale. As can be seen, a unique plant fiber is composed of numerous elementary fibers bonded to each other by pectin, comprised of layers surrounding lumens. Nanoindentation tests proved that each layer has a specific thickness, structure, microfibril orientation, and properties (Li et al., 2017). It is important to note that vegetable fibers from the same source can present diverse properties depending on their growth environment and soil, the process of fibers' extraction, age and chemical composition, etc. Thus, the literature is a good starting point when designing NFC, but an in-situ test must be taken. As an example, Table 2.1 presents some properties of several natural fibers.

2.2.1.1 Abaca

Musa textilis is its scientific name, but Abaca, or Manila hemp, is how this plant from the *Musaceae* family is known as (R R Franck, 2005). Native to the Philippines, Abaca crops play a vital role in the country's economy and the global advocacy for environmental protection and forest conservation, according to the Department of Agriculture of the Republic of the Philippines. The fibers' primary origin comes from the plant's pseudostem, which has an average diameter of 15 to 20 cm and reaches up to 6 m in highness (Wong et al., 2002). This fiber is a waste product of banana cultivation, and this feedstock can be obtained without significant investment.

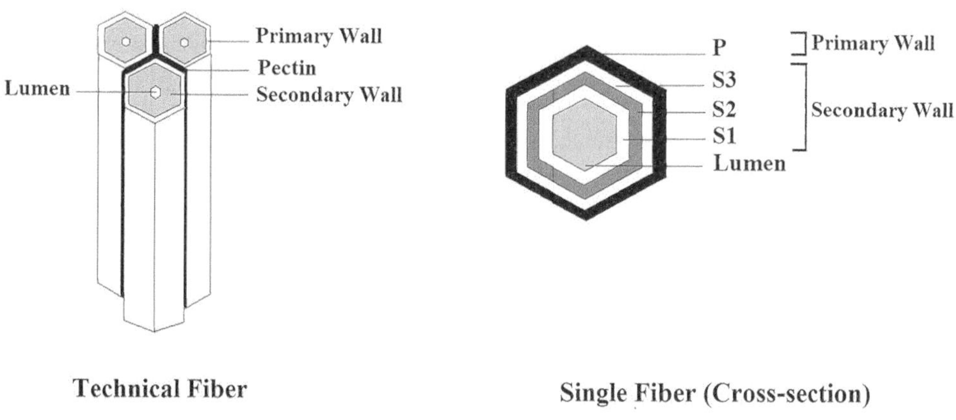

FIGURE 2.1 Diagram for the structure of a single sisal fiber.

TABLE 2.1
Properties of Different Natural Fibers

Fibers	Cellulose (wt%)	Density (g/cm³)	Moisture Content (wt%)	Young's Modulus (GPa)	Tensile Strength (MPa)	Elongation at Break (%)
Abaca	55–62	1.5	14	10–12	400	3–10
Coir	31–42	1.1	10	4–6	106–175	17–47
Cotton	84–91	1.5–1.6	8	6–12	290–800	7–8
Banana	17	1.3	7	7–20	54–754	10.35
Ramie	67–75	1.5	8	61–127	400–900	1–4
Sisal	64–70	1.45	10	9–15	568–640	3–7
Jute	60–70	1.3	11	12–25	393–770	7–9
Pineapple leaf	20–80	1.44	12	34–82	413–1627	0.8–1
Hemp	67	1.48	10	60–70	690	1.6

Source: Khalid (2021).

Formed mainly of cellulose (63–68 wt%), hemicellulose (19–20 wt%), and lignin (5–6 wt%), with a small concentration of pectin (<1 wt%) and fat/wax (<1 wt%) (Mamun et al., 2015), the structure of Abaca fiber is composed of tubular structures, with single fiber bundles (rope-like cellulose structures) embedded in lignin (Mathai, 2005). The mechanical properties of these fibers are related to these tubular structures, specifically in the hydrogen bonding among cellulose molecules, Van der Waals, and hydrogen bond interactions between fibrils (nano-sized strands that originate tubular structure) (Simbaña et al., 2020).

Abaca fibers have outstanding properties, like high tensile strength, resistance to abrasion, and putrefaction, due to their high lignin and cellulose content. In addition, abaca is resistant to ultraviolet (UV) rays/saltwater degradation, and before the use of synthetic fibers, Abaca fibers were widely used by fishermen from where this plant is native. Nowadays, it has its place in the textile and paper industries and other applications. When used for natural fiber composites' fabrication, they can be utilized in different fields like mobility, construction, furniture, and even for developing nanomaterials (Simbaña et al., 2020).

2.2.1.2 Coir

Cocos nucifera, an extensively cultivated plant in tropical countries such as Thailand and India, originates from one of the most used natural fibers, second to FAOSTAT statistics.

Coconut fiber, or coir, is obtained from the mesocarp of coconut. It is a durable and resistant lignocellulosic fiber used in uncountable applications worldwide, such as rope manufacturing, and it is present in many products, such as doormats, brushes, and mattresses. Moreover, there is considerable interest in research using coir as a composite reinforcement due to particular features allied with high abundance and cheap feedstock. Coir has an uncertain origin due to its fruits possessing floating capability and high resistance to seawater, enduring long journeys until they reach sandy beaches (Teixeira et al., 2022).

Due to the high lignin content, these fibers are considered durable and weather resistant. Adeniyi et al. (2019) reported that coconut fiber's chemical composition ranges from 32 to 50% cellulose, 0.15–15% hemicellulose, 30–46% lignin, and about 3–4% pectin, and some deviations attributed to fiber extraction methodology or the coconut plant source.

The single coir fibers are also composed of elemental fibers, with one primary wall and three secondary walls, varying in thickness and structure, with the orientation of cellulose microfibrils into these walls. The walls vary with a helical arrangement of 45° (Sarmin, 2022), which contributes to the capability of being stretched beyond the elastic limit without rupture (Adeniyi et al., 2019). Furthermore, coir is considered to have the highest tensile toughness among natural fibers, which turns this natural fiber into a prominent candidate to reinforce cementitious materials (Ali et al., 2022).

2.2.1.3 Flax

One of the most iconic vegetal fibers in human history, used across ages for luxury and utility applications, comes from a plant known as Flax, whose scientific name is *Linum usitatissimum*. Traditionally flax fibers are used in textile applications, with the first records of application dating from ancient Egypt and Mesopotamia (Duke, 1983).

The physical properties of flax are associated with the elements present in the fibers. These constituents may vary depending on the plant's origin and its original environment's conditions. They generally have between 70% and 85% cellulose, 11% and 20% hemicelluloses, 2% and 12% pectin, and about 2% lignin. In smaller quantities, it also has fat, wax, protein, tannins, dyes, inorganic salts, etc. Regarding the mechanical properties, the modulus of elasticity can vary depending on its composition and the treatments applied to the fiber, with reference values between 6 and 42 GPa (Krishna et al., 2015).

2.2.1.4 Hemp

Hemp (Cannabis sativa) is a multi-use plant that traditionally provides long bast fibers for textiles, ropes, and paper pulp production. These fibers are considered for producing insulation materials or as composite reinforcement nowadays (Gabrion et al., 2022). The plant has singular characteristics, like high adaptability to a wide range of environments, capturing substantial carbon dioxide amounts from the atmosphere, 1.4–1.6 tons of CO_2 per ton (Carus, 2015; Meirheage, 2011), and needs small quantities of water to grow (Schumacher et al., 2020).

Furthermore, hemp has high disease resistance (Amaducci et al., 2015), allied to high-value cannabinoids and flower production (Calzolari et al., 2017). Hemp crops can grow fast, between 9 and 12 weeks (Paul & Chakraborty, 2019) and showed a 12% percentage growth from 2011 to 2020 (FAOSTAT Statistics), the highest growth among natural fiber sources.

Hemp fibers also present the typically vegetal-based fiber's structure, with single fibers composed of primary and secondary walls surrounding a lumen, chemically constituted of cellulose (68–81 wt%), hemicellulose (10.66–22.4 wt%), lignin (3.7–13 wt%), pectin (0.9wt%), waxes (0.8wt%), and ash (0.8wt%) (Ramesh & Bhoopathi, 2022).

Currently, the most productive methodology for obtaining hemp fiber is through mechanical systems based on hammer mills and/or decorticators, which are considered aggressive to the fiber's structural integrity, potentially decreasing their mechanical properties (Grégoire et al., 2020). These processes generate short and medium-length fibers, but in an alternative way, research has been carried out for producing long-length hemp fibers adopting the same route used for textile flax (Musio et al., 2018; Vandepitte et al., 2020). The results showed that it is possible to obtain long-length hemp fibers with superior quality, with tensile stiffness and strength reaching more than 55 GPa and 527 MPa, respectively (Bensadoun et al., 2017).

2.2.1.5 Jute

Jute belongs to the class Corchorus and the family Tiliaceae. Jute is one of the most popular vegetable fibers nowadays. In terms of world production, the fiber extracted from the stem of the jute plant stays behind only the cotton fiber, which reflects jute's highlighted popularity and the title of the most-known natural fiber. Applications involving jute fiber are placed into the automobile and aerospace industries, as well as construction and textile goods (Ramesh et al., 2020). Jute crops can be found in countries like India, Bangladesh, Brazil, Nepal, and Thailand, while the first two are responsible for more than 90% of the world's production. The fame of this fiber goes further with its commercial availability, passing through recyclability, biodegradability, and crop cycle of up to 120 days (Kumar et al., 2022). Moreover, compared to other vegetable fibers, jute presents low production cost, lower density, higher individual fiber length, and comparable mechanical properties (Ahamed et al., 2022). In addition, jute is antistatic and insulating (Hossain et al., 2023).

The jute bast fiber's chemical composition can vary depending on the location at which the plant is cultivated, and on the plant's age in the same way as other plant fibers. However, the typical composition is 45–65% cellulose, 21–26% lignin, 18–21% pentosan, and 0.5–2% Ash (Cottrell et al., 2023).

Besides using long fiber-reinforced polymer composites, which aim to increase the mechanical performance of composites, small particles of jute fiber have been used as reinforcement and filler to polymer matrix composites seeking tribological behavior improvement (Mahakur et al., 2023). Study reports have shown that adding treated jute fibers into polymer matrices improved abrasive wear resistance (Swain & Biswas, 2017) and decreased the friction coefficient (Mahakur et al., 2022).

2.2.1.6 Kapok

Surrounding the seeds of *ceiba pentandra* is found a peculiar material known as Kapok fiber. These cellulosic fibers own a special feature that guarantees a hollowness of 80–90%, the largest among natural fibers, resulting in a very low density of about 0.29 g cm^{-3} (Zheng et al., 2015). This feature can be addressed to their single fiber structure, composed of two major layers with distinct orientations, with a large lumen inside them (Macedo et al., 2020). Therefore, naturally, Kapok fibers are an interesting choice for both reinforcing low-density composites and contributing to a final high thermal/sound insulating feature, like aerogels (Ramaswamy & Rigacci, 2021). Besides, because of their softness, they are traditionally used for stuffing pillows, bedding, and soft toys (Zheng & Wang, 2014).

2.2.1.7 Kenaf

Native to east-central Africa and present in human history since around 4000 BC (Thiruchitrambalam et al., 2012), Kenaf (Hibiscus cannabinus L.), fibers had its demand highlighted during World War II, as a source of fiber for textile and rope production (Ayadi et al., 2017). The Kenaf plants own fast-growing characteristics. They can reach heights of 1.5–4.5 m within four to five months and are highly adaptable to a wide range of weather conditions. Kenaf also has an increased percentage of usable fiber in their stalks, approximately 40%, and double of jute, hemp, or flax, making them an optimal economic choice (Akil et al., 2011; Ramesh 2016).

Kenaf's outstanding mechanical properties, specific modulus, and sound absorption coefficient are comparable to glass fiber (Tamrakar et al., 2021). There is a specificity of Kenaf fiber that distinguish it from other common vegetal fibers, that is, the presence of a wax layer on its single fiber's surface, resulting in a hydrophobic behavior, which deters the absorption of water on its surface (Santos et al., 2017). Therefore, Kenaf's surface treatment is not just dispensable but often causes defects on their surface, which can lead to surface destruction (Zhao et al., 2017). Radzuan et al. (2020) reported that combing Kenaf fibers before hot pressing (190°C) in a polypropylene matrix results in a cost-effective and environmentally friendly NFC with excellent mechanical properties.

2.2.1.8 Ramie

Ramie fiber is a bast fiber extracted from Boehmeria nivea L. (White ramie) or Betula utilis (Green ramie). Its application is dated to the ancient Egyptians in 5000 BC for mummy cloths. However, it was only in the 18th century that ramie textile products reached the Western world through Chinese exports. Ramie is adapted to a wide range of climates, but in humid climates under moderate temperatures, the crops may be harvested six or more times annually (Roy & Lutfar, 2012).

Furthermore, ramie fibers are the longest and most durable bast fiber known. They are mostly used for the production of textile fabrics (Angelini et al., 2015), but they also find their use in resin matrix composites with excellent reinforcement performance. On the other hand, applications are restricted due to ramie fibers' low thermal stability and poor fire resistance (Aristri et al., 2022).

2.2.1.9 Sisal

Brazil is the biggest producer of *Agave Sisalana* on earth, or Sisal as commonly known, despite its origin in Central America. The fibers can be extracted from its leaves, which grow up to 1.5–2 meters (Devaraju & Harikumar, 2019).

Sisal fiber can be widely found in industrial and research applications. As an example, for paper manufacturing, fibers with inferior quality and high contents of cellulose and hemicellulose are the best choice. In contrast, in the cordage industry, fibers of a medium grade are chosen, and after treatment, superior-quality sisal fibers are specified for the carpet business (Choudhary et al., 2023).

Many researchers are interested in sisal fiber. They are interested in reinforcing concrete composites (Shen & Zhang, 2020; Bahja et al., 2021), construction bricks (Alene et al., 2022; Tang et al., 2022), or even polymer matrix composites, as are presented with many examples and comparisons by Sharma et al. (2022). The elevator industry can demonstrate another example of sisal fiber versatility. The fibers are widely used as the core of steel wire rope, playing a significant function in keeping lubrication, decreasing the wear among wires, and increasing the load-bearing capacity.

Sisal fibers can be divided into three types: mechanical (mainly extracted from the periphery of the leaf, possess roughly thickened-horseshoe shape, and are the most commercially useful of sisal fiber), ribbon (the longer sisal fiber, and occur in association with the conducting tissue in the median line of the leaf, which gives them considerable mechanical strength), and xylem (opposite to ribbon fibers, have an irregular shape, and are easily broken and lost during the extraction process) (Li et al., 2000). Depending on plant age, or the fiber type, their composition can vary its content of hemicellulose (10–24 wt%), cellulose (47–78 wt%), lignin (7–14 wt%), pectin (2–10 wt%), ash (0.6–1 wt%), and wax (0.2–2 wt%) (Choudhary et al., 2023). As a consequence of these deviations, the fibers' features can vary significantly, as seen in Table 2.2.

2.2.2 Matrix Materials for Eco-friendly Composites

Petroleum derivatives, such as synthetic plastics, have been widely used for various purposes in recent decades in the form of both plastic products and matrix composites. The environmental impact resulting from the widespread use of synthetic plastics has led to the creation of new laws to control their use and disposal. Furthermore, there has been a growing awareness among the general population regarding the consumption and disposal of these materials. Consequently, the development of eco-friendly materials, which are biodegradable, recyclable, and originate from renewable sources, has become increasingly necessary (Asyraf et al., 2020; Islam et al., 2022).

TABLE 2.2
Properties of Sisal Fibers Reported by Different Researchers

Density (kg/m3)	Moisture Content (%)	Tensile Strength (MPa)	Tensile Modulus (GPa)	Maximum Strain (%)	Diameter (μm)
1450	11	604	9.4–15.8	–	50–200
1450	–	530–640	9.4–22	3–7	50–300
–	–	347	14	5	–
1030	–	500–600	16–21	3.6–5.1	–
1410	–	400–700	9–20	5–14	100–300
1400	–	450–700	7–13	4–9	–
–	–	530–630	17–22	3.64–5.12	100–300
1450	–	450–700	7–13	4–9	–

Source: Li et al. (2000).

Biopolymers are made from natural or renewable resources and are a viable option for eco-friendly materials in matrix composites, compared to synthetic polymers. They are classified into two types: thermosets and thermoplastics. Thermoplastic biopolymers are solids at room temperature and can be molded when heated. The molding process does not involve any chemical reactions. Thermoplastics have greater tolerance to processing under pressure and/or temperature and generally exhibit greater impact resistance compared to thermosets. Thermosetting polymers are materials cured by heat or by a reaction with catalysts. They cannot be cast or reshaped with the application of heat. They have higher thermal and chemical stability and higher creep resistance relative to thermoplastics. On the other hand, they are more brittle at room temperature and have lower fracture toughness than thermoplastics. In addition, the materials used in the eco-friendly composite matrix can also be classified as fully biodegradable or partially biodegradable and non-biodegradable (Faruk et al., 2012; Gholampour & Ozbakkaloglu, 2020).

Green composites are composites composed of a matrix and reinforcement derived from renewable, biodegradable, and/or recyclable sources. Bio-based polymers are commonly used as the matrix material for green composites due to their good degradability, lower emissions during production, and their potential to serve as alternatives to traditional polymers (Abhiram et al., 2021; Mann et al., 2020).

2.2.2.1 Bio-Based Polymers

Bio-based polymers are derived wholly or partially from renewable resources such as plants, animals, or microorganisms. They are typically biodegradable, non-toxic, and are associated with lower emissions of greenhouse gases compared to traditional, petroleum-based polymers. Additionally, the production of bio-based polymers can be energy-efficient and can contribute to the development of sustainable industries (Asyraf et al., 2020; Ramachandran et al., 2022).

There are three types of bio-based polymers:

- Partially bio-based and non-biodegradable, such as bio-polyethylene, bio-polypropylene, and bio-polyethylene terephthalate.
- Partially bio-based and biodegradable, such as cellulose, polylactide acid (PLA), and polybutylene succinate (PBS).
- Fully bio-based and biodegradable, such as polyhydroxyalkanoates (PHA), starch, and polyhydroxybutyrate (PHB).

2.2.2.2 Cellulose

Cellulose is a partially bio-based and biodegradable polymer, which makes it an environmentally friendly option. Additionally, its production cost is relatively low (Gholampour & Ozbakkaloglu, 2020). Cellulose can be obtained in its basic form or through the esterification of various sources such as trees, sugarcane, cotton, recycled paper, algae, and bacteria that produce this biopolymer (Abhiram et al., 2021). There are at least four types of cellulose polymorphs: cellulose I, II, III, and IV. These types have different elastic moduli, with type I being the most rigid at 138 GPa, followed by type II at 88 GPa, type III at 87 GPa, and type IV at 75 GPa (Nishino et al., 1995).

Cellulose composites can be made with a cellulose matrix and undissolved cellulose fiber reinforcement. Since both the matrix and reinforcement have the same chemical composition, there is efficient stress transfer and adhesion at the interface between them (Spörl et al., 2018). Composites reinforced with 90% Bocell fibers and 10% cellulose exhibited higher Young's modulus and better tensile strength when compared to other composites reinforced with natural fibers, such as sisal-epoxy composites and flax fibers, and polypropylene composites (Baghaei & Skrifvars, 2020).

Cellulose composites have significant biodegradability and are non-cytotoxic, making them environmentally friendly. As polymers of renewable origin, these composites are fully recyclable after their useful life. Thus, cellulose is an excellent source of eco-friendly products (Baghaei & Skrifvars, 2020).

2.2.2.3 Polylactide Acid

PLA is a biothermoplastic that is completely biodegradable and sustainable. Its production has lower manufacturing costs compared to conventional polymers, as it requires 25% to 30% less energy to produce (Naik et al., 2022).

PLA can be synthesized from agricultural residues, as well as starch extracted from sources such as rice, corn, and sugarcane. These products are converted into dextrose, which is then fermented to obtain lactic acid. The lactic acid, in the presence of a catalyst, is transformed into lactide, which is purified into a monomer by vacuum distillation. The purified monomer is then converted to PLA through a polymerization process. During this process, a CH_3 group is incorporated into the polymer structure, making it hydrophobic. Due to its relatively high melting point of 160°C, PLA can undergo processing conditions similar to those of polypropylene (Islam et al., 2022; Shekar & Ramachandra, 2018).

PLA is widely used in the manufacture of disposable items such as household goods, sanitary products, plastic bags, and packaging. It is also used by the pharmaceutical and medical industry due to its good compatibility with human tissues and non-toxic degradation (Vinod et al., 2020; Zwawi, 2021).

PLA has L and D stereoisomers, which can form poly-D-lactic acid (PDLA), poly-L-lactic acid (PLLA), and poly-L-D-lactic acid (PDLLA) (Rasal et al., 2010). When PLLA is reinforced with 70% kenaf fibers, its resistance to failure can triple compared to polypropylene (Shekar & Ramachandra, 2018). The addition of 30% by volume of jute fibers in the PLA matrix improved the flexural and tensile properties of the composite (Singh et al., 2020).

In the environment, PLA degradation occurs through hydrolysis, where the polyester macromolecules are degraded into lactic acid with reduced molecular weight. When the average molecular weight of the PLA reaches 10,000 Daltons, soil microorganisms digest the molecule, producing carbon dioxide and water (Farah et al., 2016). If the ambient conditions are favorable, with adequate temperature and humidity, the degradation of the PLA can happen in weeks (Rajeshkumar et al., 2021).

2.2.2.4 Polybutylene Succinate

PBS is a thermoplastic, biodegradable bio-based polymer synthesized from succinic acid and 1,4-butanediol (BDO). Succinic acid can be produced from bacterial fermentation of agricultural by-products and carbohydrates, and BDO is produced from a petroleum-based product. PBS is synthesized through the esterification and transesterification of BDO and succinic acid (Asyraf et al., 2020; Zwawi, 2021).

This polymer has good mechanical properties and has a wide processing temperature window. Hence, it can be extruded, injection molded, and thermoformed (Shekar & Ramachandra, 2018). The tensile strength is about 30 to 35 MPa and the Young's modulus is about 0.3 to 0.5 GPa, and these properties are related to the crystallinity of the material (Aliotta et al., 2022).

PBS has been gaining ground as an eco-friendly material for shopping bags, packaging, bottles, and molds, in addition to having several types of biomedical applications (Aliotta et al., 2022). The stiffness of PBS composites was found to increase by 29% with the addition of palm fiber reinforcement, while the use of coconut fibers resulted in a 42% increase in stiffness (Chaari et al., 2020). PBS composite foams reinforced with cellulose fibers showed an increase in the percentage of elongation and tensile strength (Vorawongsagul et al., 2021).

Degradation of PBS by microorganisms mainly occurs through hydrolysis, but it can also occur through the action of enzymes. Fungi and bacteria are capable of biodegrading the polymer. Under appropriate conditions, PBS can present a mass loss of 0.2% to 0.5% after one month of biodegradation in the soil (Aliotta et al., 2022; da Silva et al., 2018).

2.2.2.5 Polyhydroxyalkanoates

PHA is a bio-based and biodegradable polymer (Asyraf et al., 2020). The PHA polymer has a high-production cost and is typically produced through a continuous fermentation process. In this process, both gram-positive and gram-negative bacteria use glucose and sucrose from agricultural products as a substrate to produce the polymer. PHA is then stored in cells and needs to be extracted.

There are several extraction methods available, with one safe, environmentally friendly, and economically favorable method being the supercritical fluid method (Raza et al., 2018; Zwawi, 2021).

PHA and its types, such as poly-3-hydroxybutyrate (PHB), poly(3-hydroxybutyrate-co-3-hydroxyhexanoate) (PHBHHx), and poly(3-hydroxybutyrate-co-3-hydroxyvalerate) (PHBV), are used as eco-friendly materials to replace synthetic polymers in the automotive, textile, food, and packaging industries (Koller, 2019; Pati et al., 2020).

The addition of wood flour to PHA has been shown to provide improved tensile strength but may result in lower breaking strength (Chan et al., 2019). PHA composites reinforced with palm fiber have a higher Young's modulus and a better biodegradation rate than pure PHA (Wu et al., 2017). The degradation rate of PHA depends on various environmental factors, such as the pH level, temperature, and water content. The polymer characteristics, such as surface area, crystallinity, and chain composition and length, also affect the degradation rate. Small molecular chain PHA may not undergo thermal degradation below 60°C, but it may undergo degradation under composting conditions (Meereboer et al., 2020; Zwawi, 2021).

2.2.2.6 Starch

Starch is a thermoplastic, semicrystalline, bio-based, biodegradable polymer with low manufacturing cost (Gholampour & Ozbakkaloglu, 2020; Shekar & Ramachandra, 2018). It is formed by a mixture of amylopectin and amylose polysaccharides. It can be extracted from several plant sources, with corn, potato, beet, wheat, and rice being the most commonly used (Asyraf et al., 2020; Shekar & Ramachandra, 2018). In its natural form, starch cannot be melted or processed. To be used as a thermoplastic, starch needs to undergo a process of plasticization and destructuring by heat and shear in the presence of glycerol or water (Mazerolles et al., 2019).

Starch has the disadvantage of having low mechanical strength and being sensitive to water. To overcome these problems, starch can be blended with other biodegradable polymers, such as PHBV, or synthetic polymers, such as low-density polyethylene (LDPE) (Sessini et al., 2019; Vinod et al., 2020). Starch foam composites reinforced with natural fibers can be used to manufacture trays and packaging fillers as an alternative to polystyrene (Shekar & Ramachandra, 2018).

Cornstarch polymers reinforced with okra cellulose fibers showed better mechanical properties (Guleria et al., 2018). The starch and sisal fiber foam composite has better resistance to tension, compression, and cushioning capacity compared to expanded polystyrene, making it an eco-friendly option for packaging manufacturing (Xie et al., 2018).

Starch films buried in soil for 180 days showed 87% degradation of their structural integrity (Tai, Adhikari, Shanks, & Adhikari, 2019). This occurs due to the hydrophilicity of starch, which facilitates its colonization by bacteria and fungi, promoting the degradation of the material (Tai et al., 2021).

2.2.2.7 PCL

Poly(ε-caprolactone) (PCL) is a thermoplastic, semicrystalline, hydrophobic, petroleum-derived, and biodegradable polymer (Asyraf et al., 2020). PCL is prepared from a lactone, caprolactone, an initiator with active hydrogen, and a catalyst such as stannous octanoate. In the presence of the initiator, the lactone ring is opened by the action of the catalyst and undergoes polymerization (Mochizuki & Hirami, 1996).

PCL is a partially crystalline polymer with low viscosity, a melting point close to 60°C, and is easy to process. It has mechanical properties comparable to those of polystyrene and polyethylene. PCL is used in the manufacture of coatings, adhesives, fabrics, and food packaging. The material's properties can be enhanced through the formation of PCL composites (Islam et al., 2022; Sachan et al., 2023).

Reinforcing PCL with waste date palm fibers was effective in increasing the material's impact and tensile strength (Dhakal et al., 2018). The addition of rice straw fiber to PCL improved the polymer's tensile strength (Wu & Liao, 2012). Porous cellulose gel composites incorporating silver and PCL can be used to fabricate shape memory polymeric devices (Jiang et al., 2022).

The degradation of PCL is slow and can occur through hydrolysis, enzyme action, or chemical agents. In the environment, degradation is facilitated by fungi, bacteria, and algae, which use lipase to degrade PCL. The slow degradation rate of PCL helps to minimize local issues related to the metabolism of PCL by-products (Bartnikowski et al., 2019).

2.3 THE EFFECT OF NATURAL FIBERS' SURFACE TREATMENTS ON NFC PROPERTIES

The use of natural fibers for reinforcing composite materials is an alternative to the production of sustainable composites. The fibers are incorporated into the material to enhance one or more properties of the composite (Khalid et al., 2021). It is necessary to have an adequate interaction between the matrix and reinforcement for the fibers to perform their function (Adekomaya & Majozi, 2019).

Natural fibers are naturally hydrophilic due to hydroxyl and polar groups. The hydrophilic nature can result in a high tendency for natural fibers to absorb moisture and compromise the compatibility between the fiber and the composite matrix. The components that make up natural fibers, such as crystalline cellulose, non-crystalline cellulose, hemicellulose, lignin, and pectin, also affect the properties of the fibers and, consequently, the compatibility between the fiber and matrix. The presence of non-cellulosic components, such as pectin, lignin, and hemicellulose, can cause an increase in moisture absorption. This results in delamination between the fiber and matrix and a reduction in the mechanical properties of the composite (Faruk et al., 2012; Shalwan & Yousif, 2013).

The fiber's mechanical strength is improved with an increase in the presence of crystalline cellulose but is reduced by the presence of hemicellulose and lignin. Hemicellulose makes the fiber more susceptible to thermal and biological degradation. On the other hand, lignin enables the fiber to resist thermal and biological degradation (Lau et al., 2018; Nayak et al., 2012).

Surface treatment of natural fibers can be used to increase compatibility between the matrix and reinforcement material, reduce the hydrophilic nature of the fiber, and improve its mechanical properties (Athijayamani et al., 2009).

Some surface treatment techniques aim to remove non-cellulosic constituents on the fiber surface, while others are used to improve the adhesion between matrix and fibers. (Pankaj & Kant, 2022). Chemical treatments alter the constituents on the material's surface, reducing the fibers' hydrophilic capacity and achieving better compatibility between the matrix and the fiber. Physical treatments improve the fiber and the matrix interaction by modifying the fiber surface without changing its chemical components (Ramachandran et al., 2022).

2.3.1 Chemical Treatments

2.3.1.1 Alkaline Treatment

One of the most commonly used chemical treatments for natural fibers is alkaline treatment, also known as mercerization (Hassan et al., 2020). In this method, a solution of sodium hydroxide (NaOH) is used to modify the structure of the natural fibers by breaking hydrogen bonds within the fiber structure. The alkaline treatment causes an increase in surface roughness and removes non-cellulosic constituents such as hemicellulose, lignin, wax, pectin, and oily substances (Kommula et al., 2013). The treatment also promotes increased interfacial adhesion between fibers and the matrix, reinforcing thermoset polymers and thermoplastics (Ali et al., 2018).

Polypropylene (PP) composites reinforced with kenaf fibers that underwent alkaline treatment significantly improved their mechanical properties compared to composites reinforced with untreated fibers. This improvement is attributed to the enhanced adhesion between the fiber and matrix, which was demonstrated (Thyavihalli et al., 2019). Alkaline treatment of coir/pineapple leaf fibers for reinforcement of PLA composites increased Young's modulus and tensile strength of the material (Siakeng et al., 2020).

The concentration of NaOH is an important factor in the alkaline treatment, and the ideal concentration must be carefully evaluated, as an excessive concentration can damage the surface of the material. Fibers that are treated with excessive NaOH concentration may result in fiber breakage and damage to the cellulose fibers, leading to the reduced mechanical resistance of the composite that incorporates these fibers. Specifically, coir fibers treated with an excessive concentration of NaOH have been shown to have reduced impact strength (Nematollahi et al., 2019).

2.3.1.2 Silane Treatment

The silane treatment is used to enhance the adhesion strength between the polymeric matrix and the fiber. During the silane treatment, silane molecules form a chemical bond between the cellulose and the polymeric resin, creating a siloxane bridge (Ahmad et al., 2019). During the silane treatment, the silanol groups react with the OH groups on the cellulose and the functional groups in the matrix, resulting in strong interfacial interaction in the composite (Campilho, 2015).

The interlocking between the fiber and matrix is demonstrated by the increase in tensile strength of composites that utilize fibers treated with this method (Gholampour & Ozbakkaloglu, 2020). The silane treatment has been shown to increase the tensile strength of cotton fibers after a one-hour treatment (Bodur et al., 2016).

PP composites reinforced with bamboo fibers subjected to silane treatment exhibited increased tensile and flexural strength compared to composites reinforced with untreated fibers (Wang, Zhang, et al., 2020). Silane coupling agents promote increased interlocking between the polymeric matrix and the natural fiber, resulting in improved interfacial adhesion and ultimately enhancing the composite material's mechanical properties (Vijay et al., 2021). Pineapple leaf fibers subjected to silane treatment have been shown to enhance the interfacial adhesion between the fiber and the polymeric matrix. As a result, the tensile and impact strength of composites reinforced with these fibers are increased compared to those that received untreated fibers (Asim et al., 2018). Composites made from polystyrene (PS) and kenaf fibers modified by silane treatment exhibit a higher elasticity modulus than composites reinforced with untreated fibers (Faruk et al., 2012).

2.3.1.3 Acetylation

Acetylation involves the use of acetic acid and acetic anhydride to replace the hydroxyl group (OH) of fibers with the acetyl group (CH_3CO). The process occurs in two stages: first, the fiber is soaked in acetic acid, and then it is treated with acetic anhydride at high temperatures for 1 to 3 hours. This replacement of the OH groups removes moisture and increases the hydrophobic character of the fibers. The surface of the fiber becomes rougher, which promotes greater interlocking between the fiber and matrix and results in greater dimensional stability of the reinforced composite (Koohestani et al., 2019; Ramachandran et al., 2022).

PBS composites reinforced with bamboo fibers that received acetylation treatment exhibited greater thermal stability and increased tensile strength compared to composites that received untreated fibers (XU et al., 2016). PP composites that were treated with alkali-pretreated and subsequently acetylated banana empty fruit bunch fibers exhibited superior mechanical properties and thermal stability compared to composites that used untreated fibers (Zaman & Khan, 2021).

2.3.1.4 Peroxide Treatment

In peroxide treatment, peroxide is decomposed into free radicals, which interact with the OH groups of the natural fiber and matrix. This process decreases the moisture absorption capacity of the fiber, improves interlocking between the matrix and reinforcement, and enhances the thermal stability of the composite. The treatment process for peroxide treatment of natural fibers typically involves two stages. In the first stage, the fibers are treated with alkaline solutions to remove impurities and expose their hydroxyl (OH) groups. In the second stage, the fibers are immersed in a benzoyl peroxide or dicumyl peroxide solution, which is then heated to decompose the peroxide (Gholampour & Ozbakkaloglu, 2020; Ramachandran et al., 2022).

The beneficial interlocking effect resulting from peroxide treatment can be observed in PLA composites reinforced with coconut fibers. Compared to composites using untreated fibers, those using peroxide-treated fibers exhibit significantly improved impact resistance (Sun et al., 2017). Peroxide treatment applied to fibers from Algarve americana leaves increases their thermal stability compared to untreated fibers (Madhu et al., 2020).

2.3.1.5 Benzoylation Treatment

In benzoylation, benzoyl chloride is used to reduce the hydrophilic nature of the fibers, increase interfacial adhesion, and improve thermal stability. The treatment involves an initial stage where the fiber is treated with an alkaline solution to remove wax and lignin and expose its surface with hydroxyl (OH) groups. Then, the fiber is treated with benzoyl chloride, and the hydroxyl groups (OH) of the fibers are replaced by benzoyl groups of the benzoyl chloride (Izwan et al., 2020; Safri et al., 2019).

Sansevieria ehrenbergii fibers that undergo benzoylation treatment exhibit improved thermal stability when used as reinforcements in composites, compared to composites reinforced with untreated fibers (Kumaran et al., 2020).

PP composites reinforced with palm and kenaf fiber that underwent benzoylation showed better flammability and increased mechanical properties (Izwan et al., 2021). The tensile strength of PP composites reinforced with Agave Americana fibers increased, and their thermal stability improved after undergoing benzoylation (Thamarai et al., 2023).

2.3.1.6 Maleated Coupling Agents

In maleated coupling maleic anhydride is used as a coupling agent to alter the fiber surface as well as the polymeric matrix. The changes result in better interaction between matrix and reinforcement and improvement of mechanical properties (Faruk et al., 2012).

Maleated coupling agents can form a carbon-carbon bond between the fiber and the matrix polymer chains. In the amorphous fiber regions, a reaction occurs between the hydroxyl (OH) groups and the maleic anhydride, which excludes the OH groups from the fiber and forms a long-chain polymeric coating on the fiber. Covalent bonds are formed between the polymeric fiber coating and the matrix polymer, providing efficient interlocking between the matrix and reinforcement and improving mechanical properties (Gholampour & Ozbakkaloglu, 2020).

The maleated coupling treatment of aspen fibers significantly improves the mechanical properties of PP composites, including an increase in tensile strength by more than 30% and a flexural strength by over 20%. The application of maleated coupling treatment to hemp fibers also leads to a remarkable increase of more than 40% in the flexural strength of the resulting PP composite. Similarly, high-density polyethylene (HDPE) composites reinforced with bamboo fibers treated with maleated coupling agents exhibit a tensile strength of 30% higher than those reinforced with untreated fibers (Koohestani et al., 2019).

2.3.1.7 Sodium Chlorite Treatment

Sodium chlorite treatment eliminates lignin and moisture from fibers, resulting in improved hydrophobic properties. The fibers are bleached in a sodium chlorite solution, which is subsequently acidified to produce chloric acid in this treatment. Chloric acid is then oxidized to chlorine dioxide, which removes lignin and reduces fiber moisture. The treatment promotes an increase in fiber flexibility and resistance to tension and flexion, as well as a modification of its stiffness (Correia et al., 2017; Gholampour & Ozbakkaloglu, 2020).

Natural rubber composites reinforced with jute fibers that have been treated with sodium chlorite exhibit greater rigidity without compromising tensile strength compared to composites that use untreated fibers. It is due to the excellent interlocking between the matrix and reinforcement, which is facilitated by removing lignin and moisture from the fibers during the treatment process (Correia et al., 2017). Treatment of soy fibers with NaOH and sodium chlorite has been shown to enhance the mechanical properties of resulting composites. Specifically, composites reinforced with treated soy fibers exhibit higher tensile strength and Young's modulus when compared to composites reinforced with untreated fibers (Martelli et al., 2017).

2.3.1.8 Acrylation and Acrylonitrile Grafting

In this treatment, the fibers undergo a pre-treatment with an alkaline solution, followed by treatment with nitric acid. The acid reacts with the OH groups in the cellulose and increases access to cellulose microradicals. It leads to ester bond formation with the OH groups of cellulose, eliminating moisture. Subsequently, the radicals interact with the matrix, forming a copolymer and initiating the grafting of matrix-acrylic acid onto the fibers. This process improves the interlocking between the fiber and matrix, increasing overall strength (Liminana et al., 2018; Sarkar et al., 2017; Zaki & Abdullah, 2015).

Treatment of date palm leaf fibers with acrylic acid results in greater mechanical strength and improved interlocking with poly(vinyl alcohol) (PVA), resulting in better mechanical properties than composites with untreated fiber in PVA (Mohanty et al., 2013). Extruded rice husk composites treated with acrylic acid exhibit higher tensile strength, improved thermal stability, and better moisture resistance than composites reinforced with untreated fiber in PLA (Wu & Tsou, 2019).

2.3.1.9 Permanganate Treatment

The first step of the permanganate treatment involves immersing the fibers in an alkaline solution. The permanganate reacts with the fiber, forming cellulose-manganate. Cellulose-manganate reacts with the OH groups of the fiber, eliminating lignin from the fiber cell wall. The removal of this component provides the fiber with better thermal stability and resistance to moisture absorption. The fiber's hydrophilic character is altered by different concentrations of potassium permanganate (Gholampour & Ozbakkaloglu, 2020; Paul et al., 2010; Sreenivasan et al., 2012).

The tensile strength of sugar palm fiber composites is improved when subjected to permanganate pre-treatment (Mohammed et al., 2017). Unsaturated polyester composites reinforced with permanganate-treated Sansevieria cylindrica leaf fibers showed better mechanical properties compared to composites reinforced with untreated fibers (Sreenivasan et al., 2012). Jute fibers treated with permanganate provided increased flexural strength to the PLA composite that received the treated fibers (Goriparthi et al., 2012).

2.3.1.10 Isocyanate Treatment

Isocyanate promotes the formation of covalent forces between the matrix and the fiber, increasing their interaction. Additionally, the treatment increases the fiber moisture resistance. During the treatment, the isocyanate functional group reacts with the hydroxyl and lignin groups of the fiber, inducing a strong bond between the fiber and the matrix. Furthermore, combining isocyanate with moisture in the fiber produces urea, which reacts with the OH groups in the fiber, and leads to increased moisture resistance (George et al., 2012; Gholampour & Ozbakkaloglu, 2020; Zaman et al., 2011).

The bonding between the fiber and matrix in PBS reinforced with isocyanate-treated banana fiber was appropriate, as verified by the increased resistance to tensile strength and impact of composites with treated fibers (Kajiyama et al., 2013).

The epoxy resin composites reinforced with isocyanate-treated flax fiber showed a 20% increase in tensile and flexural strength compared to composites with untreated fibers (Wang, Fu, et al., 2020).

2.3.1.11 Stearic Acid Treatment

Stearic acid is dissolved in ethanol and applied to the fiber in this treatment. The carboxylic groups of the acid are not ionized in the solution, allowing them to react with the hydroxyl groups of the fiber and eliminate the non-crystalline constituents of the fiber structure. It induces fiber dispersion at the matrix interface, improving composite performance (Jain et al., 2020; Karthi et al., 2019).

Short hemp fibers that received the stearic acid treatment were added to PP, forming a composite with improved thermal stability (Pickering & Sunny, 2021). The mechanical strength of an epoxy resin reinforced with Eichhornia crassipes fibers treated with stearic acid increased up to 20%wt of reinforcement. Beyond this amount, there was a reduction in mechanical properties (Palai & Sarangi, 2022).

2.3.1.12 Oleoyl Chloride Treatment

Oleoyl chloride is a fatty acid derivative commonly used to enhance natural fibers' adhesion and wettability properties. When applied to the fiber surface, the oleoyl chloride reacts with the hydroxyl groups, removing them. It results in surface esterification and a plasticizing effect on the fiber, which improves the adhesion between fiber and reinforcement. Additionally, eliminating the hydroxyl groups induces a hydrophobic character in the fiber (Shirke & Shelke, 2020; Soman et al., 2018).

Polyester composites reinforced with linen fibers treated with oleoyl chloride showed improved impact resistance and thermal stability (Hu et al., 2016). Natural rubber composites reinforced with nanocellulose fibers treated with oleic acid and vulcanized with sulfur exhibited an increase in Young's modulus (Kato et al., 2015).

2.3.2 Alternative Methods to Chemical Treatments

2.3.2.1 Triazine Treatment

The surface of natural fibers can be modified using chemical solutions that contain triazine and its derivatives. The functional groups of triazine react with the hydroxyl groups of the fibers, inducing a plasticizing effect and promoting the formation of covalent bonds between the fibers and polymer matrices. This treatment results in better anchoring of the fibers to the matrices and improved resistance to humidity (Mohammed et al., 2022; Tanasă et al., 2020). Cotton fibers treated with triazine exhibited higher thermal stability and excellent flame retardancy (Sun et al., 2020).

2.3.2.2 Enzyme Treatment

Enzymatic treatment of natural fibers has been considered an environmentally friendly option due to its low water and energy consumption. The enzymes used in this treatment must be chosen to suit the type of product to be treated, as they have high specificity (Pankaj & Kant, 2022).

Enzymes act as biological catalysts and accelerate chemical reactions on the surface of fibers. They clean the fibers, degrease them, and make the material soft. Enzymatic treatment is faster than chemical treatment, as it is carried out at atmospheric pressure, below 100°C, and with a pH between 4 and 8. However, the specificity of enzymes and their high cost may limit their use (Koohestani et al., 2019).

PP composites were reinforced with jute fibers treated with laccase enzymes. The composites that received the treated fibers showed increased breaking strength and temper, indicating improved interfacial adhesion between the fibers and the matrix (Dong et al., 2018).

2.3.2.3 Fungal Treatment

This treatment involves using enzymes produced by white rot fungi to remove non-cellulosic substances, such as waxes, lignin, and hemicellulose, from the surface of fibers. In addition to enzymes, fungal hyphae create fine holes on the surface of the fibers, increasing surface roughness and improving adhesion between the matrix and the fibers (Kabir et al., 2012; Kenned et al., 2020; Summerscales, 2021).

The use of fungi requires care such as pH, humidity, temperature, oxygen, carbon, and nitrogen saturation control. White rot fungi proliferate at temperatures between 25 and 30 degrees Celsius and a wide pH range (Tanasă et al., 2020).

Hemp fibers that received fungal treatment was applied to PP. The resulting composite showed a 22% increase in mechanical strength and improved thermal stability when compared to composites made with untreated fibers. Fungal treatment is an eco-friendly and efficient alternative to chemical methods (Kalia et al., 2013; Khoshnava et al., 2014).

2.3.3 Physical Treatments

Physical treatments seek to modify the surface of the fibers, enhancing the adhesion between matrix and fiber without promoting chemical alteration of the surface (Faruk et al., 2012).

2.3.3.1 Corona Treatment

The corona treatment promotes changes in the fiber surface by oxidizing its surface, increasing its surface energy, and reducing its contact angle, providing increased compatibility between the hydrophilic fiber and the hydrophobic matrix (Luo et al., 2019; Sreekala et al., 2000).

Corona treatment equipment usually comprises two aluminum electrodes and a dielectric spacer, commonly made of quartz. When electrical energy, either continuous or pulsed, is applied to the material through the electrodes, the air between the electrode and the material becomes ionized, forming radicals and ions that induce the formation of functional groups containing oxygen on the surface of the fiber (Pickering et al., 2008). The portion of the fibers directly exposed to the electrical discharge receives a better quality treatment compared to the portion that does not receive direct exposure (Ragoubi et al., 2012).

Composites formed by PLA and palm fibers subjected to corona treatment had a 13% increase in their modulus of elasticity (Amirou et al., 2013). The modification of hemp fiber by the corona method led to a 30% increase in Young's modulus of composites reinforced with the treated fiber (Ragoubi et al., 2010). The use of corona treatment on aloe vera fibers reduced tensile strength and elastic modulus, contrary to the expected improvement in mechanical properties observed in other fibers subjected to the same treatment (Oudrhiri et al., 2020).

2.3.3.2 Plasma Treatment

The plasma treatment is widely used in industry to modify the surface of natural fibers. This treatment causes modifications to the surface of the fibers, reducing their ability to absorb moisture and increasing the adhesion between the fiber and the polymeric matrix (Macedo et al., 2020; Rao et al., 2018).

Plasma treatment must be performed in a vacuum chamber that will receive a molecular gas or a mixture of molecular gases. During the process, the gas surrounds the material being treated, and an ionized region is formed containing electrons, ions and radicals, which then alter the material's surface (Adekomaya & Majozi, 2019; Koohestani et al., 2019).

Plasma treatment offers various surface modifications depending on the type of gas used and the process parameters. The process can promote the formation of free radicals by breaking bonds, increase or decrease the surface energy of the material, and alter the fiber surface by removing weakly bonded surface layers to form new functional groups on the material's surface (Faruk et al., 2012; Koohestani et al., 2019). Plasma treatment can break down the hydrophobic components of fibers, leading to an increased adhesion between the fiber and polymeric matrix. This treatment can modify the surface of both the natural fiber and polymeric matrix without altering their bulk properties (Sinha & Panigrahi, 2009).

Composites formed by plasma-treated jute fiber and PLA have higher tensile strength, flexural strength, and elasticity modulus than composites made from untreated jute (Gibeop et al., 2013). Plant-derived polyethylene composites were reinforced with plasma-treated bamboo fibers, increasing tensile strength compared to composites with untreated fibers (Sánchez et al., 2020).

Plasma treatment is considered eco-friendly due to its low levels of pollution and energy consumption compared to conventional methods (Haji, 2019).

2.3.3.3 Ultraviolet Irradiation Treatment

UV irradiation energy has been utilized to promote surface modifications. During this process, the material is exposed to UV electromagnetic radiation with wavelengths ranging from 10 to 400 nm. The radiation promotes photochemical reactions in the molecules on the material's surface, causing an increase in surface polarity and removing impurities. This treatment changes the material's mechanical properties (Ching et al., 2019; Gassan & Gutowski, 2000).

Cotton fibers subjected to UV radiation undergo oxidation of their cellulosic components and develop carboxylic groups on their surface. UV radiation allows for a higher rate of dye absorption as it improves their interaction with the functional groups of the fibers (Elkashouty et al., 2020). Jute fibers treated with UV radiation and added to polyester composites showed a 40% increase in flexural strength compared to composites reinforced with untreated fibers (Ramachandran et al., 2022). Composites reinforced with coconut fiber that received UV treatment showed a tensile strength 33% higher than composites reinforced with untreated fibers (Rahman & Khan, 2007).

2.3.3.4 Ozone Treatment

Ozone treatment is a high-efficiency, low-carbon, and eco-friendly method of removing non-cellulosic substances from the surface of natural fibers (Zhang et al., 2022). Ozone is generated from dry air or oxygen. Atmospheric air is captured and passed through an oxygen concentrator to separate the oxygen, which is then sent to the ozone generator, where it is converted into ozone (Maqsood et al., 2017). This gas has a short half-life, decomposing quickly without causing environmental persistence problems (Tripathi et al., 2020).

The ozone treatment promotes the surface oxidation of natural fibers, improving interfacial adhesion with the matrix. During the treatment, carbonyl and hydroperoxide groups are formed on the fiber's surface, resulting in the degradation of lignin and alteration of the morphology of the fibers. This increases the degree of crystallinity and the size of the crystallites (Jagadeesh et al., 2021; Ramachandran et al., 2022).

Wool and silk fibers treated with ozone showed increased tensile strength, elastic modulus, and reduced elongation percentage (Benli & Bahtiyari, 2018). Using cellulose crystals obtained from jute fibers pre-treated with ozone and then subjected to enzymatic hydrolysis, as reinforcement in PLA composites, improved their mechanical properties. This is attributed to the rough and irregular surface of the particles, which provided good interaction between the matrix and reinforcement (Baheti et al., 2016).

2.4 CONCLUSIONS

The use of natural fibers in engineering applications, like in natural fiber composites (NFCs), is typically recognized as a sustainable practice, and their impacts are usually smaller than the synthetic fiber composites. Such bio-based composites offer advantages to users and the environment. They are renewable, can be biodegradable and recyclable, reduce energy consumption, and pose less risk to human and animal health.

Several factors influence the fiber's characteristics, as well as the soil, geographic position (air humidity, altitude, temperature, and solar incidence), packaging, processing methods (which can cause damage to the fibers), morphology, length, defects, and orientation of fibers, for example. Such factors will result in the final properties of natural fiber composites. In addition, most of the fibers are hydrophilic, and they need surface treatments to be carried out to avoid water absorption. The water absorption could rot the material, and that will also cause contractions and swelling of the fibers, resulting in the fiber displacement in the matrix, which will cancel the boost effect.

Despite the advantages of NFCs, they need precepts of sustainability. Therefore, being better does not necessarily mean being the best, and there is potential for research and technology development to improve this type of product's sustainability. When considering the environmental aspect, renewable or natural-origin products do not necessarily imply a smaller environmental or sustainability impact. The impacts can be even more significant than the synthetical equivalent material for specific categories, such as eutrophication. In fact, research that improves important features in natural fibers has to be encouraged, such as fibers that need less water to grow, require fewer chemicals, can absorb nitrogen and phosphorus from the soil, reduce eutrophication, upgrade recyclability and biodegradability, or measure the impact of chemicals.

Consequently, the simple natural fibers used to obtain composites in engineering are insufficient for the material to become sustainable. There is no one-size-fits-all solution, which requires attention and sustainable concerns throughout the product's life cycle. Furthermore, another issue is that natural fiber composites may not escape greenwashing; while convenient, the term "green" may not be adequate and is much less sustainable.

Moreover, research on green composites tends to focus on sustainability's environmental and economic aspects, leaving aside the social aspect. NFCs offer great potential as a waste management or poverty reduction tool, especially in developing communities that depend on exploiting or planting natural fibers. Concerns with the communities involved in planting and producing different fiber types, the use of fertilizers, the chemical treatments necessary to improve properties, the recyclability and biodegradability of NFCs, the reduction of greenhouse gas emissions and eutrophication, for example, must be considered since there are some risks involved in the production of NFCs.

Strategies for using and manufacturing smarter products or those extending their useful life are more appropriate when considering sustainability. Furthermore, improving the NFCs' recyclability and biodegradability are important features and should be seen as a purpose. Finally, perspectives indicate that natural fibers will prevail over synthetic fibers (glass, carbon, etc.) in the future because of their low cost and environmental safety.

REFERENCES

Abhiram, Y., Das, A., Sharma, K. K. (2021). Green composites for structural and non-structural applications: A review. *Materials Today: Proceedings*, 44, 2658–2664. https://doi.org/10.1016/j.matpr.2020.12.678

Adekomaya, O., Majozi, T. (2019). Sustainability of surface treatment of natural fibre in composite formation: Challenges of environment-friendly option. *International Journal of Advanced Manufacturing Technology*, 105(7-8), 3183–3195. https://doi.org/10.1007/s00170-019-04581-6

Adeniyi, A. G., Onifade, D. V., Ighalo, J. O., Adeoye, A. S. (2019). A review of coir fiber reinforced polymer composites, *Composites Part B: Engineering*, 176, 107305, ISSN 1359-8368, https://doi.org/10.1016/j.compositesb.2019.107305

Ahamed, B., Hasan, M., Azim, A. Y. M. A., Saifullah, A., Alimuzzaman, S., Dhakal, H. N., Sarker, F. (2022). High performance short jute fibre preforms for thermoset composite applications. *Composites Part C: Open Access*, 9, 100318, ISSN 2666-6820. https://doi.org/10.1016/j.jcomc.2022.100318

Ahmad Kamal, A. A., Noriman, N. Z., Sam, S. T., et al. (2019). Tensile properties and impact strength of RHDPE/BF composites: The effects of chemical treatment. *IOP Conference Series: Materials Science and Engineering.*, 557(1), 012041. https://doi.org/10.1088/1757-899X/557/1/012041

Akil, H. M., Omar, M. F., Mazuki, A. A. M., Safiee, S., Ishak, Z. A. M., Abu Bakar, A. (2011). Kenaf fiber reinforced composites: A review. *Materials and Design*, 32, 4107–4121. https://doi.org/10.1016/j.matdes.2011.04.008

Alene, T. E., Mohammed, T. A., Gualu, A. G. (2022). Use of sisal fiber and cement to improve load bearing capacity of mud blocks. *Materials Today Communications*, 33, 104557, ISSN 2352-4928, https://doi.org/10.1016/j.mtcomm.2022.104557

Ali, A., Shaker, K., Nawab, Y., et al. (2018). Hydrophobic treatment of natural fibers and their composites—A review. *Journal of Industrial Textiles*, 47(8), 2153–2183. https://doi.org/10.1177/1528083716654468

Ali, B., Hawreen, A., Kahla, N. B., Amir, M. T., Azab, M., Raza, A. (2022). A critical review on the utilization of coir (coconut fiber) in cementitious materials. *Construction and Building Materials*, 351, 128957, ISSN 0950-0618. https://doi.org/10.1016/j.conbuildmat.2022.128957

Aliotta, L., Seggiani, M., Lazzeri, A., Gigante, V., Cinelli, P. (2022). A brief review of poly (butylene succinate) (PBS) and its main copolymers: Synthesis, blends, composites. *Biodegradability, and Applications. Polymers*, 14(4). https://doi.org/10.3390/polym14040844

Amaducci, S., Scordia, D., Liu, F. H., Zhang, Q., Guo, H., Testa, G., et al. (2015). Key cultivation techniques for hemp in Europe and China. *Industrial Crops and Products*, 68, 2–16.

Amirou, S., Zerizer, A., Haddadou, I., Merlin, A. (2013). Effects of corona discharge treatment on the mechanical properties of biocomposites from polylactic acid and Algerian date palm fibres. *Scientific Research and Essays*, 8(21), 946–952.

Angelini, L. G., Scalabrelli, M., Tavarini, S., Cinelli, P., Anguillesi, I., Lazzeri, A. (2015). Ramie fibers in a comparison between chemical and microbiological retting proposed for application in biocomposites. *Industrial Crops and Products*, 75, Part B, 178–184, ISSN 0926-6690. https://doi.org/10.1016/j.indcrop.2015.05.004

Aristri, M. A., Lubis, M. A. R., Laksana, R. P. B., Sari, R. K., Iswanto, A. H., Kristak, L., Antov, P., Pizzi, A. (2022). Thermal and mechanical performance of ramie fibers modified with polyurethane resins derived from acacia mangium bark tannin. *Journal of Materials Research and Technology*, 18, 2413–2427, ISSN 2238-7854. https://doi.org/10.1016/j.jmrt.2022.03.131

Arockiam, N. J., Jawaid, M., Saba, N. (2018). *Sustainable Bio Composites for Aircraft Components*. [S.l.]: Elsevier Ltd. Available in: http://dx.doi.org/10.1016/B978-0-08-102131-6.00006-2

Asim, M., Jawaid, M., Abdan, K., Ishak, M. R. (2018). The effect of silane treated fibre loading on mechanical properties of pineapple leaf/kenaf fibre filler phenolic composites. *Journal of Polymers and the Environment*, 26(4), 1520–1527. https://doi.org/10.1007/s10924-017-1060-z

Asyraf, M. R. M., Ishak, M. R., Sapuan, S. M., Yidris, N., Ilyas, R. A., Rafidah, M., Razman, M. R. (2020). Potential application of green composites for cross arm component in transmission tower: A brief review. *International Journal of Polymer Science*, 2020, 1–15. https://doi.org/10.1155/2020/8878300

Athijayamani, A., Thiruchitrambalam, M., Natarajan, U., Pazhanivel, B. (2009). Effect of moisture absorption on the mechanical properties of randomly oriented natural fibers/polyester hybrid composite. *Materials Science and Engineering A*, 517(1-2), 344–353. https://doi.org/10.1016/j.msea.2009.04.027

Ayadi, R., Hanana, M., Mzid, R., Hamrouni, L., Khouja, M. I., Hanachi, A. S. (2017). Hibiscus cannabinus L. -kenaf: A review paper. *Journal of Natural Fibers*, 14 (4), 466–484.

Baghaei, B., Skrifvars, M. (2020). All-cellulose composites: A review of recent studies on structure, properties and applications. *Molecules*, 25(12). https://doi.org/10.3390/molecules25122836

Baheti, V., Maqsood, H. S., Wiener, J., Militky, J. (2016). Reinforcement of ozone pre-treated and enzyme hydrolyzed longer jute micro crystals in poly lactic acid composite films. *Composites Part B: Engineering*, 95, 9–17. https://doi.org/10.1016/j.compositesb.2016.03.093

Bahja, B., Elouafi, A., Tizliouine, A., Omari, L. H. (2021). Morphological and structural analysis of treated sisal fibers and their impact on mechanical properties in cementitious composites. *Journal of Building Engineering*, 34, 102025.

Baillie, C., Feinblatt, E. (2013). Ethical practices in the processing of green composites. *Natural Fibre Composites: Materials, Processes and Applications*. [S.l: s.n.].

Balla, V. K. et al. (2019). Additive manufacturing of natural fiber reinforced polymer composites: Processing and prospects. *Composites Part B: Engineering*, 174(March), 106956. Available in: https://doi.org/10.1016/j.compositesb.2019.106956

Bartnikowski, M., Dargaville, T. R., Ivanovski, S., Hutmacher, D. W. (2019). Degradation mechanisms of polycaprolactone in the context of chemistry, geometry and environment. *Progress in Polymer Science*, 96, 1–20. https://doi.org/10.1016/j.progpolymsci.2019.05.004

Bellil, S., Pantaloni, D., Shah, D. U., Le Duigou, A., Baley, C., Beaugrand, J., Bourmaud, A., Guessasma, S. (2023). Prediction of interfacial behaviour of single flax fibre bonded to various matrices by simulation of microdroplet test. *Composites Part C: Open Access*, 11, 100351, ISSN 2666-6820. https://doi.org/10.1016/j.jcomc.2023.100351

Benli, H., Bahtiyari, M. İ. (2018). Combination of dyeing method and ozone after-treatment to apply natural dyes on to cotton fabrics. *Ozone: Science and Engineering*, 40(1), 44–53. https://doi.org/10.1080/01919512.2017.1336926

Bensadoun, F., Verpoest, I., Baets, J., Müssig, J., Graupner, N., Davies, P., et al. (2017). Impregnated fibre bundle test for natural fibres used in composites. *Journal of Reinforced Plastics and Composites*, 36(13), 942–957.

Bledzki, A. K., Mamun, A. A., Jaszkiewicz, A., Erdmann, K. (2010). Polypropylene composites with enzyme modified abaca fibre. *Composites Science and Technology*, 70(5), 854–860, ISSN 0266-3538. https://doi.org/10.1016/j.compscitech.2010.02.003

Bodur, M. S., Bakkal, M., Sonmez, H. E. (2016). The effects of different chemical treatment methods on the mechanical and thermal properties of textile fiber reinforced polymer composites. *Journal of Composite Materials*, 50(27), 3817–3830. https://doi.org/10.1177/0021998315626256

Calzolari, D., Magagnini, G., Lucini, L., Grassi, G., Appendino, G. B., Amaducci, S. (2017). High added-value compounds from Cannabis threshing residues. *Industrial Crops and Products*, 108, 558–563.

Campilho, R. D. S. G., ed. (2015). *Natural Fiber Composites*. 1st ed. CRC Press. https://doi.org/10.1201/b19062

Carus, M. (2015). Overview on the European hemp industry. In: *12th International Conference of the European Industrial Hemp Association (EIHA)*, Wesseling, Germany.

Chaari, R., Khlif, M., Mallek, H., Bradai, C., Lacoste, C., Belguith, H., … Dony, P. (2020). Enzymatic treatments effect on the poly (butylene succinate)/date palm fibers properties for bio-composite applications. *Industrial Crops and Products*, 148(February), 112270. https://doi.org/10.1016/j.indcrop.2020.112270

Chan, C. M., Vandi, L. J., Pratt, S., Halley, P., Ma, Y., Chen, G. Q., … Laycock, B. (2019). Understanding the effect of copolymer content on the processability and mechanical properties of polyhydroxyalkanoate (PHA)/wood composites. *Composites Part A: Applied Science and Manufacturing*, 124(October 2018), 105437. https://doi.org/10.1016/j.compositesa.2019.05.005

Charlet, K., Jernot, J. P., Eve, S., Gomina, M., Bréard, J. (2010). Multi-scale morphological characterisation of flax: From the stem to the fibrils. *Carbohydrate Polymers*, 82(1), 54–61, ISSN 0144-8617, https://doi.org/10.1016/j.carbpol.2010.04.022

Ching, Y. C., Gunathilake, T. U., Ching, K. Y., et al. (2019). Effects of high temperature and ultraviolet radiation on polymer composites. In: Jawaid, M., Thariq, M., Saba, N., eds. *Durability and Life Prediction in Biocomposites, Fibre-Reinforced Composites and Hybrid Composites*. Elsevier, 407–426. https://doi.org/10.1016/B978-0-08-102290-0.00018-0

Choudhary, S. et al. (2023). Systematic literature review on thermal and acoustic characteristics of natural fibre polymer composites for automobile applications. *Materials Today: Proceedings*, 1–8. Available in: https://doi.org/10.1016/j.matpr.2023.01.349

Correia, C. A., de Oliveira, L. M. Valera, T. S. The influence of bleached jute fiber filler on the properties of vulcanized natural rubber. *Materials Research.* 2017; 20(suppl 2): 466–471. https://doi.org/10.1590/1980-5373-mr-2017-0126

Cottrell, J. A., Ali, M., Tatari, A., Brett Martinson, D. (2023). Effects of fibre moisture content on the mechanical properties of jute reinforced compressed earth composites, *Construction and Building Materials*, 373, 130848, ISSN 0950-0618, https://doi.org/10.1016/j.conbuildmat.2023.130848

Da Silva, D., Kaduri, M., Poley, M., Adir, O., Krinsky, N., Shainsky-Roitman, J., & Schroeder, A. (2018). Biocompatibility, biodegradation and excretion of polylactic acid (PLA) in medical implants and theranostic systems. *Chemical Engineering Journal*, 340(January), 9–14. https://doi.org/10.1016/j.cej.2018.01.010

Devaraju, A., Harikumar, R. (2019). "Life cycle assessment of sisal fiber," in *Reference Module in Materials Science and Materials Engineering* (Elsevier Ltd.). https://doi.org/10.1016/B978-0-12-803581-8.10552-1

Dhakal, H., Bourmaud, A., Berzin, F., Almansour, F., Zhang, Z., Shah, D. U., & Beaugrand, J. (2018). Mechanical properties of leaf sheath date palm fibre waste biomass reinforced polycaprolactone (PCL) biocomposites. *Industrial Crops and Products*, 126(June), 394–402. https://doi.org/10.1016/j.indcrop.2018.10.044

Dong, A., Li, F., Fan, X., et al. 2018. Enzymatic modification of jute fabrics for enhancing the reinforcement in jute/PP composites. *Journal of Thermoplastic Composite Materials*, 31(4), 483–499. https://doi.org/10.1177/0892705717706538

Duflou, J. R. et al. (2012). Do fiber-reinforced polymer composites provide environmentally benign alternatives? A life-cycle-assessment-based study. *MRS Bulletin*, 37(4), 374–382.

Duke, J. A. (1983). *Handbook of Energy Crops*. Purdue University, Center for New Crops & Plants Products.

Elkashouty, M., Elsyaed, H., Twaffiek, S., Salem, T., Elhadad, S. S. M. (2020). An overview: Textile surface modification by using sol-gel technology. *Egyptian Journal of Chemistry*, 63(9): 3301–3311. https://doi.org/10.21608/ejchem.2020.24441.2464

Elseify, Lobna A. et al. (2021). *Manufacturing Automotive Components from Sustainable Natural Fiber Composites*. Cham: Springer International Publishing. Available in: https://doi.org/10.1007/978-3-030-83025-0. (Springer Briefs in Materials).

Farah, S., Anderson, D. G., & Langer, R. (2016). Physical and mechanical properties of PLA, and their functions in widespread applications—A comprehensive review. *Advanced Drug Delivery Reviews*, 107, 367–392. https://doi.org/10.1016/j.addr.2016.06.012

Faris, M. Al-Oqla, S. M. Sapuan, M. R. Ishak, N. A. A. (2015). Selecting natural fibers for bio-based materials with conflicting criteria. *American Journal of Applied Sciences*, 12(1), 64–71, https://doi.org/10.3844/ajassp.2015.64.71

Faruk, O., Bledzki, A. K., Fink, H. P., Sain, M. (2012). Biocomposites reinforced with natural fibers: 2000-2010. *Progress in Polymer Science*, 37(11), 1552–1596. https://doi.org/10.1016/j.progpolymsci.2012.04.003

Franck, R. R. (2005). 8—Abaca, Editor(s): Robert E. Franck, In *Woodhead Publishing Series in Textiles, Bast and Other Plant Fibres*, Woodhead Publishing, 315–321, ISBN 9781855736849, https://doi.org/10.1533/9781845690618.315

Gabrion, X., Koolen, G., Grégoire, M., Musio, S., Bar, M., Botturi, D., Rondi, G., de Luycker, E., Amaducci, S., Ouagne, P., Van Vuure, A., Placet, V. (2022). Influence of industrial processing parameters on the effective properties of long aligned European hemp fibres in composite materials. *Composites Part A: Applied Science and Manufacturing*, 157, 106915, ISSN 1359-835X, https://doi.org/10.1016/j.compositesa.2022.106915

Gassan, J., Gutowski, V. S. (2000). Effects of corona discharge and UV treatment on the properties of jute-fibre expoxy composites. *Composites Science and Technology*, 60(15), 2857–2863. https://doi.org/10.1016/S0266-3538(00)00168-8

George, G., Tomlal Jose, E., Jayanarayanan, K., Nagarajan, E. R., Skrifvars, M., Joseph, K. (2012). Novel bio-commingled composites based on jute/polypropylene yarns: Effect of chemical treatments on the mechanical properties. *Composites Part A: Applied Science and Manufacturing*, 43(1): 219–230. https://doi.org/10.1016/j.compositesa.2011.10.011

Gholampour, A., Ozbakkaloglu, T. (2020). A review of natural fiber composites: Properties, modification and processing techniques, characterization, applications. *Journal of Materials Science*, 55(3), 829–892. https://doi.org/10.1007/s10853-019-03990-y

Gibeop, N., Lee D. W., Prasad, C. V., Toru, F., Kim, B. S., Song, J. I. (2013). Effect of plasma treatment on mechanical properties of jute fiber/poly (lactic acid) biodegradable composites. *Advanced Composite Materials*, 22(6): 389–399. https://doi.org/10.1080/09243046.2013.843814

Goriparthi, B. K., Suman, K. N. S., Mohan Rao, N. (2012). Effect of fiber surface treatments on mechanical and abrasive wear performance of polylactide/jute composites. *Composites Part A: Applied Science and Manufacturing*, 43(10), 1800–1808. https://doi.org/10.1016/j.compositesa.2012.05.007

Grégoire, M., Barthod-Malat, B., Labonne, L., Evon, P., De Luycker, E., Ouagne, P. (2020). Investigation of the potential of hemp fibre straws harvested using a combine machine for the production of technical load-bearing textiles. *Industrial Crops and Products*, 145.

Gukendran, R., Sambathkumar, M., Prakash Raj, T., Nalinkumar, S., Nivaas, B., Sai Pranesh, B. K. (2022). Effect of various stitching patterns on mechanical properties of jute fiber reinforced composite, *Materials Today: Proceedings*, ISSN 2214-7853. https://doi.org/10.1016/j.matpr.2022.11.323

Guleria, A., Singha, A. S., Rana, R. A. J. K. (2018). Mechanical, thermal, morphological, and biodegradable studies of okra cellulosic fiber reinforced starch-based biocomposites. *Advances in Polymer Technology*, 37(1), 104–112. https://doi.org/10.1002/adv.21646

Haji, A. (2019). Dyeing of cotton fabric with natural dyes improved by mordants and plasma treatment. *Progress in Color, Colorants and Coatings*, 12(3): 191–201.

Hassan, M. Z., Roslan, S. A., Sapuan, S. M., et al. (2020). Mercerization optimization of bamboo (Bambusa vulgaris) fiber-reinforced epoxy composite structures using a Box-Behnken design. *Polymers*, 12(6), 1–19. https://doi.org/10.3390/POLYM12061367

Hossain, A. K. N. U., Sela, S. K., Hasan, N., Rahman, H., Hassan, M. N., Alam, S. M. M. (2023). Surface modification of naturally dyed jute fabric to ameliorate its multifunctional properties and electrical conductivity, *Vacuum*, 207, 111612, ISSN 0042-207X, https://doi.org/10.1016/j.vacuum.2022.111612

Hu, Q, Yan, X., Dong, C., Hu, W. (2016). Chemical modifications on linen for unsaturated polyester composites. *Chemical Research in Chinese Universities*, 32(6), 1057–1062. https://doi.org/10.1007/s40242-016-6101-y

Islam, M. Z., Sarker, M. E., Rahman, M. M., Islam, M. R., Ahmed, A. T. M. F., Mahmud, M. S., & Syduzzaman, M. (2022). Green composites from natural fibers and biopolymers: A review on processing, properties, and applications. *Journal of Reinforced Plastics and Composites*, 41(13–14), 526–557. https://doi.org/10.1177/07316844211058708

Ita-Nagy, D. et al. (2020). Life cycle assessment of bagasse fiber reinforced biocomposites. *Science of the Total Environment*, 720.

Izwan, S. M., Sapuan, S. M., Zuhri, M. Y. M., Mohamed, A. R. (2020) Effects of benzoyl treatment on NaOH treated sugar palm fiber: Tensile, thermal, and morphological properties. *Journal of Materials Research and Technology*, 9(3), 5805–5814. https://doi.org/10.1016/j.jmrt.2020.03.105

Izwan, S. M., Sapuan, S. M., Zuhri, M. Y. M., Mohamed, A. R. (2021). Thermal stability and dynamic mechanical analysis of benzoylation treated sugar palm/kenaf fiber reinforced polypropylene hybrid composites. *Polymers*, 13(17). https://doi.org/10.3390/polym13172961

Jaafar, J. et al. (2019). Important considerations in manufacturing of natural fiber composites: A review. *International Journal of Precision Engineering and Manufacturing-Green Technology*, 6(30), 647–664, 16 Jul. Available in: https://doi.org/10.1007/s40684-019-00097-2

Jagadeesh, P., Puttegowda, M., Mavinkere Rangappa, S., Siengchin, S. (2021). A review on extraction, chemical treatment, characterization of natural fibers and its composites for potential applications. *Polymer Composites*, 42(12), 6239–6264. https://doi.org/10.1002/pc.26312

Jain, D., Sekhon, H., Bera, T. K., Jain, R. (2020). Comparison of different hydrophobic treatments for the durability improvement of palmyra natural fiber composites under hydrothermal ageing environments. *Journal of Natural Fibers*, 17(11), 1668–1682. https://doi.org/10.1080/15440478.2019.1588828

Jiang, Z., Wang, Y., Huang, Z., Ma, W., Gao, S., Dong, W., & Xu, M. (2022). Green and sustainable cellulose-based shape memory composites with excellent conductivity for temperature warning. *Carbohydrate Polymers*, 276 (October 2021), 118767. https://doi.org/10.1016/j.carbpol.2021.118767

Joshi, S. V., et al. (2004). Are natural fiber composites environmentally superior to glass fiber reinforced composites? *Composites Part A: Applied Science and Manufacturing*, 35(3), 371–376.

Kabir, M. M., Wang, H., Lau, K. T., Cardona, F. (2012). Chemical treatments on plant-based natural fibre reinforced polymer composites: An overview. *Composites Part B: Engineering.* ; 43(7): 2883–2892. https://doi.org/10.1016/j.compositesb.2012.04.053

Kajiyama, T., Yasuda, T., Mimoto, S., et al. (2013). Effect of hexamethylene diisocyanate as compatibilizer on the mechanical properties of banana fiber/poly(butylene succinate) composites. *International Polymer Processing*, 28(1): 58–63. https://doi.org/10.3139/217.2661

Kalia, S., Thakur, K., Celli, A., Kiechel, M. A., Schauer, C. L. (2013). Surface modification of plant fibers using environment friendly methods for their application in polymer composites, textile industry and antimicrobial activities: A review. *Journal of Environmental Chemical Engineering*, 1(3), 97–112. https://doi.org/10.1016/j.jece.2013.04.009

Karim, N., Sarker, F., Afroj, S., Zhang, M., Potluri, P., Novoselov, K. S.. (2021). Sustainable and multifunctional composites of graphene-based natural jute fibers. *Advanced Sustainable Systems*, 5(3), https://doi.org/10.1002/adsu.202000228

Karthi, N., Kumaresan, K., Sathish, S., Gokulkumar, S., Prabhu, L., Vigneshkumar, N. (2019). An overview: Natural fiber reinforced hybrid composites, chemical treatments and application areas. *Materials Today: Proceedings*, 27: 2828–2834. https://doi.org/10.1016/j.matpr.2020.01.011

Kato, H., Nakatsubo, F., Abe, K., Yano, H. (2015). Crosslinking via sulfur vulcanization of natural rubber and cellulose nanofibers incorporating unsaturated fatty acids. *RSC Advances*, 5(38), 29814–29819. https://doi.org/10.1039/C4RA14867C

Kenned, J. J., Sankaranarayanasamy, K., Binoj, J. S., Chelliah, S. K. (2020). Thermo-mechanical and morphological characterization of needle punched non-woven banana fiber reinforced polymer composites. *Composites Science and Technology*, 185 (October 2019): 107890. https://doi.org/10.1016/j.compscitech.2019.107890

Khalid, M. Y., Rashid, A. A., Arif, Z. U., Ahmed, W., Arshad, H., Zaidi, A. A. (2021). Natural fiber reinforced composites: Sustainable materials for emerging applications, *Results in Engineering*, 11(April), 100263, ISSN 2590-1230, https://doi.org/10.1016/j.rineng.2021.100263

Khoshnava, S. M., Rostami, R., Ismail, M., Valipour, A. (2014). The using fungi treatment as green and environmentally process for surface modification of natural fibres. *Applied Mechanics and Materials*, 554, 116–122. https://doi.org/10.4028/www.scientific.net/AMM.554.116

Koller, M. (2019). Switching from petro-plastics to microbial polyhydroxyalkanoates (PHA): The biotechnological escape route of choice out of the plastic predicament? *The EuroBiotech Journal*, 3(1), 32–44. https://doi.org/10.2478/ebtj-2019-0004

Kommula, V. P., Reddy, K. O., Shukla, M., Marwala, T., Rajulu, A. V. (2013). Physico-chemical, tensile, and thermal characterization of napier grass (Native African) fiber strands. *International Journal of Polymer Analysis and Characterization*, 18(4), 303–314. https://doi.org/10.1080/1023666X.2013.784935

Koohestani, B., Darban, A. K., Mokhtari, P., Yilmaz, E., Darezereshki, E. 2019. Comparison of different natural fiber treatments: A literature review. *International Journal of Environmental Science and Technology*, 16(1): 629–642. https://doi.org/10.1007/s13762-018-1890-9

Krishna, B. G., Sahoo, A. K., Ranveer, R. C. (2015). Bioactive components of flaxseed and its health benefits, *International Journal of Pharmaceutical Sciences Review and Research*, 31(1), 42–51.

Kumar, T. P. S., Satheesh Kumar, S., Rajesh Kumar, L. (2022). Chapter 11—Jute fibers, their composites and applications, Editor(s): Sanjay, M. R., Jyotishkumar, P., Suchart, S., Togay, O., Hao, W., In *The Textile Institute Book Series, Plant Fibers, their Composites, and Applications*, Woodhead Publishing, 253–282, ISBN 9780128245286, https://doi.org/10.1016/B978-0-12-824528-6.00020-5

Kumaran, S., Srinivasan, K., Ponmariappan, M., Yashwhanth, S., Akshay, S., Hu, Y. C. (2020). Study of raw and chemically treated Sansevieria ehrenbergii fibers for brake pad application. *Materials Research Express*, 7(5), 055102. https://doi.org/10.1088/2053-1591/ab8f48

Lau, K. T., Hung, P. Y., Zhu, M. H., Hui, D. (2018). Properties of natural fibre composites for structural engineering applications. *Composites Part B: Engineering*, 136 (September 2017), 222–233. https://doi.org/10.1016/j.compositesb.2017.10.038

Le Duigou, A., Davies, P., Baley, C. (2012). Replacement of glass/unsaturated polyester composites by Flax/PLLA biocomposites: Is it justified? *Journal of Biobased Materials and Bioenergy*, 5(4), p. 466–482.

Li, M., et al. (2020). Recent advancements of plant-based natural fiber–reinforced composites and their applications. *Composites Part B: Engineering*, 200(February).

Li, Q., Li, Y., Zhou, L. (2017). Nanoscale evaluation of multi-layer interfacial mechanical properties of sisal fiber reinforced composites by nanoindentation technique. *Composites Science and Technology*, 152, 211–221.

Li, Y., Mai, Y. W., Ye, L. (2000). Sisal fibre and its composites: A review of recent developments, *Composites Science and Technology*, 60(11), 2037–2055, ISSN 0266-3538. https://doi.org/10.1016/S0266-3538(00)00101-9

Liminana, P., Garcia-Sanoguera, D., Quiles-Carrillo, L., Balart, R., Montanes, N. (2018). Development and characterization of environmentally friendly composites from poly(butylene succinate) (PBS) and almond shell flour with different compatibilizers. *Composites Part B: Engineering*, 144(February), 153–162. https://doi.org/10.1016/j.compositesb.2018.02.031

Luo, Q. Z., Huang, Q., Chen, Z., et al. (2019). Effect of the corona treatment on the microstructure of PVDF probed by electrochemical impedance spectroscopy. *Materials Research Express*, 6(1). https://doi.org/10.1088/2053-1591/aae87e

Macedo, M. J. P., Silva, G. S., Feitor, M. C., Costa, T. H. C., Ito, E. N., Melo, J. D. D. (2020). Surface modification of kapok fibers by cold plasma surface treatment. *Journal of Materials Research and Technology*, 9(2), 2467–2476, ISSN 2238-7854. https://doi.org/10.1016/j.jmrt.2019.12.077

Madhu, P., Sanjay, M. R., Jawaid, M., Siengchin, S., Khan, A., Pruncu, C. I. (2020). A new study on effect of various chemical treatments on Agave Americana fiber for composite reinforcement: Physico-chemical, thermal, mechanical and morphological properties. *Polymer Testing*, 85(January), 106437. https://doi.org/10.1016/j.polymertesting.2020.106437

Mahakur, V. K., Paul, R., Bhowmik, S., Patowari, P. K. (2022). Influence of surface modification on mechanical and tribology performance of jute filler polymer composites and prediction of the performance using artificial neural network. *Polymer Bulletin*, 1–22. https://doi.org/10.1007/s00289-022-04636-x

Mahakur, V.K., Paul, R., Bhowmik, S., Patowari, P. K. (2023). Influence of surface modification on mechanical and tribology performance of jute filler polymer composites and prediction of the performance using artificial neural network. *Polymer Bulletin*, 80(11), 11953–11974.

Mamun, A. A., Heim, H. P., Faruk, O., Bledzki, A. K. (2015). 8—The use of banana and abaca fibres as reinforcements in composites, Editor(s): Omar, F., Mohini, S., *Biofiber Reinforcements in Composite Materials*, Woodhead Publishing, 236–272, ISBN 9781782421221, https://doi.org/10.1533/9781782421276.2.236

Mann, G. S., Singh, L. P., Kumar, P., Singh, S. (2020). Green composites: A review of processing technologies and recent applications. *Journal of Thermoplastic Composite Materials*, 33(8), 1145–1171. https://doi.org/10.1177/0892705718816354

Mansor, M. R. et al. (2018). *The Environmental Impact of Natural Fiber Composites Through Life Cycle Assessment Analysis*. [S.l.]: Elsevier Ltd. Available in: http://dx.doi.org/10.1016/B978-0-08-102290-0.00011-8

Maqsood, H. S., Bashir, U., Wiener, J., Puchalski, M., Sztajnowski, S., Militky, J. (2017). Ozone treatment of jute fibers. *Cellulose*, 24(3): 1543–1553. https://doi.org/10.1007/s10570-016-1164-y

Martelli, T. M., Assis, O. B. G., Silva, N. C., Esposto, B. S., Martins, M. A., Tapia-Blácido D. R. (2017). Chemical treatment and characterization of soybean straw and soybean protein isolate/straw composite films. *Carbohydrate Polymers*, 157, 512–520. https://doi.org/10.1016/j.carbpol.2016.10.013

Mathai, P. M. (2005), Coir, in Franck, R. R., *Bast and Other Plant Fibres*, Woodhead Publishing Limited, London, 275.

Mazerolles, T., Heuzey, M. C., Soliman, M., Martens, H., Kleppinger, R., & Huneault, M. A. (2019). Development of co-continuous morphology in blends of thermoplastic starch and low-density polyethylene. *Carbohydrate Polymers*, 206(November 2018), 757–766. https://doi.org/10.1016/j.carbpol.2018.11.038

Meereboer, K. W., Misra, M., & Mohanty, A. K. (2020). Review of recent advances in the biodegradability of polyhydroxyalkanoate (PHA) bioplastics and their composites. *Green Chemistry*, 22(17), 5519–5558. https://doi.org/10.1039/d0gc01647k

Meirheage, C. (2011). *Evaluation de la disponibilité et de l'accessibilité de fibres végétales à usages matériaux en France*. In: FRD, editor.: Etude r´ealis´ ee pour le compte de l'ADEME par FRD, Mars, 84.

Mochizuki, M., Hirami, M. (1996). Structural effects on biodegradation of aliphatic polyesters. *Sen'i Gakkaishi*, 52(5), P200–P208. https://doi.org/10.2115/fiber.52.5_P200

Mohammed, A. A., Bachtiar, D., Rejab, M. R. M., Hasany, S. F. (2017). Effect of potassium permanganate on tensile properties of sugar palm fibre reinforced thermoplastic polyurethane. *Indian Journal of Science and Technology*, 10(7): 1–5. https://doi.org/10.17485/ijst/2017/v10i7/111453

Mohammed, M., Rahman, R., Mohammed, A. M., et al. (2022). Surface treatment to improve water repellence and compatibility of natural fiber with polymer matrix: Recent advancement. *Polymer Testing*, 115(July), 107707. https://doi.org/10.1016/j.polymertesting.2022.107707

Mohanty, J. R., Das, S. N., Das, H. C., Swain, S. K. 2013. Effective mechanical properties of polyvinylalcohol biocomposites with reinforcement of date palm leaf fibers. *Polymer Composites*, 34(6), 959–966. https://doi.org/10.1002/pc.22502

Musio, S., Müssig, J., Amaducci, S. (2018). Optimizing hemp fiber production for high performance composite applications. *Frontiers in Plant Science*, 2018, 9.

Naik, N., Shivamurthy, B., Thimmappa, B. H. S., Jaladi, G., Samanth, K., Shetty, N. (2022). Recent advances in green composites and their applications. *Engineered Science*, 21, 1–16. https://doi.org/10.30919/es8e779

Nayak, S. K., Dixit, G., Appukuttan, K. K. (2012). Sisal fiber (SF) reinforced recycled polypropylene (RPP) composites. *International Journal of Plastics Technology*. 16(2), 150–165. https://doi.org/10.1007/s12588-012-9042-4

Nematollahi, M., Karevan, M., Mosaddegh, P., Farzin, M. (2019). Morphology, thermal and mechanical properties of extruded injection molded kenaf fiber reinforced polypropylene composites. *Materials Research Express*, 6(9), 095409. https://doi.org/10.1088/2053-1591/ab2fbd

Nishino, T., Takano, K., & Nakamae, K. (1995). Elastic modulus of the crystalline regions of cellulose polymorphs. *Journal of Polymer Science Part B: Polymer Physics*, 33(11), 1647–1651. https://doi.org/10.1002/polb.1995.090331110

Oudrhiri, H. F., Merbahi, N., Oushabi, A., Elfadili, M. H., Kammouni, A., Oueldna, N. (2020). Effects of corona discharge treatment on surface and mechanical properties of Aloe Vera fibers. *Materials Today: Proceedings*, 24, 46–51. https://doi.org/10.1016/j.matpr.2019.07.527

Palai, B. K., Sarangi, S. K. (2022). Characterization of chemical treated Eichhornia Crassipes fibers and its composites. *Journal of Natural Fibers*, 19(15), 10883–10896. https://doi.org/10.1080/15440478.2021.2002777

Pankaj, J. C. S., Kant, S. (2022). Critical review on chemical treatment of natural fibers to enhance mechanical properties of bio composites. *Silicon*, 14(10), 5103–5124. https://doi.org/10.1007/s12633-021-01194-1

Pati, S., Maity, S., Dash, A., Jema, S., Mohapatra, S., Das, S., & Samantaray, D. P. (2020). Biocompatible PHB production from Bacillus species under submerged and solid-state fermentation and extraction through different downstream processing. *Current Microbiology*, 77(7), 1203–1209. https://doi.org/10.1007/s00284-020-01922-7

Paul, S. A., Joseph, K., Mathew, G. D. G., Pothen, L. A., Thomas S. (2010). Influence of polarity parameters on the mechanical properties of composites from polypropylene fiber and short banana fiber. *Composites Part A: Applied Science and Manufacturing*, 41(10), 1380–1387. https://doi.org/10.1016/j.compositesa.2010.04.015

Paul, S. K., Chakraborty, S. (2019). Mixing effects on the kinetics of enzymatic hydrolysis of lignocellulosic Sunn hemp fibres for bioethanol production. *Chemical Engineering Journal*, 377. https://doi.org/10.1016/j.cej.2018.10.040

Pickering, K. L., Rowell, R. M., Thomas, S., et al. (2008). *Properties and Performance of Natural-Fibre Composites*. (Pickering, K. L., ed.).

Pickering, K. L., Sunny, T. (2021). Effect of stearic acid treatment on the properties of aligned short hemp fibre mats and their potential use as reinforcement in polypropylene matrix composites. *Cellulose*, 28(9), 5611–5624. https://doi.org/10.1007/s10570-021-03884-w

Potting, J. et al. (2017). *Circular Economy: Measuring Innovation in the Product Chain*. PBL Netherlands Environmental Assessment Agency, n. 2544, 42.

Radzuan, N. A. Mohd, D. T., Sulong, A. B., Muhamad, N., Haron, C. H. C. (2020). New processing technique for biodegradable kenaf composites: A simple alternative to commercial automotive parts, *Composites Part B: Engineering*, 184, 107644, ISSN 1359-8368, https://doi.org/10.1016/j.compositesb.2019.107644

Ragoubi, M., Bienaimé, D., Molina, S., George, B., Merlin, A. (2010). Impact of corona treated hemp fibres onto mechanical properties of polypropylene composites made thereof. *Industrial Crops and Products*, 31(2), 344–349. https://doi.org/10.1016/j.indcrop.2009.12.004

Ragoubi, M., George, B., Molina, S., et al. (2012). Effect of corona discharge treatment on mechanical and thermal properties of composites based on miscanthus fibres and polylactic acid or polypropylene matrix. *Composites Part A: Applied Science and Manufacturing*, 43(4) 675–685. https://doi.org/10.1016/j.compositesa.2011.12.025

Rahman, M. M., Khan, M. A. (2007). Surface treatment of coir (Cocos nucifera) fibers and its influence on the fibers' physico-mechanical properties. *Composites Science and Technology*, 67(11-12), 2369–2376. https://doi.org/10.1016/j.compscitech.2007.01.009

Rajeshkumar, G., Arvindh Seshadri, S., Devnani, G. L., Sanjay, M. R., Siengchin, S., Prakash Maran, J., … Ronaldo Anuf, A. (2021). Environment friendly, renewable and sustainable poly lactic acid (PLA) based natural fiber reinforced composites—A comprehensive review. *Journal of Cleaner Production*, 310(January), 127483. https://doi.org/10.1016/j.jclepro.2021.127483

Ramachandran, A. R., Mavinkere Rangappa, S., Kushvaha, V., Khan, A., Seingchin, S., Dhakal, H. N. (2022). Modification of fibers and matrices in natural fiber reinforced polymer composites: A comprehensive review. *Macromolecular Rapid Communications*, 43(17), 1–38. https://doi.org/10.1002/marc.202100862

Ramaswamy, A. P., Rigacci, A. (2021). Superinsulating composite aerogels from polymethylsilsesquioxane and kapok fibers, *Materials Chemistry and Physics*, 261, 124252, ISSN 0254-0584. https://doi.org/10.1016/j.matchemphys.2021.124252

Ramesh, M. (2016). Kenaf (Hibiscus cannabinus L.) fibre based bio-materials: A review on processing and properties. *Progress in Materials Science*, 78–79, 1–92. https://doi.org/10.1016/j.pmatsci.2015.11.001

Ramesh, M., Bhoopathi, R. (2022). Chapter 10—Hemp fibers, their composites and applications, Editor(s): Sanjay, M. R., Jyotishkumar, P., Suchart, S., Togay, O., Hao, W., In *The Textile Institute Book Series, Plant Fibers, their Composites, and Applications*, Woodhead Publishing, 233–252, ISBN 9780128245286, https://doi.org/10.1016/B978-0-12-824528-6.00013-8

Ramesh, M., Deepa, C., Kumar, L. R., Sanjay, M. R., Siengchin, S. (2020). Life-cycle and environmental impact assessments on processing of plant fibres and its bio-composites: A critical review, *Journal of Industrial Textiles*. https://doi.org/10.1177/1528083720924730

Rao, J., Bao, L., Wang, B., Fan, M., Feo, L. (2018). Plasma surface modification and bonding enhancement for bamboo composites. *Composites Part B: Engineering*, 138(June 2017), 157–167. https://doi.org/10.1016/j.compositesb.2017.11.025

Rasal, R. M., Janorkar, A. V., Hirt, D. E. (2010). Poly(lactic acid) modifications. *Progress in Polymer Science*, 35(3), 338–356. https://doi.org/10.1016/j.progpolymsci.2009.12.003

Raza, Z. A., Abid, S., & Banat, I. M. (2018). Polyhydroxyalkanoates: Characteristics, production, recent developments and applications. *International Biodeterioration & Biodegradation*, 126(January 2017), 45–56. https://doi.org/10.1016/j.ibiod.2017.10.001

Roy, S., Lutfar, L. B. (2012). Bast fibres: Ramie. *Handbook of Natural Fibers*, Woodhead Publishing, ISBN: 978-0-12-818398-4.

Sachan, R., Warkar, S. G., Purwar, R. (2023). An overview on synthesis, properties and applications of polycaprolactone copolymers, blends & composites. *Polymer-Plastics Technology and Materials*, 62(3), 327–358. https://doi.org/10.1080/25740881.2022.2113890

Safri, S. N. A., Sultan, M. T. H., Jawaid, M., Abdul Majid, M. S. (2019). Analysis of dynamic mechanical, low-velocity impact and compression after impact behaviour of benzoyl treated sugar palm/glass/epoxy composites. *Composite Structures*, 226(June). https://doi.org/10.1016/j.compstruct.2019.111308

Sánchez, M. L., Patiño, W., Cárdenas, J. (2020). Physical-mechanical properties of bamboo fibers-reinforced biocomposites: Influence of surface treatment of fibers. *Journal of Building Engineering*, 28(May 2019). https://doi.org/10.1016/j.jobe.2019.101058

Santos, W. F., Quattrone, M., John, V. M., Angulo, S. C. (2017). Roughness, wettability and water absorption of water repellent treated recycled aggregates. *Construction and Building Materials*, 146, 502–513. https://doi.org/10.1016/j.conbuildmat.2017.04.012

Sarkar, N., Sahoo, G., Khuntia, T., Priyadarsini, P., Mohanty, J. R., Swain, S. K. (2017). Fabrication of acrylic modified coconut fiber reinforced polypropylene biocomposites: Study of mechanical, thermal, and erosion properties. *Polymer Composites*, 38(12): 2852–2862. https://doi.org/10.1002/pc.23887

Sarmin, S. N. (2022). Chapter 3—Anatomical structure of coir fibers, Editor(s): Mohammad, J., In *Woodhead Publishing Series in Composites Science and Engineering, Coir Fiber and its Composites*, Woodhead Publishing, 43–54, ISBN 9780443151866, https://doi.org/10.1016/B978-0-443-15186-6.00033-3

Schumacher, D. A. G., Pequito, S., Pazour, J. (2020). Industrial hemp fiber: A sustainable and economical alternative to cotton. *Journal of Cleaner Production*, 268.

Sessini, V., Arrieta, M. P., Raquez, J., Dubois, P., Kenny, J. M., & Peponi, L. (2019). Thermal and composting degradation of EVA/thermoplastic starch blends and their nanocomposites. *Polymer Degradation and Stability*, 159, 184–198. https://doi.org/10.1016/j.polymdegradstab.2018.11.025

Shalwan, A., Yousif, B. F. (2013). In state of art: Mechanical and tribological behaviour of polymeric composites based on natural fibres. *Materials and Design*, 48, 14–24. https://doi.org/10.1016/j.matdes.2012.07.014

Sharma, A. K., Bhandari, R., Sharma, C., Dhakad, S. K., Pinca-Bretotean, C. (2022). Polymer matrix composites: A state of art review. *Materials Today: Proceedings*, 57, 2330–2333.

Shekar, H. S. S., Ramachandra, M. (2018). Green composites: A review. *Materials Today: Proceedings*, 5(1), 2518–2526. https://doi.org/10.1016/j.matpr.2017.11.034

Shen, J., Zhang, Y. (2020). Fiber-reinforced mechanism and mechanical performance of composite fibers reinforced concrete. *Journal of Wuhan University of Technology*, 35, 121–130.

Shirke, M. B., Shelke, S. N. (2020). Significance of chemical treatment on the performance of natural fiber: A review. *I-Manager's Journal on Material Science*, 8(2): 23. https://doi.org/10.26634/jms.8.2.16291

Siakeng, R., Jawaid, M., Asim, M., et al. (2020). Alkali treated coir/pineapple leaf fibres reinforced pla hybrid composites: Evaluation of mechanical, morphological, thermal and physical properties. *Express Polymer Letters*, 14(8): 717–730. https://doi.org/10.3144/expresspolymlett.2020.59

Simbaña, E. A., Ordóñez, P. E., Ordóñez, Y. F., Guerrero, V. H., Mera, M. C., Carvajal, E. A. (2020). 6—Abaca: Cultivation, obtaining fibre and potential uses, Editor(s): Ryszard, M. K., Maria, M.-T. In *Woodhead Publishing Series in Textiles, Handbook of Natural Fibres* (2nd ed.), Woodhead Publishing, 197–218, ISBN 9780128183984, https://doi.org/10.1016/B978-0-12-818398-4.00008-6

Singh, J. I. P., Singh, S., Dhawan, V. (2020). Influence of fiber volume fraction and curing temperature on mechanical properties of jute/PLA green composites. *Polymers and Polymer Composites*, 28(4), 273–284. https://doi.org/10.1177/0967391119872875

Singh, M. K. et al. (2023). A comprehensive review of various factors for application feasibility of natural fiber-reinforced polymer composites. *Results in Materials*, 17(December), 100355, 2022. Available in: https://doi.org/10.1016/j.rinma.2022.100355

Sinha, E., Panigrahi, S. (2009). Effect of plasma treatment on structure, wettability of jute fiber and flexural strength of its composite. *Journal of Composite Materials*, 43(17), 1791–1802. https://doi.org/10.1177/0021998309338078

Soman, S., Chacko, A. S., Prasad, V. S., Anju, P., Surya, B. S., Vandana, K. (2018). Self-assembly of oleylamine modified nano-fibrillated cellulose from areca husk fibers into giant vesicles. *Carbohydrate Polymers*, 182 (November 2017): 69–74. https://doi.org/10.1016/j.carbpol.2017.11.010

Spörl, J. M., Batti, F., Vocht, M. P., Raab, R., Müller, A., Hermanutz, F., & Buchmeiser, M. R. (2018). Ionic liquid approach toward manufacture and full recycling of all-cellulose composites. *Macromolecular Materials and Engineering*, 303(1), 1–8. https://doi.org/10.1002/mame.201700335

Sreekala, M. S., Kumaran, M. G., Joseph, S., Jacob, M., Thomas, S. (2000). Oil palm fibre reinforced phenol formaldehyde composites: Influence of fibre surface modifications on the mechanical performance. *Applied Composite Materials*, 7(5-6): 295–329. https://doi.org/10.1023/A:1026534006291

Sreenivasan, V. S., Ravindran, D., Manikandan, V., Narayanasamy, R. (2012). Influence of fibre treatments on mechanical properties of short Sansevieria cylindrica/polyester composites. *Materials and Design*, 37, 111–121. https://doi.org/10.1016/j.matdes.2012.01.004

Summerscales, J. (2021). A review of bast fibres and their composites: Part 4 ~ organisms and enzyme processes. *Composites Part A: Applied Science and Manufacturing*, 140(August 2020): 106149. https://doi.org/10.1016/j.compositesa.2020.106149

Sun, L., Wang, S., Zhang, J, et al. (2020). Preparation of a novel flame retardant containing triazine groups and its application on cotton fabrics. *New Journal of Chemistry*, 44(18), 7386–7394. https://doi.org/10.1039/C9NJ06268H

Sun, Z., Zhang, L., Liang, D., Xiao, W., Lin, J. (2017). Mechanical and thermal properties of PLA biocomposites reinforced by coir fibers. *International Journal of Polymer Science*, 2017, 1–9. https://doi.org/10.1155/2017/2178329

Swain, P. T. R., Biswas, S. (2017). Abrasive wear behaviour of surface modified jute fiber reinforced epoxy composites. *Materials Research*, 20, 661–674. https://doi.org/10.1590/1980-5373-MR-2016-0541

Tai, N. L., Adhikari, R., Shanks, R., Adhikari, B. (2019). Aerobic biodegradation of starch–polyurethane flexible films under soil burial condition: Changes in physical structure and chemical composition. *International Biodeterioration & Biodegradation*, 145, 104793.

Tai, N. L., Ghasemlou, M., Adhikari, R., & Adhikari, B. (2021). Starch-based isocyanate- and non-isocyanate polyurethane hybrids: A review on synthesis, performance and biodegradation. *Carbohydrate Polymers*, 265, 118029. https://doi.org/10.1016/j.carbpol.2021.118029

Tamrakar, S., Kiziltas, A., Mielewski, D., Zander, R., (2021). Characterization of kenaf and glass fiber reinforced hybrid composites for underbody shield applications, *Composites Part B: Engineering*, 216, 108805, ISSN 1359-8368, https://doi.org/10.1016/j.compositesb.2021.108805

Tanasă, F., Zănoagă, M., Teacă, C. A., Nechifor, M., Shahzad, A. (2020). Modified hemp fibers intended for fiber-reinforced polymer composites used in structural applications—A review. I. Methods of modification. *Polymer Composites*, 41(1), 5–31. https://doi.org/10.1002/pc.25354

Tang, L., Liu, T., Sun, P., Wang, Y., Liu, G. (2022). Sisal fiber modified construction waste recycled brick as building material: Properties, performance and applications, *Structures*, 46, 927–935, ISSN 2352-0124, https://doi.org/10.1016/j.istruc.2022.10.126

Teixeira, R. S., Bufalino, L., Tonoli, G. H. D., dos Santos, S. F., Savastano, J. H. (2022). 32—Coir fiber as reinforcement in cement-based materials, Editor(s): Sanjay, M. R., Madhu, P., Jyotishkumar, P., Suchart, S., Sergey, G., In *The Textile Institute Book Series, Advances in Bio-Based Fiber*, Woodhead Publishing, 707–739, ISBN 9780128245439, https://doi.org/10.1016/B978-0-12-824543-9.00008-6

Téraube, O., Gratier, L., Agopian, J.-C., Pucci, M. F., Liotier, P.-J., Hajjar-Garreau, S., Petit, E., Batisse, N., Bousquet, A., Charlet, K., Tomasella, É., Dubois, M. (2023). Elaboration of hydrophobic flax fibers through fluorine plasma treatment, *Applied Surface Science*, 611, Part A, 155615, ISSN 0169-4332, https://doi.org/10.1016/j.apsusc.2022.155615

Thamarai, S. S., Sunitha, R., Ammayappan, L., Prakash, C. (2023). Impact of chemical treatment on surface modification of agave Americana fibres for composite application–A futuristic approach. *Journal of Natural Fibers*, 20(1). https://doi.org/10.1080/15440478.2022.2142726

Thiruchitrambalam, M., Alavudeen, A., Venkateshwaran, N. (2012). Review on kenaf fiber composites. *Reviews on Advanced Materials Science* 32 (2), 106–111.

Thyavihalli Girijappa, Y. G., Mavinkere Rangappa, S., Parameswaranpillai, J., Siengchin, S. (2019). Natural fibers as sustainable and renewable resource for development of eco-friendly composites: A comprehensive review. *Frontiers in Materials*, 6(September), 1–14. https://doi.org/10.3389/fmats.2019.00226

Townsend, T. (2020). World natural fibre production and employment. In *Handbook of Natural Fibres*. Woodhead Publishing. 15–36.

Tripathi, S. K., Bhardwaj, N. K., Roy Ghatak, H. (2020). Developments in ozone-based bleaching of pulps. *Ozone: Science & Engineering*, 42(2): 194–210. https://doi.org/10.1080/01919512.2019.1647407

Vandepitte, K., Vasile, S., Vermeire, S., Vanderhoeven, M., Van der Borght, W., Latré, J., et al. (2020). Hemp (Cannabis sativa L.) for high-value textile applications: The effective long fiber yield and quality of different hemp varieties, processed using industrial flax equipment. *Industrial Crops and Products*, 158: 112969.

Vigneshwaran, S. et al. (2020). Recent advancement in the natural fiber polymer composites: A comprehensive review. *Journal of Cleaner Production*, 277, 124109. Available in: https://doi.org/10.1016/j.jclepro.2020.124109

Vijay, R., Manoharan, S., Arjun, S., Vinod, A., Singaravelu, D. L. (2021). Characterization of silane-treated and untreated natural fibers from stem of leucas aspera. *Journal of Natural Fibers*, 18(12), 1957–1973. https://doi.org/10.1080/15440478.2019.1710651

Vinod, A., Sanjay, M. R., Suchart, S., & Jyotishkumar, P. (2020). Renewable and sustainable biobased materials: An assessment on biofibers, biofilms, biopolymers and biocomposites. *Journal of Cleaner Production*, 258, 120978. Available in: https://doi.org/10.1016/j.jclepro.2020.120978

Vorawongsagul, S., Pratumpong, P., & Pechyen, C. (2021). Preparation and foaming behavior of poly (lactic acid)/poly (butylene succinate)/cellulose fiber composite for hot cups packaging application. *Food Packaging and Shelf Life*, 27(November 2020). https://doi.org/10.1016/j.fpsl.2020.100608

Wang, Q., Zhang, Y., Liang, W., Wang, J., Chen, Y. (2020). Effect of silane treatment on mechanical properties and thermal behavior of bamboo fibers reinforced polypropylene composites. *Journal of Engineered Fibers and Fabrics*, 15(33). https://doi.org/10.1177/1558925020958195

Wang, W., Fu, R., Deng, Q., et al. (2020). Surface modification of flax fibers with isocyanate and its effects on fiber/epoxy interfacial properties. *Fibers and Polymers*, 21(12), 2888–2895. https://doi.org/10.1007/s12221-020-9722-1

Wong, C., Kiew, R., Argent, G., Set, O., Lee, S. K., Gan, Y. Y. 2002. Assessment of the validity of the sections in musa (musaceae) using AFLP. *Annals of Botany*, 90 (2), 231–238. https://doi.org/10.1093/aob/mcf170

Wu, C.-S., Liao, H.-T. (2012). Polycaprolactone-based green renewable ecocomposites made from rice straw fiber: characterization and assessment of mechanical and thermal properties. *Industrial & Engineering Chemistry Research*, 51(8), 3329–3337. https://doi.org/10.1021/ie202002p

Wu, C. S., Liao, H. T., & Cai, Y. X. (2017). Characterisation, biodegradability and application of palm fibre-reinforced polyhydroxyalkanoate composites. *Polymer Degradation and Stability*, 140, 55–63. https://doi.org/10.1016/j.polymdegradstab.2017.04.016

Wu, C. S., Tsou, C. H. (2019). Fabrication, characterization, and application of biocomposites from poly(lactic acid) with renewable rice husk as reinforcement. *Journal of Polymer Research*, 26(2). https://doi.org/10.1007/s10965-019-1710-z

Xie, Q., Li, F., Li, J., Wang, L., Li, Y., Zhang, C., Xu, J. (2018). A new biodegradable sisal fiber—Starch packing composite with nest structure. *Carbohydrate Polymers*, 189, 56–64. https://doi.org/10.1016/j.carbpol.2018.01.063

Xu, X. L., Zhang, M., Qiang, Q., Song, J. Q., He, W. Q. (2016). Study on the performance of the acetylated bamboo fiber/PBS composites by molecular dynamics simulation. *Journal of Composite Materials*, 50(7), 995–1003. https://doi.org/10.1177/0021998315615690

Zaki, F. A., Abdullah, I. (2015). Graft copolymerization of acrylonitrile onto torch ginger cellulose. *Sains Malaysiana*, 44(6), 853–859. https://doi.org/10.17576/jsm-2015-4406-11

Zaman, H. U., Khan, M. A., Khan, R. A. (2011). Effect of ultraviolet radiation and potassium permanganate on the physico-mechanical properties of jute polymer composites. *Polymer - Plastics Technology and Engineering*, 50(7): 651–659. https://doi.org/10.1080/03602559.2010.543237

Zaman, H. U., Khan, R. A. (2021). Acetylation used for natural fiber/polymer composites. *Journal of Thermoplastic Composite Materials*, 34(1), 3–23. https://doi.org/10.1177/0892705719838000

Zhang, L., Meng, C., Fu, J., et al. 2022. Effect of ozone treatment on the chemical and mechanical properties of flax fibers. *Industrial Crops and Products*, 189(October), 115694. https://doi.org/10.1016/j.indcrop.2022.115694

Zhao, Z., Teng, K., Li, N., Li, X., Xu, Z., Chen, L., et al. (2017). Mechanical, thermal and interfacial performances of carbon fiber reinforced composites flavored by carbon nanotube in matrix/interface. *Composite Structures*, 159, 761–772. https://doi.org/10.1016/j.compstruct.2016.10.022

Zheng, Y., Wang, A. (2014). Kapok fiber: Applications, In Khalid, R. H., Mohammad, J., Umer, R., *Biomass and Bioenergy Applications*, Springer, 251–266.

Zheng, Y., Wang, J., Zhu, Y., Wang, A. (2015). Research and application of kapok fiber as an absorbing material: A mini review, *Journal of Environmental Sciences* 27 21–32.

Zwawi, M. (2021). A review on natural fiber bio-composites, surface modifications and applications. *Molecules*, 26(2), 404. https://doi.org/10.3390/molecules26020404

3 Fabrication of NFCs
Techniques and Procedures

J. S. G. Fracaro and P. C. Gonçalves

3.1 INTRODUCTION

Composite materials result from the amalgamation of two or more materials with disparate properties, wherein one component serves as the matrix and the others act as reinforcement. This blending imparts distinctive characteristics and specific properties to the end product. In recent times, there has been a notable surge in the utilization of polymer composites fortified with plant fibers, yielding a blend of heightened performance, adaptability, and cost advantages. Plant fibers have emerged as a significant category of reinforcing materials, exhibiting attributes highly valued in the realm of polymer composites, including low density, cost-effectiveness, biodegradability, processing flexibility, and origins from renewable and non-toxic sources. A detailed analysis on natural fiber-based composites such as sansevieria/jute fiber, jute/sisal fiber, glass/sisal fiber, and bamboo/jute fiber has been carried out by Kumar et al. (2022).

Numerous natural fibers are already employed in composite materials, showcasing strong mechanical properties when utilized as reinforcement in polymer resins. These fibers find application in various industries, including automotive, furniture, packaging, and aerospace, highlighting their versatility. Notably, natural fibers surpass glass fibers in efficiency as thermal and acoustic insulators. Furthermore, they exhibit non-abrasive qualities to equipment, are non-irritating to the skin, and are obtained with an energy-saving advantage of up to 80%.

In the investigation conducted by De Paula et.al (2014), the focus was on the creation of a thermoplastic matrix utilizing virgin polypropylene (PP) and sugarcane bagasse fibers as the dispersed phase (reinforcement), obtained through manual separation. The study delved into examining the mechanical properties of the sugarcane fiber-reinforced composite, with fiber lengths cut to 3 and 8 mm. The analysis explored variations in mechanical properties in relation to the volumetric fraction within the composite, specifically at 15%, 20%, and 25%.

A multitude of fibers worldwide offers diverse possibilities as reinforcement materials, yet often, they go unused and are lost in the process. Take banana fiber, for instance, derived from the pseudostem of the plant, which remains unused after fruit harvesting. Motaleb et al. (2021) conducted a study on the mechanical and physical properties of a composite comprising an epoxy and polyester resins matrices reinforced with banana fiber. Various combinations of fabricated composites, were tested. The results underscored the impacts of fiber alignment, treatment, volumetric fraction on tensile strength, modulus of elasticity in tension, flexural strength, and modulus of elasticity in flexure of the test specimens.

For natural fibers and the polymer matrix to synergize effectively in a particular application, it is essential to establish adequate interfacial contact between them. Surface modification of the fiber serves to diminish its hydrophilic character, typically leading to enhanced adhesion with the polymer matrix. Numerous studies suggest that the pre-incorporation surface modification of natural fibers in polymer matrices mitigates polarity differences and enhances compatibility between the fiber and the matrix.

DOI: 10.1201/9781032665337-3

The evolution of composites reinforced with natural fibers aligns seamlessly with the contemporary ethos of sustainable development, fostering the creation of eco-friendly materials. The crux of rendering these innovative materials both economically viable and environmentally sound lies in tailoring established composite manufacturing techniques to the distinctive and inherent properties of natural fibers. One such technique is the infusion process, recognized for its closed mold processing, which contributes significantly to curbing volatile emissions during manufacturing. In a study by Rodrigues (2015), the impact of key variables in the infusion process on composite properties was assessed, particularly when natural fibers served as reinforcement for an unsaturated polyester matrix. Plates fortified with curaua, jute, and costa straw fibers, arranged either as aligned threads or in a flat-style fabric, were manufactured under varying vacuum levels during processing. The escalation of the vacuum level induced changes in fiber compaction, thereby altering porosity, the fiber/matrix interface, and, consequently, the overall properties of the composite.

This chapter aims to showcase studies illustrating diverse manufacturing techniques for composite materials incorporating natural fibers. Additionally, the applications of these materials across various industrial sectors will be explored (Figure 3.1).

The choice of manufacturing methods typically hinges on the rheological properties of the polymer, with the primary distinction being whether the material is thermoplastic or thermosetting. Additional crucial factors include the softening temperature, stability, size, and shape of the intended final product.

The application of these polymers as matrices in composites varies depending on the fiber type and the manufacturing process utilized. Thermoplastics are primarily employed with short fibers in injection molding. In this process, thermoplastics come into contact with the reinforcement when in a molten state. However, challenges may arise with specific thermoplastic matrices due to the high viscosity of the polymer.

FIGURE 3.1 Samples for techniques for composite material incorporation of natural fibers.

The properties of thermoplastics are directly impacted by factors such as the degree of crystallinity, morphology, and orientation of the polymer network, all of which are influenced by processing conditions (Levy Neto and Pardini, 2006). In the case of reinforced polymers, the resulting composite exhibits properties contingent on the specific characteristics of both the resin and fibers (Bunsell and Renard, 2005).

In the production of these composites, traditional extrusion or injection processes can be employed, commencing with the composition of the reinforcement/matrix (Levy Neto and Pardini, 2006). Fundamentally, the manufacturing process of fiber-reinforced thermoplastics involves an initial pre-mixing step in a blending chamber, where the polymer is introduced first, followed by the addition of fibers after melting. This pre-mix can undergo homogenization through an extruder or be directly molded through injection. Moreover, these composites can also be manufactured using compression molding.

Thermoset polymers find extensive application in structural composite materials, offering distinct advantages over thermoplastics. These include high rigidity, elevated thermal stability, dimensional stability, effective electrical and thermal insulation properties, and resistance to creep and relaxation. Among the widely used and readily available thermoset resins are polyesters, polyurethanes, vinyl esters, and phenolic resins, commonly utilized in the formulation of fiber-reinforced composites.

Thermoset polymers, featuring cross-linked chain macromolecules, undergo polymerization during processing and are non-recyclable due to thermal decomposition. Generally, these polymers are promoted for composite manufacturing, as well as for applications in adhesives and coatings. The assembly of these composites can occur in molds with simple construction, and the organization of components can be achieved manually or automatically. The resin curing or hardening process may occur with or without the application of pressure, temperature, or vacuum.

The creation of natural fiber composite (NFC) materials involves introducing natural fibers, along with additional agents like compatibilizers and plasticizers, into the polymer or gel matrix for reinforcement. When incorporating natural fibers as reinforcement with thermoplastics, it is crucial to carefully consider processing conditions to enhance the composite's performance, taking into account the inherent characteristics of both materials (Allan et al., 2002; Ngo et al., 2018). The substitution of synthetic fibers with natural fibers requires attention to various factors, including alterations in chemical composition (comprising complex polysaccharides), material origin, and potential instabilities in thermal and moisture absorption properties (Le Duigou et al., 2020; Baley et al., 2019).

3.2 MANUFACTURING PROCESSES OF NATURAL FIBER COMPOSITE MATERIALS

The ultimate properties of materials are shaped not only by their inherent characteristics but also by the production process and their integration into compounds. All manufacturing techniques for composites, ranging from basic to advanced, come with various constraints, including the fabrication capacity of parts, size limitations, geometric shape considerations, structural quality, production uniformity, potential for automation, production scalability, and additional costs (Simões, 2024).

The initial emphasis on cost in composite construction within the maritime industry has traditionally led to manual processes or alternative techniques in open molds. However, environmental concerns, including harmful gas emissions and potential harm to workers, have prompted regulatory restrictions on open mold construction by the Environmental Protection Agency. In response, there has been substantial research and development focused on closed mold techniques to reduce costs and emissions. The resin transfer molding (RTM) method, a conventional choice for composite manufacturing, offers advantages such as improved thickness tolerances and smoother laminated surfaces but may incur high material costs for larger parts. Emerging techniques like infusion, a variation of RTM, involve replacing the closed mold with a bag, resulting in a significant reduction in process costs (Franco, 2008).

The manufacturing processes for composites are a significant cost factor, often constituting 50–60% of the total expense, drawing considerable attention from both the industrial and scientific communities. The primary goal is to minimize this portion of the final product cost. Composite processing can be classified based on the matrix used or the process employed to obtain the composite. The selection of the manufacturing process considers factors such as part size, geometry, desired microstructure, performance requirements, and market evaluation. Another classification method is based on tooling, categorizing processes into open molds or closed molds, which influences surface finish quality and mold requirements. Various processes like contact lamination, projection lamination, and filament winding use open molds, while resin transfer molding and pressing molding utilize closed molds. With increasing demands on composite materials, manual manufacturing processes faced challenges in controlling essential factors like fiber orientation and volumetric fraction, crucial for mechanical strength. While fiber orientation remained unchanged, precise control over fiber volumetric fraction became imperative for optimization. This led to the development of prepreg, an intermediate product comprising reinforcing fibers saturated with a thermosetting or thermoplastic polymer. Prepregs, created through fibers impregnated with resin in an initial curing stage, must be stored at low temperatures to prevent premature curing, ensuring ideal conditions for molding the composite (Adrian P. Mouritz, 2012).

Composites are materials composed of two or more distinct phases, typically a matrix and a reinforcement. When these components are combined, they yield properties superior to those of each individual element. Various manufacturing processes exist for composites, each with its unique advantages and limitations. The selection of a particular process hinges on the specifications of the final product. Therefore, it is crucial to comprehend the distinctions between these processes.

3.3 COMPRESSION MOLDING

During this process, a mixture of resin and reinforcement is subjected to heat and pressure within a sealed mold. The resin undergoes curing, forming a robust matrix, and the reinforcement adds to the structural strength. Compression molding, known for its simplicity, is effective in producing large components.

Compression molding serves as a fundamental manufacturing technique for NFCs. In this method, fibers are organized within a matrix and subjected to heightened pressures and temperatures. This process enables the production of components known for their impressive mechanical strength, durability, and lightweight properties (Thakur et al., 2022).

In the production of NFCs, raw materials like sisal, jute, curaua, coconut, and others are employed. These fibers are blended with a thermoplastic or thermosetting matrix, such as PP, polyester, or epoxy (Thakur et al., 2022).

Compression molding of NFCs is conducted using a hydraulic press or vacuum press. Initially, the fibers are layered, and subsequently, the thermoplastic or thermosetting matrix is added. Finally, the application of suitable pressure and temperature facilitates the fusion of the matrix and the compression of the fibers.

After the process is completed, the parts are removed from the press and undergo a cooling and finishing process, where they are cut and sanded to obtain the desired shape and appropriate finish.

Compression molding of NFCs presents numerous benefits, including weight reduction, elevated mechanical strength, and lower production costs in comparison to alternative materials. Moreover, the incorporation of natural fibers in this process contributes to a diminished environmental impact, given the renewable and biodegradable nature of these fibers (Venugopal et al., 2022).

3.4 MANUAL LAMINATION

This process involves the manual application of alternating layers of reinforcing fabric and resin in a mold. The layers are carefully placed and laminated together to form a composite part. Manual lamination is a low-cost process and can be used to produce small- to medium-sized parts.

Manual lamination (hand layup) of NFCs is a manufacturing process that involves the manual application of layers of natural fibers into a resin matrix (Lotfi et al., 2021). NFCs are composite materials that consist of fibers of natural origin (such as coconut, sisal, jute, and bamboo, among others) incorporated into a resin matrix.

The process of manual lamination of NFCs is usually performed on a small scale, where fiber layers are manually placed into a resin matrix and compacted to form a laminate. This process can be repeated several times to build up additional layers and increase the thickness of the laminate.

While manual lamination may be labor-intensive, it offers numerous advantages, such as the ability to tailor the thickness and quantity of natural fiber layers, resulting in the development of composites with elevated strength and stiffness. Additionally, NFCs are not just lighter but also more environmentally sustainable when compared to fiberglass or carbon composites.

However, it is essential to note that manual lamination may introduce variability in the quality and mechanical properties of composites, especially with inconsistent execution. Therefore, strict adherence to a meticulous manual lamination process is imperative to ensure the quality and durability of NFCs (Rubino et al., 2020).

3.5 VACUUM INFUSION

This process involves placing a dry reinforcement fabric in a mold and applying a liquid resin through vacuum. The vacuum is used to extract the air from the mold and the resin, allowing the resin to completely penetrate the reinforcement fabric. Vacuum infusion is a precise process and can be used to produce large-sized parts with superior fabrication accuracy (Gajjar et al., 2020).

Vacuum infusion of NFCs is a manufacturing process that involves applying layers of natural fibers into a resin matrix through a vacuum infusion process. This process is an alternative to hand lay-up, which is a more labor-intensive process and can result in variations in composite quality.

During the vacuum infusion process, fiber layers are positioned within a mold, accompanied by a barrier film layer that aids in sealing the mold and preventing the ingress of air. Subsequently, the mold is enveloped with a vacuum bag and linked to a vacuum pump system.

With reduced pressure, the resin is pulled into the fiber layers through an infusion process, in which the resin is injected under vacuum. This process helps to ensure that the resin is uniformly distributed throughout all fiber layers, which can result in stronger and more durable composites.

NFCs manufactured through vacuum infusion are more uniform in terms of quality and mechanical properties than composites manufactured manually. Additionally, the vacuum infusion process allows for greater control over resin content, which can help reduce composite weight and improve fatigue resistance.

Nonetheless, vacuum infusion demands specialized equipment and materials, potentially rendering the process more costly compared to manual lay-up. Moreover, the successful execution of vacuum infusion necessitates a certain level of expertise and skill (Kim et al., 2013).

The vacuum infusion process (VIP) molding technique is used in the manufacturing of large composite structures. This method is popular due to its low tooling cost and environmental safety, as it eliminates over 90% of the volatile organic compounds emitted by unsaturated polyester resins. In contrast to open mold techniques like hand-lay-up or spray-up, the VIP exhibits higher reproducibility, involves minimal operator participation, and generates components with a relatively high fiber content, reaching up to 60% by volume (Ragondet, 2005).

Rodrigues (2015) employed the resin infusion technique in the production of composites reinforced with curaua, jute, and coast straw fibers, organized in flat-style fabrics and aligned threads. The study aimed to explore manufacturing process variables and the state of fiber compaction concerning their arrangement and geometry. The resin infusion process emerged as a favorable choice for manufacturing natural fiber-reinforced composites, offering environmental advantages as a closed mold process. The research successfully yielded products with elevated fiber content and commendable mechanical properties.

3.6 FILAMENT WINDING

This process involves placing resin-impregnated reinforcing filaments onto a cone-shaped mandrel. The filaments are spirally wound around the mandrel to create a continuous reinforcement layer. The resin is cured to form the matrix, and the composite part is removed from the mandrel. Filament winding is a highly precise process and can be used to produce small to medium-sized cylindrical parts (Blachut et al., 2023).

The filament winding process for NFCs entails applying layers of natural fibers onto a resin matrix through a winding technique around a mandrel. This manufacturing method is employed to produce cylindrical components, such as tubes, characterized by both high-strength and lightweight properties.

Within the filament winding process, natural fibers are guided through a feeding device, and the resin matrix is applied concurrently as the fibers are wound around a rotating mandrel. Typically, the resin is administered through an application device, such as a brush or roller, during the winding of the fibers around the mandrel (Garkina and Danilov, 2019).

The filament winding process is highly controlled, which helps ensure that the fibers are evenly distributed along the part. This results in a high-strength and stiffness composite, which is ideal for applications where lightness is important.

NFCs manufactured through filament winding may be more expensive than composites manufactured through hand lay-up or vacuum infusion due to the need for specific equipment and materials. Additionally, the filament winding process is more suitable for manufacturing cylindrical parts, which limits its versatility in terms of the shapes and sizes of parts that can be produced.

However, filament winding offers high precision and control in the manufacturing process, which can result in high-quality and high-performance parts. Furthermore, NFCs produced via filament winding may exhibit greater sustainability compared to fiberglass or carbon fiber composites, thanks to the utilization of renewable natural fibers (Blachut et al., 2023).

3.7 INJECTION MOLDING

In this approach, liquid resin is injected into a sealed mold housing reinforcement. The resin undergoes curing to form the matrix, and the resulting composite part is then extracted from the mold. Injection molding is a high-production process well-suited for the manufacturing of small to medium-sized parts (Kuo et al., 2009).

The injection molding of NFCs is a manufacturing process where a blend of natural fibers and resin is injected into a mold under elevated pressure. This technique is employed to manufacture components with intricate geometries, precise dimensional accuracy, and commendable mechanical strength.

In the injection molding process, natural fibers are mixed with resin in a container, and the mixture is heated to reduce the viscosity of the resin. The mixture is then injected into a mold under high pressure, which is kept closed until the mixture solidifies and the part is formed.

Injection molding offers clear advantages over alternative manufacturing processes for NFCs. Specifically, it stands out as a highly automated method that enables efficient and high-quality mass production. Additionally, injection molding provides a high degree of control over the distribution of natural fibers within the part, leading to a substantial enhancement in its strength and stiffness.

However, injection molding can be more expensive than other manufacturing processes of NFCs, due to the need for specialized equipment and molds. Furthermore, the resin employed in injection molding may present recycling challenges compared to other resins, potentially exacerbating the environmental impact of the process.

In essence, injection molding for NFCs is a highly automated manufacturing process, enabling efficient and high-quality mass production. Despite the potential for higher costs compared to alternative manufacturing methods, injection molding remains the preferred choice for producing components with intricate geometries and superior dimensional accuracy (Kumar et al., 2023).

3.8 PULTRUSION

Pultrusion is a process that involves the passage of resin-impregnated reinforcing filaments through a heated mold. The resin is cured as the filaments are pulled through the mold to form a continuous piece. Pultrusion is a high-production process and can be used to produce continuous profiles of great length. Pultrusion technique provides a superior productivity/cost ratio.

The pultrusion of NFCs is a continuous manufacturing process that involves the extrusion of a mixture of natural fibers and thermosetting resin through a heated die. This process is used to produce parts with a constant cross-section and continuous length, such as profiles, tubes, and bars (Minchenkov et al., 2021).

In the pultrusion process, natural fibers are saturated with a thermosetting resin in a resin bath and subsequently guided through a heated die for resin curing. The resulting part is then drawn by a traction system that regulates the part's speed and maintains constant tension to ensure the formation of the desired cross-section.

Pultrusion presents numerous advantages compared to alternative manufacturing processes for NFCs. Notably, it is a highly automated method enabling the mass production of parts with superior strength and stiffness. Furthermore, pultrusion provides high-dimensional precision and a substantial level of control over the orientation of fibers within the part, contributing significantly to enhanced mechanical properties.

Despite this, pultrusion may involve higher expenses when contrasted with other NFC manufacturing processes, mainly due to the necessity for specialized equipment and custom dies designed for each profile or tube. Additionally, pultrusion could encounter restrictions concerning the range of shapes and sizes of parts it can produce, as its optimal performance is in crafting components with a uniform cross-section and continuous length.

In summary, pultrusion for NFCs stands out as a continuous and highly automated manufacturing process that facilitates efficient and high-quality mass production of parts featuring a consistent cross-section and continuous length. Despite the possibility of higher costs compared to other manufacturing processes, pultrusion is particularly well-suited for crafting profiles, tubes, and bars characterized by exceptional strength and stiffness (Vedernikov et al., 2020).

In their article, Maximilian (2022) addressed the possibility of advancing toward more sustainable composite manufacturing through pultrusion. Nevertheless, ensuring compatibility between fibers and the matrix remains a significant challenge that directly influences the mechanical properties and overall performance of natural composites. The challenge of ensuring compatibility between fibers and the matrix in natural composites can be addressed by employing compatibilizing agents, such as surface treatments for reinforcing fibers. The literature on pultrusion suggests that it is feasible to apply secondary materials and surface agents in a straightforward and continuous manner within a pultrusion production line. This technique has already been demonstrated for thermoplastic pultrusion, and its extension to natural material pultrusion has the potential to lower barriers to the broader adoption of sustainable composite processing.

3.9 VEGETABLE FIBERS USED IN POLYMER COMPOSITES

Vegetable fibers can be viewed as composites composed of cellulose fibrils bound together by a matrix of lignin and hemicellulose. This matrix serves as a natural barrier against microbial degradation and provides mechanical protection. The structural characteristics of these fibers are linked to the nature and crystallinity of cellulose. The typical organization of a vegetable fiber includes a complex layered structure with a thin primary wall formed during cell growth and surrounding a secondary wall. The secondary wall comprises three layers (S1, S2, and S3), where the mechanical properties are influenced by the middle layer (S2). This layer consists of microfibrils, helically formed by long cellulose chains, organized along the fiber axis (Pereira et al., 2015). The microfibrils, with a diameter of 10 to 30 nm, result from the packing of 30 to 100 extended cellulose chains (Figure 3.2).

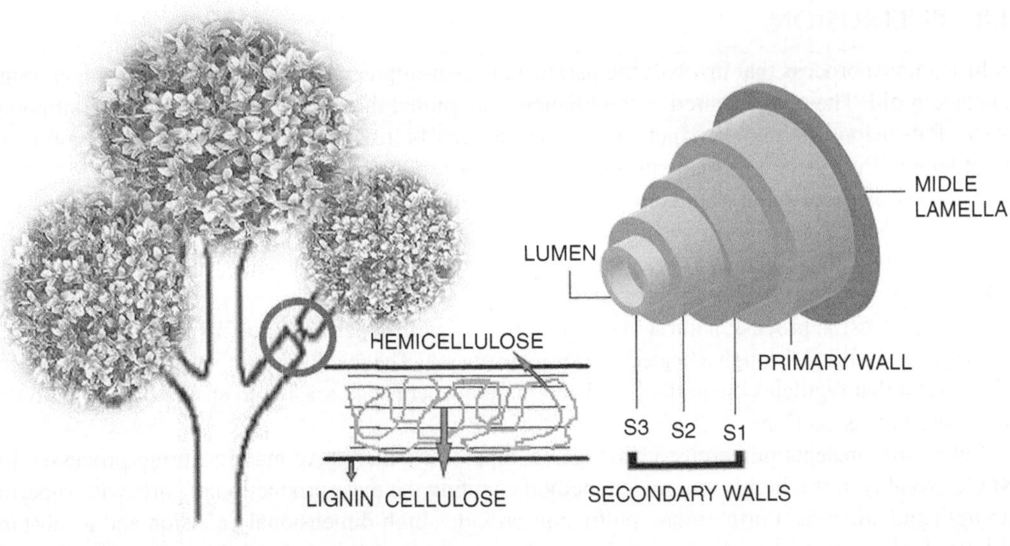

FIGURE 3.2 Structure and composition of a plant fiber.

Natural fibers are categorized based on their origin, namely plant, mineral, and animal. In the context of this work, the fibers utilized are of plant origin, meaning they are derived from the stems, leaves, fruits, or seeds of plants (Mohanty et al., 2005).

The use of plant fibers instead of synthetic fibers is a very important choice because they are biodegradable, have low cost compared to synthetic fibers, are from renewable sources, and, most importantly, cause less environmental impact. Plant fibers have lower density than synthetic fibers and cause less wear on composite processing equipment (Yimin, 2016).

In recent years, there has been a growing adoption of lignocellulosic fibers as reinforcement for polymeric materials, displacing synthetic fibers, notably glass fibers, in composites applied across various industrial sectors such as automotive and construction. This transition offers several advantages over conventional materials (Beg and Pickering, 2008). Some of these advantages are as follows:

- Derived from renewable sources
- Continuous availability
- Biodegradability
- Low cost
- Low density
- Interesting specific properties

Plant fibers exhibit a significantly lower abrasive nature compared to glass fibers, providing advantages in terms of processing and recycling (Malkapuram et al., 2009).

Plant fibers can be classified according to the specific part of the plant from which they are extracted. Fibers derived from the stem or leaves of the plant are the most commonly employed dispersed phase in polymeric composites.

Various plant fibers, including ramie, jute, sisal, bamboo, and banana, are employed as substitutes for traditional fibers like glass fibers. Flax, kenaf, hemp, and wood flour are also recognized for their suitability as fillers and impact-absorbing components in vehicles.

The physical, mechanical, and chemical properties of these plant fibers are significantly influenced by the harvest, which is in turn affected by various factors such as climate, location, soil

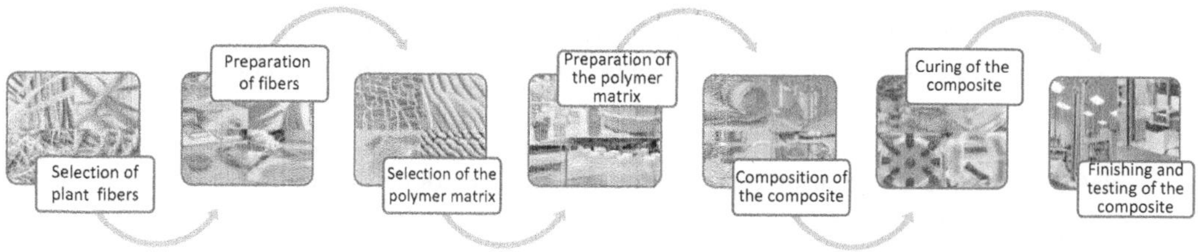

FIGURE 3.3 Steps for preparing polymer matrix composites reinforced with plant fibers.

characteristics, and weather conditions. Additionally, fiber processing and its integration into composites, including handling, impregnation, and consolidation, also play a crucial role in determining its properties.

Moreover, natural fibers exhibit certain drawbacks, including limited wettability by the polymer, compatibility issues with certain polymeric matrices, and high moisture absorption (Machado et al., 2003).

These challenges can be mitigated through physical treatments using methods like plasma and corona, as well as chemical treatments involving substances such as maleic anhydride, organosilanes, isocyanates, sodium hydroxide, permanganates, and peroxide, among other options (Marconcini et al., 2007).

From a general understanding and from the investigations on the mechanical properties of PP composites reinforced with various natural fibers, including kenaf, coconut, sisal, hemp, and jute, it was understood that (Figure 3.3) Kenaf, hemp, and sisal composites demonstrated similar tensile strength and Young's modulus, while sisal exhibited superior impact performance compared to other fibers. Composites reinforced with coconut fiber exhibited lower mechanical properties but demonstrated higher impact resistance than those reinforced with jute and kenaf.

Composite materials reinforced with plant fibers have been an attractive area for researchers and industries, where it has become a necessity to add value to the waste generated by the population, creating efficient and low-cost alternative materials, utilizing natural resources, and preserving the environment.

3.10 POLYMERIC MATRIX COMPOSITES ENHANCED WITH VEGETABLE FIBERS

The application of fiber-reinforced composite materials has generated significant interest among engineers across diverse industries, including automotive, construction, mechanics, metallurgy, pharmaceuticals, naval, aerospace, and more. This interest stems from the remarkable tensile strength and high modulus of elasticity exhibited by materials in fibrous form. Consequently, these materials find extensive use as engineering materials when combined with a matrix that safeguards the fibers and distributes stress, thus impeding the propagation of failures.

Here are some general steps for preparing polymer matrix composites reinforced with plant fibers (Figure 3.3):

1. Selection of Plant Fibers: Choose the plant fibers you want to use, such as sisal, jute, bamboo, and coconut. Consider their physical, chemical, and mechanical properties, as well as their availability and price.
2. Preparation of Fibers: Plant fibers need to be prepared before being used to reinforce the polymer matrix. This may include processes such as cutting, grinding, defibrillation, bleaching, and chemical treatment.
3. Selection of the Polymer Matrix: Choose a suitable polymer matrix for the intended application. Some common examples include polyester, epoxy, polyamide, and PP. Consider the properties of the matrix, such as tensile strength, modulus of elasticity, toughness, glass transition temperature, and melting temperature.

4. Preparation of the Polymer Matrix: The polymer matrix needs to be prepared according to the manufacturer's instructions. This may include processes such as mixing, heating, cooling, and adding curing agents.
5. Composition of the Composite: Mix the prepared plant fibers with the prepared polymer matrix according to the desired proportion. This can be done through techniques such as lamination, compression molding, injection molding, or vacuum infusion.
6. Curing of the Composite: Allow the composite to cure according to the manufacturer's instructions. This may involve the application of heat or a curing agent.
7. Finishing and Testing of the Composite: Once the composite is cured, it can be cut, polished, and tested to evaluate its mechanical and physical properties. This may include tests for tension, bending, impact, and hardness, among others.

The choice of a suitable fiber type is significantly influenced by the modulus of elasticity of the fiber. Although tropical fibers have characteristic strength values that are only half of those of fiberglass, the density of natural fibers is approximately 45% lower, resulting in comparable specific strength levels to that of fiberglass (Thyavihalli Girijappa et al., 2019).

Polymer composites reinforced with plant fibers are viewed as economically feasible alternatives for carbon sequestration in nature, actively mitigating CO_2 emissions across their entire production, processing, and utilization life cycle (Puttegowda, 2025). This increases their economic potential, as carbon credits can be commercialized in the production chain.

The quest for materials aligning with global trends of economic viability and environmental sustainability has prompted the utilization of renewable natural resources. Numerous researchers have expressed interest in employing polymer materials reinforced with natural fibers as a sustainable alternative (Maleki et al., 2025).

Polymer composites strengthened with natural fibers have garnered significant attention owing to the advantages they provide over synthetic fibers. These include low density, cost-effectiveness, renewable sources, non-toxicity, biodegradability, potential for incineration, and, most importantly, sustainability (Li et al., 2020).

Many researchers have repeatedly highlighted the potential for polymer matrix composites strengthened with natural fibers to attain cost competitiveness in comparison to wood products. This holds especially true when integrating recycled plastics into the composites, capitalizing on their affordability and advantageous processability.

The blending and formulation procedures for polymer matrix composites enhanced with plant fibers typically occur in two phases: the initial stage entails integrating the fibers into the matrix, while the subsequent stage involves shaping the composite into its ultimate product form.

The compatibility between the matrix and fibers is vital for the composite's ultimate properties, as the transfer of loads from the matrix to the fibers happens at the interface. Incompatibility can jeopardize the composite's reinforcement, leading to a weak interface and material failure (Li et al., 2008). Numerous studies suggest that surface modification of natural fibers before integration into the composite can enhance compatibility with the matrix.

The distribution of applied stress within the composite is managed by the polymer matrix. However, the susceptibility of natural fibers to degradation at elevated temperatures imposes constraints on polymer selection, particularly concerning processing temperatures. Polyethylene stands out as the preferred polymer matrix, primarily due to its widespread availability, cost-effectiveness, and ease of processing at moderate temperatures (Velde and Kiekens, 2001).

The incorporation of vegetable fibers in composites is pivotal for the advancement of "green" materials, characterized by a biodegradable matrix and commendable performance. This approach holds promise as a solution to contemporary environmental and ecological challenges. Introducing these materials to the market can alleviate agricultural waste management problems, mitigate environmental pollution, and find applications across engineering, electronics, and the automotive industry. Noteworthy advantages of these composites include their low weight, diminished machine wear, low abrasiveness, and decreased health risk during the production process (Goda et al., 2006).

3.11 3D PRINTING OF NATURAL FIBER COMPOSITES

The 3D printing process for NFCs adheres to conventional 3D printing principles, with the incorporation of a fibrous reinforcing material. The key steps in 3D printing NFCs are as follows (Figure 3.4):

1. Material Preparation: First, the material to be used in 3D printing needs to be prepared. This usually involves mixing natural fibers with a thermoplastic polymer, such as polylactic acid (PLA) or acrylonitrile butadiene styrene (ABS). The fibers are added to the resin in a mixer or extruder to ensure they are evenly distributed in the mixture.
2. Printer Setup: The next step is to set up the 3D printer for the NFC printing process. This may involve configuring the printing parameters, such as extruder temperature and printing speed.
3. Printing the Piece: With the 3D printer configured and the material prepared, the piece is printed layer by layer, following the defined settings. During the printing process, the extruder melts the fibrous reinforcing material and deposits it layer by layer, creating a continuous piece.
4. Post-Processing: After the piece is printed, post-processing is required. This may include removing support, if any, and grinding or finishing the piece to improve its surface quality and dimensional accuracy.

3D printing of NFCs presents several advantages compared to conventional 3D printing. These include heightened strength and rigidity, along with a limited scrap generation and consumption of energy (Zeid et al., 2024). Moreover, the utilization of natural fibers provides a more sustainable and environmentally friendly solution by minimizing the amount of plastic employed in the final product.

Composites consisting of degradable polymers reinforced with natural fibers represent an innovative category of biomaterials poised to substitute traditional petroleum-based polymers like polyethylene, PP, and polystyrene. Their commendable performance and environmentally friendly characteristics have garnered considerable recognition in recent years.

The interface region is a significant concern in composites due to the potential for stress concentrations leading to defects and damage. Weak interfaces, often caused by insufficient bonding between fibers and the polymer matrix or pores resulting from gas diffusion, contribute to internal defects. In 3D printed fiber-polymer composites, achieving adhesion between deposited filaments and their faces poses an even greater challenge compared to conventional composites (Liu et al., 2021).

The primary factors leading to a decline in the overall mechanical properties of fiber-polymer composites are the weak interfacial bonding and subsequent less stress transfer, as indicated by Yiou (2019). Typically, natural fibers are hydrophilic, and the polymeric matrix is hydrophobic, resulting in adhesion challenges due to their polarity difference. Surface modification is a prevalent approach for enhancing interfacial adhesion properties. Moreover, selecting an appropriate polymer matrix

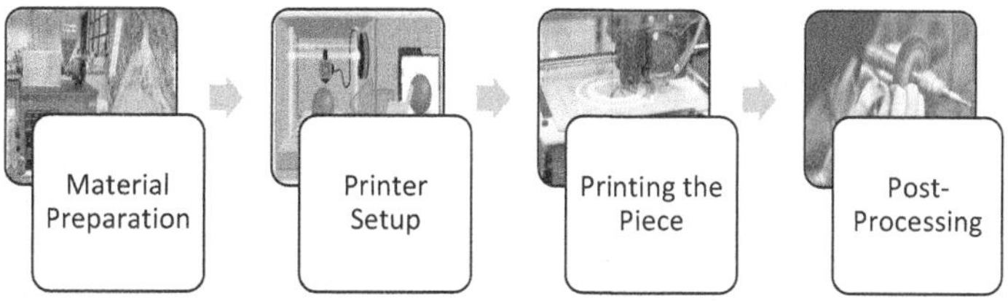

FIGURE 3.4 Steps in 3D printing natural fiber composites.

and managing the fiber content can further enhance both interfacial and mechanical properties (Haque et al., 2009; Le Troedec et al., 2008).

In 2022, Huang addressed key aspects of preparing composite filaments and gels using natural fibers, their derivatives, and polymers through techniques like fused deposition modeling (FDM) and direct ink writing (DIW). Two significant issues were discussed: the correlation between interfacial adhesion conditions and mechanical properties, with surface modification enhancing bonding strength, and the impact of thermal properties on composite inks, where adding natural or derivative fibers alters crystallization temperature and improves thermal stability. For gel printing with cellulose nanofibers (CNC), regulating ink rheology ensures a uniform product, while additional freeze-drying enhances overall structural strength and provides a highly specific ultra-high surface area (Huang, 2022).

Through thorough research and analysis of crucial factors, various advanced applications have emerged. These encompass medical devices with robust mechanical stability and antibacterial properties, biomass inks exhibiting high cellular compatibility and low cytotoxicity for tissue engineering, efficient bacteriostasis, high cross-linking resistance, and adequate mechanical strength for wound dressings. Additionally, lyophilized hydrogels are employed to achieve a 3D porous support structure with high specific surface area and porosity, ideal for drug delivery. These materials are also utilized as support for thermoelectric conversion elements, lithium-ion battery electrodes, and as a viscosity regulator for electronic inks, among other applications.

Despite extensive research in smart manufacturing, biomedicine, and electronics, the widespread practical application of 3D printing technology for fibers and their derivatives faces limitations. Challenges include the complexity, high cost, and frequent maintenance requirements of 3D printing equipment, making it less accessible to ordinary consumers. The preparation of high-performance printing inks or linear consumables involves repeated exploration and adjustment of rheological properties to ensure high resolution while maintaining excellent mechanical properties. Furthermore, the 3D printing process is time-consuming, more complex than traditional manufacturing techniques, and presents difficulties in preparing large part structures.

To guide future research directions, emphasis should be placed on two key areas: materials and structures. Regarding materials, two critical steps are essential. First, conducting a thorough quantitative characterization of materials before their use as ink or printing filament is crucial. This enables the exploration of a comprehensive set of material rheological models, standardizing industry patterns, and facilitating advanced planning for future development. Second, it is important to delve into research on the fundamental properties of the printed product, including interface adhesion, mechanical properties, and thermal properties. This enables the correction of unreasonable aspects of composite materials and the proper optimization of printing proportions.

Concerning structure, an appropriate design for the printing structure is anticipated to address any deficiency in bonding performance between material layers. Additionally, the natural microstructure achieved by integrating nanocellulose with other materials through the selection of suitable processing methods has the potential to enhance the overall performance of printed products (Huang, 2022).

In conclusion, the full exploration of the potential applications of fibers and their derivatives in 3D printing has not been realized. This has the potential to elevate 3D printing technology to new heights, driven by the synergistic progress of both technology and materials.

3.12 APPLICATIONS OF NATURAL FIBER COMPOSITES IN VARIOUS INDUSTRIAL SECTORS

NFCs find extensive applications across diverse sectors, including automotive, construction, furniture, packaging, and sports. They are utilized in the production of various items such as wall panels, coatings, sustainable packaging (trays, boxes, containers), lightweight automotive parts (door and roof panels, interior components), furniture, decorative objects, and sporting goods like surfboards and snowboards. In the construction industry, these composites are employed in the manufacturing of fence panels, roofing, windows, doors, and flooring, benefiting from their durable and resistant properties (Bressiani et al., 2020).

NFC materials are increasingly being used in the automotive industry due to their high-strength and low-density properties. They offer an attractive alternative to traditional materials such as metals, plastics, and fiberglass. The most common natural fibers used in the automotive industry include bamboo, hemp, sisal, jute, and flax (Salim et al., 2025). Generally, these fibers are combined with thermoplastic or thermosetting resins to create composite materials.

NFCs offer numerous benefits when compared to conventional materials. Their lightweight nature contributes to the reduction of vehicle weight, thereby enhancing fuel efficiency and lowering greenhouse gas emissions. Furthermore, the renewable and biodegradable characteristics of natural fibers make them a more sustainable alternative compared to synthetic materials.

These materials find extensive application in the interior components of automobiles, including door panels, seats, and carpets, as well as structural elements like bumpers and body panels (Khatri et al., 2023). While NFCs are not yet commonplace in mass-produced vehicles, the increasing demand for more sustainable solutions in the automotive industry is fueling research and development efforts for the integration of these materials.

The aerospace industry frequently employs NFC materials because of their advantageous properties, such as high strength and low density. These characteristics aid in reducing the overall weight of aircraft, consequently enhancing fuel efficiency. Notable natural fibers utilized in the production of these composites include bamboo, sisal, jute, flax, and coconut.

As an illustration, bamboo, characterized by its light weight, strength, and rigidity, proves to be an excellent choice for aerospace applications. Additionally, it serves as a renewable and sustainable resource. Sisal fiber, derived from the Agave sisalana plant, is recognized for its notable tensile and compressive strength, frequently finding application in the structural components of aircraft, including floor and interior wall panels.

Derived from the Corchorus capsularis plant, jute fiber is esteemed for its high tensile strength and low density. This makes it a frequent choice in aerospace applications where the aim is to enhance strength while minimizing component weight. On the other hand, flax fiber, obtained from the Linum usitatissimum plant, is recognized for its notable strength and rigidity. It is commonly utilized in the manufacturing of structural components in aircraft, including wings and fuselages.

NFCs are widely used in constructing both the structural and interior components of aircraft. In addition to their positive impact on weight reduction and fuel efficiency, these composites present an environmentally friendly alternative to synthetic materials. Their renewable and biodegradable qualities, combined with reduced greenhouse gas emissions in the manufacturing process, underscore their sustainability. The aerospace industry's commitment to sustainability continues to drive ongoing research and development in harnessing the potential of NFCs (Carosena et al., 2017).

3.13 CONCLUSION

The manufacturing of NFC materials offers notable advantages, utilizing renewable and biodegradable raw materials that are less polluting compared to conventional synthetics. The inherent high strength and low density of natural fibers contribute to weight reduction and enhanced fuel efficiency in products. Standard manufacturing processes involve blending the fibers with a polymer matrix and molding. Despite existing challenges, ongoing industry investment in research and development indicates a potential rise in the future utilization of these materials.

BIBLIOGRAPHY

Allan, H. Fawcett, McNally, Tony, & McNaly, Gerry, An attempt at engineering the bulk properties of blends of a bitumen with polymers. *Advances in Polymer Technology*, 21 (4), 2002, 275–286. https://doi.org/10.1002/adv.10032

Baley, C., Gomina, M., Breard, J., Bourmaud, A., Drapier, S., Ferreira, M., Le Duigou, A., Liotier, P. J., Ouagne, P., Soulat, D., & Davies, P. Specific features of flax fibres used to manufacture composite materials, *International Journal of Material Forming* 12 (6), 2019, 1023–1052.

Beg, M. D. H., & Pickering, K. L., Reprocessing of wood fibre reinforced polypropylene composites: Part I: Effects on physical and mechanical properties. *Composites Part A: Applied Science and Manufacturing*, 39(7) Jul. 2008, 1091–1100. https://doi.org/10.1016/j.compositesa.2008.04.013

Bi, Xiaoyu, & Huang, Runzhou 3D printing of natural fiber and composites: A state-of-the-art review. *Materials & Design*, 222, 2022. https://doi.org/10.1016/j.matdes.2022.111065

Bijlwan, P. P., Prasad, L., & Sharma, A. Recent advancement in the fabrication and characterization of natural fiber reinforced composite: A review, *Materials Today Proceedings*, 44, 2021, 1718–1722, https://doi.org/10.1016/j.matpr.2020.11.878

Blachut, A., Wollmann, T., Panek, M., Vater, M., Kaleta, J., Detyna, J., Hoschutzky, S., & Gude, M. Influence of fiber tension during filament winding on the mechanical properties of composite pressure vessels. *Composite Structures* 304 2023, 116337.

Bledzki, A. K., & Gassan, J. Composites reinforced with cellulose based fibres. *Progress in Polymer Science*, 1999, 221–274.

Bressiani, I. J., Keinert, A. C., Ellenberger, A., & Belini, U. L. Vegetal fibers and composites in automotive industry. *Mix Sustentável|Florianópolis*, 6(4), 2020, 129–138|ago.|. http://dx.doi.org/10.29183/2447-3073.MIX2020.v6.n4.129-138 ISSN: 2447-0899 (IMPRESSA) | 2447-3073 (ONLINE).

Bunsell, A. R., & Renard, J., *Fundamentals of Fibre Reinforced Composites Materials*. IOP Publishing Ltda, London, 2005.

Carosena, M., Boccardi, S., & Carlomagno, G. *Composite Materials in the Aeronautical Industry, Infrared Thermography in the Evaluation of Aerospace Composite Materials*. Chapter 1, 2017, Woodhead Publishing. https://doi.org/10.1016/B978-1-78242-171-9.00001-2

Casas, G. R. *A Biotecnologia a Serviço da Indústria Automotiva*, 2010. Disponível em: http://www.portaldomeioambiente.org.br/. Acesso em 28 de novembro de 2019.

Correia, J. R. Pultrusion of advanced composites. In *Advanced Fiber-reinforced Polymer (FRP) Composites for Structural Applications*, 2023, 137–177. Woodhead Publishing. https://doi.org/10.1016/B978-0-12-820346-0.00007-1

De Paula, Paula, Sanchez Rodriguez, Ruben, Duarte, Luís, Candido, Veronica, & Monteiro, Sergio. Formulation and characterization of polypropylene composites alkali treated bagasse fiber. *Materials Science Forum* 2014, 775–776. 319–324. https://doi.org/10.4028/www.scientific.net/MSF.775-776.319

Ereno, D., *Revista Pesquisa FAPESP*. Disponível em https://revistapesquisa.fapesp.br/fibra-para-toda-obra/, 2004. Acesso em 18 de novembro de 2019.

Franco, R. *Production of Composite Material Components by Resin Infusion*, 2008. http://fenix.tecnico.ulisboa.pt/Resumo, Accessed on 22 November 2025

Gajjar, T., Shah, D. B., Joshi, S. J., & Patel, K. M. Analysis of process parameters for composites manufacturing using vacuum infusion process. *Materials Today: Proceedings*, 2020, 21, 1244–1249.

Garkina, I., & Danilov, A. Composite materials: Identification, control, synthesis. *IOP Conference Series: Materials Science and Engineering*, 471(3). IOP Publishing; 2019, February, 032005.

George, J., Sreekala, M. S., & Thomas, S., A review on interface modification and characterization of natural fiber reinforced plastic composites. *Polymer Engineering and Science*, 41(9), 2001, 1471–1485, https://doi.org/10.1002/pen.10846

Goda, K., Sreekala, M. S., Gomes, A., Kaji, T., & Ohgi, J. Improvement of plant based natural fibers for toughening green composites—Effect of load application during mercerization of ramie fibers. *Composites. Part A, Applied Science and Manufacturing*, 37, 2006, 2213–2220.

Goh, G. D., & Yeong, W. Y. Mode I interlaminar fracture toughness of additively manufactured carbon fibre thermoplastic. *3rd International Conference on Progress in Additive Manufacturing (Pro-AM), Singapore, Singapore*, 2018, 505–510.

Gonçalves, A. R., Luz, S. M., & Del'Arco, A. P. Mechanical behavior and microstructural analysis of sugarcane bagasse fibers reinforced polypropylene composites. *Composites Part A: Applied Science and Manufacturing*, 38, 2007, 1455–1461.

Hagnell, M. K., & Akermo, M. A composite cost model for the aeronautical industry: Methodology and case study. *Composite Part B* 79, 2015, 254–261.

Haque, M. M., Hasan, M., Islam, M. S., & Ali, M. E. Physico-mechanical properties of chemically treated palm and coir fiber reinforced polypropylene composites, *Bioresource Technology*, 100 (20), 2009, 4903–4906.

Khatri, H., Naveen, J., Jawaid, M., Jayakrishna, K., Norrahim, M. N. F., & Rashedi, A. Potential of natural fiber based polymeric composites for cleaner automotive component production-a comprehensive review. *Journal of Materials Research and Technology*, 25, 2023, 1086–1104. https://doi.org/10.1016/j.jmrt.2023.06.019

Kim, H., Okubo, K., Fujii, T., & Takemura, K. Influence of fiber extraction and surface modification on mechanical properties of green composites with bamboo fiber. *Journal of Adhesion Science and Technology*, 27, 2013, 1348–1358.

Kumar, A., Behura, A. K., Rajak, D. K., & Kumar, P.. Thermal properties of natural based fibers composites. *Natural and Synthetic Fiber Reinforced Composites: Synthesis, Properties and Applications*, 2022, 211–239. https://doi.org/10.1002/9783527832996.ch12

Kumar, S., Gupta, D. Sharma, V. et al. Recent development in natural fiber composites, testing and fabrication methods: A review, *Materials Today: Proceedings*, 2023. https://doi.org/10.1016/j.matpr.2023.02.073

Kuo, P. Y., Wang, S. Y., Chen, J. H., Hsueh, H. C., & Tsai, M. J. Effects of material compositions on the mechanical properties of wood–plastic composites manufactured by injection moulding. *Materials and Design*, 30, 2009, 3489–3496.

Le Duigou, Antoine, Correa, David, Ueda, Masahito, Matsuzaki, Ryosuke, & Castro, Mickael, A review of 3D and 4D printing of natural fibre biocomposites. *Materials and Design*, 194, 2020, 108911. https://doi.org/10.1016/j.matdes.2020.108911

Le Troedec, M., Sedan, D., Peyratout, C., Bonnet, J. P., Smith, A., Guinebretiere, R., Gloaguen, V., & Krausz, P. Influence of various chemical treatments on the composition and structure of hemp fibres, *Composites Part A: Applied Science and Manufacturing* 39 (3),2008, 514–522.

Leong, Y. W., Thitithanasarn, S., Yamada, K., & Hamada, H. Compression and injection molding techniques for natural fiber composites, *Natural Fibre Composites Materials, Processes and Applications*, 2014, 216–232, https://doi.org/10.1533/9780857099228.2.216

Levy Neto, F., & Pardini, L. C. *Compósitos Estruturais: Ciência e Tecnologia.* São Paulo: Edgar Blucher, 2006.

Li, M., Pu, Y., Thomas, V.M., Yoo, C.G., Ozcan, S., Deng, Y., Nelson, K. and Ragauskas, A.J.. Recent advancements of plant-based natural fiber–reinforced composites and their applications. *Composites Part B: Engineering*, 200, 2020, 108254. https://doi.org/10.1016/j.compositesb.2020.108254

Li, X., Tabil, L. G., & Panigrahi, S. Chemical treatments of natural fiber for use in natural fiber-reinforced composites: A review. *Journal of Polymers and the Environment*, 15, 25–33, 2007.

Li, Yan, Chunjing, Hu, & Yehong, Yu Interfacial studies of sisal fiber reinforced high density polyethylene (HDPE) composites. *Composites Part A: Applied Science and Manufacturing*. 39, 2008, 570–578.

Liu, G., Xiong, Y., & Zhou, L. Additive manufacturing of continuous fiber reinforced polymer composites: Design opportunities and novel applications, *Composites Communications*, 27, 2021.

Liu, Zhengyi, Wang, Honguang, & Yang, Lanjie, Du, Jinbo Research on mechanical properties and durability of flax/glass fiber bio-hybrid FRP composites laminates. *Composite Structures*, 290, 2022, 115566.

Lotfi, A., Li, H., Dao, D. V., & Prusty, G. Natural fiber–reinforced composites: A review on material, manufacturing, and machinability. *Journal of Thermoplastic Composite Materials*, 34(2), 2021, 238–284.

Lukmon, O. A., Puteri, S. M. Y., Zulkifli, M. A., & Muhammad, S. H. Fabrication of pandanus tectorius (screw-pine) natural fiber using vacuum resin infusion for polymer composite application. *Journal of Materials Research and Technology*, 8(3) 2019, 3102–3113.

Luo, L., Zhang, F., & Leng, J. Shape memory epoxy resin and its composites: From materials to applications, *Research*, 2022, 2022, 1–25. https://doi.org/10.34133/2022/9767830

Machado, M. A. L., Biagiotti, J., Arroyo, M., & Kenny, J. M. Ternary composites based on PP-EPDM blends reinforced ith flax fibers. Part I: Processing and thermal behavior. *Polymer Engineering and Science*, 43, 2003, 1018.

Maleki, B., Kargar, P. G., & Ashrafi, S. S. Biobased composites: A sustainable pathway for modern applications. *Cellulose - Biobased Solutions for Society*, 2025. https://doi.org/10.5772/intechopen.1009464

Malkapuram, R., Kumar, V., & Singh Negi, Y. Recent development in natural fiber reinforced polypropylene composites. *Journal of Reinforced Plastics and Composites*, 28, 2009, 1169–1189. https://doi.org/10.1177/0731684407087759

Marconcini, J. M., Ito, E. N., Hage Jr E., & Mattoso, L. H. C. Studies on polypropylene/cellulose microfiber composites. *6th International Symposium on Natural Polymers and Composites*, 2007.

Maximilian, V., Onur, Y., Ismet, B., Hattel, J. H., Spangenberg, J., Sandberg, M. Cost-efficient, automated, and sustainable composite profile manufacture: A review of the state of the art, innovation, and future of pultrusion technologies. *Composite Part B* 246 2022.

Minchenkov, K., Vedernikov, A., Safonov, A., & Akhatov, I. Thermoplastic pultrusion: A review. *Polymers*, 13, 2021, 180.

Mohanty, A. K., Misra, M., & Drzal, L. T. *Natural Fibers, Biopolymers, and Biocomposites*. Boca Raton, FL: CRC Press, Taylor & Francis Group. 2005. https://www.crcpress.com/Natural-Fibers-Biopolymers-and-Biocomposites/Mohanty-Misra-Drzal/p/book/9780849317415

Motaleb, K. A., Ahad, A., Laureckiene, G., & Milasius, R.. Innovative banana fiber nonwoven reinforced polymer composites: Pre-and post-treatment effects on physical and mechanical properties. *Polymers*, 13(21), 2021, 3744. https://doi.org/10.3390/polym13213744

Mouritz, Adrian P. Manufacturing of fibre–polymer composite materials, *Introduction to Aerospace Materials*, 2012. https://doi.org/10.1533/9780857095152.303

Mulinari, D. R., Voorwald, H. J. C., Cioffi, M. O. H., & Silva, M. L. C. P. Preparation and properties of HDPE/sugarcane bagasse cellulose composites obtained for thermokinetic mixer. *Carbohydrate Polymers*, 75, 2009, 317–321.

Ngo, T. D., Kashani, A., Imbalzano, G., Nguyen, K. T. Q., & Hui, D. Additive manufacturing (3D printing): A review of materials, methods, applications and challenges, *Composites Part B: Engineering* 143, 2018, 172–196.

Parameswaranpillai, J., Gopi, J. A., Chavali, M., Midhun Dominic, C. D., Radoor, S., Jayakumar, A., Siengchin, S., Rangappa, S. M., Krishnasamy, S., Hameed, N., & Thomas, S. Advances in epoxy/synthetic/natural fiber composites, in: *Handb. Epoxy/Fiber Compos.*, Springer Nature Singapore, Singapore, 2022: pp. 1093–1120. https://doi.org/10.1007/978-981-19-3603-6_52

Patel, N. K., Mishra, V., & Choudhary, T. Fabrication and characterization of epoxy composites reinforced with jute fibers and coconut fiber: A mechanical study. *Materials Today: Proceedings*, 67, 2022, 495–500. https://doi.org/10.1016/j.matpr.2022.06.471

Pereira, Paulo Henrique Fernandes et al. Vegetal fibers in polymeric composites: A review. *Polímeros*, 25(1), 2015.

Puttegowda, Madhu. Eco-friendly composites: Exploring the potential of natural fiber reinforcement. *Discover Applied Sciences*, 7, 2025, 401, https://doi.org/10.1007/s42452-025-06981-8

Ragondet, A., *Experimental Characterization of the Vacuum Infusion Process*, Tese de D. Sc., University of Nottingham, Nottingham, 2005.

Rathore, Deepak Singh, & Jain, K. K. Experimental analysis of mechanical properties of coconut coir fiber & wheat straw composite material. *International Journal of Engineering Sciences & Research Technology*, ISSN: 2277-9655, Impact Factor: 5.164, IcTM Value: 3.00, 2019. 15. Coden: Ijess7. https://doi.org/10.5281/Zenodo.2837729

Rodrigues, J., Souza, J. A., & Fujiyama, R. Polimeric composites reinforced with natural fibers from Amazon manufactured by infusion. *Revista Matéria*, 20, 2015, 946–960. ISSN 1517-7076. https://doi.org/10.1590/S1517-707620150004.0099

Rubino, F., Nistico, A., Tucci, F., & Garlone, P. Marine application of fiber reinforced composites: A review. *Journal of Marine Science and Engineering*, 8, 2020, 26.

Salim, M. Y., Rizal, M. A. M., Nordin, A. H., Harjar, M. N. N., Hafila, K. Z., Ma, Q., … Sebaey, T. A. Natural fibre-reinforced thermoplastic composites: A bibliometric analysis and review of eco-friendly solutions in exterior automotive components. *Journal of Materials Research and Technology*, 2025, 2025, 6755–6774, https://doi.org/10.1016/j.jmrt.2025.10.230

Simões, S. High-performance advanced composites in multifunctional material design: state of the art, challenges, and future directions. *Materials*, 17, 2024, 5997. https://doi.org/10.3390/ma17235997

Tesfay, D., Balakrishnan, S., Ashine, F., & Sivaprakasam, P. Sisal fibre/polypropylene composites properties by plunger injection moulding. *Materials Today: Proceedings* 62, 2022, 448–453.

Thakur, Akhilesh Kumar, Schuster, Jens, & Shaik, Yousuf Pasha, Fabrication and characterization of hybrid bio-composites using braided natural fibers and aligned thermoplastic filaments. *Journal of Composites Science*, 6, 2022. https://doi.org/10.3390/jcs6100291

Thyavihalli Girijappa, Y. G., Mavinkere Rangappa, S., Parameswaranpillai, J., & Siengchin, S. Natural fibers as sustainable and renewable resource for development of eco-friendly composites: A comprehensive review. *Frontiers in Materials*, 6, 2019, 226.

Vedernikov, A., Safonov, A., Tucci, F., Carlone, P., & Akhatov, I. Pultruded materials and structures: A review. *Journal of Composite Materials*, 54, 2020, 4081–4117.

Velde, K., & Kiekens, P. Termplastic polymers: Overview of several properties and their consequences in flax fibre reinforced composites. *Polymer Testing*, 8 2001, 885–893.

Venugopal, J., Dhanasakkaravarthi, B., Surakasi, R., Rinawa, M. L., Manjunatha, L. H., Alshgari, R. A., Wabaidur, S. M., Islam, M. A., & Jenish, I. Effect on compression molding parameters in mechanical properties of MWCNT/glass fiber/epoxy composites, *Advances in Polymer Technology*, 2022, 2022. https://doi.org/10.1155/2022/9295407

Yimin, Qin, A brief description of textile fibers, *Medical Textile Materials*, Woodhead Publishing Series in Textiles, 2016, 23–42, https://doi.org/10.1016/B978-0-08-100618-4.00003-0

Yiou, Shen, Tan, Jiayi, Fernandes, Luis, Zehua, Qu, & Li, Yan. Dynamic mechanical analysis on delaminated flax fiber reinforced composites, *Materials*, 12(16), 2019, 2559. https://doi.org/10.3390/ma12162559

Zeid, M. A., Awad, M., & Elsabbagh, A.. 3D printing of natural fiber-reinforced composites. In *Encyclopedia of Green Materials*, 13–23. 2024. Singapore: Springer Nature Singapore. https://doi.org/10.1007/978-981-16-4921-9_202-1

Zin, M. H., Razzi, M. F., Othman, S., Liew, K., Abdan, K., & Mazlan, N. A review on the fabrication method of bio-sourced hybrid composites for aerospace and automotive applications, *IOP Conference Series: Materials Science and Engineering*, 152, 2016. https://doi.org/10.1088/1757-899X/152/1/012041

4 Surface Analysis and Characterization of Chemically Modified Natural Fibers and Composites

M. J. John and N. Nomadolo

4.1 INTRODUCTION

A composite is a material structure comprising two or more macroscopically distinguishable components separated by an interface. The amalgamation of the components results in a system with enhanced structural or functional properties that will otherwise be unattainable by any component alone. Therefore, the emergence of composite materials was motivated by the need to mitigate the mechanical limitations of monolithic materials, that is, clay, ceramics, polymers, and metals. Consequently, composites are extensively employed in various industries and applications, including aircraft, chemicals, transportation, civil engineering, sport, and other consumer products, because of benefits such as light weight and excellent resistance to corrosion and fatigue, among others.

Composites consist of three main elements: the matrix (binder), the reinforcement (filler), and the interface region. The orientation of these three components allows optimal synergism, where the reinforcement is the load-bearing component that provides the composite with strength and rigidity, and the matrix, on the other hand, provides structural support for the reinforcement and allows load transfer to the reinforcement. Finally, the interface region governs the coordination between the binder and the filler. Generally, any material can serve as a matrix material for a composite; the commonly used materials include ceramics, metals, and polymers. Most utilized matrices are polymers due to their lightweight, robustness, ease of manufacturing, and economical properties. Frequently used polymers include polypropylene, polyvinyl chloride, silicone, polylactic acid, phenolic, some polyesters, epoxy, and polyurethane. The reinforcement phase is typically the stiffer and a discontinuous phase in the composite. There are two groups of reinforced polymeric composites: fiber-reinforced polymeric composites (FRPs) and particle-reinforced polymeric composites. Traditionally, synthetic fibers are utilized the most due to their outstanding properties, such as specific strength and stiffness, durability, creep resistance, light weight, heat, corrosion, and fatigue resistance. Examples of synthetic fibers include glass, carbon, graphite, graphene, aramid, and Kevlar.

Consequently, FRPs can be found in civil engineering through their use as reinforcement in concrete, automotive and aerospace industries, medical devices, and so forth. However, FRPs typically have poor transverse or shear resistance, and they exhibit brittle behavior. Thus, they are susceptible to the effects of stress rupture and lose a significant amount of strength when bent, and FRPs are relatively expensive and difficult to adapt. In addition, synthetic fibers, such as glass and carbon fibers, are from fossil sources and, thus, raise environmental concerns. The sustainability and availability of materials currently drive the primary focus of all engineering and material science in every application due to the exhaustion of non-renewable resources, emission of carbon dioxide gases, and other adverse environmental issues. Plant fiber composites (PFRCs) are increasingly used instead of

 DOI: 10.1201/9781032665337-4

synthetic fibers as reinforcement in fiber-reinforced composites because they are of low cost due to their abundance in nature; they can biodegrade into unharmful products, thus are nontoxic and possess excellent mechanical strength. Other advantages of plant fibers over synthetic fibers are good impact resistance and flexibility. They also have low specific gravity, which makes them less abrasive to equipment, less dangerous for human health, process-friendly, and CO_2 neutral. The low-density characteristic of natural plant fibers allows for high fiber loading onto the polymeric matrices, which minimizes the use of the harmful polymer base.

Interestingly, the first PFRC application can be traced to 1500 BC, when mud was reinforced with straw to create strong and durable composite bricks for building purposes (Shah, 2013). Since then, civilization, research advancement, and industrialization have extended this concept to various applications. For example, by the 20th century, Ford saw the capacity of plant fibers as structural reinforcing agents and produced the first "green car" with an all-plastic body made up of 70% lignocellulosic fibers and attributed this seemingly drastic decision to the impressive durability and strength characteristics of the lignocellulosic fibers by famously demonstrating using a sledgehammer to hit the trunk of the car (Bekemeyer, 2021). Simultaneously, Aero Research Ltd replaced light alloy with a flax fiber-reinforced phenolic composite called Gordon Aerolite to make the structural elements of Spitfire fuselages for British military aircraft for World War II (Soutis et al., 2019). Following this, Britain experienced aluminum shortages, and the use of Gordon Aerolite significantly increased for aviation application due to the immense cellulose content of bast fibers (such as flax, hemp, and jute), which possessed exceptional tensile stiffness and strength. These properties are primary markers of the structural potential of plant fibers. Currently, more than 95% of plant-fibre-reinforced polymers (PFRPs) manufactured in the EU are used for non-structural automotive components, which are produced mainly via compression molding (Bourmaud et al., 2013). Moreover, PFRPs are being appraised for applications in construction and infrastructure, sports and leisure, furniture and consumer goods, pipes and tanks, and wind energy applications, thereby extending their application across all fields.

Nevertheless, despite these benchmarking qualities of plant fibers, optimizing the interface region between plant fibers and polymer matrices is one of the most crucial steps in ensuring the best formulation of PFRCs. The fiber-polymer interface is a critical zone of diffusion where the two constituents are physically, mechanically, and chemically combined. The interfacial adhesion between the fiber and matrix governs the composite's mechanical characteristics. The key elements influencing the interfacial bonding between the fiber and polymer are the mechanical interlocking, the attractive molecular forces, and the chemical bonds. However, barriers between the plant fiber and the polymer, like impurities, pectin, and waxy substances, hinder optimal interlocking between the two components. Intrinsically, plant fibers are incompatible with polymers on account of the hydrophilic and hydrophobic nature of the two, respectively. Consequently, this prevents chemical bonds from being formed between the two. To produce an effective composite with enhanced interfacial bonding and good stress distribution across the interface, it is crucial to alter the surface properties of plant fiber and polymer matrix. Numerous methods are utilized to enhance the compatibility and bonding between plant fiber and polymer, including physical, physicochemical, and chemical treatment methods.

4.2 PLANT FIBER COMPOSITES

4.2.1 Polymer Matrix

The critical functions of the polymeric matrix are to maintain the reinforcing fibers in place, protect them from mechanical and chemical damage, and ensure that the stress is appropriately distributed between the fibers. Since natural fibers have a high propensity to degrade at temperatures above 200°C, selecting a polymeric matrix is often governed by processing temperature. Traditionally, thermosets are preferred over thermoplastics in electronics/electrical applications, motor, and construction industries, mainly because thermosets have high-temperature resistance, which allows the preservation of plant fiber mechanical properties from high processing temperatures. Additionally, thermosets offer better mechanical qualities than thermoplastics because curing creates a stable, massive, cross-linked three-dimensional molecular structure. Moreover, they allow ease of adherence of the fiber during resin impregnation, resulting in better dispersion of fibers and thus interfacial properties. Lastly, thermosets provide superior interface and shear resistance compared with plant fibers that are typically polar. Examples of thermosetting matrix polymers are epoxy resin, vinyl esters, polyesters, and polyurethane.

Nonetheless, there is a general shift toward the use of thermoplastic matrices due to faster processing time, cleaner manufacturing process (dry systems with no hazardous by-products), and lower cost (for high-volume production). Thermoplastics have outstanding impact characteristics and may be molded and recycled by reheating. It is preferred for manufacturing high-performance structures using processes such as melt compounding and molding, resulting in low-cost intermediate composite goods. Polypropylene (PP) and polyethylene (PE) are the thermoplastic resin matrices that are most frequently used. Ultimately, the selection of the polymeric matrix lies with its application.

The rampant use of petroleum-based polymers has led to the exhaustion of non-renewable resources, emission of greenhouse gases, and the accumulation of plastics in the environment, which infiltrate the food chain, posing a threat to the environment, animals, and human life. Resultantly, biobased biodegradable polymers have emerged as viable alternatives to petroleum-based polymers, which have shown comparable mechanical performance. Currently, the primary barrier to the widespread application of bioplastic polymers is their high initial cost. Biobased, biodegradable polymers are three to five times more expensive than commonly used resins like polyvinyl chloride (PVC), polypropylene (PP), low-density polyethylene (LDPE), and high-density polyethylene (HDPE). The more practical alternative is to blend petroleum and biobased polymers to create an economical product with broad application. This motivation has led to the evolution of different classes of reinforced polymer composites (NFRPs), including but not limited to green composites, hybrid composites, and textile composites.

4.2.2 Plant Fibers

4.2.2.1 Chemical Compositions and Classification

Generally, plant-producing fibers are classified into primary and secondary groups. The primary group refers to plants grown to produce fiber content, that is, kenaf, sisal, and cotton. The secondary group comprises plants whose fibers are secondary by-products, like oil palms, pineapple, and cereal stalks. Fibers are commonly categorized according to origin (i.e., bast, leaf, seed, core, grass, wood, and roots). Then, following this, different systems are also used to categorize fibers further, such as in industrial applications (examples are textile fibers, papermaking fibers, and composite fibers). Table 4.1 summarizes plant fibers according to their chemical composition, species, and botanical type.

4.2.2.2 Structure of Natural Fibers

Natural plant fiber is the multi-layered heterogeneous membrane of the cell wall found in different plant parts (i.e., leaves, stems, roots). The structure of the cell wall greatly influences the mechanical properties of the fiber, and it is primarily organized in a framework of semi-crystalline cellulose encapsulated in the amorphous matrix of hemicellulose, pectin, lignin, and proteins. Figure 4.1 shows the structure of the plant fiber. The first deposited layer during cell growth is the primary wall, a hydrated thin layer containing cellulose encased in the hemicellulose and pectin polysaccharide matrix. The prominent role of the primary cell wall is to encourage cell wall extension, thereby promoting plant growth, and to provide strength to bear the load of the secondary wall as it encircles it. The secondary wall is a heavily gel-like compound comprising primarily cellulose in a network

TABLE 4.1
Chemical Composition of Plant Fiber according to Species and Fiber Origin (Birniwa et al., 2023; Bourmaud et al., 2018; John & Anandjiwala, 2008)

Fiber	Chemical Composition (wt.%)				Species	Botanical Type
	Cellulose	Hemicellulose	Lignin	Wax		
Abaca	56–63	20–25	7–12	3.0	Musa textilis	Leaf
Banana	55.2	16.8	25.3	–	Musa indica	
Curaua	73.6	9.9	7.5	–	Ananas erectifolius	
Henequen	60	28	8	1.09	Agave fourcroydes	
Mauritius hemp	68	15	10	0.8	Cannabis sativa	
Palm	65	–	29	–	Phoenix dactylifera	
Piassava	28.6	25.8	45	–	Attalea funifera	
Pineapple	81	–	12.7	–	Ananus comosus	
Sisal	65	12	9.9	2	Agave sisalana	
Alfa	45.4	38.5	14.9	2	Strippa tenacissima	Grass/reed
Bagasse	55.2	16.8	25.3	–	–	
Bamboo	26.43	30	21–31	–	>1,250 species	
China Jute	61–71	14–20	12–13	0.5	Abutilon theophrasti	Stem
Flax	71	18.6–20.16	2.2	1.5	Linum usitatissimum	
Hemp	68	15	10	0.8	Cannabis sativa	
Isora	74	–	23	1.09	Helicteres isora	
Jute	61–71	14–20	12–13	0.5	Corchorus capsularis	
Kenaf	72	20.3	9	–	Hibiscus cannabinus	
Kudzu	33	11.6	14	–	Puerarla thunbergiana	
Ramie	68.6–76.2	13–16	0.6–1.7	–	Boehmeria nivea	
Sunn hemp	41–48	8.3–13	22.7	–	Crotalaria juncea	
Coir	32–43	0.15–0.25	40–45	–	Cocos nucifera	Fruit
Sponge	63	19.4	11.2	3	Luffa cylindrica	
Cotton	82.7–98.0	4.0–5.7	0.7	0.6	Gossypium sp.	Seed

of hydrophobic polyphenol lignin and hemicellulose polysaccharides. The presence of lignin in the secondary wall restricts water permeation; in doing so, the secondary wall is stiffer. The wall has three distinct layers, namely, S1, S2, and S3. S1 is the thin outer layer of the secondary cell wall with cellulose microfibrils organized in a diagonal orientation network. S2 is the thickest layer of the three and has the lowest microfibril angle (microfibril angle forms between the fiber axis and microfibril). It is fundamentally made up of helically wound cellular microfibrils formed from long-chain cellulose molecules.

S2 constitutes about 80% of the secondary cell wall. Consequently, it determines the mechanical properties of the fiber as its primary function is to provide high axial stiffness while also ensuring high collapse and burst resistance, that is, yield strength. Overall, the mechanical properties of the fiber are heavily bolstered by the nature and crystallinity of cellulose, which is symbiotically supported by hemicellulose and lignin. Finally, the inner layer of the secondary wall is referred to as S3, and like the S1 layer, it also has the diagonal orientation of cellulose microfibril and thus presents higher microfibril angles. The S3 layer, often called the tertiary wall, is situated at the lumen-to-cell wall transition and acts as a barrier. The S3 layer in the secondary wall has the highest concentration of lignin when compared to the other secondary wall layers. The microfibrillar angle (MFA) of the microfibrils in the S3 layer ranges from 50° to 90°, almost perpendicular to the microfibrils in the S2

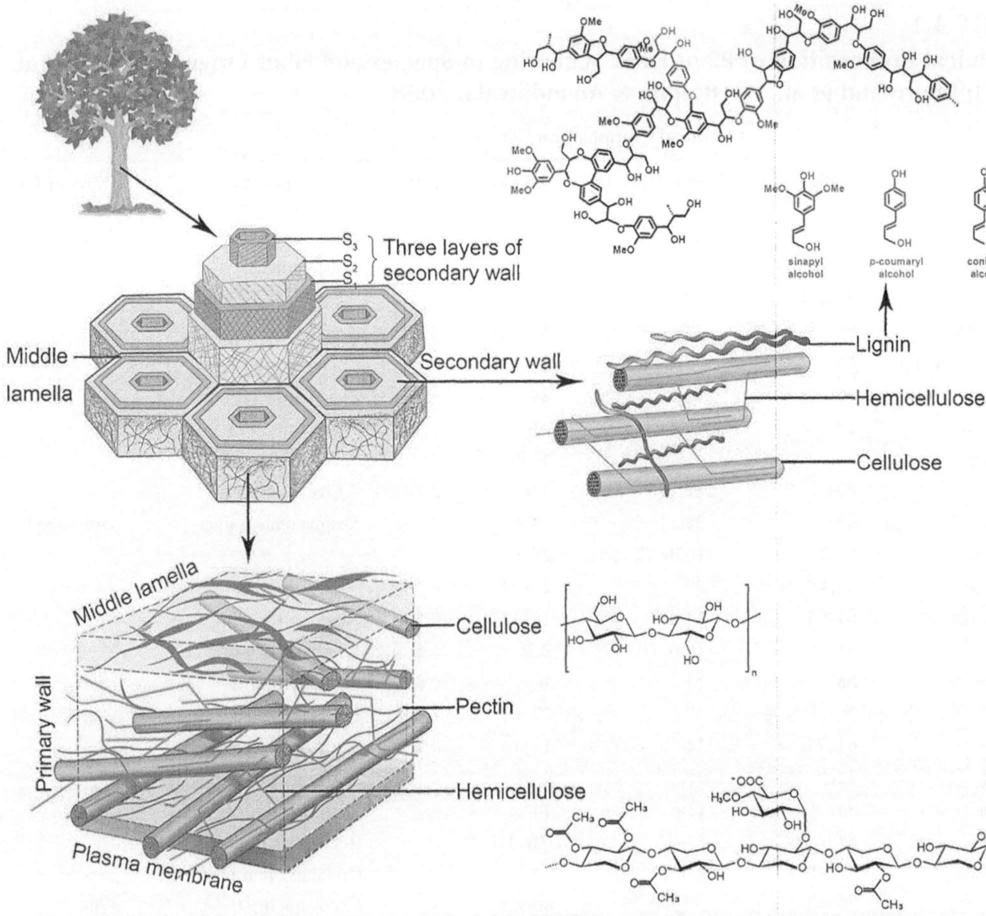

FIGURE 4.1 Typical plant fiber structural arrangement (Zhao et al., 2019).

layer. Finally, there is the so-called warty layer, found in the innermost part of the cell wall, most likely created from protoplasmic waste called the lumen. Overall, S1 and S3 allow the plant fiber to restrict deformation when exposed to compression and tension, and S2 provides the fiber with tensile strength and stiffness.

Table 4.1 illustrates that the three main chemical constituents of plant fibers are cellulose, hemicellulose, and lignin from a molecular perspective. Cellulose is a natural polymer comprising D-anhydrous-glucose ($C_6H_{11}O_5$) repeating units linked by the β-(1,4) glycosidic bonds. The repeating units contain three hydroxyl groups that interact to form hydrogen bonds, which are particularly essential for the physical properties of cellulose, as they govern the crystalline packing of the polymer. Cellulose is a semi-crystalline polymer with a microcrystalline structure of high order (crystalline region) and low order (amorphous regions). The crystalline chains form microfibrils of Iα orientation, which convert to the more thermodynamically stable Iβ orientation upon chemical treatment. The primary and secondary cell walls contain cellulose chain lengths of 2000–6000 and 10,000, respectively. The crystalline nature of cellulose renders it more chemically stable and is fundamental in its extraction, modification, and application. Although cellulose is rapidly hydrolyzed by acids to produce water-soluble sugars, it is extremely resistant to strong alkali and oxidizing agents.

Unlike cellulose, hemicellulose is a nonlinear, heterogeneous, and complex polymer. Hemicellulose contains a range of highly branched and short chains of polysaccharides composed of 5- and 6-carbon ring sugars with a degree of polymerization of about 50–300. It is amorphous and very hydrophilic.

Hemicellulose is soluble in alkali and easily hydrolyzed in acid. Lignin is a class of hydrocarbon polymers with both aromatic and aliphatic properties. The high molecular weight, three-dimensional amorphous lignin polymer comprises hydroxyl and methoxy-substituted phenylpropane units, which are converted to phenylpropane units. It is often referred to as the cementing agent that binds the individual cellulose fibers together, thereby strengthening the rigidity of the plant cell wall. Due to its hydrophobic nature, lignin also plays a crucial role in protecting hydrophilic cellulose and hemicellulose. Other lignin functions include encouraging mineral transport and protecting the plant against environmental stress. The fundamental issue in lignin chemistry is that no method has been developed for isolating lignin in its original condition from the fiber. Lignin is a thermoplastic polymer with a glass transition temperature of roughly 90°C and a melting temperature of about 170°C. It is soluble in hot alkali, rapidly oxidized, and readily condensable with phenol; however, acids do not hydrolyze it.

4.2.2.3 Mechanical Properties of Plant Fibers

Table 4.2 summarizes various plant fibers' morphological and mechanical properties data. It is observed that the densities of different plant fibers do not significantly vary from each other. However, both the diameters and MFA substantially vary between plant types, and these variables affect the mechanical characteristics of the fibers. Young's modulus and tensile strength of fibers are directly proportional to the cellulose concentration. For example, flax and natural bamboo fibers have cellulose concentrations of about 71% and 26%, respectively, and their tensile strength and Young's modulus are reported to be 345–1035 MPa and 27.6 GPa for flax fibers and 140–230 MPa and 11–17 GPa for natural bamboo fibers. Cotton versus coir fiber is another example that illustrates the dependence of mechanical properties on the cellulose content, with 98% and 43% cellulose concentration for the plant fibers, respectively. Another factor that is pivotal to the mechanical properties of natural fibers is the aspect ratio. A smaller fiber diameter with a higher fiber length is crucial as it allows high fiber content loading with complete coverage across the polymeric matrix. This enables optimal load transfer between fiber and matrix. The microfibrillar angle, MFA, is also said to influence the stiffness of the natural fibers. Plant fibers will be more ductile if the microfibrils are

TABLE 4.2
Morphological and Mechanical Properties of Various Plant Fibers (Bourmaud et al., 2018; Chichane et al., 2023; Imraan et al., 2023; Mokhothu & John, 2015)

Fiber	Density (g/cm³)	Diameter (mm)	Length (mm)	Microfibrillar Angle (Degrees)	Tensile Strength (MPa)	Young's Modulus (GPa)	Elongation at Break (%)
Abaca	1.5	18.2	2–12	5–17	400	12	3–10
Alfa	0.89	5–10	2–5	–	350	22	5.8
Bagasse	1.25	20	–	–	290	17	–
Bamboo	0.6–1.1	5–40	0.5–5	8–11	140–230	11–17	–
Banana	1.35	80–250	–	10–12	500	12	5.9
Coir	1.2	100–450	0.28–1.1	30–49	175	4–6	30
Cotton	1.5–1.6	12–38	10–60	20–30	287–800	5.5–12.6	7–8
Flax	1.5	12.4–23.9	6–80	5–11	345–1035	27.6	2.7–3.2
Hemp	1.4–1.6	10.9–42.0	5–55	2–11	690	70	1.6
Jute	1.3–1.4	15–35	1–6	7–12	393–773	26.5	1.5–1.8
Kenaf	1.5	2.1–36	2.1–2.7	9–15	930	53	1.6
Pineapple	0.8–1.6	80–250	–	8–15	1.44	400–627	14.5
Ramie	1.5	10–50	120–150	7.5	560	24.5	2.5
Sisal	1.5	50–200	0.5–8	10–25	511–635	9.4–22	2.0–2.5

oriented in a spiral direction relative to the fiber axis. The microfibrils will be rigid, inflexible, and have a high tensile strength if orientated parallel to the fiber axis.

4.2.3 Challenges of Plant Fiber Composites

Using lignocellulosic fiber as reinforcement in polymer composites undoubtedly has numerous benefits; however, the effectiveness of the composite materials is determined by several criteria. The primary criterion is the interfacial compatibility and adhesion between the fiber and matrix, which is fundamental to the mechanical behavior of the PFRC. The critical function of the fiber in the polymeric composite is to enhance the mechanical properties of the composite material by carrying most of the compressive strength exerted into the composite, to bridge polymer matrix cracks, and to decrease crack propagation by dissipating the energy around the fracture tip. Therefore, the interfacial bonding between the fiber and the matrix is of great significance as it allows the optimal distribution of load stress between the two components of the composite.

However, the inherent aquaphilic attribute of the plant fiber, which depends on fiber content loading, degree of crystallinity, and time spent in a humid environment, leads to poor stress distribution from matrix to fibers, culminating in composite mechanical failure. Generally, the hydroxyl moieties on the fiber's surface form hydrogen bonds with polar molecules, that is, water. The water is then permeated from the surface to the bulk material of the fiber by capillary action or diffusion, especially when the amorphous cellulose content is high. This then leads to the swelling of fibers, the creation of cavities in the interface, the leaching of the water-soluble compounds from the surface of the fiber, and ultimately causes the interfacial bonds between the matrix and the natural fibers to crumble, leading to poor mechanical strength and reduction of dimensional stability.

In addition, fibers are highly prone to rotting and microbial attack due to their hydrophilic property. Another factor that threatens the interfacial bonds between the lignocellulosic fiber and the polymeric matrix is the incompatibility of the two due to the polar and non-polar properties of the two components. This incompatibility causes heterogeneous dispersion of the fibers within the polymer during compounding, as it produces poor interfacial bonding. Moreover, poor fiber dispersion has a high proclivity to make loose bundles with a low aspect ratio, low mechanical strength, and, thus, less reinforcing ability than a single fiber. Other criteria that influence the mechanical performance of the natural fiber polymeric composite include limited processing temperature, as fibers are thermally unstable above 200°C, fiber aspect ratio (for optimum effectiveness, fiber aspect ratio must be maintained in the range of 100–200), and fibril orientation, which informs the microfibril angle and consequently, fiber strength.

Referring to these challenges, various physical and chemical alterations have been attempted to reduce the hydrophilicity of plant fiber, hence improving the interfacial adhesion. These modification techniques differ in their efficiency and methodology for optimizing the composite's interfacial properties. In general, physical treatments, such as electric discharge, alter the structural and surface properties of the fiber by introducing surface crosslinking, changing surface energy, and creating reactive free radicals and groups, ensuring improved composite mechanical strength. The chemical treatments permanently alter the character of fiber cell walls by eliminating hydroxyl groups and surface contaminants such as waxy substances, hence addressing PFRC compatibility and interfacial bonding. Recent studies have focused on the use of biobased coupling agents so as to maintain the renewable and environmentally friendly characteristics of natural fibers (John, 2022; John & Anandjiwala, 2009).

4.3 CHEMICAL TREATMENTS

4.3.1 Alkaline Treatment or Mercerization

The most common chemical process utilized to enhance the interfacial bonds between the fiber and the polymer is alkaline treatment. This method is also called a mercerization process, named after a dye and fabric chemist from England, John Mercer (Heines, 1944). There are two possible outcomes of alkaline

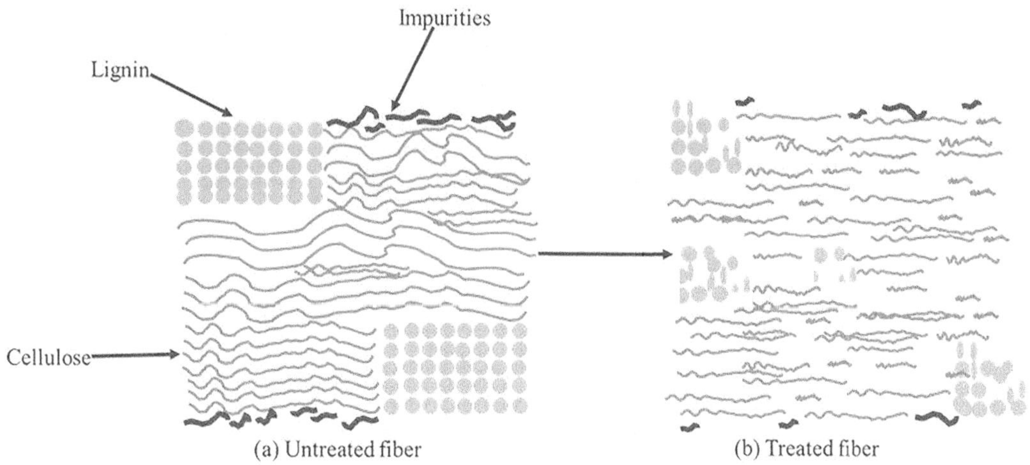

FIGURE 4.2 Scheme depicts effect of alkalization on the plant fiber (Mwaikambo & Ansell, 2002).

treatment. The first is that alkaline treatment shortens the length of crystallites, increasing amorphous regions of cellulose, which improves the number of reaction sites. The second outcome is that this treatment removes surface impurities. In doing so, the surface roughness increases, and as such, better mechanical interlocking may occur; additionally, the second outcome also improves the aspect ratio that results from the fiber diameter reduction, enhancing the fiber surface area. Figure 4.2 illustrates the impact of alkaline modification on the fiber structure, and Equation 4.1 shows the alkali reaction pathway.

$$Fibre - OH + NaOH \rightarrow Fibre - O^{-}Na^{+} + H_2O \tag{4.1}$$

The loss of hydrogen bonding from the lignocellulosic network structure is a critical alteration during alkylation. The alkaline treatment stimulates a swelling reaction on cellulose fiber, relaxing the cellulose's native crystalline structure. The kind and concentration of the alkali solution (KOH, LiOH, or NaOH) will affect the level of swelling and, thus, the level of lattice transfiguration into thermodynamically stable cellulose-II. It has been demonstrated that Na has a suitable diameter that enables the fiber microstructure to expand and penetrate even the compact pores between the lattice arrangements. Consequently, treatment with sodium hydroxide (NaOH) causes more swelling. This creates a new Na-cellulose-I lattice form, with voids filled with water molecules and relatively high distances between the cellulose molecules. The cellulose's OH-groups are then changed in this structure into ON a-groups, increasing the size of the molecules. Following a subsequent water rinse, the connected Na-ions will be released, causing the cellulose to transform into cellulose-II. This new crystalline structure is thermodynamically more stable than cellulose-I (Figure 4.3). In contrast to other alkalis, NaOH can trigger a complete lattice transformation from a monoclinic crystalline lattice of cellulose-I to a polymorphous form of cellulose-II. The alkali solution affects the plant fiber's hemicellulose, lignin, and pectin components.

4.3.2 Acetylation Treatment

Acetylation of plant fiber surfaces is a desirable way to increase their hydrophobicity while improving their resilience to environmental deterioration and dimensional stability. Acetylation treatment proceeds through an esterification process, and it involves a reaction between the hydroxyl moieties of the cell wall with acetic acid at high temperatures. The addition of the acetyl functional group CH3COO – allows for this reaction to also be referred to as the plasticization of the fiber.

$$Fibre - OH + CH_3 - O - OC - CH_3 \xrightarrow[\text{Conc. } H_2SO_4]{CH_3COOH} Fibre - O - CO - CH_3 + CH_3CHOOH \tag{4.2}$$

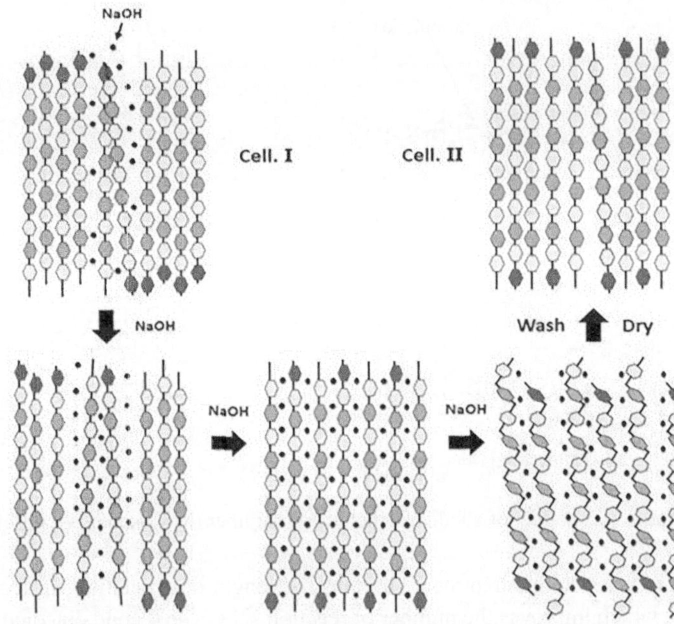

FIGURE 4.3 Schematic depicts the transfiguration of cellulose-I to cellulose-II ordered structure after NaOH treatment (Uusi-Tarkka et al., 2021).

The reaction's fundamental idea is to substitute the hydroxyl moieties (OH) in the cell wall with the acetyl group CH3COO – from acetic anhydride (CH_3COO-C=O-CH_3), making the surface of the fiber more hydrophobic. This esterification reaction has moderate reaction rates as the acid targets the hydroxyl group on the hemicellulose, lignin, and the amorphous region of cellulose only. The hydroxyl group of the crystalline cellulose is sterically inaccessible to the acid. The degree of acetylation is controlled by the catalyst and acids employed. The technique also lowers the number of void contents, evidenced by the rough surface consequence. As a result, the composite has superior mechanical performance because of the augmented interfacial bonding between the fiber and the polymer.

4.3.3 Benzoylation Treatment

Benzoylation is another chemical modification technique that reduces the hydrophilic characteristic of natural plant fiber. Benzoylation employs alkali treatment of the natural fiber first to remove the waxy, oily, and lignin compounds found on the fiber's surface so that the reactive hydroxyl group can be accessible on the surface. Subsequently, benzoyl chloride is used to condition the fibers, allowing the substitution of the hydroxyl group with the benzoyl group (C_6H_5C=O). This action bestows the natural fibers with hydrophobic characteristics, which strengthen the adherence of the fibers to the polymer, leading to the augmentation of mechanical behavior of the composite. The reaction mechanism is represented in Figure 4.4.

4.3.4 Peroxide Treatment

Peroxide treatment starts with alkali pre-treatment of the natural fibers, followed by a 30-minute treatment with low-concentrated benzoyl peroxide or dicumyl peroxide in an acetone medium at elevated temperatures. During the treatment, the peroxide group (ROOR) decomposes to highly reactive radicals (RO), which are then grafted onto the cellulose fiber through interaction with the

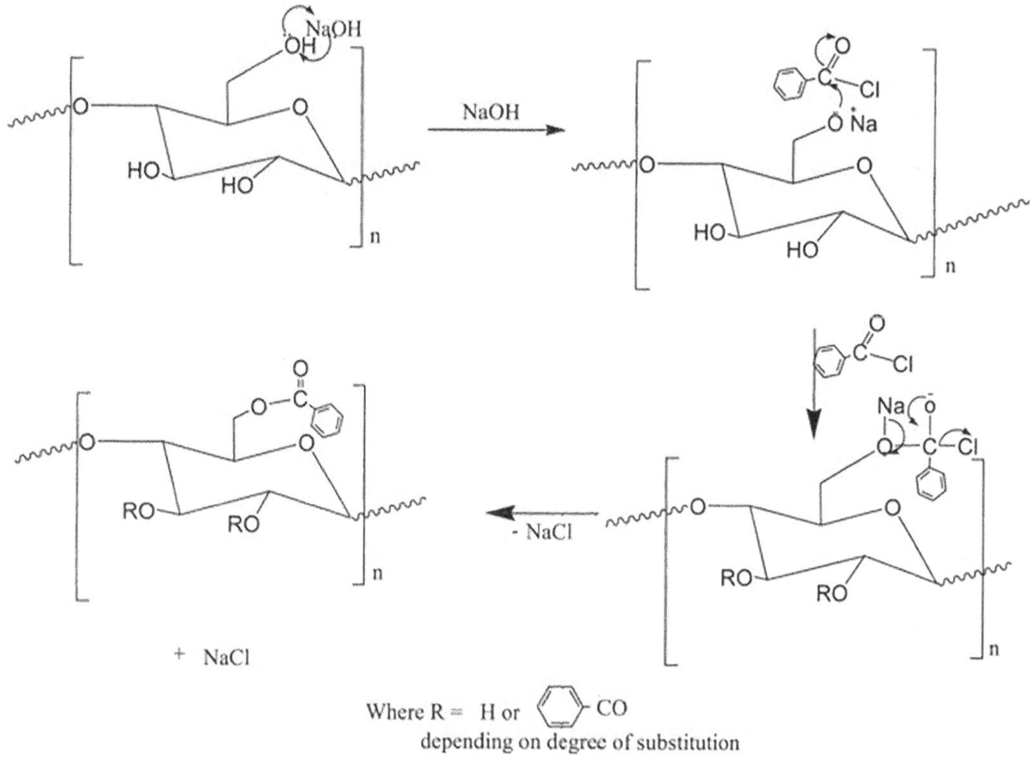

FIGURE 4.4 Benzoylation treatment reaction pathway (Ajayi et al., 2023).

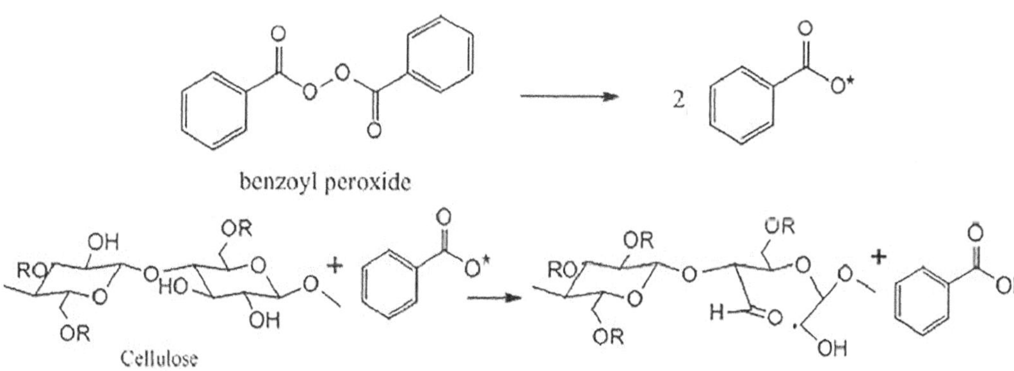

FIGURE 4.5 Peroxide chemical treatment method scheme (K. Kaushik et al., 2013).

hydroxyl moieties. The interfacial bonding of plant fiber and the matrix is improved, positively influencing the composite's thermal stability. The textile industry has named this method bleaching due to the combined effect of alkaline-peroxide treatment (Figure 4.5).

4.3.5 Silane Treatment

Silane treatment is an efficient chemical treatment that coats the surface of the lignocellulosic fiber and increases the moisture resistance of the plant fiber, leading to better adhesion between the components of the composite. Typically, the technique utilizes molecules that possess two

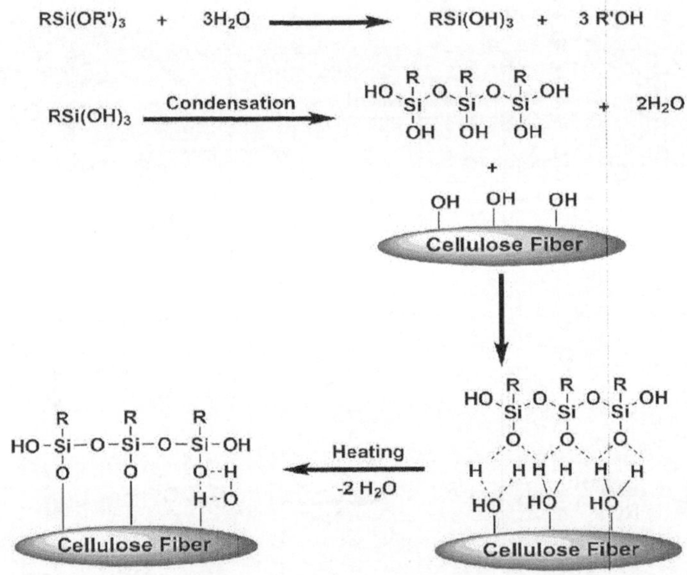

FIGURE 4.6 Silane treatment reaction mechanism (Gadhave et al., 2021).

reactive groups; one is the inorganic group, which interacts with the fiber, and the other is an organo group, which bonds to the thermoset polymer matrix. This bridging arrangement improves the adhesion between the two components, and such molecules are referred to as coupling agents. Generally, silane is a monomeric silicon compound in the form of SiH_4. It is often utilized as a coupling agent as it possesses key functional moieties; trialkoxy and alkyl chain substituents on the silane atom; the standard structure of silane is $(RO)_3$-Si-R-X. Silane treatment follows a three-step reaction mechanism: (1) hydrolysis – the trialkoxysilane hydrolyses to form silanols; (2) condensation – the silanol then condenses with each other and forms siloxane linkages oligomers; and (3) curing/Adhesion – finally, the oligomers interact with the plant fiber's hydroxyl moieties, creating strong attachments through covalent bonding to the cell wall and thus rendering the lignocellulosic surface less hydrophilic. Simultaneously, the alkyl chain diffuses into the polymeric matrix, producing an entangled network with the matrix and, thus, preventing the swelling of fibers within the composite, which then improves the interfacial bonding between the two. Figure 4.6 represents a schematic overview of silane treatment. Commonly used silanes include amino, glycidoxy, vinyl, alkyl, and azide silanes. Factors such as pH, temperature, organo functionality, and hydrolysis time influence the absorption and adsorption of silane onto the natural fiber composite. Enhanced silane treatment is often achieved with alkaline pre-treatment, providing optimal hydrolysis and condensation conditions. This technique offers wettability, good compatibility, and surface tension.

4.3.6 Maleated Coupling Agents

Another coupling agent that has found its use in the PFRC reinforcement is maleic anhydride. Like the silane coupling agent, maleic anhydride interacts with the surface of the fiber and the polymer. The organic end-group forms a strong C-C bond with the polymer, and the hydrolytically anhydride linkages end-group interacts with the hydroxyl moieties of the plant fiber, thereby reducing the hydrophilicity of the fiber. Strong adhesion is the outcome of this kind of chemical treatment because of the molecular entanglement through covalent bonding that connects the polymer and the plant fiber through the coupling agent.

4.4 CHEMICAL MODIFICATION AND CHARACTERIZATION OF NATURAL POLYMER FIBERS

Enhancing cellulose's utility as a composite component requires optimizing surface modification processes, which need a thorough grasp of the intricate structure of fiber surfaces. Strong adhesion at their interfaces is important to transfer stress and distribute bonds along an interface because the polymers and fibers have distinct chemical compositions. Adhesion can be via chemical bonding, electrostatic interaction, acid-base adhesion, and wetting diffusion in the form of mechanical or entanglement interlocking of components. Chemical modification can alter fibers due to the hydroxyl groups on the surface of cellulose, which possess the potential to participate in hydrogen bonding by either energizing existing groups or introducing new functional groups that create functional interlocks in the structure. Surface properties like as adhesion, surface tension, wetting, and porosity of fibers can all be enhanced by chemical treatment. A key element in the mechanical interlocking at the interface is the imperfections on the fiber surface. By appropriately modifying the constituents, the interfacial properties can be enhanced, leading to variations in the physical and chemical interactions at the interface. Extensive research has been conducted on fiber modification and characterization, including the current studies listed below.

4.4.1 Alkali Treatment

4.4.1.1 Scanning Electron Microscopy

The examination of the micro surface of plant fibers using scanning electron microscopy (SEM) aids in the visual assessment of the texture of the fibers, their orientation, composition, etc., subject to various chemical pre-treatments of the fibers. Overall, the effect of alkaline treatment is reported to remove impurities such as oil, lignin, and waxy compounds on the surface of the cellulose fibers, thereby allowing the cellulose fiber to have better adhesion to the polymeric matrix and improving the interfacial bonding, which will positively influence the mechanical behavior of the plant fiber polymer composite. An investigation done by Abdallah et al. (2022) explored the impact of alkaline treatment on thermal insulation application of date palm wood fibers (DPWF) and polylactic acid (PLA) green composite. The noticeable effect of alkaline treatment was depicted by SEM images (Figure 4.7), which showed pre-treated DPWF with clusters of impurities on the surface of the fibers. These clusters increased the interfacial bonding distance between the fibers and the thermoset, thus restricting the optimal load distribution from the polymer to the fiber. Treatment with NaOH resulted in removing the aggregates on the surface, exposing the cellulose surface (Figure 4.7 c–d). The exposure allows access to the cellulosic hydroxyl group, which is fundamental in interlocking the fiber to the matrix. The authors reported increased crystallinity of the composite by 41% and 51% and decreased density by 15% and 9% after treatment with NaOH and KOH, respectively. The observations were consistent with those found by Tuğluca et al. (2023) when investigating the impact of varying molarities of NaOH (1 M–10 M) and the duration of the treatment (2 hr, 4 hr, 6hr, and 12 hr) on the elimination of surface impurities and enhancement of waste wood fiber (WWF) properties and, consequently, fiber-matrix compatibility, that is, cementitious composite (Figures 4.8 and 4.9). The study revealed that the degradation of non-cellulose substances, that is, hemicellulose, oils, wax, and lignin, has resulted in the creation of a rough-textured cellulose surface, and led to fiber separation by breaking the fiber bundles into microfibrils (defibrillation), and alkaline conditioning caused the appearance of hollow regions on the surface of the fibers. Additionally, *Tugluca and team* further revealed that, as alkaline duration increased from 2 hr to 12 hr, the number of pores increased, fibrillation was most favorable, and better consistency of fiber arrangement as depicted by Figure 4.9(a–c). At the same time, the decomposition of inorganic materials was observed. After 6 hours, the microstructure of cellulose fiber also started to decompose, concluding that 4-hr treatment with 2.5 M NaOH was sufficient for optimal cellulose fiber modification. The observed changes in the microstructure of WWF due to alkaline conditioning led to the augmentation of flexural strength capacity and compressive strength.

FIGURE 4.7 Micrographs of (a, b) untreated date palm wood fiber and (c, d) NaOH-treated fiber (Abdallah et al., 2022).

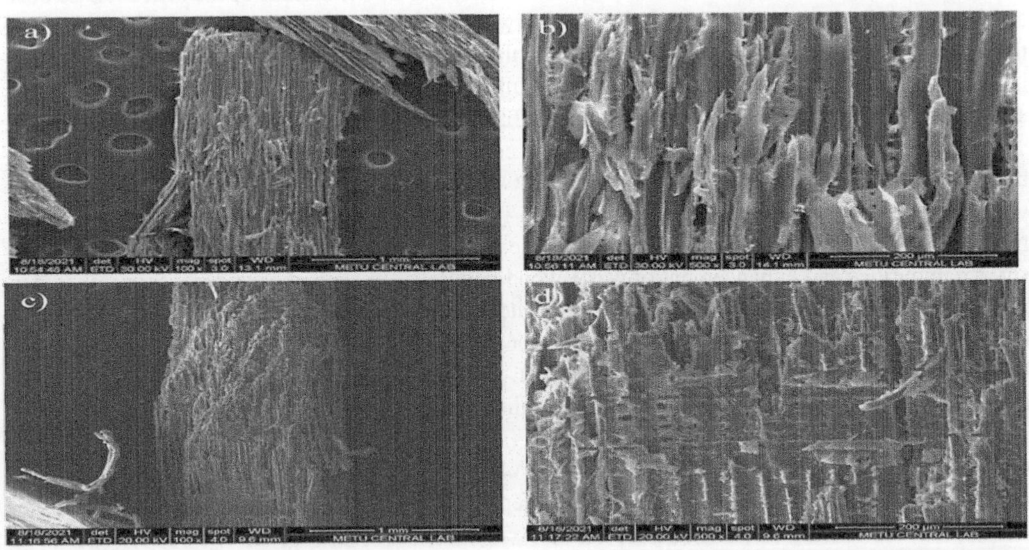

FIGURE 4.8 Micrographs of (a) and (b) untreated WWF and (c) and (d) treated WWF with 2.5 M NaOH solution for 12 h (Tuğluca et al., 2023).

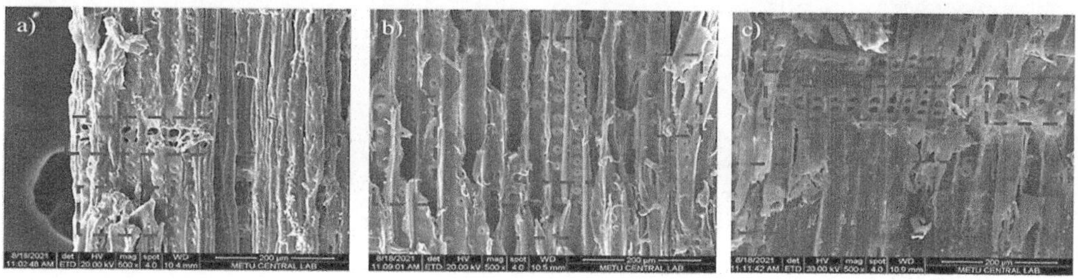

FIGURE 4.9 Micrographs illustrating the effect of treatment duration at 2.5M NaOH: (a) 2h, (b) 4h, and (c) 6h (Tuğluca et al., 2023).

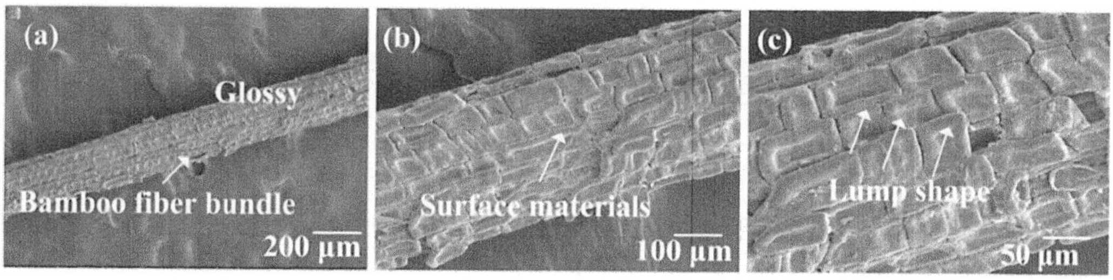

FIGURE 4.10 Micrographs of surface of 2 wt.% NaOH-treated bamboo fiber bundles (Zhang et al., 2018).

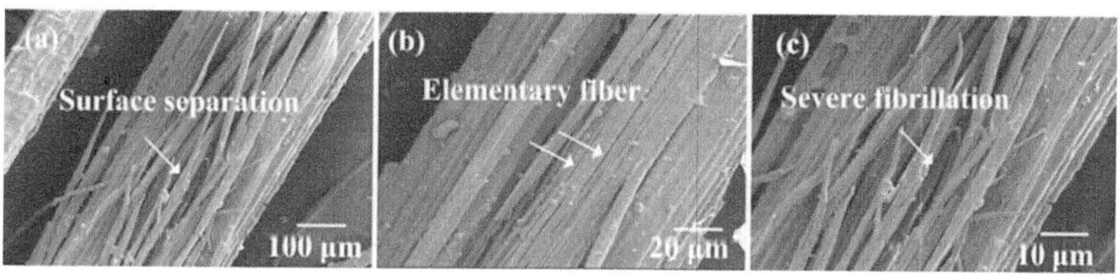

FIGURE 4.11 Micrographs of surface of the 10 wt.% NaOH-treated bamboo fiber bundles (Zhang et al., 2018).

A distinct defibrillation outcome of NaOH conditioning of fiber was also described by Zhang et al. (2018) when evaluating the impact of varied concentrations of NaOH solution on bamboo fibers and the fabrication of bamboo fiber epoxy composites. Like Abdallah *and* Tuğluca et al. (Abdallah et al., 2022; Tuğluca et al., 2023) results, the authors concluded that conditioning the plant fiber with NaOH resulted in the elimination of impurities on the surface of the fiber, forming a rough surface (Figure 4.10). Zhang and colleagues further presented Figure 4.11(a–c) SEM images showing that the evolution of individual fiber strands was attributed to the high concentration of the alkaline solution. The authors went on to assess the role of NaOH concentration on the interfacial adhesion between bamboo fibers and epoxy matrix by carrying out a pull-out test. Figure 4.12(a–c) illustrates that NaOH-treated fiber was more compatible with the epoxy matrix since more of the matrix was found on the surface of the fibers after being pulled out, and it was reportedly harder to pull out the fibers, evidenced by the necking during the test, as opposed to the unmodified fiber epoxy composite (Figure 4.12), and the fibers were reportedly profoundly embedded in the matrix.

4.4.1.2 FTIR Analysis

Fourier transform infrared spectroscopy (FTIR) analysis monitors the peak positions associated with functional groups belonging to the fibrous material and thermoset polymer. Generally, the plant fiber exhibits functional groups of O-H stretching of hydroxyl α-cellulose peak, C-H stretching of cellulose, C=O stretching of hemicellulose and lignin, symmetric C-OH benzene ring, -OCH3 lignin

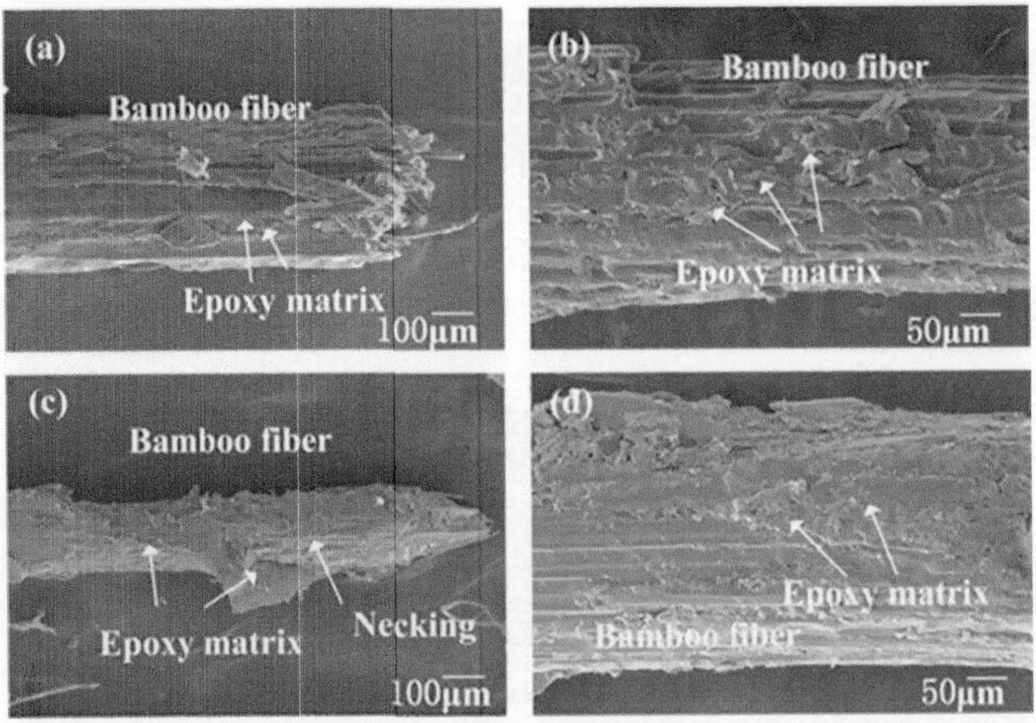

FIGURE 4.12 Pull-out test micrographs depicting the effect of NaOH treatment on the fiber epoxy: (a) unmodified fiber, and (b) 2 wt.%, (c) 6 wt.%, and (d) 10 wt.% NaOH-treated bamboo fiber (Zhang et al., 2018).

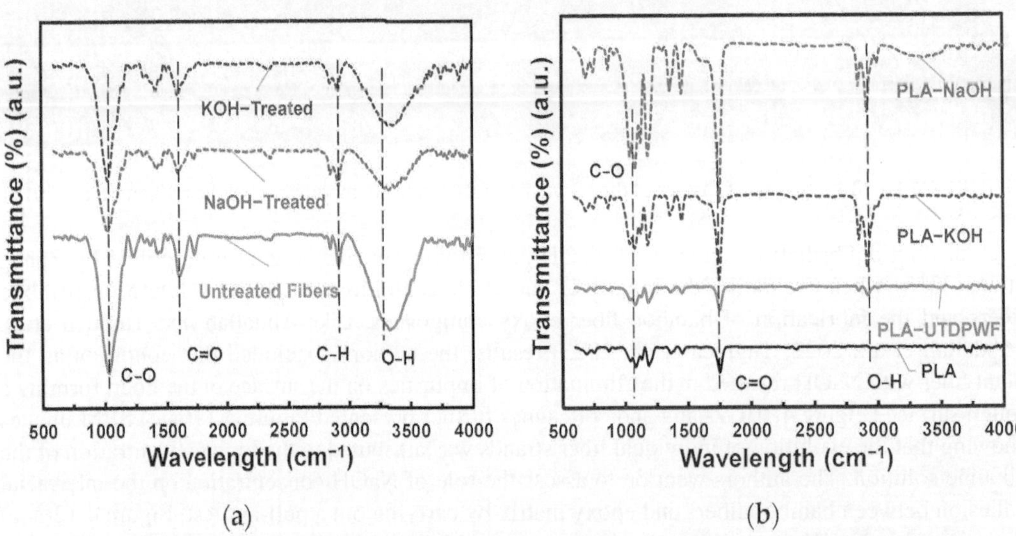

FIGURE 4.13 FTIR of (a) treated and untreated fibers and (b) neat PLA, PLA-UTDPWF, PLA-NaOH, and PLA-KOH composite (Abdallah et al., 2022).

stretches, and C-OH bending in cellulose located around 3400 cm^{-1}, 2900 cm^{-1}, 1700 cm^{-1}, 1200–1600 cm^{-1}, 1041 cm^{-1}, and 601 cm^{-1}, respectively. Interestingly, Abdallah et al. (2022) observed the reduction of peaks associated with hemicellulose and lignin due to alkaline conditioning, illustrating the delignification that has occurred (Figure 4.13a). Conversely, Figure 4.13b displayed the O-H band on the composite at a wavelength of around 3000 cm^{-1}. This peak resulted from

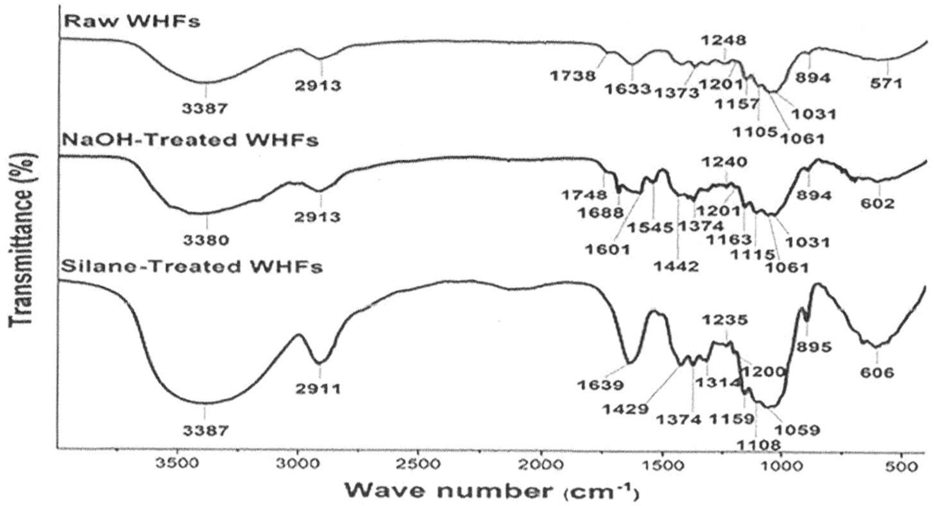

FIGURE 4.14 FTIR of alkalized versus and silanated WHFs (Sumrith et al., 2020).

the incorporation of the aquaphilic fiber into the matrix. Additionally, lignin and hemicellulose are removed from plant fiber during alkaline treatment, as demonstrated by the cellulose ratio, which was more significant in the treated samples. The observed increase in the cellulose ratio confirmed the disappearance or reduction of their respective peaks. The alkaline-conditioned composite also presented a very sharp carboxylic acid group (C=O) band, possibly due to the cellulose-PLA reaction.

Sumrith et al. (2020) reported that alkaline conditioning of water hyacinth fiber (WHF) led to the weeding out of lignin and hemicellulose, evidenced by the decreased peak intensity of the O-H functional group (Figure 4.14). Furthermore, the authors showed that the silane-treated WHF spectrum exhibited an intense peak at around 1640 cm^{-1} and a new band at 1430 cm^{-1}, both indicative of NH_2 group deformation vibrations and, thus, indicated that the coupling agent was effectively attached to WHFs. The vibrating bands at around 1160 cm^{-1} and 1110 cm^{-1} are characteristic of triethoxysilane's Si-O-Si bonds. The authors suggest that the triethoxysilane moieties reacts with hydroxyl moieties on the surface of WHFs, as indicated by the stretching of -Si-O-cellulose at 1200 cm^{-1}. The disappearance of the bands around 1739 cm^{-1}–1749 cm^{-1} assigned to the C=O moieties in silane-modified WHFs indicated that hemicellulose structures might have been removed. The peak at 1639 cm^{-1} had high intensity, indicating that more water was being absorbed by cellulose due to wax elimination following silane treatment.

4.4.1.3 Thermal and Mechanical Characterization

The chemical compositions of fibers significantly impact their crystallinity, water resistance, as well as their thermal, morphological, and mechanical qualities. Generally, alkaline treatment increases the cellulose index and dramatically impacts the fibers' mechanical properties. For example, Pokhriyal et al. (2023), Vinod, Sanjay, et al. (2021), and Umashankaran and Gopalakrishnan (2021) remarked that the cellulose percentage of alkali-conditioned fiber increased from 72.5% to 76.8%, 41.5% to 54.7%, and 62.3% to 71.3%, respectively. The observed results are ascribed to the delignification by alkylation conditioning, which allows the fiber bundle to disintegrate into separate fibers, and the propensity of alkali treatment to convert some portion of hemicellulose into native cellulose, increasing the cellulose content and perhaps improving the mechanical characteristics and crystalline index of the fiber. Pokhriyal et al. reported that tensile strength and Young's Modulus increased from 132 to 196.5 MPa and 10.26 to 19.61 GPa, respectively. While Venod et al. observed an enhanced tensile strength from 20.7 to 26.74 MPa. Umashankaran et al. also observed an increase in the tensile strength and Young's modulus from 322.0 to 343.6 MPa and 9.67-12.7 GPa, respectively. Table 4.3 summarizes the impact of alkali conditioning on the mechanical and thermal properties of the fiber.

TABLE 4.3
Comparability of Crystalline, Thermal, and Mechanical Properties of Untreated and Treated Plant Fibers

	Crystalline Properties		Thermal Properties		Mechanical Properties		
Fiber	Crystal Index (%)	Crystalline Size (mm)	Thermal Stability	Max. Degradation Temp (°C)	Tensile Strength (MPa)	Young's Modulus (GPa)	Refs
Himalayacalamus falconeri (HF)	58.92	3.39	250	356.25	132	10.26	Pokhriyal et al. (2023)
5% alkali treat HF	67.79	3.80	258	361.78	196.5	19.61	
Muntingia calabura fiber (MCF)	32.77	–	213	332	20.71	–	Vinod, Yashas Gowda, et al. (2021)
NaOH-treated MCF	37.67	–	254	379	26.74	–	
Pongamia pinnata	45.31	5.43	332	360	322	9.67	Umashankaran and Gopalakrishnan (2021)
NaOH-treated Pongamia pinnata	54.43	8.32	348	375	343.6	12.71	
Zamioculcas Zamiifolia (ZZ)	25.75	2.54	310	361	34.92	0.136	Tengsuthiwat et al. (2022)
5% alkali-treated ZZ	37.80	4.16	326	367	48.25	0.240	
Soy stem	46.33	1.31	330.91	347.3	–	–	Vinod, Sanjay, et al. (2021)
5% treated soy stem	50.43	2.02	337.40	351.0	–	–	
Catharanthus roseus (CRF)	25.09	–	203	296	27.02	1.23	Vinod et al. (2019)
NaOH-treated CRF	30.34	–	212	310	30.34	1.84	
Saharan aloe vera	52.6	5.6	225	350	621.8	40.03	Balaji and Nagarajan (2017)
NaOH-treated aloe vera	56.5	5.72	231	355	805.5	42.29	
Perotis indica	48.3	15.12	225	329.35	317–1608	8.41–69.61	Prithiviraj et al. (2016)
NaOH-treated Perotis indica	55.43	18.12	238	348.78	34.9	28.36–79.45	
Coccinia grandis	52.17	13.38	213.4	351.6	113	4.38	Senthamaraikannan and Kathiresan (2018)
Treated Coccinia grandis	57.6	8.15	220.6	360.1	273	10.17	
Tridax procumbens	34.46	25.04	270	340	25.75	0.94	Vijay et al. (2019)
Alkali Tridax procumbens	40.85	38.23	350	380	33.62	1.5	

4.4.1.4 X-Ray Photoelectron Spectroscopy

X-Ray photoelectron spectroscopy (XPS) is an excellent instrument for describing the interfacial interaction of a polymer matrix with a fiber, as it examines the chemical framework of the composite. It reveals the different binding energies for C1s peaks corresponding to the carbon of cellulose, hemicellulose, lignin, and other waxy impurities in the fibers. Moreover, XPS also shows that lignin has a low oxygen-to-carbon ratio (O/C ratio) relative to cell wall polysaccharides because lignin is a hydrocarbon polymer with aliphatic and aromatic functional moieties. Conversely, carbohydrates (cellulose, hemicelluloses, pectin) have larger O/C ratios because they have more OH, CHO, and COOH functional groups. This information allows XPS to monitor any changes to the chemical framework of the fiber attributed to chemical treatment and deduce the efficacy of various chemical treatments. Typically, XPS atomic spectra exhibit the major C, O, and N atomic binding energies at around 286.4 eV, 532.7 eV, and 399 eV for C1s, O1s, and N1s, respectively. Twebaze et al. (2022) employed XPS to compare the impact of alkaline versus ultrasonication conditioning on the roughness of banana fibers. Figure 4.15(b–d) displays associated Gaussian peaks for the untreated (UT), alkali-treated, and ultra-sonicated fibers that correspond to C1s (C1, C2, C3), O1s (O1, O2, and O3), and N1s, respectively. Interestingly, due to the difference in the binding energies of carbon, the authors could deduce that C1 peak illustrates C-C or/and C-H bonds in extractives and lignin, the C2 diagram displays C-O bonds in cellulose and hemicellulose, and the C3 diagram shows C-O bonds in hemicellulose and lignin. N1s at 399 eV are linked to the amide group ($-NH_2$), which is clearly visible in the untreated fiber but incredibly low in samples that have been treated and ultra-sonicated. The centered C1s of ultra-sonicated fibers in Figure 4.15(e) displayed comparable C1s at 284.8 eV, 286.4 eV, and 287.6 eV, in that order. While the XPS O1s-centered spectra displayed peaks at 531.2 eV in Figure 4.15(f), which were attributed to lignin and extracts. Furthermore, the authors demonstrated that longer ultrasonication times result in effective delignification, as evidenced by the C1 peaks, which diminish with increasing ultrasonication time.

Ashadujjaman et al. (2021) quantified the C/O ratio of untreated yarn fibers versus the alkaline-treated fibers, and they reported a reduction from ~3.93 to 3.44, which confirmed the removal of impurities on the surface of the fiber (Figure 4.16). The authors also assessed the use of glycine after the NaOH modification step and observed that the ratio further decreased to ~2.36. The observations were attributed to the increased oxygen and nitrogen moieties that cover up the C/O associated with the hydroxyl group.

4.4.2 Silane Treatment

4.4.2.1 SEM Characterization

Sudeshkumar et al. (2022) studied the effect of treating opuntia cladode fiber (OCF) and fumed silica (FS) with 3-aminopropyletriethoxysilane on the OCF-FS-Epoxy composite. The authors show that untreated fiber exhibited a smoother surface compared to the silane-treated fiber, as represented by Figure 4.17. While the enhanced mechanical properties were also attributed to the addition of FS on the composite, the main conclusion of the study was that silane treatment encouraged better adhesion between the OCF, FS, and the epoxy matrix and thus resulted in enhanced wear resistance, shown in Figure 4.17 (e-h). Asyraf et al. (2021) reported on a study that concluded that silane-treated fiber adhered better to the unsaturated polyester (UPE) matrix, as depicted by Figure 4.18, with small gaps observed on the images.

Wang et al. (2020) investigated the effect of three silane coupling agents, that is, amino (KH550), epoxy (KH560), and methyl (KH570) silane, on bamboo fiber surface modification and, consequently, the fiber-matrix adhesion properties. Spectra for all silane-treated bamboo fibers displayed the Si–O and Si–O–C stretching vibrations at 1040 cm^{-1} and 1113 cm^{-1}, illustrating that the coupling agents are grafted into the fibers (Figure 4.19). Interestingly, the FTIR spectra did not appear to be any different for the three agents but upon mechanical assessment. The effectiveness of the modifications was found to be in the following order: KH570 > KH550 > KH560. The fiber composite treated with KH570 showed the best mechanical performance.

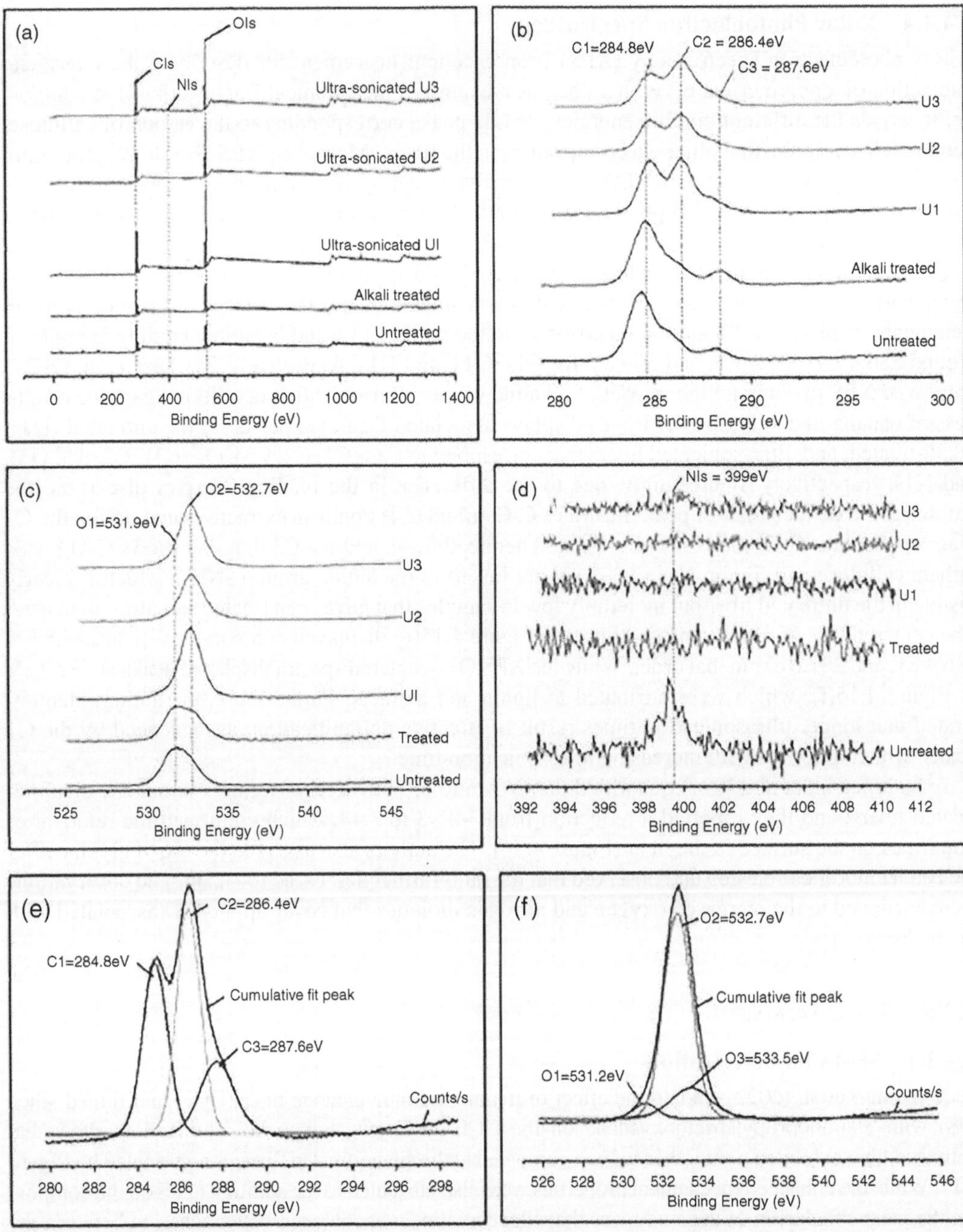

FIGURE 4.15 (a–d) X-ray photoelectron spectroscopy (XPS) spectra of untreated (UT), alkali-treated (T), and ultra-sonicated (U) fibers showing C, O, and N peaks. Centered peaks of (e) ultra-sonicated fiber C1s peaks and (f) ultra-sonicated fiber O1s (Twebaze et al., 2022).

4.4.2.2 Mechanical Characterization

Huda et al. (2008) investigated the effect of 5% silane-treated pineapple leaf fiber on the mechanical properties and showed that impact and flexural strength improved by 86% and 45%, respectively. Interestingly, Asim et al. (2016) compared the 2% silane and 6% NaOH treatment efficacies on pineapple and kenaf fiber. The authors revealed that silane-treated fibers were the most mechanically

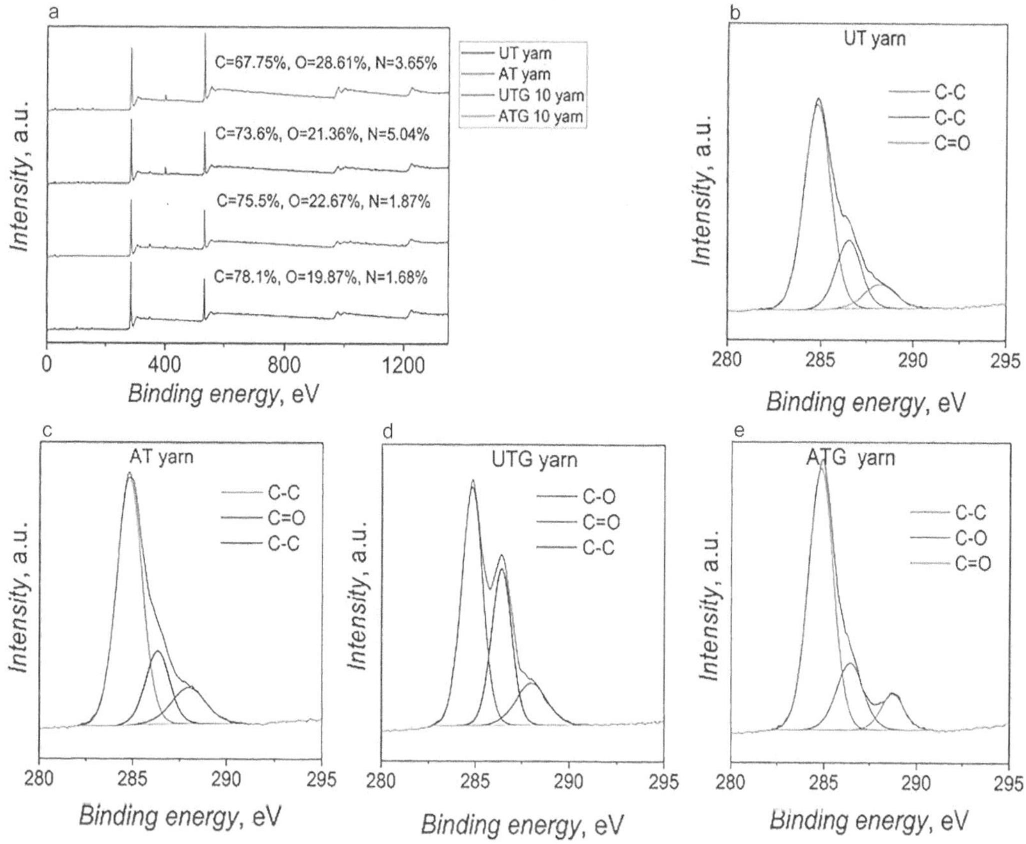

FIGURE 4.16 XPS spectra of (a) untreated (UT), alkali (AT), untreated-glycine, and alkaline-glycine treated yarn fibers and their respective C-centered peaks (Ashadujjaman et al., 2021).

FIGURE 4.17 SEM images of (a–d) untreated and silane-treated FS and OCF and (e–h) worn surface morphology treated OCF-FS epoxy composite (Sudeshkumar et al., 2022).

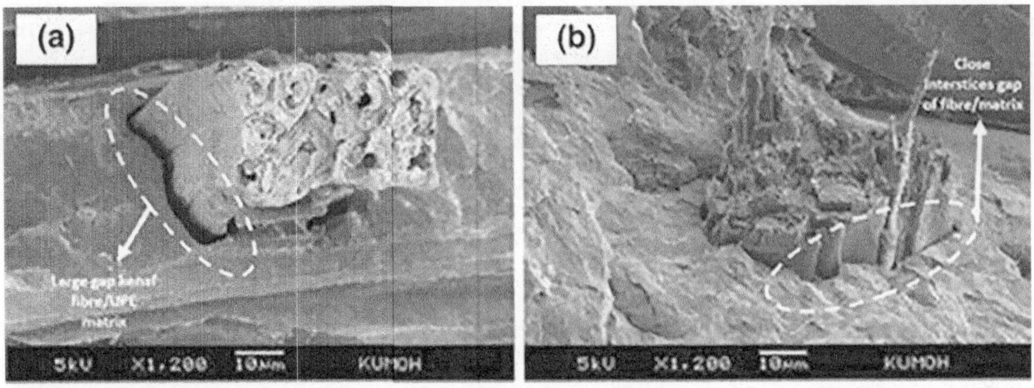

FIGURE 4.18 SEM images of kenaf/unsaturated polyester surface of (a) untreated and (b) 0.5 wt.% silane treatment (Asyraf et al., 2021).

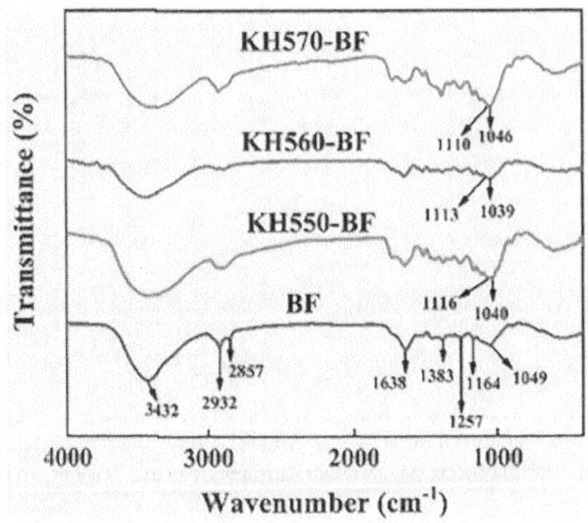

FIGURE 4.19 FTIR spectra of BF and 5 wt.% treated BF (Wang et al., 2020).

improved, with 105% and 175% improvement for pineapple leaf and kenaf fiber, respectively. Additionally, about 400% enhancement of the interfacial shear stress was obtained after 2% silane treatment, further indicating the efficiency of silane treatment on the mechanical properties of the fiber. Zahari et al. (2015) evaluated the effectiveness of the vinyl trimethoxy reactive group in the organofunctional silane treatment on the mechanical properties and water absorption behavior of the jute fiber-polypropylene composite. The authors illustrated that silane treatment improved the adhesion between the ijuk fiber and PP matrix, evidenced by the higher strength and Young's modulus while maintaining the elongation at break compared to untreated ones. Yet, the treatment did not affect the water absorption properties of the composite. Amino-propyl-triethoxysilane, a silane coupling agent, was used by Liu et al. (2019) to treat maize stalk fiber at 1%, 5%, 9%, and 13% concentrations. Silane treatment demonstrated higher interfacial adhesion by creating a slightly rougher, cleaner surface that encourages better mechanical interlocking. In comparison to untreated corn stalk fiber, it was shown that composites containing fiber with a 5 wt.% silane treatment had the highest maximum tensile strength and Young's modulus, measuring approximately 97.72% and 91.26% higher, respectively. Often, silane treatment is carried out as a part of pre-treatment methods. As such, researchers like Sepe et al. (2018) investigated hemp fiber treated with NaOH followed by (3-Glycidyloxypropyl) trimethoxy silane coupling agent versus separate silane and alkali

TABLE 4.4
Comparison Data on the Effect of Silane Treatment on Fiber Composite on Mechanical Properties

Natural Fiber/Matrix	Silane Agent	Concentration	Effect of Treatment	Refs
Pineapple/polylactic acid	NaOH+ Amino silane	5% + 5%	23%, 45%, and 86% increase in tensile strength, modulus, and impact strength	Huda et al. (2008)
Henequen/HDPE	NaOH+ xylene+ vinyl silane	2% + 1.5% + 0.015%	10% and 43% increase in longitudinal and transverse tensile strength	Herrera-Franco and Valadez-González (2004)
Jute/Epoxy	NaOH + Xiameter OFS 6040	2g + 2%	8% tensile strength	Cavalcanti et al. (2019)
Jute + Curaua/Epoxy hybrid			22% flexural strength	
Jute/polyester	NaOH + oligomer siloxane	1% + 96% alcohol solution	28.80%, 59.57%, and 35.41% increase in tensile strength, flexural strength, and flexural modulus	Seki (2009)
Jute/Epoxy			32.26%, 12%, and 10.67% increase in tensile strength, flexural strength, and flexural modulus	

treatments. The authors observed that silane treatment enhanced properties better than alkaline and silane-alkaline treatments. The 2,2,6,6-tetramethylpiperidine-1-oxy radical (TEMPO)-mediated oxidation pre-treatment step is frequently employed for enhanced silane effect. Fathi et al. (Hasan et al., 2022) modified flax fiber by TEMPO oxidation, followed by direct amino silane application without pre-hydrolysis. A comparison revealed that the composites containing TEMPO-oxidized fiber showed better properties than the alkali-treated fiber. Other studies are summarized in Table 4.4.

4.4.3 Acetylation

4.4.3.1 SEM

Chung et al. (2018) employed field emission scanning electron microscopy (FESEM) to assess the level of adhesion between kenaf and PLA after acetylation treatment (Figure 4.20). The authors depicted that the untreated kenaf/PLA composite exhibited a rough surface with several unfilled spaces, illustrating that during the flexural test, fibers were forced out of the PLA lattice. After treatment, the images show decreased voids or cavities, better fiber dispersion on the polymer matrix, and the composite surface is smoother with more prolonged exposure to acetyl treatment. The observations illustrate improved interfacial adhesion and compatibility manifested through the improved mechanical strength of the composite.

4.4.3.2 FTIR

Tserki et al. (2005) monitored FTIR spectra to examine the effect of acetylation and propionylation treatments on the fiber structure and surface properties of flax, hemp, and wood fibers. The authors showed that the emergence of new peaks appeared at 1740 cm^{-1}, 1366 cm^{-1}, and 1215–1230 cm^{-1}, arising from the C=O, O-C-H, and C-O acetyl stretch, respectively (Figure 4.21). The intense carbonyl band suggests that the hydroxyl groups of the fibers and the acetyl and propionyl groups are participating in an ester bond. The development of a new peak at 1200 cm^{-1}, which is ascribed to the ester

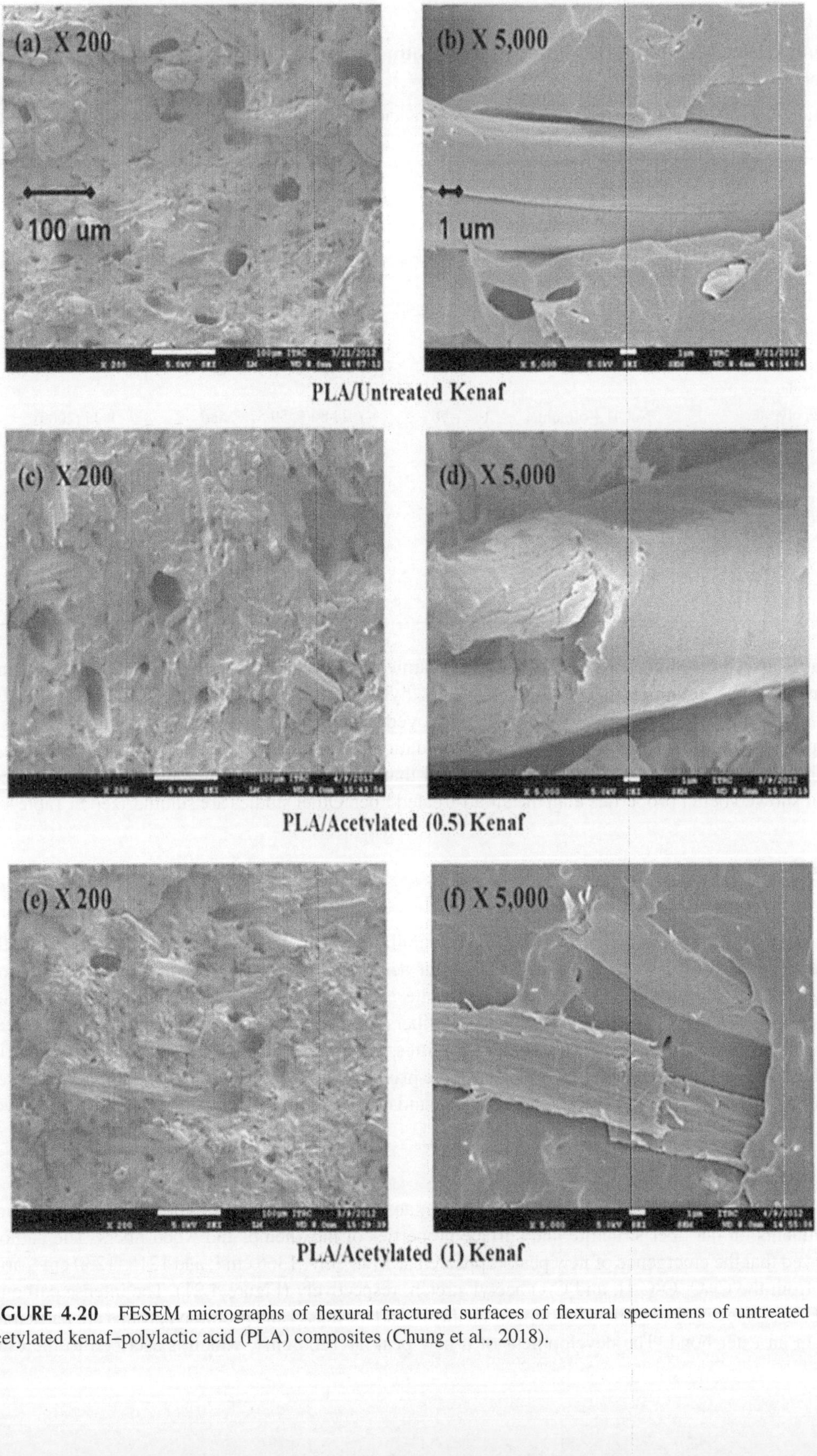

FIGURE 4.20 FESEM micrographs of flexural fractured surfaces of flexural specimens of untreated and acetylated kenaf–polylactic acid (PLA) composites (Chung et al., 2018).

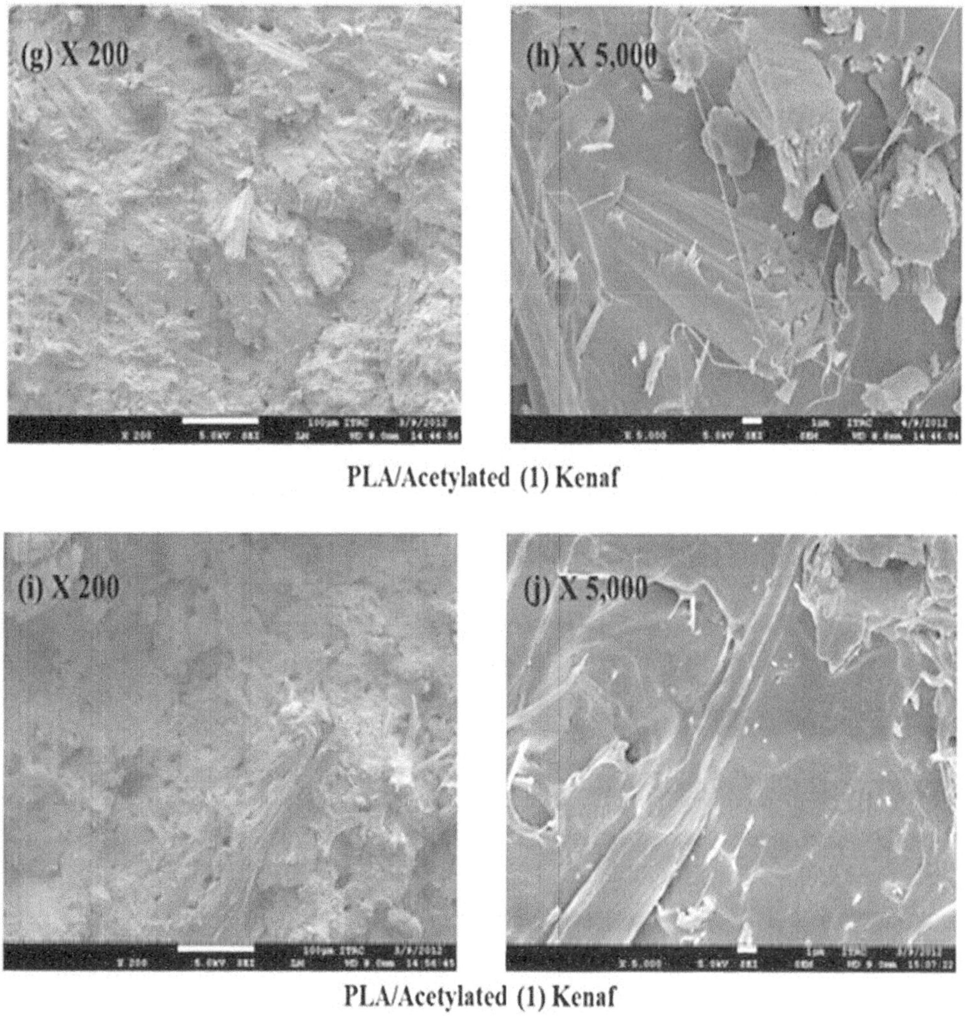

FIGURE 4.20 (Continued)

carboxyl group's C–O stretching, further supports the esterification reaction. The authors also observed an absorption band around 3500 cm^{-1} and ascribed it to the residual OH groups in the fiber components.

4.4.3.3 Mechanical Properties

Acetylation promotes the lowering of the plant fibers' hygroscopic nature, thereby enhancing the natural fibers' dimensional stability. Bledzki and Gassan et al. (Bledzki et al., 2008) reported that acetylation of jute and pine fibers reduced 50% and 65% of moisture uptake, respectively. Nair et al. discovered that after applying 18% NaOH for the sisal fiber pre-treatment and acetylation, the fiber's surface was extremely rough, which encouraged improved mechanical compatibility and increased thermal sturdiness. Bledzki et al. (2008) further observed that 18% degree of acetylation lowered the optimal tensile and flexural strengths. Kabir et al. (2012) investigated the mechanical properties of modified hemp fibers with glacial acetic acid and acetic anhydride and discovered improved mechanical strengths. According to Kabir et al. (2012), the fiber that had acetylation alteration performed better than the fiber that underwent silane modification. Furthermore, Oladele et al. (2020) showed that a 4% acetylation concentration is ideal for enhancing the tensile strength of practically all plant fibers. These results were consistent with those reported by Chung et al., Oladele et al., and Pickering et al. Table 4.5 shows a summary of the enhancement of mechanical properties of PFRC by acetylation chemical treatment.

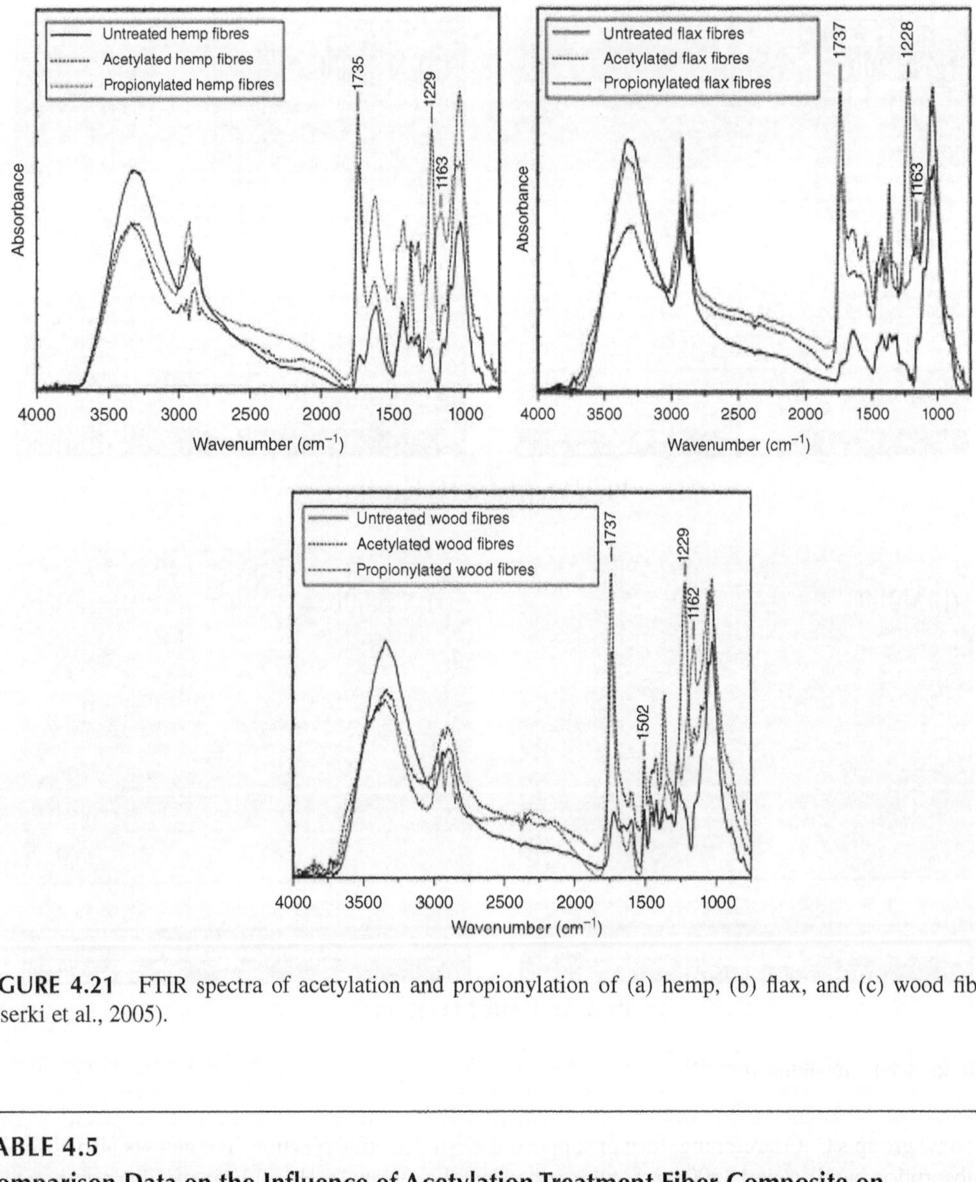

FIGURE 4.21 FTIR spectra of acetylation and propionylation of (a) hemp, (b) flax, and (c) wood fibers (Tserki et al., 2005).

TABLE 4.5
Comparison Data on the Influence of Acetylation Treatment Fiber Composite on Mechanical Properties

Natural Fiber/Matrix	Acetylating Agent	Concentration	Effect of Treatment	Refs
Banana/polypropylene	NaOH+ Toluene, Acetic acid, perchloric acid catalyst	5% + 250 ml toluene +125 acetic anhydride	17%, 16%, and 11.8% increase in tensile strength, modulus, and impact strength	Zaman and Khan (2021)
Flax/polypropylene	Acetic anhydride	18%	11.8%, 28.8%, and 72.78% increase in tensile strength, modulus, and impact strength	Bledzki et al. (2008)
Banana fiber	NaOH + acetic acid	4% + 50%	35% and 20% increase in tensile strength and modulus	Joseph et al. (2005)
Banana fiber/phenol formaldehyde	NaOH + acetic acid	4% + 50%	36% and 800% increase in tensile strength and modulus	Joseph et al. (2006)

4.5 CONCLUSION

This chapter outlined the types and characteristics of natural fibers, the common chemical modification techniques used for improving properties in natural fiber-reinforced composites, as well as the analytical techniques employed to characterize the modification changes. Alkaline chemical modification of the fibers is the most popular of the various chemical treatments, mainly because it results in shortened crystallite length, thereby increasing the amorphous regions of cellulose and consequently increasing reaction sites. Sodium hydroxide is the most preferred alkali solution as it favors complete lattice transformation from a monoclinic crystalline lattice to a polymorphous form of cellulose. The maleated coupling agents are also frequently utilized, particularly with polypropylene matrices, since the incompatibility issue between the fiber and matrix can be resolved by interactions between the hydroxyl moieties of natural fibers and the anhydride groups of maleated coupling agents. Silane treatment is another chemical treatment that stands out as it employs coupling agents that create bridges linking the fibers and matrix; in doing so, the interfacial adhesion is enhanced. The common analytical techniques utilized for characterizing natural fiber composites include FTIR, XPS, and SEM. FTIR provides critical data of moieties found on the surface of modified fibers and composites, while XPS is used to quantify C/O ratios and to distinguish between different carbon, oxygen, and nitrogen atoms according to the binding energies, and this key information is used to monitor the changes that occur because of chemical treatment. In addition, SEM provides an indirect measure of adhesion between fiber and matrix in the form of fiber pull-out and fiber breakage.

REFERENCES

Ajayi, S. M., Olusanya, S. O., Sodeinde, K. O., Didunyemi, A. E., Atunde, M. O., Fapojuwo, D. P., Olumayede, E. G., & Lawal, O. S. (2023). Hydrophobic modification of cellulose from oil palm empty fruit bunch: Characterization and application in Pickering emulsions stabilization. *Carbohydrate Polymer Technologies and Applications*, *5*. https://doi.org/10.1016/j.carpta.2023.100282

Al Abdallah, H., Abu-Jdayil, B., & Iqbal, M. Z. (2022). The effect of alkaline treatment on poly(lactic acid)/date palm wood green composites for thermal insulation. *Polymers*, *14*(6). https://doi.org/10.3390/polym14061143

Ashadujjaman, M., Saifullah, A., Shah, D. U., Zhang, M., Akonda, M., Karim, N., & Sarker, F. (2021). Enhancing the mechanical properties of natural jute yarn suitable for structural applications. *Materials Research Express*, *8*(5). https://doi.org/10.1088/2053-1591/abfd5e

Asim, M., Jawaid, M., Abdan, K., & Ishak, M. R. (2016). Effect of alkali and silane treatments on mechanical and fibre-matrix bond strength of kenaf and pineapple leaf fibres. *Journal of Bionic Engineering*, *13*(3), 426–435. https://doi.org/10.1016/S1672-6529(16)60315-3

Asyraf, M. R. M., Rafidah, M., Azrina, A., & Razman, M. R. (2021). Dynamic mechanical behaviour of kenaf cellulosic fibre biocomposites: A comprehensive review on chemical treatments. In *Cellulose*, *28*(5), 2675–2695. Springer Science and Business Media B.V. https://doi.org/10.1007/s10570-021-03710-3

Balaji, A. N., & Nagarajan, K. J. (2017). Characterization of alkali treated and untreated new cellulosic fiber from Saharan aloe vera cactus leaves. *Carbohydrate Polymers*, *174*, 200–208. https://doi.org/10.1016/j.carbpol.2017.06.065

Bekemeyer, J. (2021). *Ford Working on Natural Fiber Materials for Automotive Applications*. https://www.slashgear.com/ford-working-on-natural-fiber-materials-for-automotive-applications-28249593

Birniwa, A. H., Abdullahi, S. S., Ali, M., Mohammad, R. E. A., Jagaba, A. H., Amran, M., Avudaiappan, S., Maureira-Carsalade, N., & Flores, E. I. S. (2023). Recent trends in treatment and fabrication of plant-based fiber-reinforced epoxy composite: A review. *Journal of Composites Science*, *7*(3), 120. https://doi.org/10.3390/jcs7030120

Bledzki, A. K., Mamun, A. A., Lucka-Gabor, M., & Gutowski, V. S. (2008). The effects of acetylation on properties of flax fibre and its polypropylene composites. *Express Polymer Letters*, *2*(6), 413–422. https://doi.org/10.3144/expresspolymlett.2008.50

Bourmaud, A., Beaugrand, J., Shah, D. U., Placet, V., & Baley, C. (2018). Towards the design of high-performance plant fibre composites. In *Progress in Materials Science* (Vol. 97, pp. 347–408). Elsevier Ltd. https://doi.org/10.1016/j.pmatsci.2018.05.005

Bourmaud, A., Morvan, C., Bouali, A., Placet, V., Perré, P., & Baley, C. (2013). Relationships between microfibrillar angle, mechanical properties and biochemical composition of flax fibers. *Industrial Crops and Products*, *44*, 343–351. https://doi.org/10.1016/j.indcrop.2012.11.031

Cavalcanti, D. K. K., Banea, M. D., Neto, J. S. S., Lima, R. A. A., da Silva, L. F. M., & Carbas, R. J. C. (2019). Mechanical characterization of intralaminar natural fibre-reinforced hybrid composites. *Composites Part B: Engineering*, *175*. https://doi.org/10.1016/j.compositesb.2019.107149

Chichane, A., Boujmal, R., & El Barkany, A. (2023). Bio-composites and bio-hybrid composites reinforced with natural fibers: Review. *Materials Today: Proceedings*, *72*, 3471–3479. https://doi.org/10.1016/j.matpr.2022.08.132

Chung, T. J., Park, J. W., Lee, H. J., Kwon, H. J., Kim, H. J., Lee, Y. K., & Yin Tze, W. T. (2018). The improvement of mechanical properties, thermal stability, and water absorption resistance of an eco-friendly PLA/kenaf biocomposite using acetylation. *Applied Sciences (Switzerland)*, *8*(3). https://doi.org/10.3390/app8030376

Gadhave, R. V., Dhawale, P. V., & Sorate, C. S. (2021). Surface Modification of Cellulose with Silanes for Adhesive Application: Review. *Open Journal of Polymer Chemistry*, *11*(02), 11–30. https://doi.org/10.4236/ojpchem.2021.112002

Hasan, A., Rabbi, M. S., & Maruf Billah, M. (2022). Making the lignocellulosic fibers chemically compatible for composite: A comprehensive review. In *Cleaner Materials* (Vol. 4). Elsevier Ltd. https://doi.org/10.1016/j.clema.2022.100078

Heines, S. V. (1944). John Mercer and Mercerization, 1844. *Journal of Chemical Education*, 430–433. https://doi.org/10.1021/ed021p430

Herrera-Franco, P. J., & Valadez-González, A. (2004). Mechanical properties of continuous natural fibre-reinforced polymer composites. *Composites Part A: Applied Science and Manufacturing*, *35*(3), 339–345. https://doi.org/10.1016/j.compositesa.2003.09.012

Huda, M. S., Drzal, L. T., Mohanty, A. K., & Misra, M. (2008). Effect of chemical modifications of the pineapple leaf fiber surfaces on the interfacial and mechanical properties of laminated biocomposites. *Composite Interfaces*, *15*(2–3), 169–191. https://doi.org/10.1163/156855408783810920

Imraan, M., Ilyas, R. A., Norfarhana, A. S., Bangar, S. P., Knight, V. F., & Norrrahim, M. N. F. (2023). Sugar palm (Arenga pinnata) fibers: New emerging natural fibre and its relevant properties, treatments and potential applications. *Journal of Materials Research and Technology* 24, 4551–4572). Elsevier Editora Ltda. https://doi.org/10.1016/j.jmrt.2023.04.056

John, M. J. (2022). Biobased alginate treatments on flax fibre reinforced PLA and PHBV composites. *Current Research in Green and Sustainable Chemistry*, *5*. https://doi.org/10.1016/j.crgsc.2022.100319

John, M. J., & Anandjiwala, R. D. (2008). Recent developments in chemical modification and characterization of natural fiber-reinforced composites. *Polymer Composites* 29, 2, 187–207. https://doi.org/10.1002/pc.20461

John, M. J., & Anandjiwala, R. D. (2009). Chemical modification of flax reinforced polypropylene composites. *Composites Part A: Applied Science and Manufacturing*, *40*(4), 442–448. https://doi.org/10.1016/j.compositesa.2009.01.007

Joseph, S., Koshy, P., & Thomas, S. (2005). The role of interfacial interactions on the mechanical properties of banana fibre reinforced phenol formaldehyde composites. *Composite Interfaces*, *12*(6), 581–600. https://doi.org/10.1163/1568554054915183

Joseph, S., Oommen, Z., & Thomas, S. (2006). Environmental durability of banana-fiber-reinforced phenol formaldehyde composites. *Journal of Applied Polymer Science*, *100*(3), 2521–2531. https://doi.org/10.1002/app.23680

Kabir, M. M., Wang, H., Lau, K. T., & Cardona, F. (2012). Chemical treatments on plant-based natural fibre reinforced polymer composites: An overview. *Composites Part B: Engineering*, *43*(7), 2883–2892. https://doi.org/10.1016/j.compositesb.2012.04.053

Kaushik, V. K., Kumar, A., & Kalia, S. (2013). Effect of mercerization and benzoyl peroxide treatment on morphology, thermal stability and crystallinity of sisal fibers. *International Journal of Textile Science*, *1*(6), 101–105. https://doi.org/10.5923/j.textile.20120106.07

Liu, Y., Lv, X., Bao, J., Xie, J., Tang, X., Che, J., Ma, Y., & Tong, J. (2019). Characterization of silane treated and untreated natural cellulosic fibre from corn stalk waste as potential reinforcement in polymer composites. *Carbohydrate Polymers*, *218*, 179–187. https://doi.org/10.1016/j.carbpol.2019.04.088

Mokhothu, T. H., & John, M. J. (2015). Review on hygroscopic aging of cellulose fibres and their biocomposites. In *Carbohydrate Polymers* (Vol. 131, pp. 337–354). Elsevier Ltd. https://doi.org/10.1016/j.carbpol.2015.06.027

Mwaikambo, L. Y., & Ansell, M. P. (2002). Chemical modification of hemp, sisal, jute, and kapok fibers by alkalization. *Journal of Applied Polymer Science*, *84*(12), 2222–2234. https://doi.org/10.1002/app.10460

Oladele, I. O., Michael, O. S., Adediran, A. A., Balogun, O. P., & Ajagbe, F. O. (2020). Acetylation treatment for the batch processing of natural fibers: Effects on constituents, tensile properties and surface morphology of selected plant stem fibers. *Fibers*, *8*(12), 1–19. https://doi.org/10.3390/fib8120073

Pokhriyal, M., Rakesh, P. K., Rangappa, S. M., & Siengchin, S. (2023). Effect of alkali treatment on novel natural fiber extracted from Himalayacalamus falconeri culms for polymer composite applications. *Biomass Conversion and Biorefinery*. https://doi.org/10.1007/s13399-023-03843-4

Prithiviraj, M., Muralikannan, R., Senthamaraikannan, P., & Saravanakumar, S. S. (2016). Characterization of new natural cellulosic fiber from the Perotis indica plant. *International Journal of Polymer Analysis and Characterization*, *21*(8), 669–674. https://doi.org/10.1080/1023666X.2016.1202466

Seki, Y. (2009). Innovative multifunctional siloxane treatment of jute fiber surface and its effect on the mechanical properties of jute/thermoset composites. *Materials Science and Engineering A*, *508*(1–2), 247–252. https://doi.org/10.1016/j.msea.2009.01.043

Senthamaraikannan, P., & Kathiresan, M. (2018). Characterization of raw and alkali treated new natural cellulosic fiber from Coccinia grandis. L. *Carbohydrate Polymers*, *186*, 332–343. https://doi.org/10.1016/j.carbpol.2018.01.072

Sepe, R., Bollino, F., Boccarusso, L., & Caputo, F. (2018). Influence of chemical treatments on mechanical properties of hemp fiber reinforced composites. *Composites Part B: Engineering*, *133*, 210–217. https://doi.org/10.1016/j.compositesb.2017.09.030

Shah, D. U. (2013). Developing plant fibre composites for structural applications by optimising composite parameters: A critical review. *Journal of Materials Science*, 48, 18, 6083–6107. https://doi.org/10.1007/s10853-013-7458-7

Soutis, C., Yi, X. S., & Bachmann, J. (2019). How green composite materials could benefit aircraft construction. In *Science China Technological Sciences* (Vol. 62, Issue 8, pp. 1478–1480). Springer Verlag. https://doi.org/10.1007/s11431-018-9489-1

Sudeshkumar, M. P., Ramadoss, R., Jayaseelan, V., & Vijayananth, S. (2022). Preparation and characterization of amino-silanized opuntia cladode fibre and fumed silica toughened epoxy composite. *Silicon*. https://doi.org/10.1007/s12633-022-02228-y

Sumrith, N., Techawinyutham, L., Sanjay, M. R., Dangtungee, R., & Siengchin, S. (2020). Characterization of alkaline and silane treated fibers of 'water hyacinth plants' and reinforcement of 'water hyacinth fibers' with bioepoxy to develop fully biobased sustainable ecofriendly composites. *Journal of Polymers and the Environment*, *28*(10), 2749–2760. https://doi.org/10.1007/s10924-020-01810-y

Tengsuthiwat, J., Vinod, A., Srisuk, R., Techawinyutham, L., Rangappa, S. M., & Siengchin, S. (2022). Thermo-mechanical characterization of new natural cellulose fiber from zmioculus zamiifolia. *Journal of Polymers and the Environment*, *30*(4), 1391–1406. https://doi.org/10.1007/s10924-021-02284-2

Tserki, V., Zafeiropoulos, N. E., Simon, F., & Panayiotou, C. (2005). A study of the effect of acetylation and propionylation surface treatments on natural fibres. *Composites Part A: Applied Science and Manufacturing*, *36*(8), 1110–1118. https://doi.org/10.1016/j.compositesa.2005.01.004

Tuğluca, M. S., Özdoğru, E., İlcan, H., Özçelikci, E., Ulugöl, H., & Şahmaran, M. (2023). Characterization of chemically treated waste wood fiber and its potential application in cementitious composites. *Cement and Concrete Composites*, *137*. https://doi.org/10.1016/j.cemconcomp.2023.104938

Twebaze, C., Zhang, M., Zhuang, X., Kimani, M., Zheng, G., & Wang, Z. (2022). Banana fiber degumming by alkali treatment and ultrasonic methods. *Journal of Natural Fibers*, *19*(16), 12911–12923. https://doi.org/10.1080/15440478.2022.2079581

Umashankaran, M., & Gopalakrishnan, S. (2021). Effect of sodium hydroxide treatment on physico-chemical, thermal, tensile and surface morphological properties of Pongamia Pinnata L. bark fiber. *Journal of Natural Fibers*, *18*(12), 2063–2076. https://doi.org/10.1080/15440478.2019.1711287

Uusi-Tarkka, E. K., Skrifvars, M., & Haapala, A. (2021). Fabricating sustainable all-cellulose composites. In *Applied Sciences (Switzerland)* (Vol. 11, Issue 21). MDPI. https://doi.org/10.3390/app112110069

Vijay, R., Lenin Singaravelu, D., Vinod, A., Sanjay, M. R., Siengchin, S., Jawaid, M., Khan, A., & Parameswaranpillai, J. (2019). Characterization of raw and alkali treated new natural cellulosic fibers from Tridax procumbens. *International Journal of Biological Macromolecules*, *125*, 99–108. https://doi.org/10.1016/j.ijbiomac.2018.12.056

Vinod, A., Sanjay, M. R., Siengchin, S., & Fischer, S. (2021). Fully bio-based agro-waste soy stem fiber reinforced bio-epoxy composites for lightweight structural applications: Influence of surface modification techniques. *Construction and Building Materials*, *303*. https://doi.org/10.1016/j.conbuildmat.2021.124509

Vinod, A., Vijay, R., Singaravelu, D. L., Sanjay, M. R., Siengchin, S., & Moure, M. M. (2019). Characterization of untreated and alkali treated natural fibers extracted from the stem of Catharanthus roseus. *Materials Research Express*, *6*(8). https://doi.org/10.1088/2053-1591/ab22d9

Vinod, A., Yashas Gowda, T. G., Vijay, R., Sanjay, M. R., Gupta, M. K., Jamil, M., Kushvaha, V., & Siengchin, S. (2021). Novel Muntingia Calabura bark fiber reinforced green-epoxy composite: A sustainable and green material for cleaner production. *Journal of Cleaner Production*, *294*. https://doi.org/10.1016/j.jclepro.2021.126337

Wang, Q., Zhang, Y., Liang, W., Wang, J., & Chen, Y. (2020). Effect of silane treatment on mechanical properties and thermal behavior of bamboo fibers reinforced polypropylene composites. *Journal of Engineered Fibers and Fabrics*, *15*. https://doi.org/10.1177/1558925020958195

Zahari, W. Z. W., Badri, R. N. R. L., Ardyananta, H., Kurniawan, D., & Nor, F. M. (2015). Mechanical properties and water absorption behavior of polypropylene/ijuk fiber composite by using silane treatment. *Procedia Manufacturing*, *2*, 573–578. https://doi.org/10.1016/j.promfg.2015.07.099

Zaman, H. U., & Khan, R. A. (2021). Acetylation used for natural fiber/polymer composites. *Journal of Thermoplastic Composite Materials*, *34*(1), 3–23. https://doi.org/10.1177/0892705719838000

Zhang, K., Wang, F., Liang, W., Wang, Z., Duan, Z., & Yang, B. (2018). Thermal and mechanical properties of bamboo fiber reinforced epoxy composites. *Polymers*, *8*(6). https://doi.org/10.3390/polym10060608

Zhao, Y., Man, Y., Wen, J., Guo, Y., & Lin, J. (2019). Advances in imaging plant cell walls. In *Trends in Plant Science* (Vol. 24, Issue 9, pp. 867–878). Elsevier Ltd. https://doi.org/10.1016/j.tplants.2019.05.009

5 Mechanical and Metallurgical Testing of Natural Fibre Composites

K. Georgy

5.1 INTRODUCTION

Natural fibre composites are biodegradable, cheaper, renewable, eco-friendly, non-toxic, low energy-consuming, flexible, lighter in weight, and have superior sound and thermal insulation. Hence, research on natural fibre composites is essential in the present scenario of climate change arising from greenhouse gas emissions (Gholampour and Ozbakkaloglu, 2020). In a composite material, stiff, as well as strong, fibre is used as the major load carrier, and it distributes load to the ductile matrix/resin, which has good elongation (Rahman and Putra, 2019). Conventional fibres (synthetic fibres, such as carbon, aramid, glass) could be sustainably replaced by natural fibres in fibre-reinforced polymer composites. Natural fibres could be derived from plants (banana, kapok, henequen, coir, sisal, jute, hemp, flax, bamboo, kenaf, ramie, bagasse, abaca, pina, rice husk, rice straw, wheat straw, curaua, oil palm) or animals (wool, feather, camel hair, silk). Polymer matrix could be of two classes: thermoplastics or thermosetting. Polyvinyl chloride, polyethylene, polypropylene, and polystyrene are the thermoplastics, and epoxy, polyester, vinyl ester, and phenolic resins are the thermosetting polymers (Malkapuram et al., 2009).

Natural fibre composites have been used in structural applications (automotive industry, construction sector, storage devices, sports equipment, aerospace) because of their attractive mechanical properties. Hence, mechanical testing of natural fibre composites is of great significance. Also, metallurgical testing of natural fibre composites is required for understanding the material extensively (Faruk et al., 2014). The mechanical properties of natural fibre composites have been improved considerably using physical modification or chemical modification of fibres. Physical modification could be performed by techniques such as fibre plasma treatment, fibre corona treatment, fibre alkaline treatment, and fibre-beating treatment. Chemical modification could be performed by techniques such as fibre acetylation treatment, fibre silane treatment, fibre peroxide treatment, fibre benzoylation treatment, fibre acrylation and acrylonitrile grafting, fibre treated by maleated coupling agent, fibre bleaching, fibre isocyanate treatment, fibre permanganate treatment, fibre stearic acid treatment, and fibre fungal treatment. Also, resin treatment by glutaraldehyde and vinyl alcohol aids in lowering moisture, and hence it could also be implemented to improve fibre/matrix bonding (Gholampour and Ozbakkaloglu, 2020).

A vast amount of literature is available on the different tests performed by researchers according to their specific requirements or applications. When natural fibre composites are used in real-life applications, different types of loads might be acting on the composite. This book chapter focuses on mentioning all the possible mechanical and metallurgical tests that could be useful in estimating the performance of natural fibre composites.

DOI: 10.1201/9781032665337-5

5.2 MECHANICAL TESTING

Various destructive tests performed for understanding the mechanical aspects of natural fibre composites are discussed in the following sections. Mechanical properties of the composites depend on a lot of factors, such as crystallinity, which indeed is governed by cooling rate. Durability of natural fibre composites depends on properties such as water absorption, wear resistance, and weathering, and lower durability will result in lowering mechanical properties. In addition to the testing, modelling/simulation could also be used for predicting the properties. For example, the Halpin-Tsai condition could be used to predict the Young's modulus of composites containing various sorts of natural fibres (Ku et al., 2011). Other analytical models available in literature are the shear lag model, Hirsch model, Bowyer-Bader model, rule of mixture, series and parallel, etc. Finite element methods (such as representative volume element) could be used for predicting the type of fibre reinforcement, fibre orientation, etc. (Gholampour and Ozbakkaloglu, 2020).

5.2.1 Tensile Test

During tensile testing, the sample clamped between grips (operated mechanically or pneumatically) of the universal testing machine is subjected to a tensile force, and a desired cross-head speed should be selected. Test specimens should be prepared according to the standards for tensile test (ASTM D638-15 or ASTM D3039 M-95 or DIN EN 61) (Herrera-Franco and Valadez-Gonzalez, 2008). Figure 5.1 depicts the specimen used for tensile testing. The specimens should be subjected to heating (~80°C) to remove the moisture, and end tabs could be attached to the sample for safeguarding the sample (Huang and Young, 2019). While performing the test, a load versus elongation curve (further transformed to stress-strain curve) is generated. From the stress-strain curve, tensile strength, modulus, and elongation at break (%) could be determined. Elastic modulus (stiffness) is obtained from the initial straight-line region of the stress-strain curve. The slope measured in this initial region

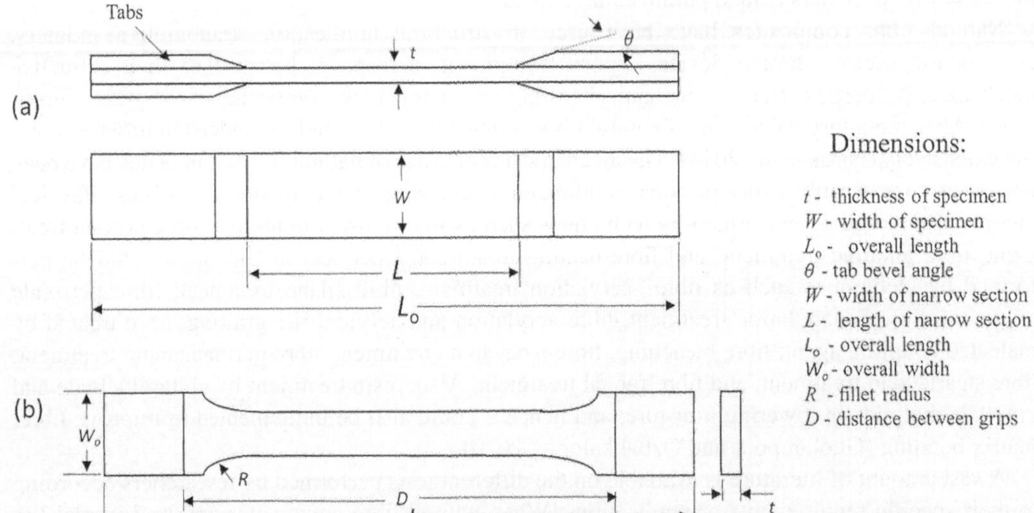

FIGURE 5.1 Sample geometry used for tensile test: (a) rectangular specimen and (b) dumbbell-shaped specimen

Reprinted from Properties and Performance of Natural Fibre Composites, P J Herrera-Franco and A Valadez-Gonzâlez, Mechanical testing of natural-fibre composites, pp. 375– 401, Copyright (2008) with permission from Elsevier.

TABLE 5.1
Tensile Properties of Different Natural Fibre Composites (Oksman et al., 2003; Bax and Mussig 2008; Bledzki and Jaszkiewicz 2010; Sawpan et al., 2011)

Fibre	Tensile Strength (MPa)	Young's Modulus (GPa)	Elongation to Break (%)
Flax	53	8.3	1
Cordenka	58	4.85	–
Jute	81.9	9.6	1.8
Hemp	75.5	8.18	–

gives a quantification of the modulus value in the load application direction (Herrera-Franco and Valadez-Gonzalez, 2008). Table 5.1 shows the tensile properties of natural fibre composites using polylactic acid as the resin, with 30 wt.% of fibre. It could be observed from Table 5.1 that depending on the fibre material, the tensile properties vary.

If the material is orthotropic, tensile tests have to be conducted once again using samples prepared with an orthogonal orientation as compared to the initial testing (Herrera-Franco and Valadez-Gonzalez, 2008). The fibre orientation dictates the tensile strength of the composite. Higher strength is achieved while applying the tensile load in the direction parallel to the fibre orientation (Ku et al., 2011). Depending on the processing route of the composite, there will be variations in fibre orientation. Various fibre surface treatments could be used for improving tensile strength of natural fibre composites. The surface-treated fibre has better bonding with the matrix and hence provides better tensile properties (Herrera-Franco and Valadez-Gonzalez, 2008). Physical modifications (by subjecting to heat treatment), altering the crystallinity of the polymer, could also lead to improvement in modulus and tensile strength (Barone, 2005). The chemical as well as physical modification of the fibre or resin will improve the tensile property of the composite.

The fibre volume fraction present in the natural fibre composites also significantly contributes to the tensile properties. An optimal amount of volume fraction of fibre can result in high tensile strength, and excess fibre content might cause brittle failure of fibres, eventually leading to poor tensile strength (Gholampour and Ozbakkaloglu, 2020). A higher amount of fibre content can increase the elastic modulus of the composite and decrease the elongation of the natural fibre composites. The fibre length also plays a major role in the tensile properties of the composites.

With respect to the loading direction, the tensile samples will fail in a transverse plane. They could also fracture because of the matrix yielding, fibre pull out, shear failure, matrix tearing, etc., depending on the adhesion of fibre with the matrix. Therefore, it could be understood that the higher the interface exists between fibre and matrix, the better the tensile properties. Also, the stiffness of the matrix material might also dictate the failure mode. For example, a low stiffness material could fail because of shear failure resulting from matrix yielding and fibre splitting (Herrera-Franco and Valadez-Gonzalez, 2008).

5.2.2 Compressive Test

Compression test could be performed in a universal testing machine. A load cell is connected with the machine, and the cross-head speed (measured in mm/min) is maintained constant during the test. Polytetrafluoroethylene (PTFE) paste is applied to reduce the friction between the sample and compression platens (Fan and Hui, 2014). The sample details and the equipment for testing as per the ASTM D3410 standard are provided in Figure 5.2 (Herrera-Franco and Valadez-Gonzalez, 2008).

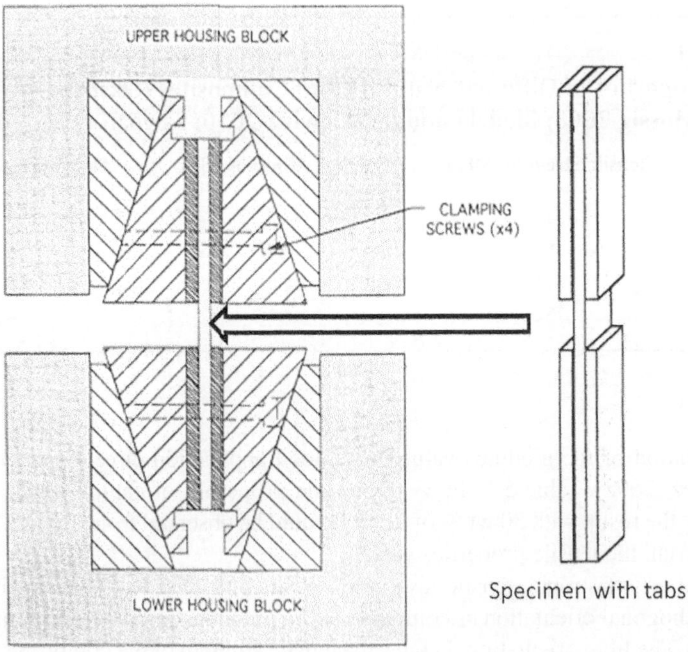

FIGURE 5.2 Compression test sample along with its fixture

Adapted from Standard Test Method for Compressive Properties of Polymer Matrix Composite Materials with Unsupported Gage Section by Shear, 2003.

From the stress-strain curve generated, elastic properties (such as modulus) could be calculated. The test has to be performed in two orthogonal directions for anisotropic material. Depending on the stiffness (modulus) of the matrix material, the samples under compression could exhibit different failure modes:

(i) If the matrix material has low modulus values, the composite would undergo buckling during loading.
(ii) For matrix material having medium stiffness, the composite would fail by transverse rupture.
(iii) If the matrix material has high stiffness, the composite would fail because of a shear failure of fibres.

The different possible failure modes are provided in Figure 5.3. Compressive studies on natural fibre composite tubes have reported other failure modes such as micro-buckling, progressive crushing, concertina-shaped buckling, and diamond-shaped buckling. Micro-buckling (initiating at an individual fibre) occurs when a huge amount of stress acts on the sample, and it results in sudden failure. In the case of diamond shape buckling and concertina shape buckling, the load is sustained until the entire system fails (Fan and Hui, 2014). Compression strength depends on the fibre volume fraction, density, bonding between the fibre and matrix, and fibre orientation (Sarikaya et al., 2019; Gholampour and Ozbakkaloglu, 2020).

5.2.3 Flexural Test

The structural capability of a composite is found by carrying out a flexural test. A universal testing machine is utilized for performing a three-point bending test as per ASTM D790 (Rajesh et al., 2018). Bending tests performed as per ISO 178 are also reported (Sarikaya et al., 2019). While performing the three-point bend test, a specific testing speed has to be selected (Rajesh et al., 2018).

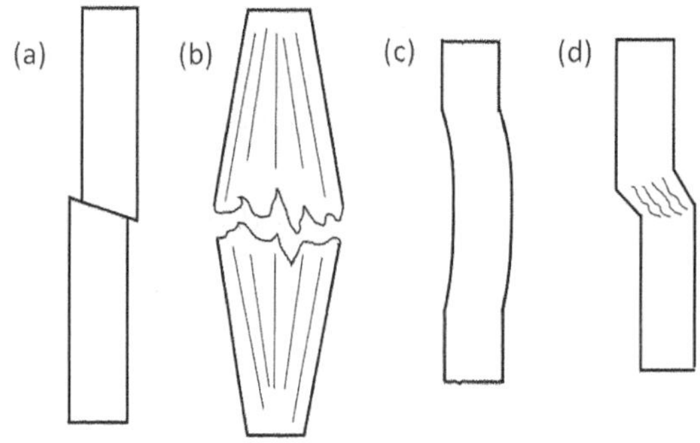

FIGURE 5.3 Possible modes of failure during compression testing: (a) shear failure, (b) axial splitting, (c) buckling, and (d) kink zone

Reprinted from Properties and Performance of Natural Fibre Composites, P J Herrera-Franco and A Valadez-Gonzâlez, Mechanical testing of natural-fibre composites, pp. 375- 401, Copyright (2008) with permission from Elsevier.

Even though three-point bending test is referred to as the three-point loading test, a single load is only applied on the sample supported at either end. Alternatively, four-point loading also could be used for evaluating flexural properties. In four-point loading, two loads will be applied on the sample in a symmetrical manner. Four-point loading could be applied either as a third-point loading or quarter-point loading. The different possible loading patterns are provided in Figure 5.4 (Herrera-Franco and Valadez-Gonzalez, 2008). Third-point loading refers to the case in which two-point loads are applied at a distance of (1/3)*beam span length away from each other. During quarter-point loading, two-point loads are provided at (1/4)*beam span length away from the supports. The specimen will be subjected to maximum tensile or compressive force at the outer surfaces, and the bending force will be maximum at the middle plane (Adams and Whitney, 2017). The compressive and tensile stresses within the beam at any point could be calculated using simple beam theory by using the following equation (Herrera-Franco and Valadez-Gonzalez, 2008):

$$\sigma_{\max} = \frac{M_{\max} \cdot \left(t/2 \right)}{I}$$

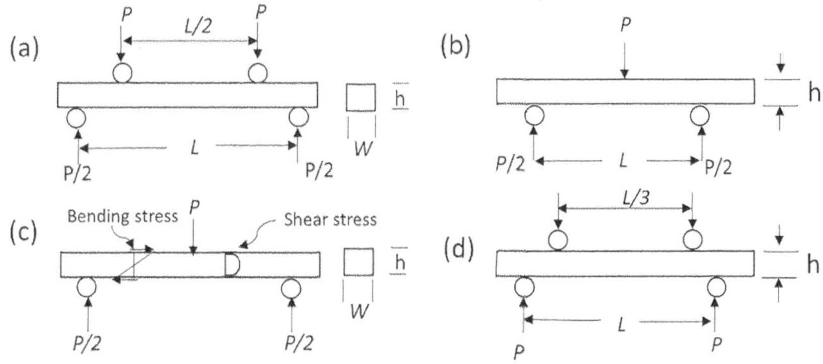

FIGURE 5.4 Loading configurations for bending test: (a) four-point bending, (b) three-point bending, (c) stresses in beam subjected to three-point bending, and (d) third-point loading

Reprinted from Properties and Performance of Natural Fibre Composites, P J Herrera-Franco and A Valadez-Gonzâlez, Mechanical testing of natural-fibre composites, pp. 375- 401, Copyright (2008) with permission from Elsevier.

TABLE 5.2
Flexural Properties of Natural Fibre Composites (Rong et al. 2001; Islam et al., 2010)

Matrix	Fibre	Fibre Content	Flexural Strength (MPa)	Flexural Modulus (GPa)	Treatment
Epoxy	Sisal (aligned)	73 vol %	320	27	Alkali treated
Hemp (carded)	Polylactic acid	30 wt.%	143	7	Alkali treated
Jute (woven)	Unsaturated polyester	30 vol %	103	7	–
Flax (random)	Unsaturated Polyester	30 vol %	91	5	–

where t is the beam thickness, I is the moment of inertia of the beam having width 'w' given by $I = wt^3/12$, M_{max} is the maximum bending moment. For a three-point bending test having load P and a span length of L, $M_{max} = PL/4$. The flexural strength of composite increases up to an optimum amount of fibre addition.

Flexural modulus of beam with deflection of δ could be calculated using the equation:

$$E_f = \frac{L^3}{4wt^3}\frac{\Delta P}{\Delta \delta}$$

Strain gauges bonded to the specimen could be used for measuring the deflection. Alternatively, linear variable differential transducers (LVDTs) could be used for automated measurements in deflection (Adams and Whitney, 2017). The flexural modulus increases with an increase in fibre content. The fibre orientation as well as fibre surface treatment could also affect the flexural properties. Joseph et al. have reported that for a composite prepared from short fibre, the flexural properties showed improvement with an increase in fibre length (Joseph et al., 2002).

Table 5.2 shows the flexural properties of different matrix-fibre combinations.

5.2.4 Impact Test

Impact tests of natural fibre composites could be carried out by performing Izod or Charpy impact tests. ASTM D256-10 could be used as the guidelines for the Izod impact testing. The samples should be given a V notch as per the DIN 847 standard, and Figure 5.5 shows an impact test sample specification (Nascimento et al., 2012). Also, there are reports on impact tests performed without using a V notch (Rajesh et al., 2018). The sample is placed on the machine, and a pivoted arm is released from a height. The energy absorbed by sample (W) is obtained after the testing, and it is used for the calculation of impact strength using the equation:

$$\text{Impact strength} = W/(\text{sample thickness} * \text{base width at the notch})$$

Charpy impact test could be conducted as per DIN EN ISO 179 standard. In Charpy tests, rectangular samples without notches have to be used. The impact strength could be represented in kJ/m^2 (Bax and Mussig 2008).

The impact strength of natural fibre composites depends on the fibre-matrix bonding. The better the bonding, higher the impact strength and hence higher the energy absorption of material. Physical and chemical modifications of fibres used for increasing the fibre-matrix interactions result in higher

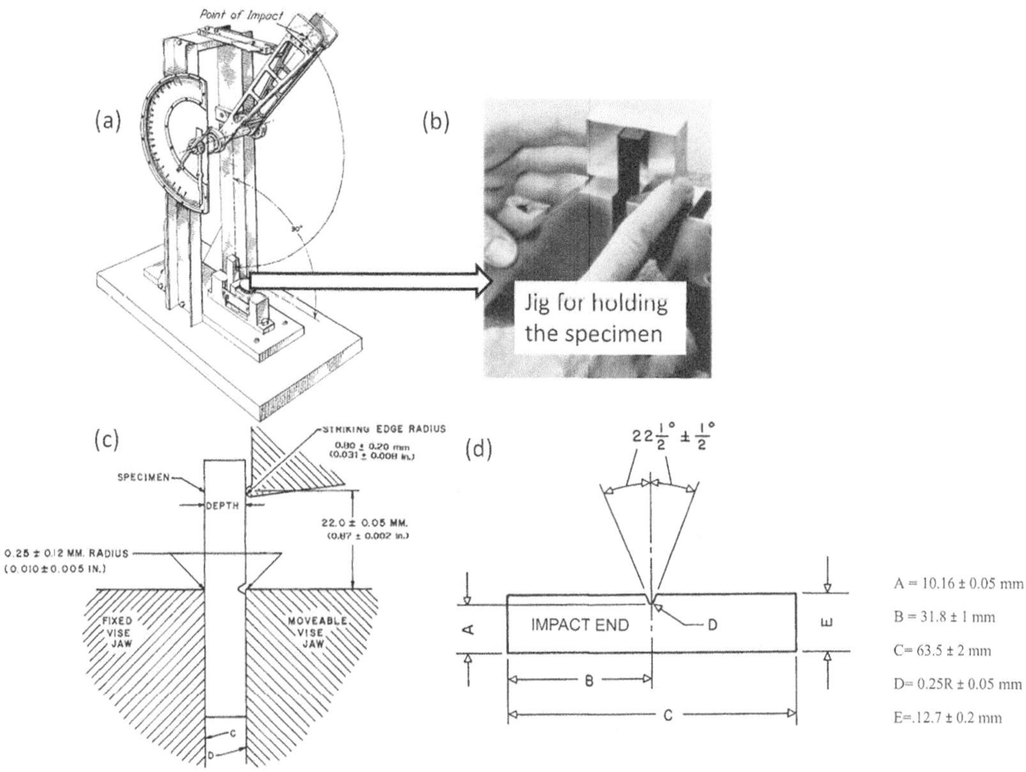

FIGURE 5.5 (a) Impact testing machine, (b) Jig, (c) relation between sample and the striking edge, and (d) sample dimensions

Adapted from Standard Test Methods for Determining the Izod Pendulum Impact Resistance of Plastics, ASTM D256, 2010.

impact properties (Gholampour and Ozbakkaloglu, 2020). Nascimento et al. have reported that a higher volume fraction of fibre would improve the impact energy capacity (Nascimento et al., 2012).

5.2.5 Fracture Toughness

The composites could undergo varying loads during their lifetime, and hence cracks might occur at a particular instant. Thus, eventually it might lead to failure because of the increase in the stress values near the crack. The stress intensity at the point of failure is termed as fracture toughness, and it is also referred to as fracture toughness or critical stress intensity factor $\left(K_{Ic}\right)$. Fracture toughness is evaluated depending on the mode of failure. There are three different possible modes of failure, and the fracture toughness tests based on the failure modes are shown in Figure 5.6 (Chittimenu et al., 2021; Prasad et al., 2011). The ASTM standard for testing is also selected depending on the material and the failure mode. ASTM D5528, ASTM 7905, and ASTM D6671 should be used in Mode I, Mode II, and Mode III failures, respectively (Chittimenu et al., 2021). The fracture toughness depends on the type of fibre used, architecture of fibre, rate of displacement, orientation of fibre, stacking pattern, fibre modification, aging, and hybridization of fibres.

5.2.6 Shear Test

Direct shear or pure shear is not ideally achievable in a composite. Direct shear refers to the state in which other forms of shear stresses do not appear in the sample length (Adams and Whitney, 2017).

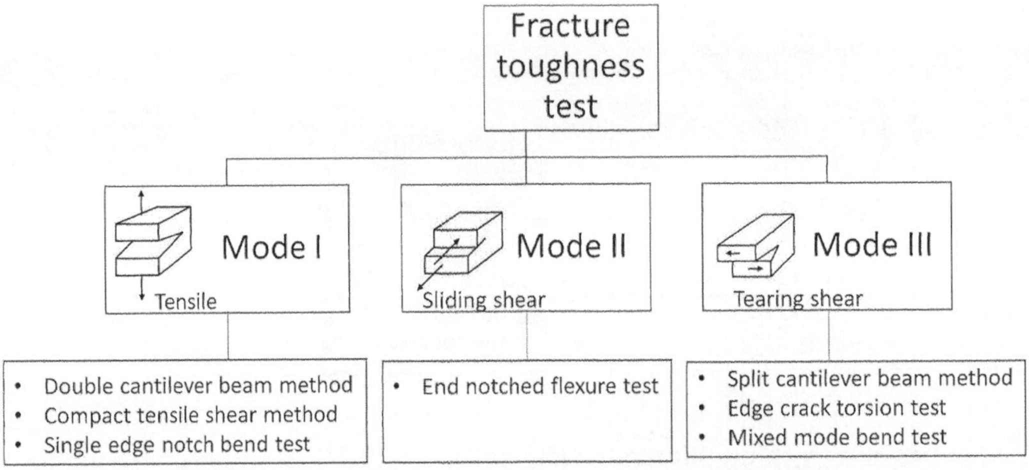

FIGURE 5.6 Categorization of fracture toughness tests depending on the failure modes

Reprinted from Journal of Minerals and Materials Characterization and Engineering, Vol. 10, M. S. S. Prasad, C. S. Venkatesha, and T. Jayaraju, Experimental Methods of Determining Fracture Toughness of Fibre Reinforced Polymer Composites under Various Loading Conditions, pp. 1263-1275, Copyright (2011), with permission from Elsevier.

5.2.6.1 Direct Shear

The Iosipescu shear test could be used for understanding the shear properties. In this test, a double V-notched sample (figure is provided in Ref. (Walrath and Adams, 1983; Herrera-Franco and Valadez-Gonzalez, 2008)) is placed in a fixture such that a compressive force is applied on the sample and the loading axis is in line with the notches. The test is performed as per the ASTM D5379 standard. The failure induced by shear occurs in the matrix depending on the geometry of fibre and fibre location. The adhesion between matrix and fibre influences the failure occurring due to shear (Herrera-Franco and Valadez-Gonzalez 2008).

5.2.6.2 Induced Shear

In laminated composites, two of the induced shears that appear are (a) in-plane shear strength and (b) inter-laminar shear strength. For determining these quantities, in-plane shear test (+45 tensile testing) and inter-laminar shear test (short beam shear method) are used, respectively.

(a) **In-plane shear test**
The sample should be selected such that fibre orientation is ±45°, and it will be subjected to a uniaxial tensile loading. ASTM D3518 standard could be used for performing the test. The tensile load applied induces shear stress in the same orientation as that of ply orientation (Adams and Whitney, 2017). Load cell will be used for tracking force, and the displacement could be traced using a linear variable differential transformer. The shear modulus of the sample would be obtained by plotting shear stress-shear strain diagram, and the slope of the curve gives the shear modulus (Ahmed and Vijayarangan, 2006). Shear stress is calculated as $\tau = P/2A$ (P is the load applied, A is the area of cross-section), and shear strain is obtained from the difference of longitudinal and transverse strains (Gowda et al., 1999).

(b) **Inter-laminar shear test**
The delamination properties of the layers are understood from the inter-laminar shear strength. The sample is subjected to three-point bending during the test, and the shear is induced in proportion to the load applied. The test is conducted by following the ASTM D2344-16 standard and is used for short beam shear testing. In this test, the sample length

is kept short so that the failure occurs because of shear (longer sample length will result in bending failure). The sample is kept on a horizontal fixture so that the loading axis is parallel to the fibres. The shear strength is calculated using $S = 0.75^{*} P_b/\text{Area}$, where P_b is the breaking load (Gowda et al., 1999). Atmakuri et al. have reported testing as per ASTM D-3846-02, and the transverse loading was applied on a sample having two opposite notches at the mid portion (Atmakuri et al., 2020).

5.2.7 Creep

For performing creep test, a constant load is applied, and the variation of strain with time is noted. Creep tests could be conducted as per ASTM D2290 standard. The creep response obtained could be classified into three different regions: (i) primary creep, (ii) secondary creep, and (iii) tertiary creep. The factors affecting the creep in natural fibre composites are matrix material, amount of fibre and their orientation, stress applied, time, temperature used for testing, etc. (Misra et al., 2011). Creep resistance showed improvement with an increase in fibre content. When the fibres are aligned in the loading direction, the creep of fibre dictates the creep property of the composite. On the contrary, the creep of the matrix dictates the composite material's creep behaviour when the fibres are oriented randomly. The properties such as compliance and viscosity could be obtained by performing the creep test (Misra et al., 2011).

5.2.8 Fatigue

A fatigue test could be performed on a universal testing machine by following the ASTM D 638-03 standard. While performing a fatigue test, a cyclic (tensile-tensile) load would be applied at a particular fatigue stress ratio (fatigue stress ratio is the ratio of minimum fatigue load to that of the maximum fatigue load) for a certain amount of frequency (Fotouh et al., 2014). During the test, an extensometer is used to monitor the strain, and the stress-strain curve generated is recorded. The S-N curve is used to depict the fatigue life. While applying a cyclic stress (S), the number of cycles to failure (N) is shown in the S-N curve. The fatigue properties depend on the type of matrix (brittle/ductile), architecture of fibre, interfacial strength, fibre modulus, amount of fibre, etc. Better the fibre properties, higher the load for initiation of crack/failure (Misra et al., 2011; Gassan, 2002).

5.2.9 Buckling Test

The structural performance of the composite subjected to load (compressive) could be evaluated by determining its buckling load and free vibration characteristics. The dynamic behaviour of the structure could be monitored by conducting a free vibration test. The sample will be placed within the grips of the universal testing machine, and the top cross-head starts to move downwards at a fixed feed rate. Thus, the sample is subjected to a compressive load, and the corresponding deflection is noted based on the cross-head movement (Rajesh and Pitchaimani, 2017). The critical buckling load would be predicted using the load versus deflection curve generated, and the graphical methods used for the prediction are mentioned in Ref. (Waddar et al., 2019).

5.2.10 Free Vibration Test

Free vibration test is carried out in a similar setup (refer to Figure 5.7) mentioned in the buckling test. The compressive load on the sample placed in the universal testing machine is increased in short intervals (say, 50 N). Between the incremental loading step, impulse hammer method is used to give small excitations to the sample, and the corresponding vibration responses from the sample are recorded using an accelerometer. The time-dependent input force and vibration response are given as input to DEWESOFT (software for modal analysis) by using a data acquisition system. The load is interrupted for a small amount of time while the modal analysis is carried out. The Fast

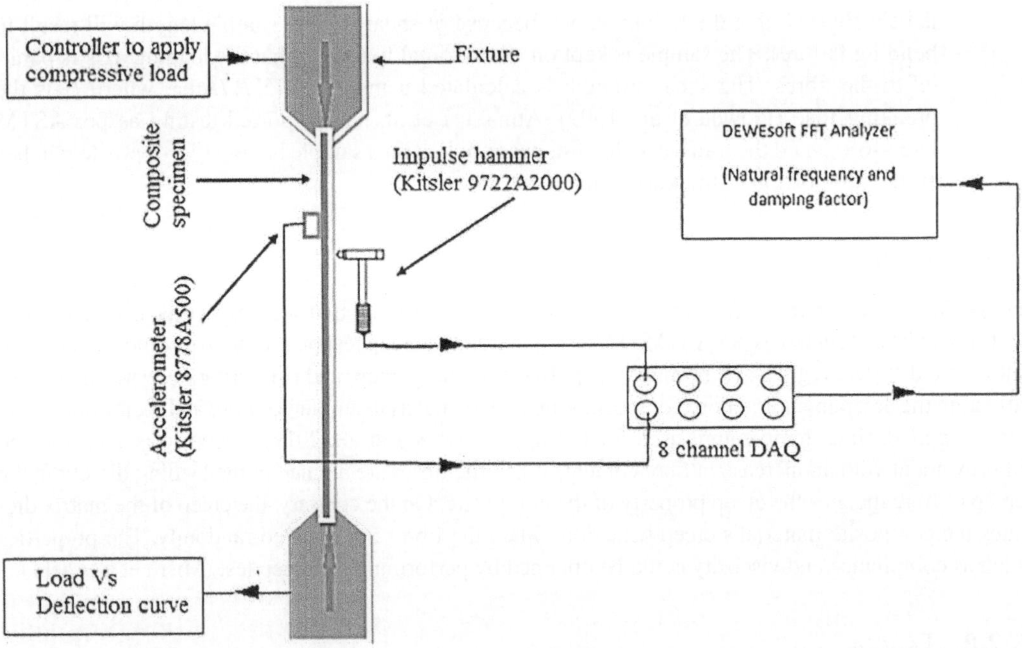

FIGURE 5.7 Schematic of buckling and free vibration setup

Reprinted from Composite Structures, Vol. 163, M. Rajesh and Jeyaraj Pitchaimani, Experimental investigation on buckling and free vibration behaviour of woven natural fibre fabric composite under axial compression, pp. 302–311, Copyright (2017), with permission from Elsevier.

Fourier Transformation method in the software is used for converting the time domain information into frequency domain. Hence, the damping ratio, natural frequency, and free vibration shape mode are obtained as the response of the test. (Rajesh and Pitchaimani, 2017).

5.2.11 Dynamic Mechanical Analysis

The viscoelastic properties (such as damping factor, loss modulus, storage modulus) of the composites are characterized using dynamic mechanical analysis. This analysis is significant to understand the effect of unbalanced loading in a structure. The viscoelastic properties depend on the volume fraction, shape, and size, as well as the orientation of the fibre and temperature (Ashok et al., 2019). Dynamic mechanical testing is possible using a double cantilever system, and the test is conducted in a selected temperature range. Storage modulus, damping factor, and loss factor of the composites could be determined by performing the test in a selected temperature range at a fixed heating rate (Rajesh et al., 2018). A sinusoidal force varying with time is applied during dynamic mechanical analysis, and the response from the sample will have a phase shift of δ with respect to the initial force. The complex modulus of the composite could be represented in a complex form with storage modulus $\left(E'\right)$ as the real part and loss modulus $\left(E''\right)$ as the imaginary part.

$$\text{Complex modulus, } E = E' + iE''$$

Storage modulus $\left(E' = E\cos\delta\right)$ signifies the load-carrying capacity of the material, whereas the loss modulus $\left(E'' = E\sin\delta\right)$ gives the measure of energy lost because of heat loss. The storage modulus conveys the stiffness or elastic constant of the sample. Damping of the material is in proportion with the loss modulus, and this modulus value shows the viscous response. Then the flexibility of the composite could be understood by calculating the damping factor $\left(\tan\delta = E'/E''\right)$.

Complex modulus discussed here is not the same as that of Young's modulus obtained from load versus deformation plot (Ashok et al., 2019).

5.2.12 Micromechanical Studies

Micromechanical studies are performed using micro-indentation test, microbond test, fragmentation test, pull-out test of single fibre, etc. Pull-out test is performed by applying an axial load on a fibre encapsulated in a matrix. The load is applied until the fibre ruptures or until the fibre comes out of the matrix. The shear strength at the interface between the matrix and fibre material could be obtained using pull-out test, shown in Figure 5.8 (Huang and Young, 2019). The tensile force applied is balanced by the shear strength at the interface between the matrix and fibre. Therefore, shear strength at the interface is found out using $F/\Pi DL$, where 'F' is applied pulling force, 'L' is fibre length, and 'D' is fibre diameter. During the test, the fibre is pulled at a constant displacement rate, and the corresponding pulling force is obtained using a sensor.

Microbond test could be contemplated as a modification of single-fibre pull-out test (Figure 5.9). An ellipsoidal droplet of resin is attached around the fibre in the microbond test, and therefore the

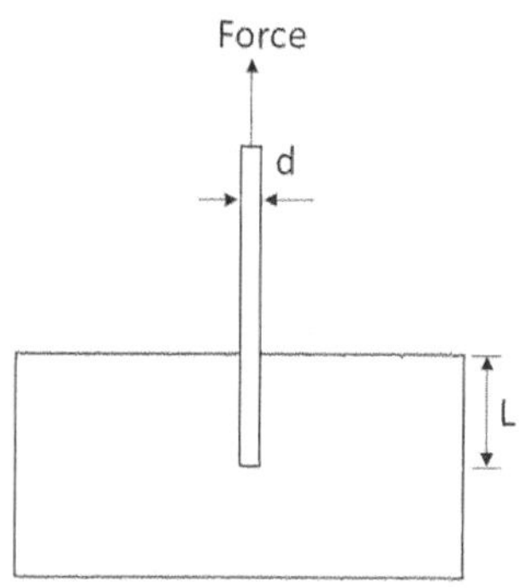

FIGURE 5.8 Pull-out test of single fibre

Reprinted from Composites Part B: Engineering, Vol. 166, Jyun-Kai Huang and Wen-Bin Young, The mechanical, hygral, and interfacial strength of continuous bamboo fibre reinforced epoxy composites, pp. 272–283, Copyright (2019), with permission from Elsevier.

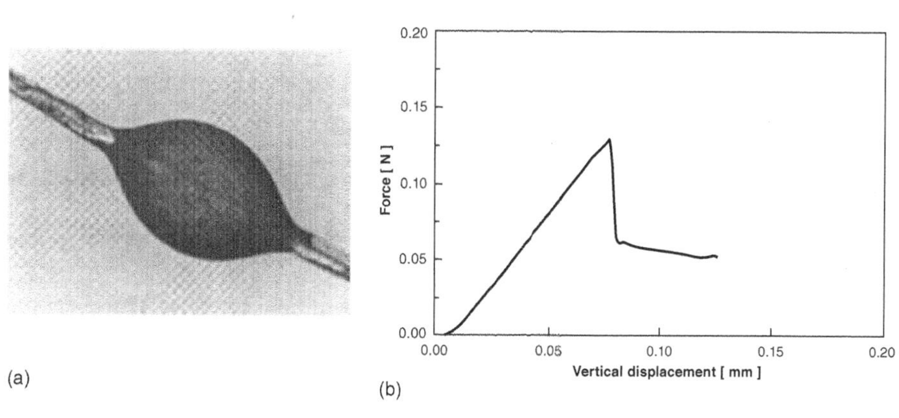

FIGURE 5.9 (a) Microbond test sample and (b) force displacement curve obtained from microbond test

Reproduced from Polymer Engineering and Science, Vol. 41, J. George, M.S. Sreekala, S. Thomas, A review on interface modification and characterization of natural fiber reinforced plastic composites, pp. 1471-1485, 2001 (George et al., 2001).

shear strength at the interface is distributed uniformly along the fibre length. The interfacial shear strength between the fibre and matrix could be calculated as the ratio of shear force at debonding with that of fibre length bonded with resin (George et al., 2001). The reliability of microbond test results depends on the resin droplet shape. If the length of resin droplet exceeds a critical value, the fibre will fail. Fragmentation test is another possibility for understanding the micromechanical characteristics of the composites. In the fragmentation test, a fibre is impregnated within the matrix, and it is subjected to tensile load until the fibre breaks. The stronger the fibre, the higher the strain required for rupturing the fibre. Micro-indentation tests are performed on actual composites. While performing micro-indentation test, a compressive load is applied on an individual fibre (on a cross-section which is polished) until the fibre debonds, and the shear strength at the interface is calculated using debond length (George et al., 2001).

5.2.13 Hardness Test

Resistance to scratch could be understood by measuring hardness. Nano-indentation tests could be performed to measure the hardness of the composites. Indentations at a depth (d) with sufficient space between successive measurements would be taken. Hardness value was calculated to be the ratio of load to that of projected area. Projected area could be calculated as $24.5d^2$ (Lee et al., 2007). Also, it is reported that the Rockwell hardness method could also be used as per ASTM 785-98 (Binoj et al., 2016).

5.2.14 Wear Test

The sample held in the machine's jaws was subjected to erosion by keeping it in contact with a disc, which is rotating at a fixed speed. The initial weight of the sample was measured before testing. Later, the weight loss, because of abrasion, was calculated by measuring the final weight. The test could be performed by applying different loads. Each test will be run for a fixed amount of time (Singha and Thakur, 2009).

5.2.15 Moisture Absorption Test

Moisture absorption will reduce the mechanical properties of the composites. Hence, it is discussed here. Tests were performed as per ASTM D-570 standard. Initially, the samples were placed in an oven at a particular temperature for a certain amount of time, and the initial weight was noted. Then the samples were placed inside distilled water, and the weight was periodically monitored for the required period. The moisture intake of the composite could be calculated by finding the ratio of the difference in weight before and after immersing with respect to the initial sample weight. The moisture absorbed depends on the voids present, temperature of the immersed liquid, contact area with the liquid, amount of fibre, etc. (Atmakuri et al., 2020).

The various standards to be followed while performing mechanical tests are summarized in Table 5.3.

5.3 METALLURGICAL TESTING

Since the mechanical and metallurgical aspects are dependent on each other, they are usually not discussed separately in literature. In this section, efforts are made to focus on metallurgical tests performed to understand the material behaviour.

5.3.1 Non-Destructive Testing

X-ray computer tomography is a non-destructive testing method that could be used for understanding the porosity, distribution of particles, cracks, fibre architecture, and tensile/compressive/fatigue/impact failures in the natural fibre composites. Depending on the resolution required (micrometres

TABLE 5.3
Standards Used for Conducting Mechanical Tests

Mechanical Tests	Standards	
Tensile test	ASTM D638 (arbitrarily oriented composite)	
	ASTM D3039 (systematically oriented composite)	
Compressive test	ASTM D3410	
Flexural test	ASTM D790, ISO 178	
Creep test	ASTM D2290	
Fatigue	ASTM D 638-03	
Impact test (Izod test)	ASTM D256	
Fracture toughness	Mode I	ASTM D5528
	Mode II	ASTM 7905
	Mode III	ASTM D6671
Shear test	Direct shear	ASTM D5379M
	In-plane shear test	ASTM D3518
	Inter-laminar shear test	ASTM D2344

or nanometres), the acquisition time will vary from a few minutes to hours. Synchrotron-based X-ray computer tomography could be used for real-time acquisition of sample behaviour. Three-dimensional images obtained from the computer tomography data could be used for generating finite element models for simulating experiments [25]. The contrast difference between the fibre and matrix will be a limiting factor while using the computer tomography technique.

Dye penetrant testing (chemical method) could be used for quick inspection with the application of a dye on the sample surface, and it helps in revealing surface defects. Ultrasonic testing is another non-destructive testing technique that makes use of high-frequency waves that are propagated through the sample for detecting defects in the sample. Eddy current testing allows the detection of defects in ferromagnetic as well as non-ferromagnetic materials. In the thermography technique, the sample is heated, and the heat waves are used to detect the defect. While performing this test, an infrared camera is required to visualize the heat flow. Eddy current thermography is another upcoming testing available, which makes use of the advantages of eddy current testing and thermography (Abidin et al., 2019).

5.3.2 Microscopy

The bonding and morphology between fibre and matrix could be studied using scanning electron microscopy or optical microscopy. Optical microscopy could be useful for a quick inspection, whereas scanning electron microscopy could be advantageous for understanding the composition (using energy dispersive X-ray analysis) and to visualize the material at a microscale. Better bonding achieved through physical or chemical modification of fibres is visualized using microscopy techniques. Also, the fractured samples obtained after performing mechanical tests (tensile tests) could be used for the same [4]. Pits, voids, and pores will affect the mechanical properties of composites (Binoj et al., 2016). Hence, microscopic techniques could be used to figure out the defects present in composites. When laminates are made from two different fibres, the curing temperature used might result in uneven surfaces. The uneven surface (pits/voids) arises because of the difference in the hygroscopic property of the fibres. For a particular curing temperature, the lesser hygroscopic fibre will have excess moisture loss and eventually result in corrugations (Vijayan and Thiagarajan, 2022). Atomic force microscopy could be used to visualize the modified fibre at a nanoscale. Since the microscopic analysis is already discussed in the previous book chapter on *Surface Analysis and Characterization of NFCs*, it is not discussed in detail here.

5.3.3 X-Ray Diffraction

X-ray diffraction technique is useful for understanding the material at a qualitative level. The crystallinity index (affecting mechanical properties) of fibres could be found out using X-ray diffraction (Sanjay et al., 2017).

5.3.4 Corrosion Testing

The corrosion behaviour of the composites depends on the dimensions of fibre, composition of solution, and fibre pH, temperature, ageing duration, etc. Non-destructive (refer to Section 5.3.1) and destructive (refer to Section 5.3.2) techniques could be used for detecting the corrosive regions. The chemical resistance of composites can be identified by immersing the sample in an acidic (HCl, H_2SO_4) or basic (NaOH) solution and then the weight loss is calculated.

5.3.5 Wettability Test

The wettability of natural fibre composite could be understood by measuring the contact angle of a liquid in contact with the composite. The tests were performed as per ASTM D-7334. The water droplet could be introduced onto the composite surface using a pipette, and the contact angle could be measured using a camera attached with an image processing unit (Atmakuri et al., 2020).

5.4 CONCLUSIONS

The mechanical as well as metallurgical aspects of the natural fibre composites are influenced by the fibre/matrix interfacial bonding, fibre length, fibre orientation, volume fraction of fibre, fibre and matrix material used, etc. The quasistatic behaviour of the natural fibre composites could be evaluated by tensile test, compression test, flexural test, and the dynamic behaviour of the natural fibre composites could be understood by performing impact test, fatigue tests, etc. The interfacial properties could be estimated using hardness testing and micromechanical studies such as microbond test, fragmentation test, and fibre pull-out test. The mechanical and metallurgical testing of natural fibre composites is essential in producing composites with enhanced properties.

REFERENCES

Abidin, Nor Ishida Zainal, Mohd Faizul Mohd Sabri, Katayoon Kalantari, M. A. Amalina, and Roslina Ahmad. 2019. "Corrosion Detection for Natural/Synthetic/Textiles Fiber Polymer Composites." In *Structural Health Monitoring of Biocomposites, Fibre-Reinforced Composites and Hybrid Composites*, Elsevier Ltd: 93–112. https://doi.org/10.1016/B978-0-08-102291-7.00006-X

Adams, Donald F., and Thomas J. Whitney. 2017. "Test Methods for Mechanical Properties." In *Comprehensive Composite Materials II*, edited by Peter W. R. Beaumont and Carl H. Zweben. Elsevier Ltd: 4–37. https://doi.org/10.1016/B978-0-12-803581-8.10032-3

Ahmed, K. Sabeel, and S. Vijayarangan. 2006. "Elastic Property Evaluation of Jute-Glass Fibre Hybrid Composite Using Experimental and CLT Approach." *Indian Journal of Engineering & Materials Sciences* 13: 435–442.

Ashok, R. B., C. V. Srinivasa, and B. Basavaraju. 2019. "Dynamic Mechanical Properties of Natural Fiber Composites - A Review." *Advanced Composites and Hybrid Materials* 2 (4). Springer: 586–607. https://doi.org/10.1007/s42114-019-00121-8

Atmakuri, Ayyappa, Arvydas Palevicius, Madhusudan Siddabathula, Andrius Vilkauskas, and Giedrius Janusas. 2020. "Analysis of Mechanical and Wettability Properties of Natural Fiber-reinforced Epoxy Hybrid Composites." *Polymers* 12. MDPI: 1–15. https://doi.org/10.3390/polym12122827

Barone, J. R. 2005. "Polyethylene/Keratin Fiber Composites with Varying Polyethylene Crystallinity" *Composites Part A: Applied Science and Manufacturing* 36. Elsevier: 1518–1524. https://doi.org/10.1016/j.compositesa.2005.03.006

Bax, Benjamin, and Jorg Mussig. 2008. "Impact and Tensile Properties of PLA/Cordenka and PLA/Flax Composites." *Composites Science and Technology* 68. Elsevier: 1601–1607. https://doi.org/10.1016/j.compscitech.2008.01.004

Binoj, J. S., R. Edwin Raj, B. S. S. Daniel, and S. S. Saravanakumar. 2016. "Optimization of Short Indian Areca Fruit Husk Fiber (Areca Catechu L.) Reinforced Polymer Composites for Maximizing the Mechanical Property." *International Journal of Polymer Analysis and Characterization* 5341. Taylor & Francis: 112–122. https://doi.org/10.1080/1023666X.2016.1110765

Bledzki, A. K., and A. Jaszkiewicz. 2010. "Mechanical Performance of Biocomposites Based on PLA and PHBV Reinforced with Natural Fibres - A Comparative Study to PP." *Composites Science and Technology* 70 (12). Elsevier: 1687–1696. https://doi.org/10.1016/j.compscitech.2010.06.005

Chittimenu, Haasith, Monesh Pasupureddy, Chandrasekar Muthukumar, Senthilkumar Krishnasamy, Senthil Muthu Kumar Thiagamani, and Suchart Siengchin. 2021. "Fracture Toughness of the Natural Fiber-Reinforced Composites: A Review." In *Mechanical and Dynamic Properties of Biocomposites*, edited by Senthilkumar Krishnasamy, Nagarajan Rajini, Senthil Muthu Kumar Thiagamani, and Suchart Siengchin. WILEY-VCH GmbH: 293–304. https://doi.org/10.1002/9783527822331

Fan, Mizi, and David Hui. 2014. "Compressive Behaviour of Natural Fibre Composite." *Composites: Part B* 67. Elsevier: 183–191. https://doi.org/10.1016/j.compositesb.2014.07.014

Faruk, Omar, Andrzej K. Bledzki, Hans Peter Fink, and Mohini Sain. 2014. "Progress Report on Natural Fiber Reinforced Composites." *Macromolecular Materials and Engineering* 299 (1). Wiley Online Library: 9–26. https://doi.org/10.1002/mame.201300008

Fotouh, Ahmed, John D. Wolodko, and Michael G. Lipsett. 2014. "Fatigue of Natural Fiber Thermoplastic Composites." *Composites Part B: Engineering* 62. Elsevier: 175–182. https://doi.org/10.1016/j.compositesb.2014.02.023

Gassan, Jochen. 2002. "A Study of Fibre and Interface Parameters Affecting the Fatigue Behaviour of Natural Fibre Composites." *Ccg* 33. Elsevier: 369–374.

George, Jayamol, M. S. Sreekala, and Sabu Thomas. 2001. "A Review on Interface Modification and Characterization of Natural Fiber Reinforced Plastic Composites." *Polymer Engineering and Science* 41 (9): 1471–1485. https://doi.org/10.1002/pen.10846

Gholampour, Aliakbar, and Togay Ozbakkaloglu. 2020. A Review of Natural Fiber Composites: Properties, Modification and Processing Techniques, Characterization, Applications. *Journal of Materials Science* 55. Springer US. https://doi.org/10.1007/s10853-019-03990-y

Gowda, T. Munikenche, A. C. B. Naidu, and Rajput Chhaya. 1999. "Some Mechanical Properties of Untreated Jute Fabric-Reinforced Polyester Composites." *Composites Part A: Applied Science and Manufacturing* 30. Elsevier: 277–284.

Herrera-Franco, P. J., and A. Valadez-Gonzalez. 2008. "Mechanical Testing of Natural-Fiber Composites." In *Properties and Performance of Natural-Fibre Composites*, edited by Kim L. Pickering. Woodhead Publishing: 375–401. https://doi.org/10.1533/9781845694593.3.375

Huang, Jyun-Kai, and Wen-Bin Young. 2019. "The Mechanical, Hygral, and Interfacial Strength of Continuous Bamboo Fiber Reinforced Epoxy Composites." *Composites Part B* 166. Elsevier: 272–283. https://doi.org/10.1016/j.compositesb.2018.12.013

Islam, M. S., K. L. Pickering, and N. J. Foreman. 2010. "Influence of Alkali Treatment on the Interfacial and Physico-Mechanical Properties of Industrial Hemp Fibre Reinforced Polylactic Acid Composites." *Composites: Part A* 41 (5). Elsevier: 596–603. https://doi.org/10.1016/j.compositesa.2010.01.006

Joseph, Seena, M. S. Sreekala, Z. Oommen, P. Koshy, and Sabu Thomas. 2002. "A Comparison of the Mechanical Properties of Phenol Formaldehyde Composites Reinforced with Banana Fibres and Glass Fibres." *Composites Science and Technology* 62 (14). Elsevier: 1857–1868. https://doi.org/10.1016/S0266-3538(02)00098-2

Ku, H., H. Wang, N. Pattarachaiyakoop, and M. Trada. 2011. "A Review on the Tensile Properties of Natural Fiber Reinforced Polymer Composites." *Composites Part B: Engineering* 42 (4). Elsevier: 856–873. https://doi.org/10.1016/j.compositesb.2011.01.010

Lee, Seung-hwan, Siqun Wang, and George M. Pharr. 2007. "Evaluation of Interphase Properties in a Cellulose Fiber-Reinforced Polypropylene Composite by Nanoindentation and Finite Element Analysis." *Composites Part A: Applied Science and Manufacturing* 38. Elsevier: 1517–1524. https://doi.org/10.1016/j.compositesa.2007.01.007

Malkapuram, Ramakrishna, Vivek Kumar, and Yuvraj Singh Negi. 2009. "Recent Development in Natural Fiber Reinforced Polypropylene Composites." *Journal of Reinforced Plastics and Composites* 28 (10). Sage Publications: 1169–1189. https://doi.org/10.1177/0731684407087759

Misra, M., S. S. Ahankari, A. K. Mohanty, and A. D. Nga. 2011. "Creep and Fatigue of Natural Fibre Composites." In *Interface Engineering of Natural Fibre Composites for Maximum Performance*, 289–340. Woodhead Publishing Limited. https://doi.org/10.1533/9780857092281.2.289

Nascimento, Denise Cristina O., Ailton S. Ferreira, Sergio N. Monteiro, Regina Coeli M. P. Aquino, and Satyanarayana G. Kestur. 2012. "Studies on the Characterization of Piassava Fibers and Their Epoxy Composites." *Composites Part A: Applied Science and Manufacturing* 43. Elsevier: 353–362. https://doi.org/10.1016/j.compositesa.2011.12.004

Oksman, K., M. Skrifvars, and J.-F. Selin. 2003. "Natural Fibres as Reinforcement in Polylactic Acid (PLA) Composites." *Composites Science and Technology* 63. Elsevier: 1317–1324. https://doi.org/10.1016/S0266-3538(03)00103-9

Prasad, M. S. Sham, C. S. Venkatesha, and T. Jayaraju. 2011. "Experimental Methods of Determining Fracture Toughness of Fiber Reinforced Polymer Composites under Various Loading Conditions." *Journal of Minerals and Materials Characterization and Engineering* 10 (13): 1263–1275. https://doi.org/10.4236/jmmce.2011.1013099

Rahman, Rozyanty, and Syed Zhafer Firdaus Syed Putra. 2019. "Tensile Properties of Natural and Synthetic Fiber-Reinforced Polymer Composites." In *Mechanical and Physical Testing of Biocomposites, Fibre-Reinforced Composites and Hybrid Composites*, edited by Mohammad Jawaid, Mohamed Thariq, and Naheed Saba, 81–102. Elsevier Ltd. https://doi.org/10.1016/C2016-0-04437-6

Rajesh, M., and Jeyaraj Pitchaimani. 2017. "Experimental Investigation on Buckling and Free Vibration Behavior of Woven Natural Fiber Fabric Composite under Axial Compression." *Composite Structures* 163. Elsevier: 302–311. https://doi.org/10.1016/j.compstruct.2016.12.046

Rajesh, M., Savendra P. Singh, and Jeyaraj Pitchaimani. 2018. "Mechanical Behavior of Woven Natural Fiber Fabric Composites: Effect of Weaving Architecture, Intra-Ply Hybridization and Stacking Sequence of Fabrics." *Journal of Industrial Textiles* 47 (5). Sage Publications Sage UK: London, England: 938–959. https://doi.org/10.1177/1528083716679157

Rong, Min Zhi, Ming Qiu Zhang, Yuan Liu, Gui Cheng Yang, and Han Min Zeng. 2001. "The Effect of Fiber Treatment on the Mechanical Properties of Unidirectional Sisal-Reinforced Epoxy Composites." *Composites Science and Technology* 61. Elsevier: 1437–1447. https://doi.org/10.1016/S0266-3538(01)00046-X

Sanjay, M. R., P. Madhu, Mohammad Jawaid, P. Senthamaraikannan, S. Senthil, and S. Pradeep. 2017. "Characterization and Properties of Natural Fiber Polymer Composites: A Comprehensive Review." *Journal of Cleaner Production* 172. Elsevier: 566–581. https://doi.org/10.1016/j.jclepro.2017.10.101

Sarikaya, Engin, Hasan Çallioğlu, and Hakan Demirel. 2019. "Production of Epoxy Composites Reinforced by Different Natural Fibers and Their Mechanical Properties." *Composites Part B: Engineering* 167. Elsevier: 461–466. https://doi.org/10.1016/j.compositesb.2019.03.020

Sawpan, Moyeenuddin A., Kim L. Pickering, and Alan Fernyhough. 2011. "Improvement of Mechanical Performance of Industrial Hemp Fibre Reinforced Polylactide Biocomposites." *Composites Part A* 42 (3). Elsevier: 310–319. https://doi.org/10.1016/j.compositesa.2010.12.004

Singha, A. S., and Vijay Kumar Thakur. 2009. "Chemical Resistance, Mechanical and Physical Properties of Biofibers-Based Polymer Composites." *Polymer-Plastics Technology and Engineering* 48. Taylor & Francis: 736–744. https://doi.org/10.1080/03602550902824622

"Standard Test Method for Compressive Properties of Polymer Matrix Composite Materials with Unsupported Gage Section by Shear." 2003. *ASTM D3410*, 1–16.

Vijayan, D., and Rajmohan Thiagarajan. 2022. "Influence of Stacking Sequence on Mechanical and Metallurgical Properties of Ramie/Areca Laminates Using B4C Nano Filled Epoxy Resin." *Journal of Natural Fibers* 19 (10). Taylor & Francis: 3598–3614. https://doi.org/10.1080/15440478.2020.1848713

Waddar, Sunil, Jeyaraj Pitchaimani, Mrityunjay Doddamani, and Ever Barbero. 2019. "Buckling and Vibration Behaviour of Syntactic Foam Core Sandwich Beam with Natural Fiber Composite Facings under Axial Compressive Loads." *Composites Part B: Engineering* 175. Elsevier: 107133.

Walrath, D. E., and D. F. Adams. 1983. "The Iosipescu Shear Test as Applied to Composite Materials." *Experimental Mechanics* 23 (1). Springer: 105–110. https://doi.org/10.1007/BF02328688

6 Tribo-Adaptability and Tribo-Response of Natural Fiber Composites

Brailson Mansingh Bright, Narayanan Rajesh Prabha, Jerin Jose, and Joseph Selvi Binoj

6.1 INTRODUCTION

The production of ecologically sound, greener substances and methods has become imperative due to the increasing worldwide recognition of ecological challenges. One of the biggest challenges faced by investigators in this discipline is creating substances that can improve the quality of goods while adhering to environmental standards. Natural yarns have become a very attractive substitute for artificial fibers like fiberglass and carbon because of their remarkable qualities, which include accessibility, higher particular modulus, lighter weight, affordability, and decomposition (Thooyavan et al., 2022). In recent years, Natural Yarn-Augmented Polymeric Composites (NYAPCs) have gained a lot of traction in the furnishings, automotive, and building sectors. Furthermore, the use of natural strands instead of artificial ones has increased car fuel effectiveness, mostly due to the development of lighter composite materials. When NYAPCs are contrasted to pure polymeric or traditional polymeric composites, they offer a higher level of ecological sustainability because the yarns used in them come from sustainable sources (Britto et al., 2023). Yet, depending on the exact proportion of organic yarns to polymeric matrices and the manufacturing processes employed, the level of ecological tolerance and degradation of NYAPCs can vary. Therefore, it is crucial to consider the selection of resources and manufacturing methods that align with ecologically sound concepts.

In terms of tribological qualities, NYAPCs could exhibit milder levels of friction and wear than synthetic versions, which could have an effect on the composite's tribological functionality. Several strategies, such as modifying organic yarns or adding lubricant to improve tribological efficiency, can be used to solve this problem. Furthermore, friction and wear characteristics can be greatly impacted by the suitability of organic yarns with the polymeric matrix (Subramanian et al., 2021). These difficulties can be lessened by carefully applying the right surface modifications and manufacturing methods. Wear, friction, and lubrication are some of the instances of the variables that affect tribological properties and are activated when two opposing surfaces interface because of their corresponding displacement. Increased transpiration of energy from friction and wear ultimately results in material deterioration. A number of factors that affect the tribological attributes of NYAPCs are the yarn matrices proportion, binding substances, matrices features, methods of manufacture, fiber position, water captivation, and stuffing substance (Ben Samuel et al., 2021). The main issue with NYAPCs is the restricted interaction among the yarns and matrix, which results from cellulose's hydrophilic behavior compared to the hydrophobic properties of the matrix substance. Poor bonding between the matrix's surface and the strands and inadequate strand dispersal are the outcomes of this discrepancy.

DOI: 10.1201/9781032665337-6

Natural yarns need to undergo chemical processing in order to improve bonding between surfaces and reduce water captivation. The tribological durability of such composite substances may deteriorate as a result of variations from specified requirements, which are frequently caused by elements like manufacturing flaws (Nayak and Satapathy, 2020). A few instances of production flaws include the development of cavities in composite materials, fiber damage, incorrect alignment, tearing, and detaching between strands and the substrate. Therefore, extensive study into various NYAPC issues is required for the creation of robust and self-sufficient composites. This investigation might also end up in the establishment of ideal methodologies and the removal of flaws from manufacturing processes. This chapter offers a thorough investigation of the tribological properties that are found in several NYAPCs. This compilation provides a useful tool for investigators who are comparing various NYAPC characteristics, including important factors like ideal fiber loadings, sliding velocities, and other relevant variables. Though there is a wealth of literature on NYAPCs, there is a significant gap in which important aspects such as production processes, tribological characteristics, influencing aspects, real-world application, critical environmental responsibility deliberations, and related obstacles have not been thoroughly covered in tandem.

6.2 CLASSIFICATION OF NATURAL FIBERS

Natural fiber is separated into categories according to where it comes from, like plants, livestock, or minerals (Nanda and Satapathy, 2021). Animal fibers are made of proteins, whereas all plant fibers are made of cellulose. Bast fiber, leaf fiber, grain, berries, timber, grain straw, and various grass fibers are examples of plant fibers. Figure 6.1 shows a diagram that classifies the different types of fibers. Many academics were investigating the use of plant fiber in composite products during the

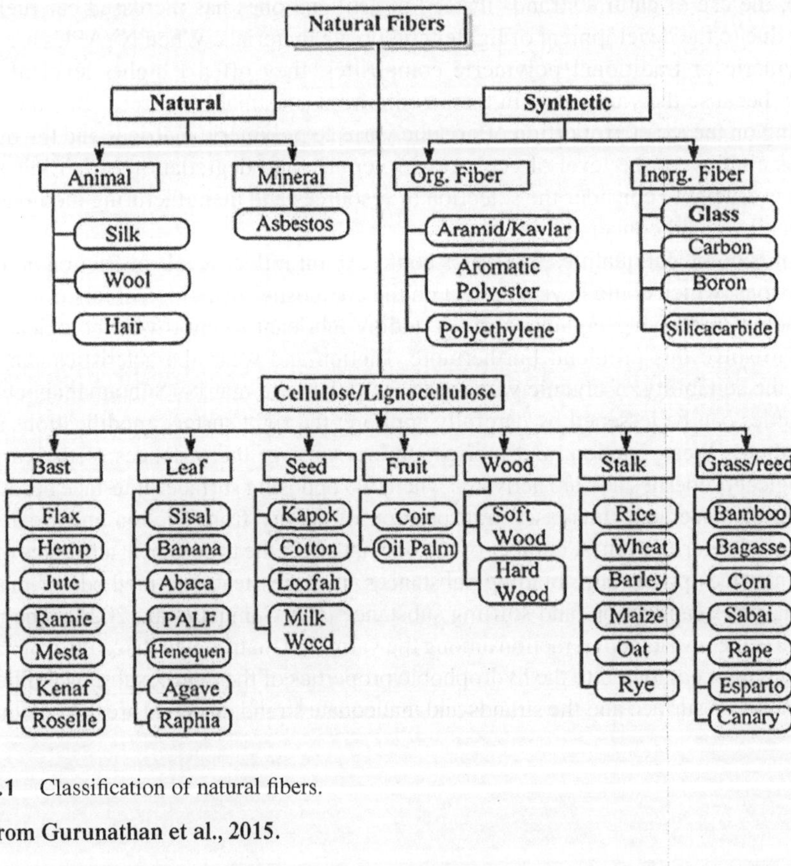

FIGURE 6.1 Classification of natural fibers.

Adapted from Gurunathan et al., 2015.

past few years. Three main chemical components that make up plant fibers are cellulose, hemicellulose, and lignin (Li et al., 2021). Furthermore, plant fibers have trace amounts of water-soluble compounds such as wax and pectin. The chemical makeup and topologies of plant fibers differ significantly according to the type of plant, maturity, soil, and environment. The qualities and consequently the uses of plant-based additives and fibers are determined by their chemical makeup, which makes it crucial to be known. The fairly intrinsic physical and mechanical characteristics of natural fiber are mostly attributed to their unique morphology and low density.

6.3 NYAPCs BASED ON YARN REINFORCEMENT

Dimensions, orientation, and shape are other categories by which organic yarn reinforcements can be categorized. Furthermore, as Figure 6.2 illustrates, they may exist as particles or fibers. Continuously and fragmentary are the two kinds of yarns, which vary in shape and size, were commonly used for augmentation in NYAPCs (Brailson Mansingh et al., 2022). Additionally, strand reinforcements might be positioned at random, multifaceted, or unilateral. The uninterrupted strands could be arranged in woven or non-woven pattern. Weaving is the continual linking of orthogonal threads in a recognizable pattern. Structures made of many strands are woven together to form yarns. Twist angle determines yarn toughness and cohesion up to a particular degree; beyond that, a surge in twist angle causes the optimum fiber strength to decrease (Singh and Sinha, 2021). In the event that the fiber is non-woven, meaning it lacks interlaced fibers, the mat may be longitudinal or arbitrarily orientated.

6.4 MANUFACTURING TECHNIQUES OF NYAPCs

The various methods for producing NYAPCs are depicted in Figure 6.3. The most used manufacturing processes among those listed are hand lay-up, spray-up, injection molding, compression molding, and resin transfer molding. In accordance with the needs of the product, which include characteristics

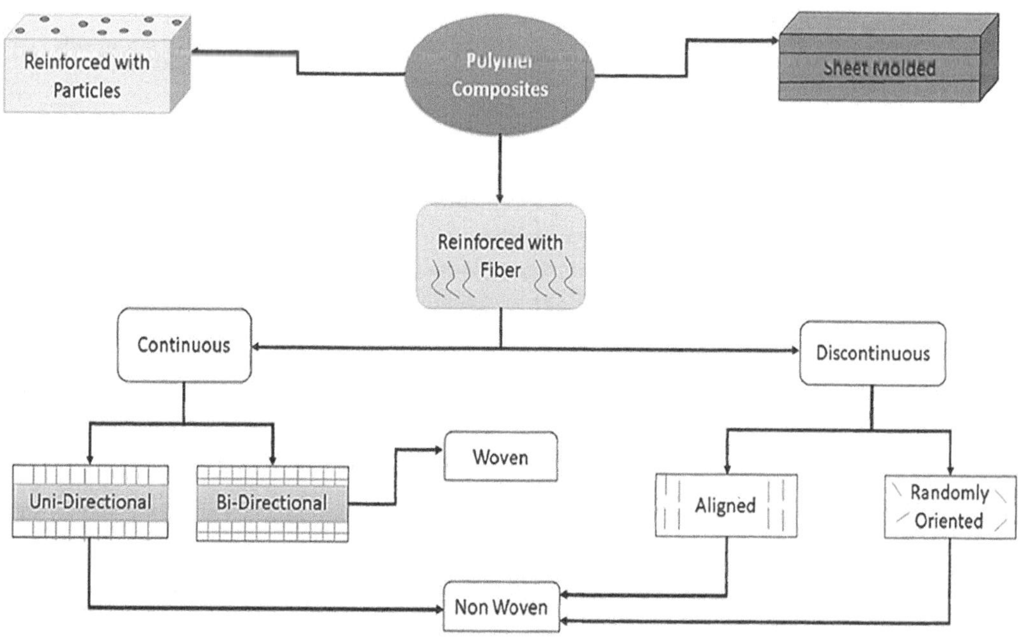

FIGURE 6.2 Composites based on yarn reinforcement.

Adapted from Bhat et al., 2023.

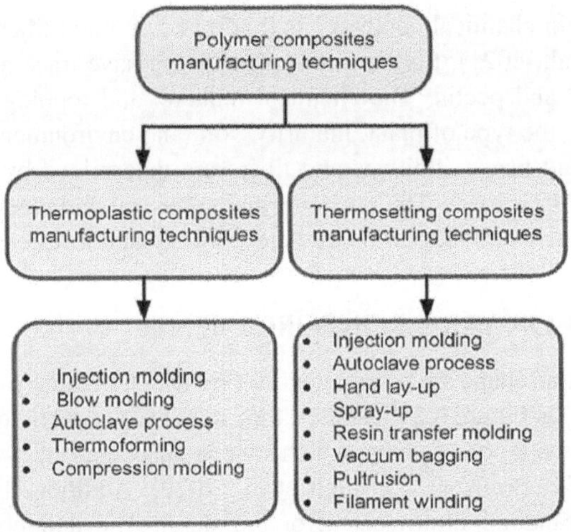

FIGURE 6.3 Classification of manufacturing techniques of NYAPCs.

Adapted from Singh et al., 2023.

like price, assets, quantity, and quality, other procedures are also employed (Ponnusamy et al., 2023). When producing composites, one of the main issues these days is the impact on the surroundings. Therefore, the cleaner methodology should be the primary consideration while choosing a manufacturing method. The aforementioned manufacturing procedures have benefits and downsides that vary depending on the individual components utilized in the composite manufacturing as well as the manufacturing circumstances. It takes more time and effort to create composites utilizing the aforementioned methods, so it's important to look into other cutting-edge methods that can speed up the composite manufacturing method. In an effort to address the problem of high production time and energy, experts and industry professionals have recently turned their focus to microwave-assisted composite fabrication (Nagaprasad et al., 2021). Thermosets and thermoplastic-based composites can both be produced with microwave-assisted composite manufacturing process.

6.5 TRIBOLOGICAL ASSETS OF NYAPCs

The investigation of wear, abrasion, and lubrication, all related phenomena that occur when objects come into contact with one another and move in relation to one another, is known as tribology. Wear is generally brought about by the energy lost as a consequence of friction. Friction and wear have an interrelated correlation (Hariharan et al., 2023). In general increase in friction leads to increase in wear. However, wear reduction is not always the outcome of decrease in friction. Wear and friction are not always bad things. In fact, there are times when they are thought to be beneficial for the best performance of the individual parts. Instances where wear is considered to be beneficial are machining, lettering with a pencil, and polishing. These are called productive wear and friction. Engineering component breakdown is known to be mostly caused by wear and friction among contacting interfaces. As a result, it is necessary to look into the tribological properties of the component pieces.

The tribological properties of NYAPCs are often the deciding factor when selecting them for various uses in industry. These NYAPCs are subjected to abrasive, erosive, and adhesive wear under conditions of use; therefore, an evaluation of their tribological characteristics is necessary (Gargol et al., 2021). A significant amount of energy utilized for overcoming friction can be conserved by following good tribological procedures. Another important component that must be taken into account for the tribological characteristics of NYAPCs is wear endurance. Improved durability from wear is crucial in situations when NYAPCs are subjected to frequent erosion during operation.

6.5.1 Tribological Testing Practices

There are notable differences in the methods used in tribology devices, which can be mostly attributed to the interaction method (point, line, or area) between the material under test and the counterface surface. Figure 6.4 illustrates the commonly employed tribological testing procedures for NYAPCs. Investigators must choose the tribo test to be used in the experimentation and ensure that the tests are carried out under suitable circumstances before beginning tribological studies (Karthik et al., 2022). In tribology studies, sensors are essential because they make it easier to collect and display data that allows investigators to see and understand how each material performs during a specific wear experiment. The investigation employs a variety of sensors and information-gathering gadgets as its equipment for temperature, speed, and force measurement. Additionally, NYAPC sample weights are measured using a weighing device both prior to and following testing. Surface damage identification is accomplished by the use of a microscope, while data collection, archiving, and analysis are accomplished through computer programs.

6.5.2 Influence of Various Variables on Tribological Properties NYAPCs

Tribological functionality of NYAPCs is dependent on a number of factors, such as the amount of matrix/yarn, the kind of yarn, and different testing circumstances, such as sliding distance, sliding velocity, and load (Selvan et al., 2023). Furthermore, the friction and wear behavior are significantly influenced by the alignment with respect to the fiber matrix interfaces, the pattern of loading, the manufacturing techniques, and the chemical procedures. Several studies have examined different fiber layouts, such as arbitrary, oppositely oriented, parallel, and standard, to confirm their impact on tribological characteristics.

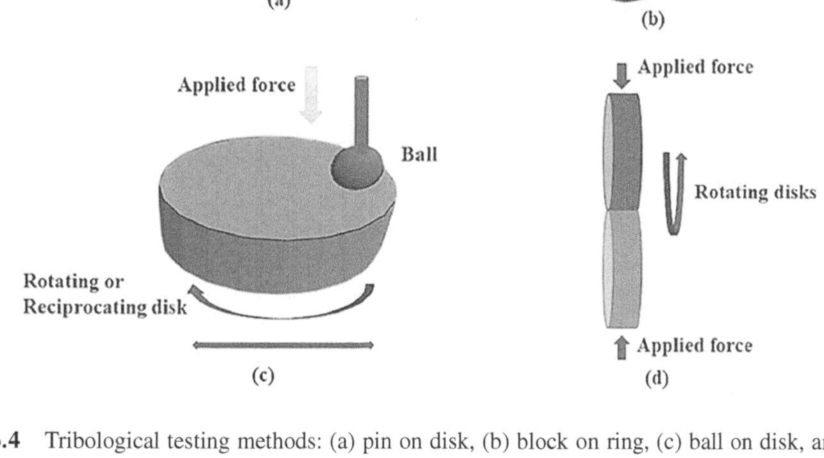

FIGURE 6.4 Tribological testing methods: (a) pin on disk, (b) block on ring, (c) ball on disk, and (d) twin disk tribometer.

Adapted from Bhat et al., 2023.

6.5.3 Influence of Fiber Augmentation and Chemical Treatments

Investigators" main attention in the area of NYAPC manufacturing has been centered on analyzing the impact of fiber reinforcement. Research indicates that adding natural fibers to plain epoxy enhances its hardness and resilience to erosion. In addition, as the amount of fiber increased, hardness and durability against wear of NYAPCs also increased (Bharath et al., 2023). In a separate study, experts saw a decrease in wear and the coefficient of friction (COF) of NYAPCs under all sliding situations when they added a hybrid combination of natural fibers at a 20% proportion. It was shown that the fibers' existence had no influence on the durability of NYAPCs at smaller loads, because of their non-brittle nature; they significantly improved NYAPCs' wear resilience at greater loads. After conducting additional research, scientists discovered that durability against wear dropped as fiber length increased.

When sisal yarn is added to the phenolic matrix up to a specific point, less substance is lost, improving the NYAPC"s durability against wear, according to studies done on the NYAPC for tribological evaluation at various temperatures, as shown in Figure 6.5 (Xin et al., 2007). At different temperatures of study, nevertheless, the frictional coefficient rose by up to 20% with mass proportion in the NYAPC. Investigators have also employed wood powder as a polypropylene augmentation (Ibrahim et al., 2019). The outcomes demonstrated a significant improvement in frictional coefficient and durability against wear. This was substantiated since there was no longer any direct interaction with the NYAPC due to the exterior development of polyamide rings.

It has also been discovered that surface alterations made by chemical treatments change tribological properties. Investigators found that when weighed against raw fibers, chemically modified fibers showed greater durability and wear resistance. Even treatment with saline solution enhanced the wear characteristics of the NYAPCs. Researchers created NYAPC, which is made of Coccinia indica strengthened with epoxy resin (Mylsamy et al., *2020*). In comparison to untreated (UT) yarn reinforcement, the mass reduction of NYAPCs treated with NaOH (NT) and silane (ST) reinforcement during wear tests showed a decline of 15% and 39%, respectively, as shown in Figure 6.6. Furthermore, for each surface modification, the frictional coefficient dropped by 41% and 39%, respectively. Scanning electron microscopy (SEM) pictures demonstrated that thc cpoxy substrate and Coccinia indica yarns in untreated NYAPCs detached, and the worn-out surface was coated in eroded epoxy substrate material and broken Coccinia indica yarns, as shown in Figure 6.7.

6.5.4 Influence of Fiber Alignment and Stacking Order

It has been determined that one important aspect affecting the tribological behavior of composites is the fiber orientation. In addition, the order in which the fibers are stacked has an impact on the tribological properties of hybrid NYAPCs, which include multiple layers of fiber infill. In one

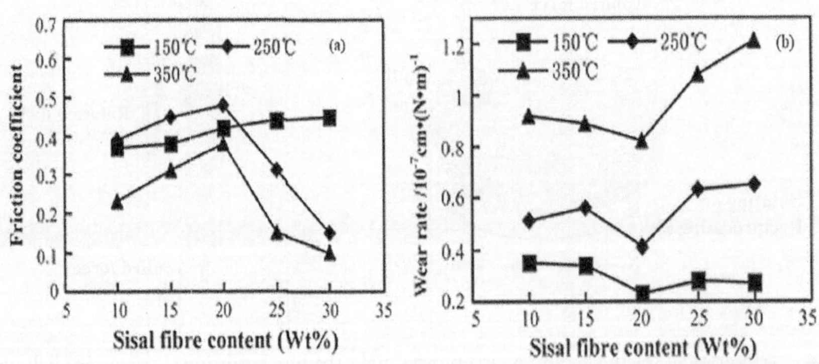

FIGURE 6.5 Influence of fiber proportion on (a) frictional coefficient and (b) wear of phenolic matrix NYAPC. **Adapted from Xin et al., 2007.**

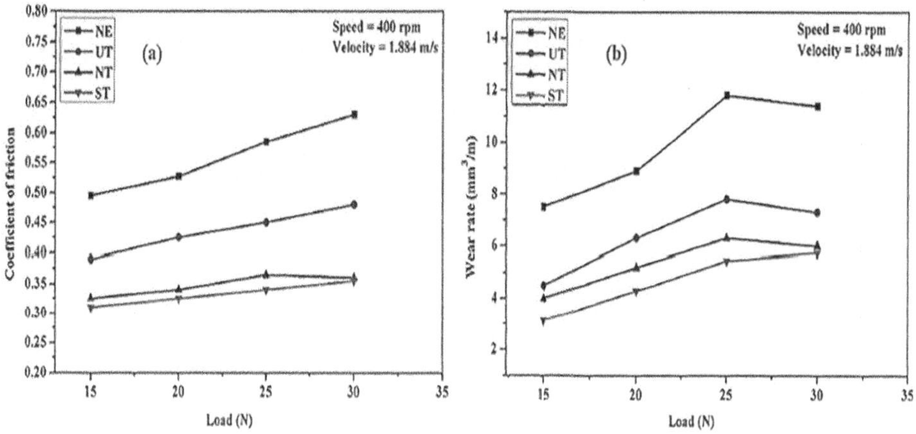

FIGURE 6.6 Influence of surface modification on Coccinia indica yarn-augmented NYAPCs: (a) frictional coefficient and (b) wear.
*NaOH treatment (NT), silane treatment (ST), untreated (UT), and neat epoxy (NE).

Adapted from Mylsamy et al., 2020.

investigation, NYAPCs with a bidirectional alignment showed the least specific wear rate and COF when several fiber alignments were taken into account. Three distinct sets of experts used isotropic, perpendicular, and anisotropic orientations to create NYAPCs. Through tribological examination, it was discovered that NYAPCs with anisotropic yarn alignment functioned as a connecting process, reducing the impacts of fissures and increasing the strand's fracture toughness (Singh et al., 2023). When working with natural and synthetic fiber-strengthened composites, investigators discovered that a specific layer order had an impact on the composite's wear properties. The ability to withstand wear and other tribological properties of the composite were enhanced as the proportion of natural fiber increased. The lower and upper limits of proportion of the reinforcement in the NYAPCs depend on the type and form of reinforcement. Due to the natural and synthetic fibers" improved interface connection, there was a decreased rate of shedding material as a result. Researchers looked at the tribological characteristics of NYAPCs using bamboo yarn augmentation in an epoxy resin with various yarn alignments (Nirmal et al., 2012). The NYAPC with anti-parallel alignment produced the most effective performance, as shown in Figure 6.8. The higher resilience to shear of bamboo yarn in the NYAPC justified the case.

6.5.5 Influence of Experimental Variables

The influence of loading on the frictional behavior of NYAPCs differs for different types of composites. In general, nevertheless, most NYAPCs exhibit a pattern of COF rising with increasing load until a certain threshold, at which it starts to decline. The COF rises with greater load due to a rise in the actual interface area, while the COF declines with a given load, which is caused by the melting of polymeric matrix under high loads (Gurunathan et al., 2015). From the perspective of composite material wear behavior, Archard's wear law states that wear loss rises with upsurge in applied load. Furthermore, durability against wear is increased by reinforcing the polymeric matrix with fibers. Numerous studies show that when sliding speed rises, the COF drops. One potential reason could be the thermal deterioration of the matrix.

The frictional behavior depends on the type of fibers and the formation of a thin polymer covering. The resin will distort plastically as speed increases. A thin polymeric film forms between the counterface and the NYAPC's touching face. This layer reduces COF and serves as a barrier to damage. Numerous investigations have established that whereas wear loss for pure polymers rises as you raise sliding speed, it decreases for NYAPCs. The weight percentage of fiber in the NYAPC does not

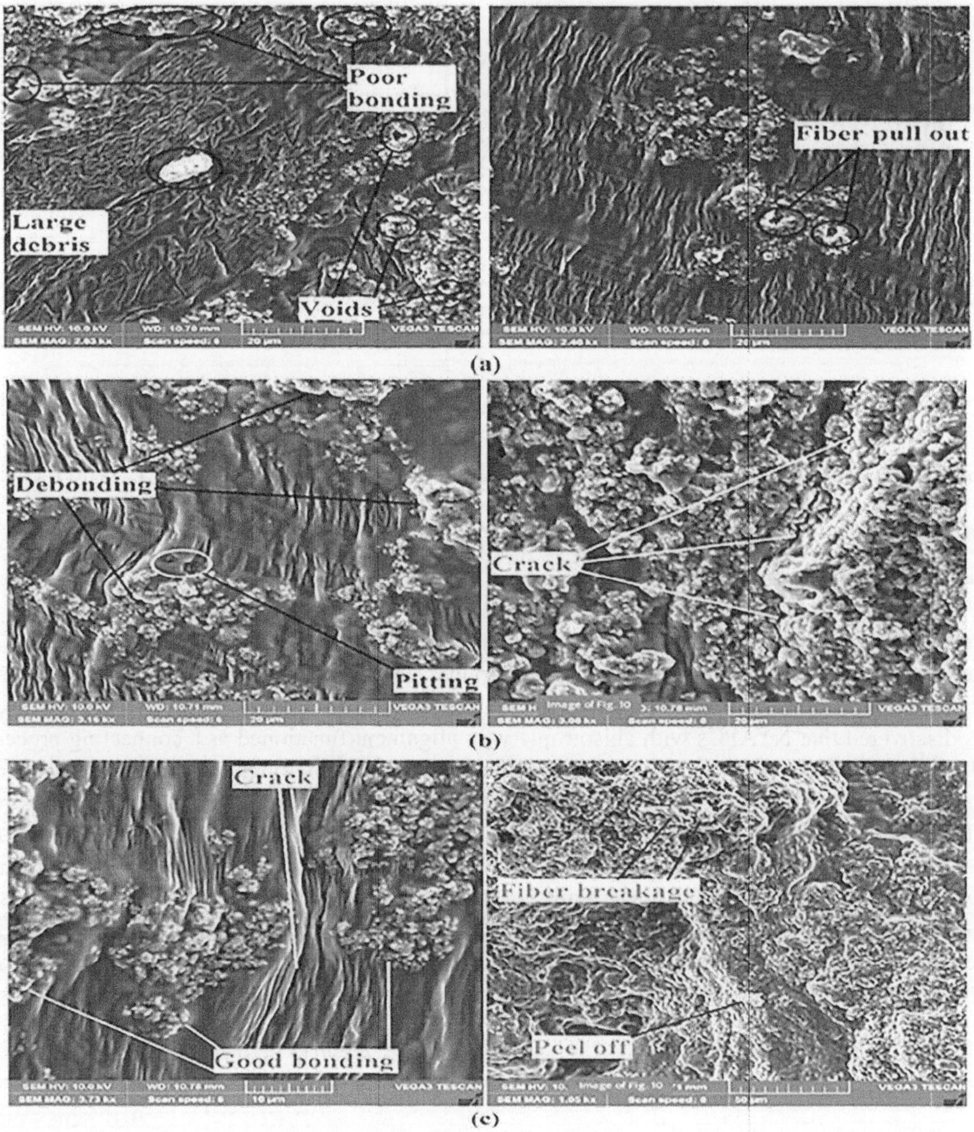

FIGURE 6.7 SEM images of damaged section modification on Coccinia indica yarn-augmented NYAPCs: (a) untreated yarn, (b) 5% alkali-modified, and (c) 3% saline water-modified Coccinia indica yarn-strengthened NYAPCs.

Adapted from Mylsamy et al., 2020.

affect the COF at greater sliding speeds. This makes sense, given that as sliding distance rises, so do the plastic deformities and delaminated region. As the length of the slide grows, so does the friction between the counter layer and the NYAPC surface, increasing the heat produced by friction between the two sides. The temperature of the NYAPC rises as a result of increased heat from friction, which aids in the material's heat softening. As a result, the resin's plastic flow increases, producing a nearly constant COF over a longer sliding distance.

The wear characteristics of epoxy augmented with Coccinia indica yarn were evaluated under diverse conditions in a distinct study (Mylsamy et al., 2020). The findings demonstrated that the frictional coefficient rose with the load until it reached 25 N, after which it decreased, as shown in

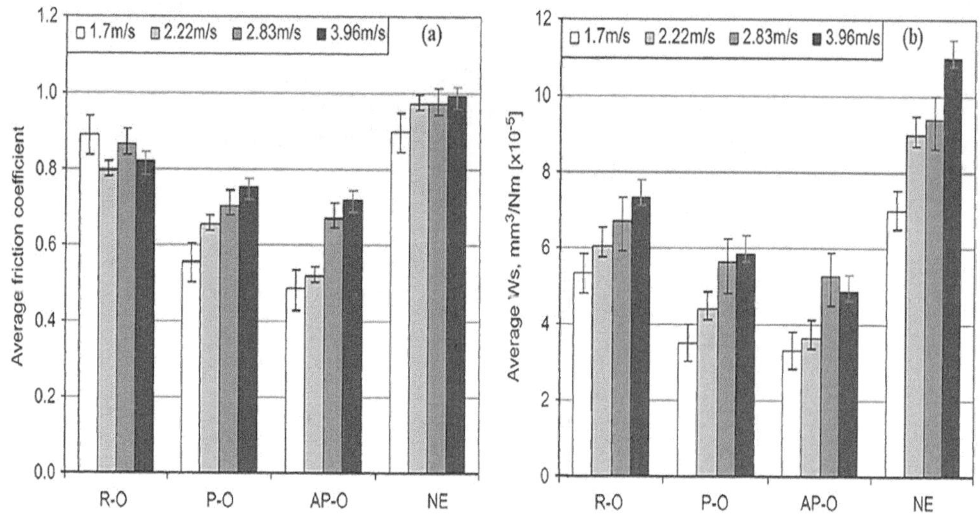

FIGURE 6.8 Influence of yarn alignment on (a) frictional coefficient and (b) wear of bamboo-armored NYAPCs *Random orientation (R-O), anti-parallel orientation (AP-O) and parallel orientation (P-O), and neat epoxy (NE).

Adapted from Nirmal et al., 2012.

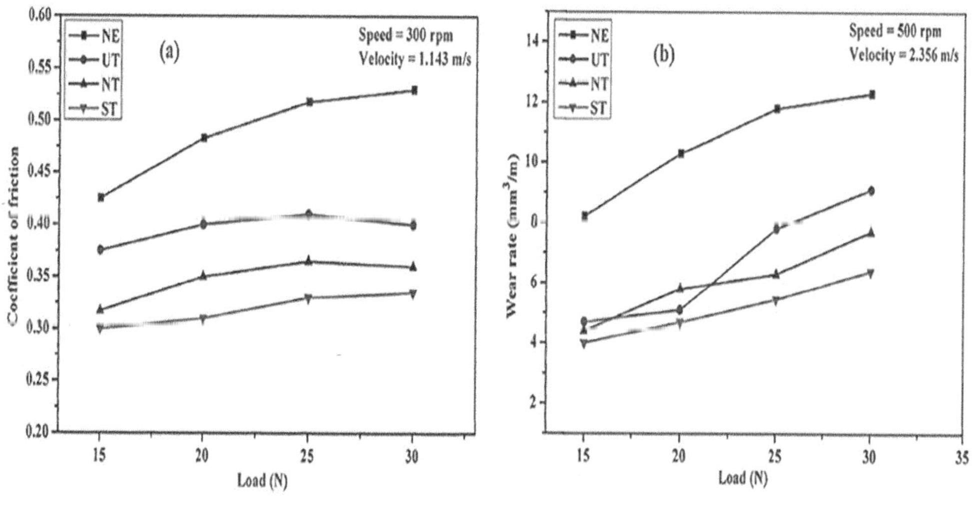

FIGURE 6.9 Influence of applied load on (a) frictional coefficient and (b) wear of Coccinia indica-armored NYAPCs.
*NaOH treatment (NT), silane treatment (ST), untreated (UT), and neat epoxy (NE).

Adapted from Mylsamy et al., 2020.

Figure 6.9 (a). As the force applied progressed, wear worsened for all produced NYAPCs, as shown in Figure 6.9 (b). The frictional coefficient value of NYAPCs at longer rubbing distances is almost identical, regardless of reinforcement weight percentage, according to studies on the tribological properties of polylactic acid augmented with organic yarns, as shown in Figure 6.10 (a) and (b) (Yadav et al., 2023). Furthermore, a decrease in frictional coefficient with rise in rubbing speed was observed for most of the NYAPCs, as shown in Figure 6.10(c).

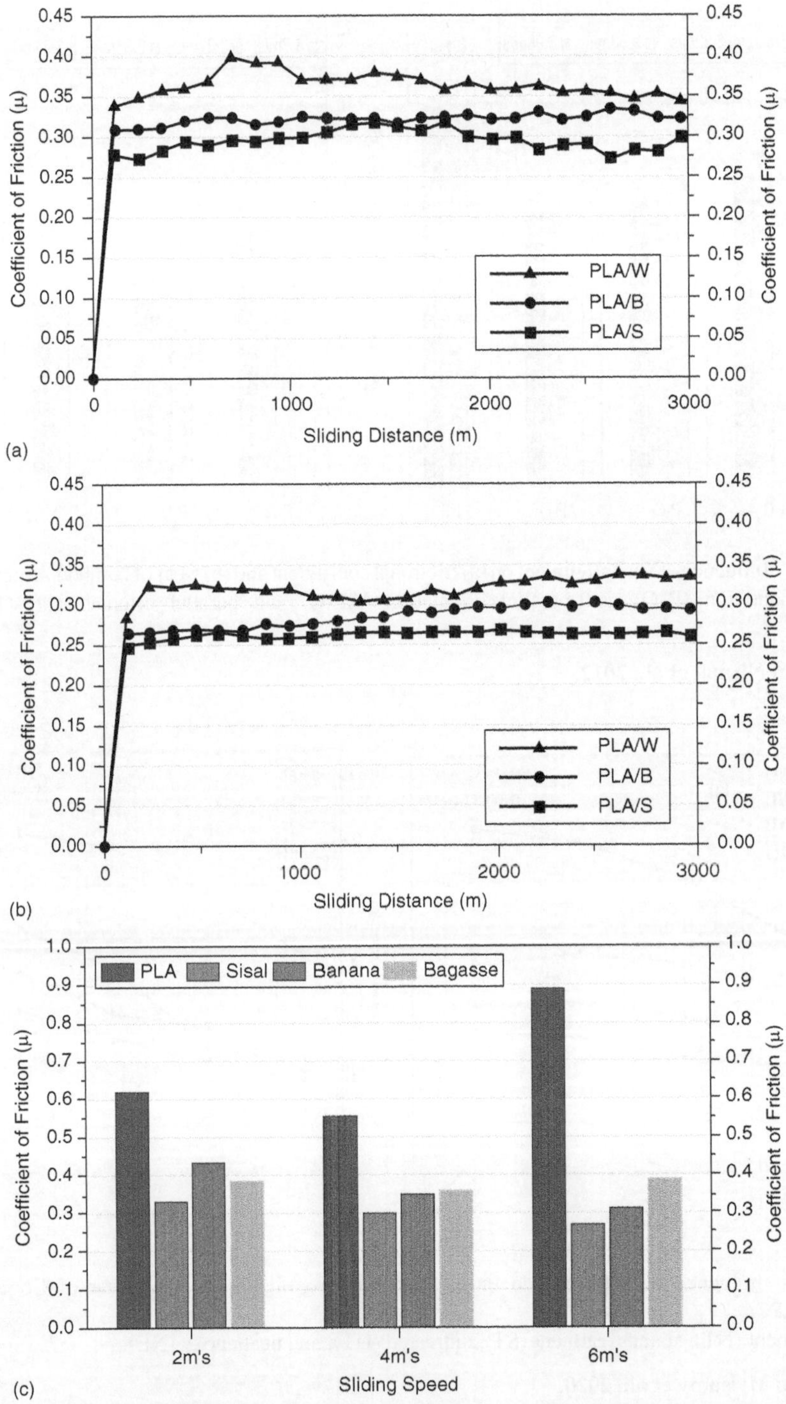

FIGURE 6.10 Frictional coefficient of polylactic acid NYAPCs: (a) 10 wt.% reinforcement, (b) 20 wt.% reinforcement, and (c) varied rubbing velocities.

*Polylactic acid with bagasse reinforcement (PLA/W), polylactic acid with banana fiber reinforcement (PLA/B), and polylactic acid with sisal fiber reinforcement (PLA/S).

Adapted from Yadav et al., 2023.

6.6 TRIBOLOGICAL APPLICATIONS, LIMITATIONS, AND FUTURE SCOPE OF NYAPC

NYAPCs have a number of benefits over other materials, but their application is limited by a few disadvantages. The incorporation of natural yarns into synthetic resin has several obstacles, such as poor mechanical qualities, retention of moisture, resilience, and insufficient bonding between the yarn and resin, as shown in Figure 6.11. These difficulties are especially noticeable when NYAPCs are subjected to outside environmental factors, which raises serious questions about their durability and effectiveness over time. Particularly, microbes have the ability to cause NYAPCs to biodegrade. First, microbes destroy the hemicellulose in the yarn and hydrolyze it into molecules that may be digested. Since NYAPCs are hydrophilic by biological terms, they attract humidity from the outside environment. It is based on the quantity of amorphous constituents and the quantity of cavities in organic yarns.

UV rays deteriorate NYAPCs when they are subjected to outside conditions. It is the lignin portion of strand that is causing the climate damage (Ibrahim et al., *2019*). The exposure of NYAPCs to higher temperatures can damage the natural yarns in the NYAPCs. The hemicellulose and lignin are the first components in the organic yarn that get damaged. The uniformity and quality of the strands used in the NYAPCs are important additional issues. NYAPCs are employed currently in industries in tribological applications like braking, bearing, and clutches. The above-discussed hurdles can be overcome to some extent by proper surface modification of natural yarn before being augmented in NYAPCs, adapting appropriate manufacturing technique, and maintaining the quality of the natural yarn during the extraction process. This may widen the possibility of utilizing the NYAPCs in other industrial tribological applications.

6.7 SUMMARY

NYAPCs' exceptional tribological qualities have potential uses in a variety of modern industries, including automotive and aerospace sectors. Due to their comparable mechanical qualities to synthetic fibers, advantageous thermal and acoustic attributes, eco-friendly processing, and economical manufacture, natural fibers are a strong option for augmentation in polymeric composites. These characteristics make NYAPCs very attractive materials, especially for applications related to tribology. Owing to its noteworthy characteristics, low frictional behavior, intrinsic self-lubricating qualities, and admirable durability against wear, NYAPCs have clearly acquired significant acceptance in an array of tribological purposes. However, there are still a lot of unanswered questions in this subject,

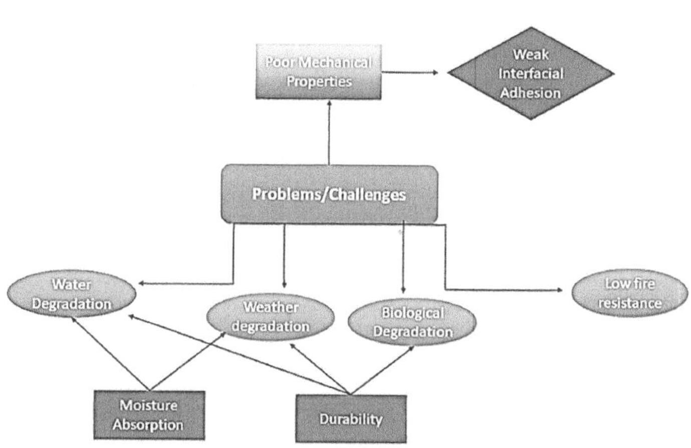

FIGURE 6.11 Hurdles in utilizing NYAPCs.

Adapted from Bhat et al., 2023.

ranging from basic science questions to difficult engineering design problems. The complexity stems from the wide range of polymeric and polymeric-based NYAPCs that are used, the variety of applications that they are applied to, and the inherent difficulties in figuring out the tribo-mechanical principles that underlie their behavior. This chapter provided a thorough discussion of the numerous tribological characteristics of NYAPCs, with an emphasis on their practicality, long-term viability, and economics. This will help the researchers continue to progress the field of NYAPCs.

REFERENCES

Ben Samuel J, Julyes Jaisingh S, Sivakumar K, Mayakannan A V and Arunprakash V R. 2021. Visco-elastic, thermal, antimicrobial and dielectric behaviour of areca fibre-reinforced nano-silica and neem oil-toughened epoxy resin bio composite. *Silicon*. 13. Springer: 1703–1712.

Bharath Kurki Nagaraja, Binoj Joseph Selvi, Mansingh Bright Brailson, Manjunath Gangaplara Basavarajappa, Raghu Gowdru Veerabhadrappa, Siengchin Suchart and Sanjay Mavinkere Rangappa. 2023, "Effect of stacking sequence and interfacial analysis of biomass sheep wool/glass fiber reinforced epoxy biocomposites." *Biomass Conversion and Biorefinery*. Springer: https://doi.org/10.1007/s13399-023-03918-2

Bhat Asrar Rafiq, Kumar Rajiv and Murali Prasanna Kumar S.. 2023. "Natural fiber reinforced polymer composites: A comprehensive review of tribo-mechanical properties." *Tribology International*, 189. Elsevier: 108978.

Brailson Mansingh B, Binoj J S, Hassan Shukur Abu, Mariatti M, Siengchin Suchart, Sanjay M R and Bharath K N. 2022. "Characterization of natural cellulosic fiber from cocos nucifera peduncle for sustainable biocomposites." *Journal of Natural Fibers*. 19(14). Taylor & Francis: 9373–9383.

Britto Antony Sagai Francis, Prabha Narayanan Rajesh, Mansingh Bright Brailson, David Rajayyan, Singh Amirthaiah Amala Mithin Minther and Binoj Joseph Selvi. 2023. "Extraction and characterization of Dypsis Lutescens peduncle fiber: Agro-waste to probable reinforcement in biocomposites - A sustainable approach." *Biomass Conversion and Biorefinery*. Springer: https://doi.org/10.1007/s13399-023-04950-y

Gargol Mateusz, Klepka Tomasz, Klapiszewski Łukasz and Podkościelna Beata. 2021. "Synthesis and thermo-mechanical study of epoxy resin-based composites with waste fibers of hemp as an eco-friendly filler." *Polymers*. 13. MDPI: 503–516.

Gurunathan T, Mohanty S and Nayak S K. 2015. "A review of the recent developments in biocomposites based on natural fibres and their application perspectives." *Composites Part A: Applied Science and Manufacturing*. 77. Elsevier: 1–25.

Hariharan Gajendiran, Krishnamachary Pattipati Chinna, Binoj Joseph Selvi and Mansingh Bright Brailson. 2023. "Influence of SiC nanoparticles reinforcement in areca/tamarind hybrid biopolymer composites: Thermo-mechanical, tribological and morphological features." *Journal of Bionic Engineering*. 20. Springer: 1723–1736.

Ibrahim M A, Hirayama T and Khalafallah D. 2019. "An investigation into the tribological properties of wood flour reinforced polypropylene composites." *Materials Research Express*. 7(1). IOP Science: 015313.

Karthik Krishnasamy, Prakash Jayavelu Udaya, Binoj Joseph Selvi and Mansingh Bright Brailson. 2022. "Effect of stacking sequence and SiC nanoparticles on properties of carbon/glass/kevlar fiber reinforced hybrid polymer composites." *Polymer Composites*, 43(9). Wiley: 6096–6105.

Li Peilong, Zhang Zhaozhu, Yang Mingming, Yuan Junya and Jiang Wei. 2021. "Synchronously improved thermal conductivity and tribological performance of self-lubricating fabric liner composites via integrated design method with copper yarn." *Tribology International*. 164. Elsevier: 107204.

Mylsamy B, Chinnasamy V, Palaniappan S K, Subramani S P and Gopalsamy C. 2020. "Effect of surface treatment on the tribological properties of Coccinia Indica cellulosic fiber reinforced polymer composites." *Journal of Materials Research and Technology*. 9(6). Elsevier: 16423–16434.

Nagaprasad Nagaraj, Stalin Balasubramaniam, Vignesh Venkataraman, Ravichandran Manickam, Rajini Nagarajan and Ismail Sikiru Oluwarotimi. 2021. "Applicability of cellulosic-based Polyalthia longigolia seed filler reinforced vinyl ester biocomposites on tribological performance." *Polymer Composites*. 42. Wiley: 791–804.

Nanda Bishnu Prasad and Satapathy Alok. 2021. "An analysis of the sliding wear characteristics of epoxy-based hybrid composites using response surface method and neural computation." *Journal of Natural Fibers*. 18. Taylor & Francis: 2077–2091.

Nayak Sandip Kumar and Satapathy Alok. 2020. "Wear analysis of waste marble dust-filled polymer composites with an integrated approach based on design of experiments and neural computation." *Proceedings of the Institution of Mechanical Engineers, Part J: Journal of Engineering Tribology*. 234. Sage: 1846–1856.

Nirmal U, Hashim J and Low K O. 2012. "Adhesive wear and frictional performance of bamboo fibres reinforced epoxy composite." *Tribology International*. 47. Elsevier: 122–133.

Ponnusamy Dharani Arasangudi, Gajendiran Hariharan, Mansingh Bright Brailson and Binoj Joseph Selvi. 2023. "Diffractional, spectroscopical, morphological and thermal analysis of pretreated/enzyme modified cellulosic Cocos nucifera L. peduncle fiber." *Biomass Conversion and Biorefinery*. Springer: https://doi.org/10.1007/s13399-023-05076-x

Selvan Martin Thangaraj Gerald Arul, Binoj Joseph Selvi, Mansingh Bright Brailson and Sajin Justin Abraham Baby. 2023. "Physico-chemical properties of alkali treated cellulosic fibers from fragrant screw pine prop root." *Journal of Natural Fibers*. 20(1). Taylor and Francis: 2129897.

Singh Neha and Sinha Sujeet K. 2021. "Tribological performances of hybrid composites of epoxy, UHMWPE and MoS_2 with in situ liquid lubrication against steel and itself." *Wear*. 204072. Elsevier: 486–487.

Singh Manoj Kumar, Tewari Renu, Zafar Sunny, Rangappa Sanjay Mavinkere and Siengchin Suchart. 2023. "A comprehensive review of various factors for application feasibility of natural fiber-reinforced polymer composites." *Results in Materials*. 17. Elevier: 100355.

Subramanian M, Arunkumar N, Rethnam George Sahaya Nixon and Balasubramanian T. 2021. Exploration of the damping characteristics of basalt hybrid composites reinforced with natural fibers and epoxy resin. *Fibers and Polymers*. 22. Springer: 1684–1692.

Thooyavan Y, Kumaraswamidhas L A, Edwin Raj R, Binoj J S and Brailson Mansingh B. 2022. "Effect of combined micro and nano SiC particles addition on mechanical, wear and moisture absorption features of basalt bidirectional mat/vinyl ester composites." *Polymer Composites*. 43(5). Wiley: 2574–2583.

Xin X, Xu C G and Qing L F. 2007. "Friction properties of sisal fibre reinforced resin brake composites." *Wear*. 262(5–6). Elsevier: 736–741.

Yadav V, Sarbjit S, Neeru C, Mohinder PG, Shubham S, Amit K, Changhe L and Elsayed Mohamed T E. 2023. "Dry sliding wear characteristics of natural fibre reinforced polylactic acid composites for Engineering applications: Fabrication, properties and characterizations." *Journal of Materials Research and Technology*. 23. Elsevier: 1189–1203.

7 Thermal Analysis and Properties of Natural Fiber Composites for a Sustainable Environment

Mariya Mathew, Rosemary Johnson, Hanna J. Maria, Karen Ter-Zakaryan, and Sabu Thomas

7.1 INTRODUCTION

The global environmental concerns and awareness regarding biocomposites necessitate the development of renewable as well as eco-friendly composites for industrial and other applications. It's the need of today to replace petroleum- or glass-based composites with biocomposites for a healthy, sustainable environment. The inclusion of natural fiber as a reinforcing tool in polymer composites seeks importance as it ensures the sustainability of the ecosystem. Basically, natural fibers are naturally occurring hair-like strands obtained from plant stems, branches, leaves, fruits, and animals, mainly used for manufacturing clothes, rope, paper, etc. The remarkable thermal and insulating characteristics of natural fiber polymer composites (NFPC) facilitate enormous applications in different sectors without harming the environment. Moreover, the fiber-loaded composites are economical, and thermal recycling is possible, which is not achievable for the glass composites.

Fibers are a group of hair-like matter that are continuous filaments or distinct stretched parts. They can be spun into ropes, threads, etc. (Chandramohan & Marimuthu, 2011). There are mainly two varieties of fibers, which are either manmade or natural, based on their source. Natural fibers are mainly from mineral, plant, and animal sources. Vegetable fibers are primarily comprised of cellulose, which is ultimately used for the production of paper and clothes. Stalk fibers mainly occurred in the straws of rice, wheat, barley, etc. The conventionally used fibers include cotton, hemp, and flax, as well as jute, coconut, kenaf, and sisal. Hemp fibers are typically utilized for aerofoils and ropes due to their greater elasticity and resilience in a hostile environment.

The natural fibers started a new era in polymer composites by elaborating their applications to various fields. Although synthetic fibers possess superior dimensional and other characteristics, the natural fibers are nonabrasive, light in weight, cost-effective, combustible, non-corrosive, and recyclable, which makes them a promising filler in the polymer blends (Kalia et al., 2009). Compared with synthetic fibers, just 17% of energy is used in the manufacture of natural fibers (Shalwan & Yousif, 2013). It is noticed that the natural fiber-reinforced polymer composites are gaining interest in engineering and construction domains owing to their minimum density, renewability, superior specific strength, biodegradability, and improved tool life (Kumar et al., 2019). The cellulose concentration, microfibrillar angle, and aspect ratio of the fiber all contribute to its strength. Strength increases with cellulose content, whereas the inverse is true for microfibrillar angle. The other factors that determine the physical parameters are matrix-filler adhesion, strength and modulus of fibers and stiffness of the matrix, dispersion of fillers, etc. The mechanical behavior, especially tensile strength, is very much affected by the volume fraction of the fibers (Kumar et al., 2019).

DOI: 10.1201/9781032665337-7

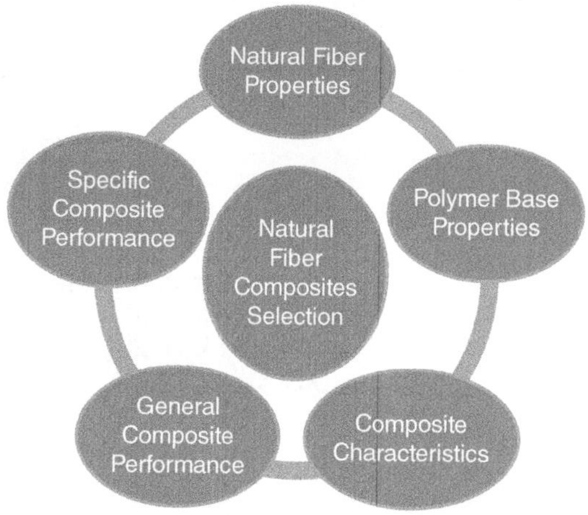

FIGURE 7.1 Schematic representation of the factors influencing the properties of the natural fiber composite materials (Al-Oqla & Sapuan, 2014).

In order to expand the application of natural fibers to industrial domains, factors such as transport properties, thermal behavior, and tribological features are also crucial. The matrix-fiber adhesion is a key parameter for superior mechanical properties. The physico-mechanical properties may deteriorate due to the poor adhesion at the phase boundary of the composites. The crucial challenge associated with natural fibers is their high value of sorption parameters, which leads to minimizing the mechanical properties of the composites (Faruk et al., 2012). The environmental factors, such as sunlight, the water uptake during the plant growth, rain, and soil, also influence the efficiency of natural fibers. In addition, the thermal behavior of the NFPC is as vital as the mechanical parameters. Figure 7.1 shows an overview of the selection criteria associated with the NFCs.

The thermal stability and the mechanical parameters are the principal elements in order to analyze its applicability in various sectors. The degradation temperature and thermal stability are determined since the manufacturing procedure largely depends on the degradation and processing temperature; the thermal studies gained more attention in the natural fiber composites (Neto et al., 2021). The fibers started getting degraded at higher temperatures, which alters the physical-mechanical properties of the elaborated composites. A wide range of thermal analysis techniques are generally exploited in order to examine the thermal behavior of the composites such as thermogravimetric analysis (TGA), dynamic mechanical thermal analysis (DMTA), thermal mechanical analysis (TMA), differential scanning calorimetry (DSC), differential thermal analysis (DTA), cone calorimetry, pyrolysis combustion flow calorimeter (PCFC), and limited oxygen index (LOI).

Studies have been conducted to enhance the mechanical, thermal, and other properties by proper surface modifications before the fabrication of the composites. According to reports, surface alterations increase the bonding, wettability, and roughness of the matrix-fiber interface while lowering the fibers' moisture absorption (Beg & Pickering, 2008; Zhou et al., 2016). Ultraviolet (UV), plasma, corona, temperature, radiation beams, and fiber beating are the most employed physical methods for surface treatments. The aforementioned process alters the surface characteristics of the NFs without affecting the structural composition, and thus it increases the adhesion between matrix and fibers (Latif et al., 2019).

Similarly, chemical treatments are widely employed to minimize fiber water absorption and boost the interactions between the fiber matrix through chemical bonding or mechanical interlocking at the interface. This is achieved by using compounds such as sodium hydroxide (de Araujo Alves Lima et al., 2020) and silane (Rout et al., 2001), which stimulate the interaction between matrix and fiber via chemical coupling.

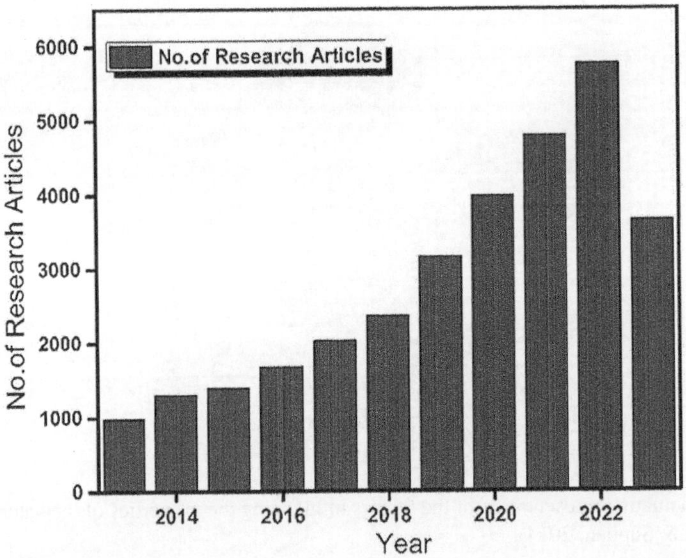

FIGURE 7.2 Graphical representations of the research articles in the past 10 years on thermal properties of natural fiber composites.

The thermal properties are considered to be an inevitable part of the study of natural fiber composites. Research has also been conducted only on the thermal parameters of polymer composites with natural fibers. Esra Erbas et al. investigated the variations in the thermal properties of natural fiber blend of kenaf, flax, and hemp along with polyamide. The crystallization, glass transition, and melting temperature were found to be increased during the DSC analysis (Kiziltas et al., n.d.). Also, the natural fiber blend enhances the storage modulus obtained from dynamic mechanical thermal analysis owing to the reinforcing efficiency of the natural fibers. A study conducted on the thermal degradation of NF-loaded polypropylene composites by evaluating TGA, DTA, and DSC, where 50% of the composite consists of flour, newsprint, rice hulls, and kenaf fibers (Tajvidi & Takemura, 2009). The effect of each component's contribution to thermal parameters was also examined. The thermal tests showed that the rice hulls were the least thermally stable among the considered natural fibers in the polypropylene matrix. Similarly, a study explored the DMTA of polymer composites reinforced with natural fibers (Costa et al., 2016). The review focused on the viscoelastic properties of composites by evaluating DMTA and explained the importance of DMTA in the correlation between structure and relationships. Another review was carried out on the thermal characterizations of natural and hybrid filler composites by Jorge et al. (Neto et al., 2021). The study broadly discussed a wide range of thermal analytical techniques in order to evaluate the thermal properties for extending their use to various applications.

Figure 7.2 depicts research publications in the years 2013–2023. The plot, however, implies the importance of thermal parameters of natural fiber composites. It is ensured that the research on the thermal properties of fibers is vital in the current research scenario. Accordingly, the chapter overview of the thermal properties, flammability factors, and thermal properties of natural fiber-reinforced composites.

7.2 PREPARATION OF NATURAL FIBER-BASED COMPOSITES

7.2.1 Compression Molding

Compression molding is considered a leading methodology compared to other molding processes. There are various advantages to this technique, such as low cycle time, high reproducibility, economic, and minimum waste generation. The processing of the natural fiber composites hangs on

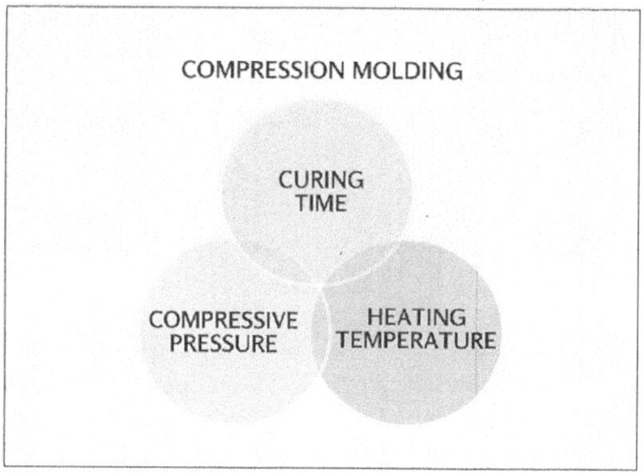

FIGURE 7.3 Compression molding parameter.

several elements, including the type and content of natural fiber, its length, and the moisture content present in it (Kerni et al., 2020). The preparation of natural fiber composites using compression molding mainly depends on the three basic process parameters (Figure 7.3):

- Curing time
- Heating temperature
- Compressive pressure

The whole process of compression molding will rely on the operating parameters, which promise the composite material much enhanced properties (Yallew et al., 2020). Barthod-Malat et al. utilized hot compression molding containing double plates for the fabrication of flax/polypropylene (PP) nonwovens. This method provides the advantage of pressing the samples with two successive cycles. The first cycle provides a higher temperature, whereas the second provides a lower temperature, preferably room temperature. The core temperature of flax/PP nonwovens was determined; after that, they were thermo-compressed. They were hot pressed under the temperature range of 200°C, 210°C, 220°C, 230°C, and 240°C, but the targeted temperature was 180°C and 200°C. This was carried out using 100 bar pressure along with a Teflon sheet on either side. The core temperature was determined using a type K thermocouple. After this, they are kept between cooled plates (room temperature) with a pressure of 100 bars for 40s (Barthod-Malat et al., 2023). Boccarusso et al. created PP composites reinforced with woven mesh textiles of flax and hemp. They prepared composites by stacking the layers of fiber and PP alternatively with a stacking sequence of $(0/90)_4$. A conventional compression molding method was applied for all samples. The heating temperature was 200°C, compressive pressure was set to 4MPa, and the time was 300s. They followed two strategies for molding their composites. First, the waiting time was considered as the mold plate temperature within the composite samples in-between the thermocouples. Second, they didn't set a molding temperature but started the process after acquiring a temperature of 200°C (Boccarusso et al., 2022).

7.2.2 Extrusion

Another fascinating method followed by researchers is extrusion. Extrusion is all about melting pellets of thermoplastic polymers, followed by mixing them with natural fiber using either one or two rotating screws, and the final composite comes out of the die (Rajeshkumar, Seshadri, et al., 2021). For the preparation of fiber-reinforced polymer composite by extrusion, a hot-melt method is followed. The thermal and chemical properties of the composite can be improved by this method of processing. In order to follow this method, there are certain process parameters that are to be taken into account, which are depicted in Figure 7.4.

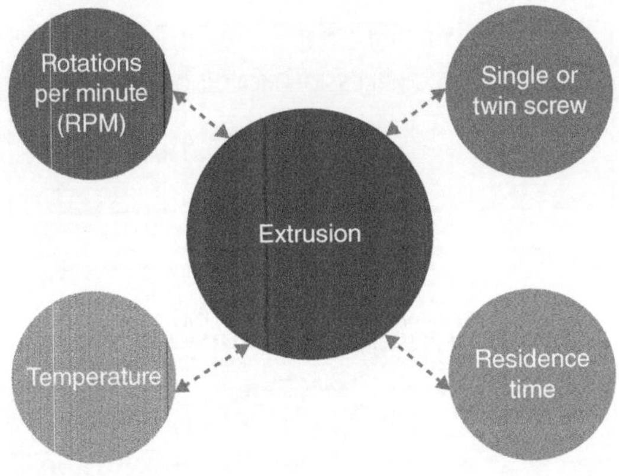

FIGURE 7.4 Process parameters of extrusion.

The structure of the natural fibers is considered significant in the properties of the final material. The dispersion, aspect ratio, and spatial orientation play a vital role in it (Gallos et al., 2017). This method of fabrication helps in the upliftment of stiffness, strength, and impact strength of the composites prepared (Adeniyi et al., 2019).

7.2.3 Injection Molding

In injection molding, as the name suggests, the hot-melt thermoplastic polymer is injected into a mold having a desired shape. This method helps in the proper dispersion of fibers into the matrix, thereby improving the tensile and flexural properties (Adeniyi et al., 2019). Nowadays, most of the processing technologies that fabricate natural fiber-reinforced thermoplastic composites are developed by using the injection molding method (Sun et al., 2023). There are various injection molding parameters, which are depicted in Figure 7.5.

Aliotta et al. developed a composite using ultra-short cellulosic arbocel fibers and polybutylene succinate (PBS). They prepared the specimens by controlling the temperature of the nozzle, and the material that came out of the mold was cooled by water. To remove the product from the mold, it was kicked with extractor pins with a specific pressure (Aliotta et al., 2022). Dolza et al. used injection molding to create environmentally acceptable wood plastic composites using eco-friendly polyethylene and short natural fibers. In this process, they followed a temperature of 140°C in the hopper

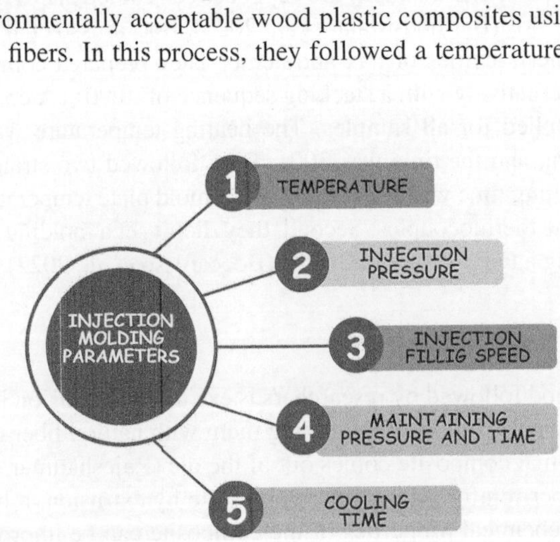

FIGURE 7.5 Injection molding parameters.

and 150°C, 155°C, and 160°C in the injection nozzle. The cavity filling and cooling times were set to 1s and 10s. A clamping force of 75 tons was applied, and finally a sample of average thickness of 4mm was obtained (Dolza et al., 2021).

7.2.4 Resin Transfer Molding

These days, the use of resin transfer molding (RTM) is most popular due to its rich fiber and fewer voids, better surface finish, required process control, and low cost. The significant advantage of the RTM is that it is capable of producing high-strength composites. One feature of this approach is that the primary processing conditions, especially pressure and temperature, may be changed (Suresh Babu et al., 2019). This method utilizes resins such as epoxy, vinyl ester, phenolic, and polyester (Sapuan & Yusoff, 2015).

7.2.5 Vacuum-Assisted Resin Transfer Molding

Vacuum-supported RTM is a closed mold process similar to RTM. Through this process, it is possible to fabricate large and complex-shaped materials. The great advantage is the low emission of volatile organic compounds (Chandgude & Salunkhe, 2022). This method is unique for the injection of epoxy resins into the molded preform and for studying various curing operations. After keeping the molds of RTM in an airtight manner, the mixed resin/hardener was injected from one side. Simultaneously, for the proper distribution of resin/hardener, a vacuum was applied on the opposite side, which effectively minimizes the air bubbles. The mixing ratio of the resin+hardener/fiber is 13/7 (w/w), respectively. The curing process was done using cartridge heaters (Sarikaya et al., 2019).

7.2.6 Solvent Casting

Solvent casting is also known as solution casting or wet processing method. The preparation of natural fiber composites is easily possible through the solvent casting method. The main highlight of this technique is the ease of fabrication does not need a specialized instrument. The main steps involved in solution casting are the mixing of the solubilized polymer matrix along with the fibers mechanically. After casting it, the solvent or solution is evaporated or dried. The solvent used in this method can be water, alcohol, or any organic solvent (Kong et al., 2015). Mostly, the solvent used is water, which is eco-friendly. Also, this method provides good dispersion of fibers inside the polymer matrix without degrading (Mochane et al., 2019).

7.3 FACTORS AFFECTING NATURAL FIBER COMPOSITES

7.3.1 Fiber Length

The most important factor that influences the characteristics of natural fiber composites is fiber length. The length of the fiber must be longer than the critical fiber length for optimal reinforcement of the fiber in the matrix. The critical fiber length $\left(L_c\right)$ is expressed by the equation:

$$\frac{L_C}{d} = \frac{\sigma_f}{2\tau}$$

where d, σ_f, τ are fiber diameter, tensile strength of fiber, and interfacial strength, respectively. Aspect ratio (length/diameter) has a significant effect on the characteristics of natural fiber-reinforced polymer composites. Cross-section of fibers has an influence on the inaccuracy of the measurements (Jariwala & Jain, 2019). The fabrication techniques like extrusion, injection, or compression molding can affect the reinforcement geometry. Good reinforcement and efficient load transfer between the fiber/matrix interface are possible if it is above the critical fiber length. But when the length of fiber is below the critical fiber length, fibers will be debonded and lack efficient load transfer (Duigou et al., 2020). The temperature and pressure that are applied during the manufacture of natural fiber-reinforced polymer composites initially cause the fiber to break. Another reason is the collision of fibers among themselves and between the tool/mold. This decreases the length of fibers, eventually leading to less strength in composites. Sometimes this may be the reason for causing a defect in the composite (Balla et al., 2019).

7.3.2 Fiber Loading

Fiber loading has a great influence on the tensile properties of the composites, such as strength, modulus, and percentage of elongation. Senthilkumar et al. have reported that the ductility of the pineapple leaf fiber/polyethylene composites can be improved by enhancing the loading of the fiber. The tensile modulus also showed a substantial rise with the fiber concentration. The compressive stress was noted to be improved on fiber loading, which indicates the interface adhesion between the fiber and matrix. The flexural strength of the composites can be increased by the addition of a lower wt% of fiber loading (Senthilkumar et al., 2019). Mohanty et al. reported that as the fiber loading increased to 30%, there was a rise in the mechanical properties, and after that, it exhibited a decline in the properties. High stress transfer was possible at the interface by the elevation of fiber loading as this boosted the storage modulus (Prakash et al., 2022).

7.3.3 Fiber Orientation

The orientation of the fiber is a crucial element in evaluating the properties of natural fiber composites. The mechanical characteristics of composites are mostly influenced by this component. The load is equally transferred to the fiber and matrix owing to its fiber orientation. There are two major classifications for fiber orientations, and they are as follows:

- Aligned orientations
- Random orientations

Aligned orientations are again classified as follows:

- Uni-directional
- Bi-directional

These two classifications have many differences, which are listed in Table 7.1:

The best method for the enhancement of mechanical properties is to create a good fiber orientation (Aravinth et al., 2022). For the injection molding process, the randomly oriented fiber composites are

TABLE 7.1
The Differences between Uni-Directional and Bi-Directional Orientations

Properties	Uni-Directional	Bi-Directional
Fiber orientation	Only in one direction	Two or more directions or stacking each other
Angle of orientation	• Longitudinal (0°) • Transverse (90°) • Inclined (45°)	• 0°/90° • 90°/45° • 0°/45°

the main flow. Resin flowing pattern, viscosity of resins, and shape of products are the factors that determine the fiber orientation. The fiber orientation changes according to the injection molding process, which eventually leads to the uneven strength of final products. Another factor that determines the fiber orientation is the surface tension and roughness of the barrel. This causes the fibers to align toward the wall rather than to concentrate on the center (Lau et al., 2018).

7.3.4 Fiber Layer Sequence

Ramesh and Anand explored the impact of fiber sequencing on flexural and tensile parameters while working on Kevlar/basalt fiber-reinforced hybrid composites. The best sequence for tensile and flexural properties was observed for Kevlar-basalt-Kevlar composite (Ramesh & Anand, 2021). The performance of natural fiber-reinforced composites is significantly affected by the order of the fiber layers. The order of the fiber layers and the fiber volume percentage both influence the composites' mechanical characteristics. Fiber sequence has a slight effect on tensile properties, but other properties such as flexural properties are greatly affected. Sarasini et al. reported that fiber sequence has a remarkable role in the impact performance. They studied two configurations of the hybrid composites using flax and carbon fibers. First, the flax fiber acts as the skin layer and carbon in the middle (FCF), and the other configuration was its opposite (CFC). The FCF configuration exhibited good impact performance due to its ability to withstand the crack that might form in the hybrid composite (Malik et al., 2021).

7.3.5 Fiber/Matrix Interfacial Bonding

The role of fiber-matrix interaction in the performance of a composite is noteworthy. Poor interfacial interaction is shown by the combination of a hydrophilic natural fiber and a hydrophobic matrix. This is because of the failure of proper stress transfer between the fiber and matrix. A good interfacial bonding brings out good results by reducing the void space. If the surface between the fiber and matrix is rough, they can create a mechanical interlock within themselves, thereby improving interfacial adhesion (Rocha et al., 2022). When natural fiber-reinforced composites come into contact with water, the interfacial interaction decreases as there is the formation of intermolecular hydrogen bonding with the fibers. The performance of the composite decreases as there is swelling in that region, causing microcracks. Therefore, the modification of fibers and matrix has equal importance in revamping the interaction between the fiber and matrix. Both physical and chemical modifications help to ameliorate the interaction between them. Fiber dispersion is another key factor that reduces the voids, leading to excellent bonding interactions (Ramachandran et al., 2022).

7.4 THERMAL PROPERTIES

The thermal characteristics of composites are crucial in defining their demand for a variety of applications. Thermal characteristics, such as DSC, thermal conductivity (K), thermogravimetric analysis (TGA), TMA, and dynamic mechanical analysis (DMA), are covered in this chapter.

7.4.1 Thermogravimetric Analysis

TGA is a conventional thermal analytical technique used for environmental, pharmaceutical, food, and petrochemical applications. In this analytical method, the mass of the sample is monitored with temperature or time. The sample is maintained at a constant temperature, or sometimes the sample is subjected to nonlinear temperature, as in sample TGA (STGA). Typically, samples are weighed at 5 to 10 mg and placed on a clean ceramic plate and heated at a constant rate and set temperature to record the thermal stability data (Yang et al., 2019). TGA calculates the mass of the sample when it is heated or cooled in a furnace. A sample is constantly weighed while heating, as an inert gas of the atmosphere is passed over it. In this method, the temperature choice depends largely on the information required for the sample. Moreover, a reactive, oxidizing, and inert atmosphere is also used in TGA experiments. There are also chances to vary the atmosphere during the analysis (Gabbott Paul, n.d.). The weight loss curves discovered by the study reveal details on alterations in the sample's composition, kinetics of chemical reactions, and thermal stability (Nurazzi Norizan et al., n.d.).

Han-Seung Yang et al. exploited TGA as an appropriate tool to evaluate the thermal stability of the microcrystalline cellulose nanofiber-based polypropylene composites. The study explained the dispersion behavior of cellulose in the polymer backbone with the help of residual mass calculation using TGA analysis (Yang et al., 2011). It was noted that the residual mass increases with the filler loading and concluded that TGA is a viable method to analyze the thermal characteristics of the cellulose-loaded thermoplastic composites.

Research was carried out in the development of green tires by incorporating cellulose nanofibers from Cuscuta refluxa in a natural rubber matrix, in which TGA was the potential tool to analyze the thermal performance of the fabricated composites by Midhun et al. (Dominic et al., 2020). It was observed that the inclusion of cellulose nanofibers enhances the onset of degradation temperature owing to the delay of the thermal degradation. Another study exploited the packaging application with Moringa oleifera fruit fibers/polyethylene terephthalate (MOF/PET) composites, where thermal degradation was obtained using the TGA curve displayed in Figure 7.3 (Nayak & Khuntia, 2019). The thermal stability seemed to be increased upon the addition of fiber, and it was noted that the inclusion of 20 wt% of fiber showed maximum thermal stability among the prepared composites. Figure 7.6 presents the TGA thermogram of neat PET and PET/MOF composite at 20 wt% of fiber loading. The thermograph clearly demonstrates that the composite degradation is a three-staged process. The initial weight loss occurred at a temperature range of 30–230°C, which corresponds to the elimination of water molecules. Weight loss may be the cause of the second stage of degradation, which begins at 230°C for composite and 250°C for matrix, respectively. Further observation shows that the degradation of the PET matrix was found to be at 480°C, while the overall degradation of the composite occurred at 540°C. Beyond 540°C, no weight loss was noticed. Consequently, it can be inferred that compared to neat PET, the PET/MOF composite exhibits greater heat stability. The thermal characteristics of natural cellulosic fibers from purple bauhinia (PBFs) were studied for prospective reinforcement in polymer composites in order to investigate their thermal properties and increase their use in

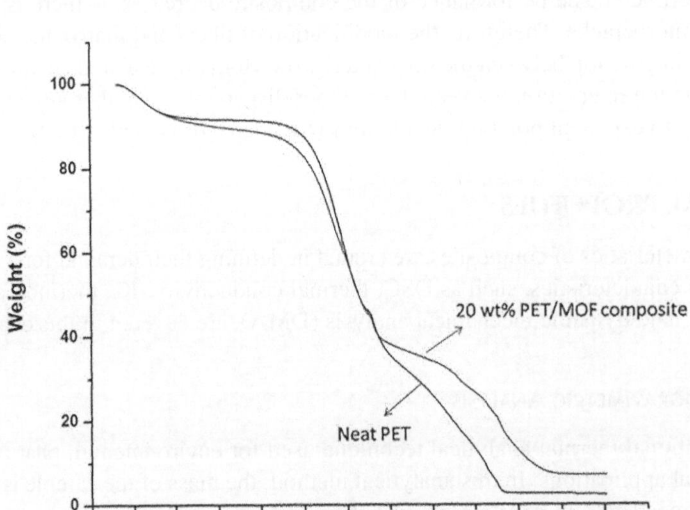

FIGURE 7.6 TGA curve of neat PET and PET/MOF composite at 20 wt% of fiber loading (Nayak & Khuntia, 2019).

composite applications (Rajeshkumar, Devnani, et al., 2021). The study inferred that the thermal stability is better, and it is also an excellent reinforcing filler for developing polymer composites.

7.4.2 Thermal Conductivity

The thermal insulation capabilities of the materials used in vehicle interior design are given specific consideration in industrial applications, especially automotive ones. The hollow tube structure of natural fibers, which offers higher insulation against heat and noise, is one of their most valued qualities. In other words, for automotive purposes, the lower the heat conductivity, the better. The studies demonstrated that the thermal conductivities of date palm fiber (DPF), hemp, and sisal are 0.083 (W/m K), 0.115 (W/m K), and 0.07 (W/m K) (Agoudjil et al., 2011), respectively, while it is 0.047 (W/m K) for coir. Consequently, for automotive applications, a lower heat conductivity is preferable (Al-Oqla & Sapuan, 2014). Figure 7.7 shows that date palm fibers have a conductivity that is lower than hemp's but yet comparable to sisal. This guarantees that date palm fibers may be used for insulation in the automobile industry. DPF may thus be a beneficial development for the automotive industry.

The thermal conductivity of different combinations of natural fiber composites of jute, banana, and pineapple was investigated by Muthukumar et al (Muthukumar et al., 2020). The study analyzed thermal analysis with the thermal conductivity parameter and accordingly its suitability for insulation applications.

For six different combinations (Table 7.2), banana fiber makes up 60% of the total amount of fiber, with the remaining 40% being shared only with pineapple and jute fibers. Additionally, each

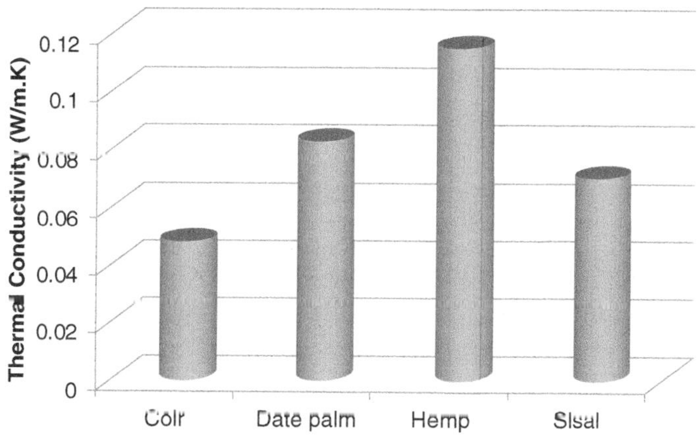

FIGURE 7.7 The thermal conductivity comparison of coir, date palm, hemp, and sisal (Al-Oqla & Sapuan, 2014).

TABLE 7.2
Formulation of Fabricated Composites (Muthukumar et al., 2020)

	% Volume of		
Samples	**Pineapple**	**Jute**	**Banana**
C 01	0	40	60
C 02	8	32	60
C 03	16	24	60
C 04	24	16	60
C 05	32	8	60
C 06	40	0	60

combination has three different thicknesses (5 mm, 10 mm, and 15 mm). The research inferred that the blend of banana, pineapple, and jute fiber composite has a significant influence on thermal qualities, including thermal conductivity as well as resistance. For each combination of fiber composite up to combination 05, the thermal conductivity is inversely proportional to the fiber thickness.

7.4.3 Dynamical Mechanical Thermal Analysis

DMTA is a versatile thermal analysis method that complements information provided by other analyses, such as DSC and TGA. This method can be used to evaluate the glass transition temperature of the material since the temperature of the sample or the frequency of the stress is repeatedly altered, which causes variations in the complex modulus. Among the composites, the dynamic properties are often required for automotive and engineering applications.

The main fundamentals that impact the dynamic mechanical characteristics of natural fiber composites include the fiber content, additive type and content, fiber geometry, fiber orientation, and modes of loading (Akil & Zamri, 2014). The DMTA study analyzes how the temperature affects variations in storage modulus, loss modulus, and tan δ. The loss modulus is related to internal friction and is related to molecular mobility, transitions, relaxation processes, morphology, and other structural parameters, whereas the storage modulus is connected to the material's stiffness. The tan δ is the proportion of the storage modulus to the loss modulus (Saba et al., 2016). Figure 7.8 depicts the dependence of these three parameters on the temperature.

An investigation on the thermal behavior of hemp fiber-reinforced polymer composites (Reddy et al., 2014) implied that the volume fraction of the fiber has a significant impact on the dynamic mechanical characteristics of hemp and hemp- and glass fiber-reinforced general purpose resin (GPR) composites. With fiber integration below the glass transition temperature, the dynamic modulus decreases, and at temperatures above T_g, it has a positive impact on the modulus. Also, the fiber addition reduces the peak height of tan δ, further demonstrating the increased fiber/matrix adherence. With the addition of fiber, the glass transition temperature is raised. Another study on the fabrication of natural fiber composites with plant oil-based resin was done by O'Donnell et al. (O'Donnell et al., 2004). The study used DMA analysis, where they observed that the storage modulus E′ of the pristine resin was 1.1 GPa, and at roughly 50 wt% fiber, E′ went up to more than 5 GPa with natural fiber reinforcements. The DMA analysis thus suggests that natural fiber can enhance the thermal parameters of polymer composites.

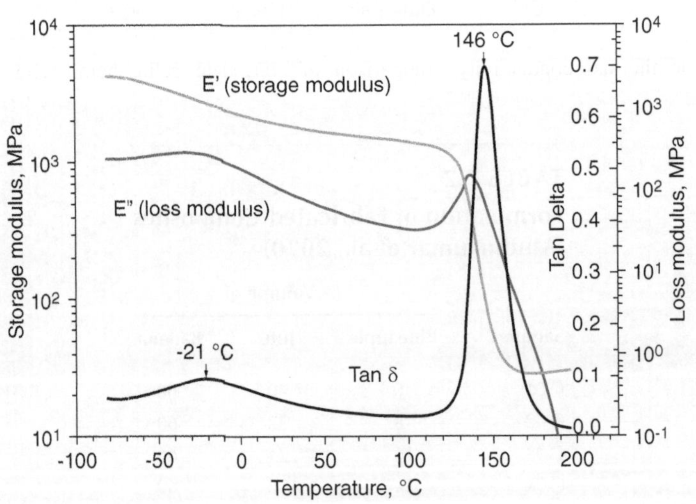

FIGURE 7.8 The variation of storage modulus, loss modulus, and tan delta with temperature (Saba et al., 2016).

7.4.4 Thermomechanical Analysis

A helpful procedure for describing the dimensional properties of various materials is thermomechanical analysis (TMA). Whether applied in isolation or in conjunction with other thermal analysis techniques such as DSC, thermogravimetric analysis, DMA, or thermal conductivity, the method provides valuable information on polymers and other materials that is challenging or even impossible to obtain by other analytical techniques. TMA is more sensitive than DSC in determining the glass transition temperature of densely packed or strongly cross-linked materials, such as brake linings, printed circuit boards, and composites (James, 2017).

7.4.5 Differential Scanning Calorimetry

By exposing both specimens to the same temperature regimes under controlled conditions, DSC measures the energy required to keep a temperature differential between a substance and an inert reference material close to zero. The fundamental tenet of this technique is that, in order to keep the sample and the reference at the same temperature during a physical transformation, like a phase transition, more (or less) heat must be applied to the sample than to the reference. This technique has been extensively used to test many thermal characteristics, including the glass transition, melting, and crystallization.

Several research used DSC analysis to study the thermal characteristics of natural fiber-reinforced polymer composites (Rajisha et al., 2011). One of the research studies examined the thermal degradation of PP composites reinforced with natural fibers, with the composite containing 50% different natural fibers (such as rice hull, newsprint, wood flour, and kenaf fibers). The comparatively large peak seen below 100°C, which is missing for pure PP, is thought to be caused by moisture content evaporation based on the DSC curves. For all composite materials, a very small drop in melting temperature is observed (Tajvidi & Takemura, 2009). With the exception of rice hulls, all natural fibers increased crystallinity as compared to pure PP. Additionally, a reduction in fractional crystallinity was seen for the formulation of uncompatibilized wood flour as compared to that of pure PP. It has been suggested that natural fibers can aid in the manufacture of crystals by functioning as nucleating agents when a coupling agent is present. The thermal and crystallization features of sisal fiber/PP composites were examined by P.V. Joseph et al. (Joseph et al., 2003). The research evaluated melting temperature (T_m), heat of fusion (ΔH_f), and percentage of crystallinity (X_c) using a DSC experiment. The results showed that there is an increase in the crystallization temperature, but there was only an insignificant increase in the Tm. However, the Tm showed a sudden increase upon the fiber modification owing to the increased matrix-fiber adhesion. Another research work was carried out for the analysis of thermal properties via DSC and TGA. The study utilized retted flax and hemp fibers in order to evaluate the thermal stability. Figure 7.9 demonstrates the temperature dependence of the mentioned fibers with the DSC technique (Wielage et al., 1999).

7.4.6 Cone Calorimetry

Cone calorimetry is considered an efficient technique to determine the flammability parameters, such as heat release rate (HRR) and other properties, via the principle of oxygen consumption. The amount of heat released by burning materials per unit area and time is called the HRR. Peak heat release rate (PHRR) measures how quickly and how far a fire spreads at its most intense. The total heat released by combustion is determined by integrating the heat release for a given time and is a function of time per unit area. The mass loss rate (MLR) affects the structural stability of the material (Kozłowski & Władyka-Przybylak, 2008). A study focused on the fire behavior of jute- and sisal-reinforced polypropylene and polylactic acid composites and compared the results with glass (Kandola et al., 2018). The fibers showed a marginal effect on time to ignition (TTI) of the PP or polylactic acid (PLA) matrix. The addition of glass reduced peak heat release to a great extent.

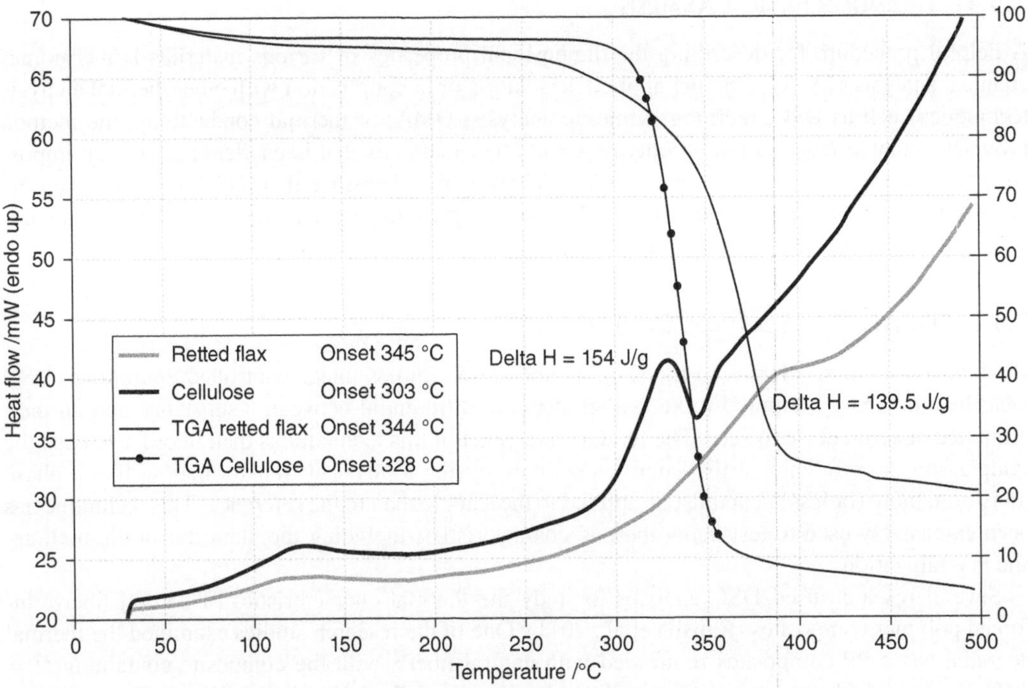

FIGURE 7.9 DSC curve of retted flax and hemp fibers (Wielage et al., 1999).

However, it is noted that the total heat release (THR) was high for natural fibers. The least thickness of sisal/PP composite might be the reason for the lowest value of THR for PP/sisal composite. Thermally and physically thin samples ignite early, burn for a shorter amount of time, and produce less total heat than thermally and physically thick samples, which burn slowly but for a longer amount of time. The depth of the heat wave penetration is less than the sample's true depth when a material is thermally thin. Figure 7.10 presents the variations in the HRR of jute-, sisal-, and glass-reinforced PP and PLA composites.

Studies reported that the type and nature of the polymer, as well as fiber, density, thermal conductivity, etc., are the factors that affect the combustibility of the polymer-fiber composites. A study carried out on the flammability factors of sansevieria fiber-loaded polyester matrix (Ramanaiah et al., 2013). The MLR, the PHRR, and the HRR of the matrix have all been significantly decreased by the addition of sansevieria fiber by 10.4%, 25.7%, and 25%, respectively. According to thermal conductivity measurements, the composite has a comparatively good thermal insulation capacity when compared to a neat polyester matrix. Although the sample burned for a longer period of time than the neat polyester resin, the THR rose by 37.8%, which was assumed to be caused by the sansevieria fiber. One of the key factors in determining fire safety is TTI.

The HRR of different fibers was exploited using cone calorimetry analysis by Dorez et al. The investigation compared the HRR of cellulose, flax, hemp, sugarcane, and bamboo with the temperature (Dorez et al., 2013). Figure 7.11 shows the differences in the HRR of the aforesaid fibers with temperature. When compared to pure cellulose, lignocellulosic fibers begin to break down at a lower temperature and release fewer combustible gases. Because of this, fibers comprising hemicellulose and lignin have a shorter TTI than those made entirely of cellulose. The production of extremely flammable levoglucosan makes pure cellulose more flammable. Levoglucosan's thermal breakdown is prevented by the presence of lignin in the fiber composition, which also makes the fibers less flammable. Levoglucosan formation and fiber flammability may be affected by additional factors such as crystallinity and fibrillar orientation. A useful indicator of the likelihood of a fire developing in a real

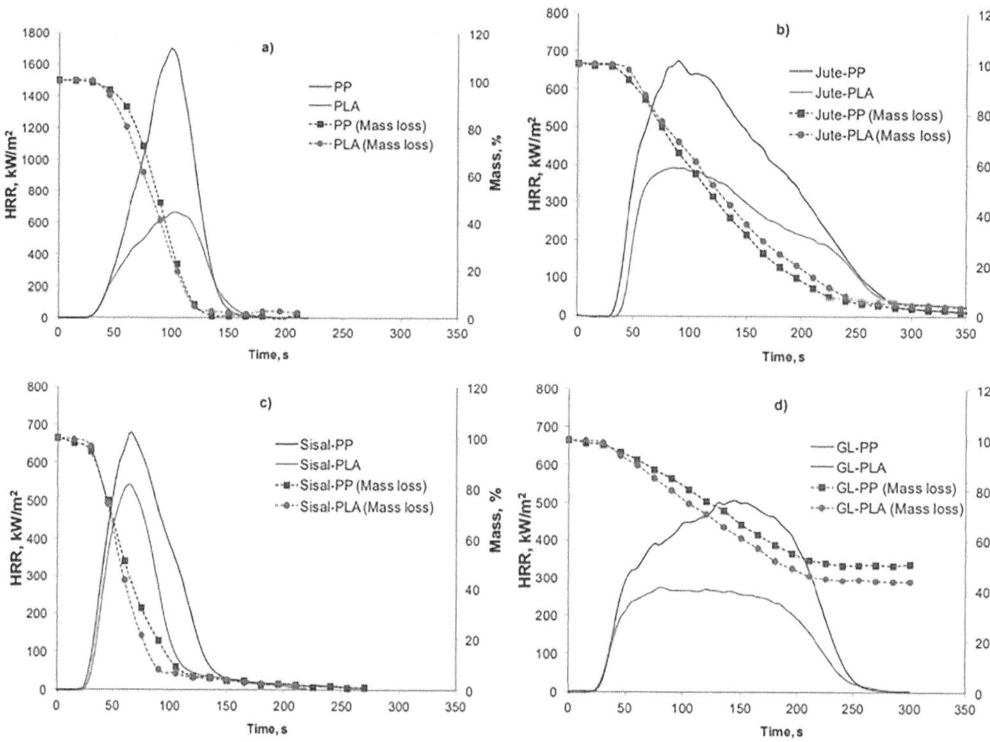

FIGURE 7.10 Variations in the HRR and mass loss of jute-, sisal-, and glass-reinforced PP and PLA composites (Kandola et al., 2018).

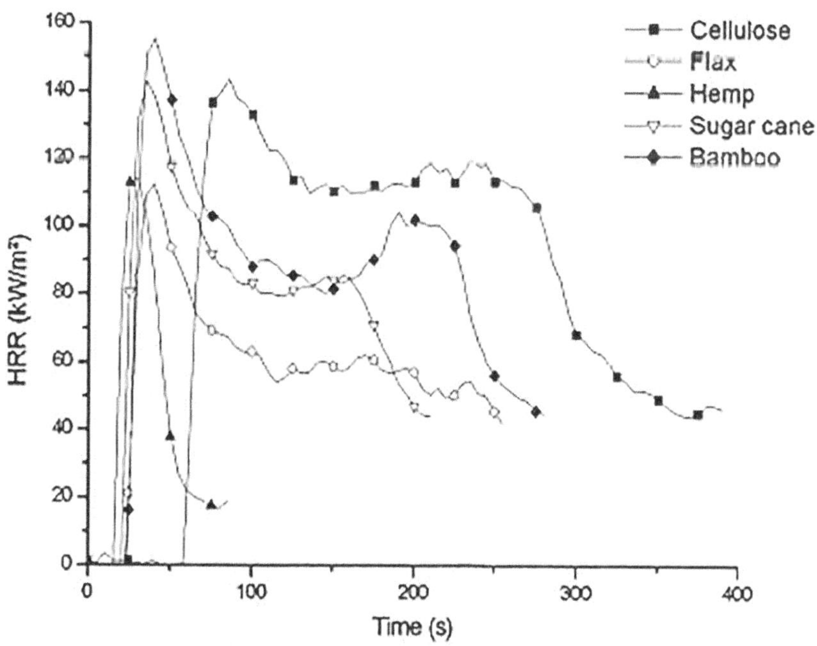

FIGURE 7.11 Plot of HRR of cellulose, flax, hemp, sugarcane, and bamboo against the temperature (Dorez et al., 2013).

environment is the maximum rate of heat emission (MARHE). The best natural fibers in this situation would be hemp and flax. It should be noted that the HRR for hemp progressively drops before extinction.

7.4.7 Limiting Oxygen Index

LOI is considered the precise as well as the easiest method in order to evaluate the fire self-extinguishment for natural fiber-incorporated polymer composites (Zhang et al., 2012). The minimum concentration of oxygen in percentage volume in order to initiate the combustion of a material is regarded as the LOI. The thermal properties upon the inclusion of coconut fiber to PLA were studied, and it was inferred that the composites showed an LOI value above 20, suggesting that it is non-flammable (Jang et al., 2012). The thermal stability and flame retardancy might be due to the strength of the coconut coir fiber, owing to its high lignin content and high heat resistance. The fire behavior of banana fiber-loaded epoxy composites was studied. Sadly, the flame-retardant properties of banana fiber epoxy composites were low compared to the neat composite. However, the composite with 20% loaded fiber showed a stability toward fire resistance (Nguyen & Nguyen, 2021). Mounika et al. examined the flammability of bamboo fiber in polyester composites where LOI is used as a tool for the same (Mounika et al., 2014). The LOI value showed that the inclusion of bamboo decreases the flammability. But the index was low when evaluated with coconut fiber/polyester composites.

Figure 7.12 displays the LOI of pristine polyester and bamboo fiber-added polyester composite.

7.4.8 Pyrolysis Combustion Flow Calorimeter

An efficient procedure for assessing the flammability parameters of natural fiber-induced composites is the PCFC. The char production and heat of combustion of the fiber composites are associated with the THR and heat release capacity (HRC) assessed using PCFC (Schartel et al., 2007). Dorez et al. explored the PCFC test to determine the flammability factors such as PHRR, HRR, and effective heat of combustion (EHC) of different fibers, including cellulose, flax, hemp, sugarcane, and bamboo in the PBS matrix. The research concluded that the inclusion of fibers reduces the thermal stability, but it enhances the mass residue owing to the generation of a char barrier (Dorez et al., 2013). Figure 7.13 represents the changes in HRR with temperature for the mentioned fibers.

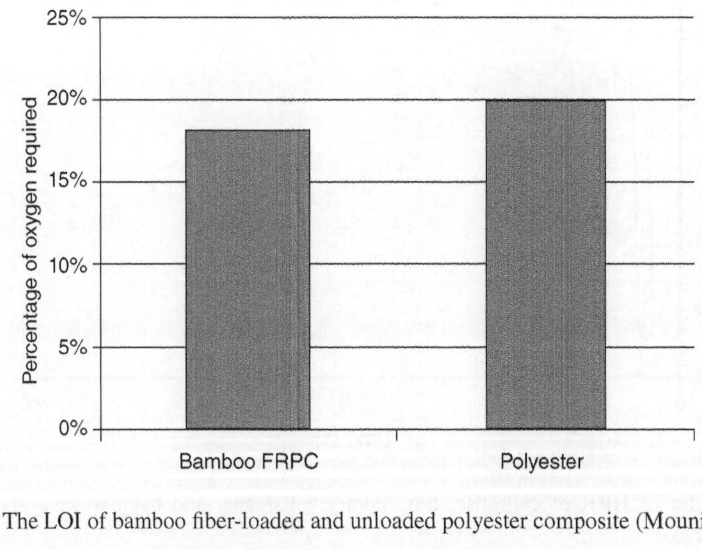

FIGURE 7.12 The LOI of bamboo fiber-loaded and unloaded polyester composite (Mounika et al., 2014).

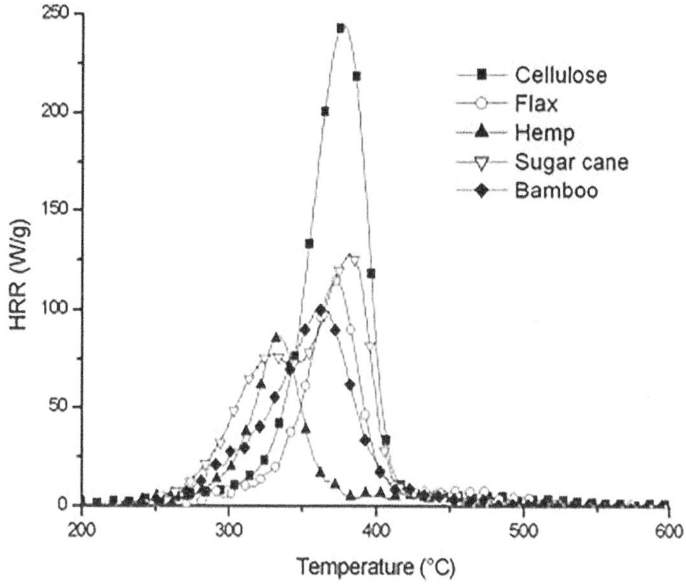

FIGURE 7.13 The variation in HRR with temperature via PCFC test (Dorez et al., 2013).

7.4.9 Thermal Applications of Natural Fiber Composites

Natural fibers have been proven to be a possible alternative to synthetic fibers exploited in engineering and industrial applications. Natural fiber-based composites have been employed in a variety of structurally designed goods, from civil engineering to the production of automobiles, since they have properties like low density, high aspect ratio, biodegradability, and ease of fabrication (Kalia et al., 2011). Beyond all of these qualities, the natural fiber's thermal stability is the key component for extending its utility to a variety of applications, particularly in structural ones like temperature transfer from end to end, load capacity at a given temperature, material behavior, and dimensional stability at high temperatures. Sisal is one of the best natural fibers for use as reinforcement in a variety of engineering applications, such as those for the transportation sector, the construction industry, the electrical sector, geotextiles, and the packaging sector. Additionally, it is used in numerous applications such as mats, clothes, bags, caps, and ropes. This is explained by its richness and inherent qualities (Veerasimman et al., 2022).

The research on five hybrid composites with bagasse fiber/bamboo charcoal proportions of 100%/0%, 70%/30%, 50%/50%, 30%/70%, and 0%/100% with 65% polyurethane foam was investigated by Abedom et al. (Abedom et al., 2021). The study exploited the reinforcing ability of sugarcane bagasse fiber and bamboo charcoal in thermal and mechanical properties for insulation applications. The result showed that the composite loaded with 30%/70% of bagasse fiber/bamboo charcoal showed better thermal insulation properties, which might be suitable for insulation applications. It is reported that the hybrid composite of flax-, glass-, and banana fiber-reinforced polymer has higher fire retardance and thermal stability than banana and flax fiber composites (Praveena et al., 2022).

Muthuraj et al. fabricated a composite made from cellulose nanofiber and elium acrylic resin with fascinating thermal properties. The degradation temperature of elium/cellulose nanofibrils (CNF) composite is around 90°C higher than that of neat CNF aerogels, increasing the material's tolerance to high temperatures. Hence, they can be effectively used for insulating purposes in the construction fields (Muthuraj et al., 2017). Similarly, Do et al. designed pineapple leaf aerogels for insulating applications. These aerogels can be used for insulating purposes as these have very low thermal conductivity. They can also be used as a thermal jacket that has an insulation capacity three times better than the commercially available product (Do et al., 2020). Hu et al. discovered the useful application of natural silk nanofibrils with the assembly of poly(vinyl alcohol) (PVA). It showed a very low thermal conductivity of 0.028 $(W{\cdot}m{\cdot}K)^{-1}$. This also paved the way to fabricate 3D aerogels with natural silk for ultra-heat retention materials (Hu et al., 2021).

7.5 CONCLUSION

The natural fibers have been proven to be promising and sustainable fillers in polymers for engineering and industrial applications. The chapter discussed the thermal properties in detail, including thermal gravimetric analysis (TGA), thermal mechanical analysis (TMA), DMA, DSC, and thermal conductivity (K). The chapter will also be dealing with flammability characteristics such as cone calorimetry analysis, LOI, and PCFC. Also, the chapter examines the preparation techniques and factors of natural fiber composites and their thermal applications. Many studies demonstrate the importance and applications of natural fibers in various industries such as automotive, building, and furniture. Consequently, the natural fibers are an efficient reinforcement in the polymer and epoxy composites for a sustainable future.

REFERENCES

Abedom, F., Sakthivel, S., Asfaw, D., Melese, B., Solomon, E., & Kumar, S. S. (2021). Development of natural fiber hybrid composites using sugarcane bagasse and bamboo charcoal for automotive thermal insulation materials. *Advances in Materials Science and Engineering, 2021*. https://doi.org/10.1155/2021/2508840

Adeniyi, A. G., Onifade, D. V., Ighalo, J. O., & Adeoye, A. S. (2019). A review of coir fiber reinforced polymer composites. *Composites Part B, 176*, 107305. https://doi.org/10.1016/j.compositesb.2019.107305

Agoudjil, B., Benchabane, A., Boudenne, A., Ibos, L., & Fois, M. (2011). Renewable materials to reduce building heat loss: Characterization of date palm wood. *Energy and Buildings, 43*(2–3), 491–497. https://doi.org/10.1016/J.ENBUILD.2010.10.014

Akil, H., & Zamri, M. H. (2014). Performance of natural fiber composites under dynamic loading. *Natural Fibre Composites: Materials, Processes and Applications*, 323–344. https://doi.org/10.1533/9780857099228.3.323

Aliotta, L., Gasenge, M., Gigante, V., & Lazzeri, A. (2022). Micromechanical deformation processes and failure of PBS based composites containing ultra-short cellulosic fibers for injection molding applications. *Polymers, 14*(21), 4499.

Al-Oqla, F. M., & Sapuan, S. M. (2014). Natural fiber reinforced polymer composites in industrial applications: Feasibility of date palm fibers for sustainable automotive industry. *Journal of Cleaner Production, 66*, 347–354. https://doi.org/10.1016/J.JCLEPRO.2013.10.050

Aravinth, K., Ramakrishnan, T., Tamilarasan, V. D., & Veeramanikandan, K. (2022). A brief review on plant fibres composites: Extraction, chemical treatment and fibre orientation. *Materials Today: Proceedings, 62*, 2005–2009. https://doi.org/10.1016/j.matpr.2022.02.291

Balla, K. V., Kate, K. H., Satyavolu, J., Singh, P., & Tadimeti, J. G. D. (2019). Additive manufacturing of natural fiber reinforced polymer composites: Processing and prospects. *Composites Part B, 174*, 106956. https://doi.org/10.1016/j.compositesb.2019.106956

Barthod-Malat, B., Hauguel, M., Behlouli, K., Grisel, M., & Savary, G. (2023). Influence of the compression molding temperature on VOCs and odors produced from natural fiber composite materials. *Coatings, 13*(371), 1–21.

Beg, M. D. H., & Pickering, K. L. (2008). Mechanical performance of Kraft fibre reinforced polypropylene composites: Influence of fibre length, fibre beating and hygrothermal ageing. *Composites Part A-Applied Science and Manufacturing, 39*, 1748–1755.

Boccarusso, L., De Fazio, D., & Durante, M. (2022). Production of PP composites reinforced with flax and hemp woven mesh fabrics via compression molding. *Inventions, 7*(5), 1–17.

Chandgude, S., & Salunkhe, S. (2022). In state of art: Mechanical behavior of natural fiber based hybrid polymeric composites for application of automobile components. *Polymer Composites*, 1–26. https://doi.org/10.1002/pc.26045

Chandramohan, D., & Marimuthu, K. (2011). A review on natural fibers. *IJRRAS, 8*(2).

Costa, C. S. M. F., Fonseca, A. C., Serra, A. C., & Coelho, J. F. J. (2016). Dynamic mechanical thermal analysis of polymer composites reinforced with natural fibers. *Polymer Reviews*, *56*(2), 362–383. https://doi.org/10.1080/15583724.2015.1108334

de Araujo Alves Lima, R., Kawasaki Cavalcanti, D., de Souza e Silva Neto, J., Meneses da Costa, H., & Banea, M. D. (2020). Effect of surface treatments on interfacial properties of natural intralaminar hybrid composites. *Polymer Composites*, *41*(1), 314–325. https://doi.org/10.1002/PC.25371

Do, N. H. N., Luu, T. P., Thai, Q. B., Le, D. K., Chau, N. D. Q., Nguyen, S. T., Le, P. K., Phan-Thien, N., & Duong, H. M. (2020). Heat and sound insulation applications of pineapple aerogels from pineapple waste. *Materials Chemistry and Physics*, *242*, 1–7.

Dolza, C., Fages, E., Gonga, E., Gomez-Caturla, J., Balart, R., & Quiles-Carrillo, L. (2021). Development and characterization of environmentally friendly wood plastic composites from biobased polyethylene and short natural fibers processed by injection moulding. *Polymers*, *13*(11), 1692. https://doi.org/10.3390/polym13111692

Dominic, M. C. D., Joseph, R., Begum, P. M. S., Joseph, M., Padmanabhan, D., Morris, L. A., Kumar, A. S., & Formela, K. (2020). Cellulose nanofibers isolated from the cuscuta reflexa plant as a green reinforcement of natural rubber. *Polymers*, *12*(4), 814. https://doi.org/10.3390/POLYM12040814

Dorez, G., Taguet, A., Ferry, L., & Lopez-Cuesta, J. M. (2013). Thermal and fire behavior of natural fibers/PBS biocomposites. *Polymer Degradation and Stability*, *98*(1), 87–95. https://doi.org/10.1016/j.polymdegradstab.2012.10.026

Duigou, L. A., Correa, D., Ueda, M., Matsuzaki, R., & Castro, M. (2020). A review of 3D and 4D printing of natural fibre biocomposites. *Materials & Design*, *194*, 108911. https://doi.org/10.1016/j.matdes.2020.108911

Faruk, O., Bledzki, A. K., Fink, H.-P., & Sain, M. (2012). Biocomposites reinforced with natural fibers: 2000–2010. *Progress in Polymer Science*, *37*(11), 1552–1596. https://doi.org/10.1016/j.progpolymsci.2012.04.003

Gallos, A., Paës, G., Allais, F., & Beaugrand, J. (2017). Lignocellulosic fibers: A critical review of the extrusion process for enhancement of the properties of natural fiber composites. *RSC Advances*, *7*, 34638–34654. https://doi.org/10.1039/c7ra05240e

Hu, Z., Yan, S., Li, X., You, R., Zhang, Q., & Kaplan, D. L. (2021). Natural silk nanofibril aerogels with distinctive filtration capacity and heat-retention performance. *ACS Nano*, *15*, 8171–8183. https://doi.org/10.1021/acsnano.1c00346

James, J. (2017). Thermomechanical analysis and its applications. *Thermal and Rheological Measurement Techniques for Nanomaterials Characterization*, *3*, 159–171. https://doi.org/10.1016/B978-0-323-46139-9.00007-4

Jang, J. Y., Jeong, T. K., Oh, H. J., Youn, J. R., & Song, Y. S. (2012). Thermal stability and flammability of coconut fiber reinforced poly(lactic acid) composites. *Composites Part B: Engineering*, *43*(5), 2434–2438. https://doi.org/10.1016/j.compositesb.2011.11.003

Jariwala, H., & Jain, P. (2019). A review on mechanical behavior of natural fiber reinforced polymer composites and its applications. *Journal of Reinforced Plastics & Composites*, *38*(10), 1–13. https://doi.org/10.1177/0731684419828524

Joseph, P. V., Joseph, K., Thomas, S., Pillai, C. K. S., Prasad, V. S., Groeninckx, G., & Sarkissova, M. (2003). The thermal and crystallisation studies of short sisal fibre reinforced polypropylene composites. *Composites Part A: Applied Science and Manufacturing*, *34*(3), 253–266. https://doi.org/10.1016/S1359-835X(02)00185-9

Kalia, S., Avérous, L., Njuguna, J., Dufresne, A., & Cherian, B. M. (2011). Editorial natural fibers, bio-and nanocomposites. *International Journal of Polymer Science*, *2011*. https://doi.org/10.1155/2011/735932

Kalia, S., Kaith, B. S., & Kaur, I. (2009). Pretreatments of natural fibers and their application as reinforcing material in polymer composites-A review. *Polymer Engineering and Science*, *49*(7), 1253–1272. https://doi.org/10.1002/pen.21328

Kandola, B. K., Mistik, S. I., Pornwannachai, W., & Anand, S. C. (2018). Natural fibre-reinforced thermoplastic composites from woven-nonwoven textile preforms: Mechanical and fire performance study. *Composites Part B: Engineering*, *153*, 456–464. https://doi.org/10.1016/j.compositesb.2018.09.013

Kerni, L., Singh, S., Patnaik, A., & Kumar, N. (2020). A review on natural fiber reinforced composites. *Materials Today: Proceedings*, *28*, 1616–1621. https://doi.org/10.1016/j.matpr.2020.04.851

Kiziltas, E. E., Kiziltas, A., & Boran Torun, S. (n.d.). *Thermal Analysis of Polyamide 6 Composites Filled by Natural Fiber Blend Han-Seung Yang Nanodyne Measurement Systems*. https://doi.org/10.15376/biores.11.2.4758-4769

Kong, I., Tshai, K. Y., & Hoque, M. E. (2015). Manufacturing of natural fibre-reinforced polymer composites by solvent casting method. In *Manufacturing of Natural Fibre Reinforced Polymer Composites* (pp. 1–15). https://doi.org/10.1007/978-3-319-07944-8_1

Kozłowski, R., & Władyka-Przybylak, M. (2008). Flammability and fire resistance of composites reinforced by natural fibers. *Polymers for Advanced Technologies*, *19*(6), 446–453. https://doi.org/10.1002/PAT.1135

Kumar, R., Ul Haq, M. I., Raina, A., & Anand, A. (2019). Industrial applications of natural fibre-reinforced polymer composites–challenges and opportunities. In *International Journal of Sustainable Engineering* (Vol. 12, Issue 3, pp. 212–220). Taylor and Francis Ltd. https://doi.org/10.1080/19397038.2018.1538267

Latif, R., Wakeel, S., Khan, N. Z., Noor Siddiquee, A., Lal Verma, S., & Akhtar Khan, Z. (2019). Surface treatments of plant fibers and their effects on mechanical properties of fiber-reinforced composites: A review. *Journal of Reinforced Plastics and Composites*, *38*(1), 15–30. https://doi.org/10.1177/0731684418802022/ASSET/IMAGES/LARGE/10.1177_0731684418802022-FIG5.JPEG

Lau, K., Hung, P., Zhu, M., & Hui, D. (2018). Properties of natural fibre composites for structural engineering applications. *Composites Part B*, *136*, 222–233. https://doi.org/10.1016/j.compositesb.2017.10.038

Malik, K., Ahmad, F., Gunister, E., Nakato, T., Mouri, E., Muhammad, M. B., & Ali, S. (2021). A review of flax fiber reinforced thermoset polymer composites: Structure and mechanical performance. *Journal of Natural Fibers*, *19*(14), 1–25. https://doi.org/10.1080/15440478.2021.1990182

Mochane, M. J., Mokhena, T. C., Mokhothu, T. H., Mtibe, A., Sadiku, E. R., Ray, S. S., Ibrahim, I. D., & Daramola, O. O. (2019). Recent progress on natural fiber hybrid composites for advanced applications: A review. *European Polymer Journal*, *13*(2), 159–198.

Mounika, M., Anusha, P., Ramanaiah, K., Ratnaprasad, A. V., Hema, K., & Reddy, C. (2014). Thermal properties of bamboo fiber reinforced polyester composite. *International Journal of Engineering Research and Science & Technology*. http://www.ijerst.com/currentissue.php

Muthukumar, K., Sabariraj, R. V., Dinesh Kumar, S., & Sathish, T. (2020). Investigation of thermal conductivity and thermal resistance analysis on different combination of natural fiber composites of banana, pineapple and jute. *Materials Today: Proceedings*, *21*, 976–980. https://doi.org/10.1016/j.matpr.2019.09.140

Muthuraj, R., Grohens, Y., & Seantier, B. (2017). Mechanical and thermal insulation properties of elium acrylic resin/cellulose nanofiber based composite aerogels. *Nano-Structures & Nano-Objects*, *12*, 68–76. https://doi.org/10.1016/j.nanoso.2017.09.002

Nayak, S., & Khuntia, S. K. (2019). Development and study of properties of Moringa oleifera fruit fibers/polyethylene terephthalate composites for packaging applications. *Composites Communications*, *15*, 113–119. https://doi.org/10.1016/J.COCO.2019.07.008

Neto, J. S. S., de Queiroz, H. F. M., Aguiar, R. A. A., & Banea, M. D. (2021). A review on the thermal characterisation of natural and hybrid fiber composites. *Polymers*, *13*(24), 4425. MDPI. https://doi.org/10.3390/polym13244425

Nguyen, T. A., & Nguyen, T. H. (2021). Banana fiber-reinforced epoxy composites: Mechanical properties and fireretardancy. *International Journal of Chemical Engineering*, *2021*. https://doi.org/10.1155/2021/1973644

Nurazzi Norizan, M., Hasnah Kamarudin, S., & Shazra Shazleen, S. (n.d.). *Thermogravimetric Analysis (TGA) and Differential Scanning Calorimetry (DSC) of PLA/Cellulose Composites Sugar Palm Fibre Reinforce Polymer Composites VIEW Project SAFETY and Health of Biocomposites View Project*. https://doi.org/10.1201/9781003160458-7

O'Donnell, A., Dweib, M. A., & Wool, R. P. (2004). Natural fiber composites with plant oil-based resin. *Composites Science and Technology*, *64*(9), 1135–1145. https://doi.org/10.1016/J.COMPSCITECH.2003.09.024

Paul G.. (n.d.). *Principles and Applications of Thermal Analysis - Google Books*. Blackwell. Retrieved May 7, 2023, from https://books.google.co.in/books?hl=en&lr=&id=1u4v_tGUEQoC&oi=fnd&pg=PA87&dq=thermogravimetric+analysis+tga&ots=lKNtgWFYlU&sig=5aYDvoJeoj2SRezQU6bYMZqLsD4&redir_esc=y#v=onepage&q=thermogravimetric%20analysis%20tga&f=false

Prakash, S. O., Sahu, P., Madhan, M., & Santhosh, A. J. (2022). A review on natural fibre-reinforced biopolymer composites: Properties and applications. *International Journal of Polymer Science*, 1–15.

Praveena, B. A., Buradi, A., Santhosh, N., Vasu, V. K., Hatgundi, J., & Huliya, D. (2022). Study on characterization of mechanical, thermal properties, machinability and biodegradability of natural fiber reinforced polymer composites and its Applications, recent developments and future potentials: A comprehensive review. *Materials Today: Proceedings*, *52*, 1255–1259. https://doi.org/10.1016/J.MATPR.2021.11.049

Rajeshkumar, G., Devnani, G. L., Maran, J. P., Sanjay, M. R., Siengchin, S., Al-Dhabi, N. A., & Ponmurugan, K. (2021). Characterization of novel natural cellulosic fibers from purple bauhinia for potential reinforcement in polymer composites. *Cellulose*, *28*(9), 5373–5385.

Rajeshkumar, G., Seshadri, S. A., Devnani, G. L., Sanjay, M. R., Siengchin, S., Maran, J. P., Al-Dhabi, N. A., Karuppiah, P., Mariadhas, V. A., Sivarajasekar, N., & Anuf, A. R. (2021). Environment friendly, renewable and sustainable poly lactic acid (PLA) based natural fiber reinforced composites–A comprehensive review. *Journal of Cleaner Production*, *310*, 1–26. https://doi.org/10.1016/j.jclepro.2021.127483

Rajisha, K. R., Deepa, B., Pothan, L. A., & Thomas, S. (2011). Thermomechanical and spectroscopic characterization of natural fibre composites. *Interface Engineering of Natural Fibre Composites for Maximum Performance*, 241–274. https://doi.org/10.1533/9780857092281.2.241

Ramachandran, A., Rangappa, S. M., Kushvaha, V., Khan, A., Seingchin, S., & Dhakal, H. N. (2022). Modification of fibers and matrices in natural fiber reinforced polymer composites: A comprehensive review. *Macromolecular Rapid Communications*, *2100862*, 1–38. https://doi.org/10.1002/marc.202100862

Ramanaiah, K., Ratna Prasad, A. V., & Hema Chandra Reddy, K. (2013). Mechanical, thermophysical and fire properties of sansevieria fiber-reinforced polyester composites. *Materials & Design*, *49*, 986–991. https://doi.org/10.1016/j.matdes.2013.02.056

Ramesh, V., & Anand, P. (2021). Evaluation of mechanical properties on Kevlar/Basalt fiber reinforced hybrid composites. *Materials Today: Proceedings*, *39*, 1494–1496. https://doi.org/10.1016/j.matpr.2020.05.406

Reddy, I., Reddy, M. I., & Reddy, V. S. (2014). Dynamic mechanical analysis of hemp fiber reinforced polymer matrix composites article in. *International Journal of Engineering Research.* www.ijert.org

Rocha, D. L., Junior, L. U. D., Marvila, M. T., Pereira, E. C., Souza, D., Rangel, A., & Azevedo, G. D. (2022). A review of the use of natural fibers in cement composites: Concepts, applications and Brazilian history. *Polymers*, *14*, 1–22.

Rout, J., Misra, M., Tripathy, S. S., Nayak, S. K., & Mohanty, A. K. (2001). The influence of fibre treatment on the performance of coir-polyester composites. *Composites Science and Technology*, *61*(9), 1303–1310. https://doi.org/10.1016/S0266-3538(01)00021-5

Saba, N., Jawaid, M., Alothman, O. Y., & Paridah, M. T. (2016). A review on dynamic mechanical properties of natural fibre reinforced polymer composites. *Construction and Building Materials*, *106*, 149–159. https://doi.org/10.1016/J.CONBUILDMAT.2015.12.075

Sapuan, S. M., & Yusoff, N. B. (2015). The relationship between manufacturing and design for manufacturing in product development of natural fibre composites. In *Manufacturing of Natural Fibre Reinforced Polymer Composites* (pp. 1–15). https://doi.org/10.1007/978-3-319-07944-8

Sarikaya, E., Çallio, H., & Demirel, H. (2019). Production of epoxy composites reinforced by different natural fibers and their mechanical properties. *Composites Part B*, *167*(15), 461–466. https://doi.org/10.1016/j.compositesb.2019.03.020

Schartel, B., Pawlowski, K. H., & Lyon, R. E. (2007). Pyrolysis combustion flow calorimeter: A tool to assess flame retarded PC/ABS materials? *Thermochimica Acta*, *462*(1), 1–14. https://doi.org/10.1016/j.tca.2007.05.021

Senthilkumar, K., Saba, N., Chandrasekar, M., Jawaid, M., Rajini, N., Alothman, O. Y., & Siengchin, S. (2019). Evaluation of mechanical and free vibration properties of the pineapple leaf fibre reinforced polyester composites. *Construction and Building Materials*, *195*, 423–431. https://doi.org/10.1016/j.conbuildmat.2018.11.081

Shalwan, A., & Yousif, B. F. (2013). In state of art: Mechanical and tribological behaviour of polymeric composites based on natural fibres. *Materials and Design*, *48*, 14–24. https://doi.org/10.1016/j.matdes.2012.07.014

Sun, Z., Duan, Y., An, H., Wang, X., Liang, S., & Li, N. (2023). Research progress and application of natural fiber composites. *Journal of Natural Fibres*, *20*(2), 2206591. https://doi.org/10.1080/15440478.2023.2206591

Suresh Babu, A., Gowthamraj, S., & Jaivignesh, M. (2019). A comparative study on mechanical properties of kenaf fiber-reinforced polyester composites prepared by VARI, RTM and CM techniques. *IOP Conference Series: Materials Science and Engineering*, *577*(1), 10. https://doi.org/10.1088/1757-899X/577/1/012189

Tajvidi, M., & Takemura, A. (2009). Thermal degradation of natural fiber-reinforced polypropylene composites. *Journal of Thermoplastic Composite Materials*, *23*(3), 281–298. https://doi.org/10.1177/0892705709347063

Veerasimman, A., Shanmugam, V., Rajendran, S., Johnson, D. J., Subbiah, A., Koilpichai, J., & Marimuthu, U. (2022). Thermal properties of natural fiber sisal based hybrid composites–A brief review. *Journal of Natural Fibers*, *19*(12), 4696–4706. Taylor and Francis Ltd. https://doi.org/10.1080/15440478.2020.1870619

Wielage, B., Lampke, T., Marx, G., Nestler, K., & Starke, D. (1999). Thermogravimetric and differential scanning calorimetric analysis of natural fibres and polypropylene. *Thermochimica Acta*, *337*(1), 169–177. https://doi.org/10.1016/S0040-6031(99)00161-6

Yallew, T. B., Kassegn, E., Aregawi, S., & Gebresias, A. (2020). Study on effect of process parameters on tensile properties of compression molded natural fiber reinforced polymer composites. *SN Applied Sciences*, *2*, 1–8. https://doi.org/10.1007/s42452-020-2101-0

Yang, H. S., Gardner, D. J., & Nader, J. W. (2011). Dispersion evaluation of microcrystalline cellulose/cellulose nanofibril-filled polypropylene composites using thermogravimetric analysis. *Journal of Thermal Analysis and Calorimetry*, *103*(3), 1007–1015. https://doi.org/10.1007/S10973-010-1147-4

Yang, Z., Li, X., Si, J., Cui, Z., & Peng, K. (2019). Morphological, mechanical and thermal properties of poly(lactic acid) (PLA)/cellulose nanofibrils (CNF) composites nanofiber for tissue engineering. *Journal Wuhan University of Technology, Materials Science Edition*, *34*(1), 207–215.

Zhang, Z. X., Zhang, J., Lu, B.-X., Xin, Z. X., Kang, C. K., & Kim, J. K. (2012). Effect of flame retardants on mechanical properties, flammability and foamability of PP/wood–fiber composites. *Composites Part B: Engineering*, *43*(2), 150–158. https://doi.org/10.1016/j.compositesb.2011.06.020

Zhou, Y., Fan, M., & Chen, L. (2016). Interface and bonding mechanisms of plant fibre composites: An overview. *Composites Part B: Engineering*, *101*, 31–45. https://doi.org/10.1016/j.compositesb.2016.06.055

8 Machining and Machinability of Natural Fiber Composites
Conventional Route

Titus Thankachan, T. Satish Kumar, and V. Kavimani

8.1 INTRODUCTION

Fiber-reinforced composites are a possible replacement for conventional metals and alloys in engineering applications owing to their superior properties, including high strength-to-weight ratio, fatigue resistance, low weight, and high corrosion resistance, making them a highly researchable material for scientists, engineers, and researchers (Ramesh et al., 2016; Alarifi, 2023). Fabrication of fiber-reinforced composites includes the selection of appropriate fibers based on application requisites. Fibers used as reinforcement are mainly classified into natural and synthetic fibers based on source, and studies have proved that fiber-reinforced composites dispersed with synthetic fibers showcase high mechanical and thermal characteristics with minimal water absorption capability (Wang et al., 2022; Vinayagamoorthy et al., 2018). However, the usage of synthetic fibers as reinforcements has certain drawbacks, including high fiber cost, recycling, and disposal issues, which makes natural fiber composites more prioritized. Biodegradability of natural fiber-based composites, along with low density and enhanced mechanical properties, makes it a potential material for various applications, including aerospace and automotive (Jani et al., 2016; Nandhakumar et al., 2020). Application of natural fiber composites in various industries is provided in Figure 8.1.

Different types of natural fibers have been utilized by researchers to develop different sets of natural fiber-based polymer composites; however, the most commonly and widely used polymers in studies are provided in Table 8.1. A detailed view of its physical and mechanical characteristics is provided so as to have a view on the characteristics of the different fibers that are used in researches. When it comes to the usage of a material for certain applications, the role of secondary manufacturing processes is significant. Manufacturing of natural fiber composites happens through many methods, such as hand layup processes, resin transfer molding, compression molding, extrusion process, and automated fiber placement to attain the required shape (Mahadevappa et al., 2020; Lee et al., 2017; Hejjaji et al., 2018). However, to join the fabricated composites to any other complex assemblies requires a secondary manufacturing process, mainly machining and joining processes. Machining of natural fiber composites happens mainly through conventional and advanced machining processes, wherein, in this study, we focus mainly on the conventional machining process, as shown in Figure 8.2.

Fabrication of natural fiber composites is carried out in such a way that the near-net shape will be attained after the manufacturing method, making the need for secondary machining operations nullified. But for certain applications with complex part assemblies, the need for the above machining process, including turning, drilling, milling, and grinding, is mandatory so as to acquire the dimensional and tolerance requirements. However, studies have proved that natural fiber composites during machining demonstrated severe delamination, de-bonding, fiber pull-out, crack development, burning, etc., due to their anisotropic and heterogeneous nature (Lotfi et al., 2021; John et al., 2021;

DOI: 10.1201/9781032665337-8

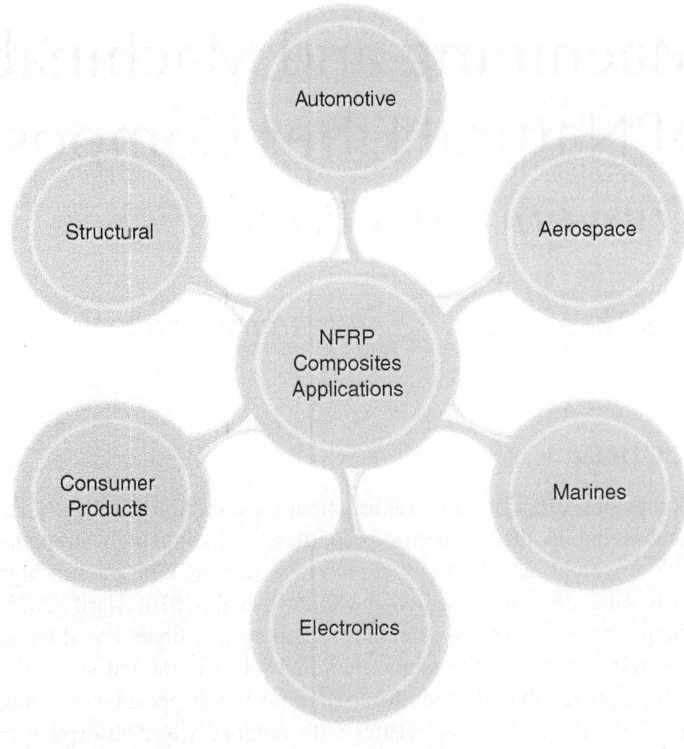

FIGURE 8.1 Application of natural fiber-based composites.

TABLE 8.1
Natural Fiber Composites and Their Properties

Fiber	Density gm/cm³	Young's Modulus GPa	Tensile Strength Mpa	Percentage Elongation %	References
Flax	1.3–1.6	27.6–	500–1500	2.7–3.2	Khan et al. (2018), Naik et al. (2022), Jawaid and Abdul Khalil (2011), Gurunathan et al. (2015), Lu et al. (2012), Prakash et al. (2014), Maya et al. (2017), Eichhorn et al. (2001), Keya et al. (2019), Boopathi et al. (2023), and Tan et al. (2015)
Abaca	1.5	9.0–28	980	10–12	
Kenaf	1.45	53	930	1.6	
Hemp	1.48	70	550–900	1.6	
Jute	1.46	10–30	400–800	1.5	
Sisal	1.45	09–12	400–700	05–14	
Cotton	1.21	6–10	287–597	2–10	
Bamboo	4.3	10–40	73–505	4.11	
Pineapple	1.526	6.21	413	1.6	
Coir	1.2	4–6	175	30	
Banana	1.02	1.63	10.854	6.085	

Delahaigue et al., 2017). Poor machinability of natural fiber composites is a major cause for researchers to investigate the different machining possibilities that can be carried out in natural fiber-based composites. Being a heterogeneous material by nature, natural fiber composites, while machining the matrix and the reinforced fibers, will react differently, like de-cohesion between the fiber and the matrix material. This will lead to uncut fibers protruding outside and reducing the surface integrity of the material.

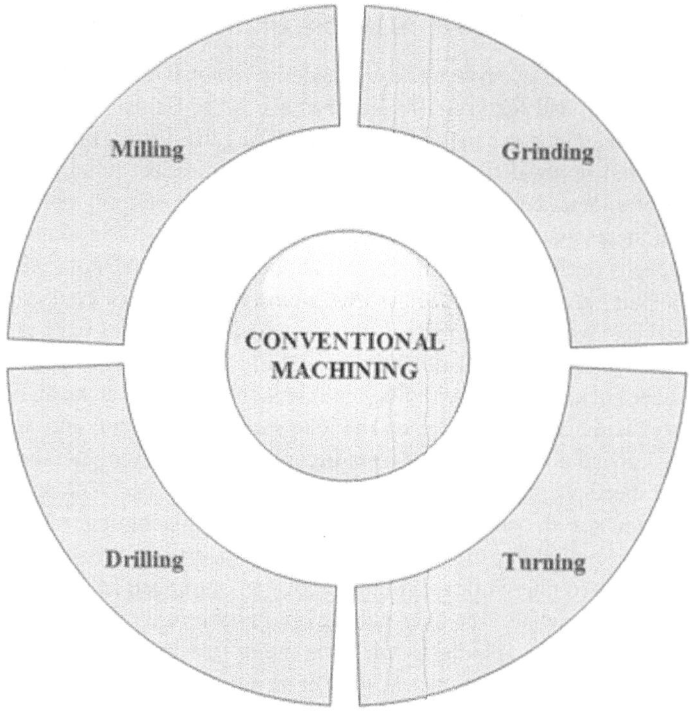

FIGURE 8.2 Conventional machining of natural fiber-based composites.

8.2 TURNING OF NATURAL FIBER COMPOSITES

As reported earlier, the need for machining processes is minimal when it comes to natural fiber composites, but to attain the tolerance and dimensional requirements, the turning operation is considered a major secondary operation for cylindrical specimens. Turning of natural fiber composites is entirely different from the turning of metals and alloys, as the material removal mechanisms happening per unit area will be wholly different from next unit area. This is mainly due to the presence of materials like matrix, fibers, and filler materials with different characteristics (Rajmohan et al., 2019; Pandya and Rathod, 2020; Raj et al., 2021). While turning natural fiber-reinforced composites, difficulties such as rapid tool wear, surface damage, or fiber pull-out may happen, and to overcome this, appropriate machining conditions and tooling have to be used.

8.2.1 Process Parameters of Turning Process

To carry out the turning operation of a natural fiber-based composite, the major machining parameters that have to be considered are depth of cut, cutting speed, and feed rate of machining. These three factors affect the machinability of a natural fiber-based composite to a great extent. Tool materials and their geometry also play a considerably higher role in defining the machinability of natural fiber-based composites. Coming to machinability of composite material, the metal removal rate and surface finish of the considered material are the major output factors that will be focused on (Júnior et al., 2022; Sujuan et al., 2022). Along with this, tool life is a key factor that has to be considered while machining a natural fiber-based composite as studies have proved that a reduction in tool life is observed. Tool life depends on hardness of the dispersed natural fiber and the matrix resin that is used in the study. Cutting edge of the tool coming in contact with hard fibers and soft matrix will lead to irregular cutting and vibration induced, thereby affecting the tool life.

8.2.2 Effects of Process Parameters on Machinability

Machinability, including the surface roughness and material removal rate, is influenced mainly by the depth of cut, cutting speed, and feed rate (Mansour et al., 2021; Equbal et al., 2020). It is evident from Figure 8.3 that with an increase in the depth of cut during the turning operation of natural polymer composites, material removal rate tends to increase. With respect to an increase in speed and feed rate, the point of contact between tool cutting point and natural polymer composite tends to increase, henceforth increasing the material removal rate per minute. The major material removal mechanism that happens during the machining of a natural fiber composite are delamination, fiber cutting, and fiber buckling. As mentioned above, the geometry of cutting tool also has a major role in defining the material removal rate of a natural fiber composite, as studies have proved that with an increase in rake angle, material removal rate happens majorly through delamination and fiber cutting, and in other case, fiber buckling along with fiber cutting is the major mechanism happening.

Surface roughness is mainly dependent on the feed rate, depth of cut, and machining speed at which the process is carried out. It is evident from Figure 8.4 that carrying out the turning operation at a medium spindle speed and feed rate provides a better surface finish. With an increase in spindle speed over an optimized speed, centrifugal force created during machining will pull out the natural fibers from the matrix radially after the machining process, increasing the surface roughness of the material. Increasing the feed rate while machining above an optimized level also affects the roughness value of machined composites, as fiber pulling out happens aggressively. Depth of cut, when combined with the feed rate and spindle speed, has a major role in defining the surface roughness value of machined natural fiber composites as with an increase in depth of cut, the roughness value also tends to increase. High depth of cut at high spindle speed and feed rate leaves high crest and valley values due to the improper material removal that happens during the machining process.

With increase in cutting speed, cutting time, feed rate, and depth of cut, tool wear increases as the tool comes in contact with the hard natural fiber composites due to continuous abrasion of fibers along the tool during machining. Research has proved that the chips generated during the turning of natural fiber composites are mainly shattered chips and splintered chips. As these chips are formed while machining a natural fiber composite, the main tool wear mechanisms, crater wear and tool mechanical chipping, are found to be minimal, while rounding wear and flank wear will be observed.

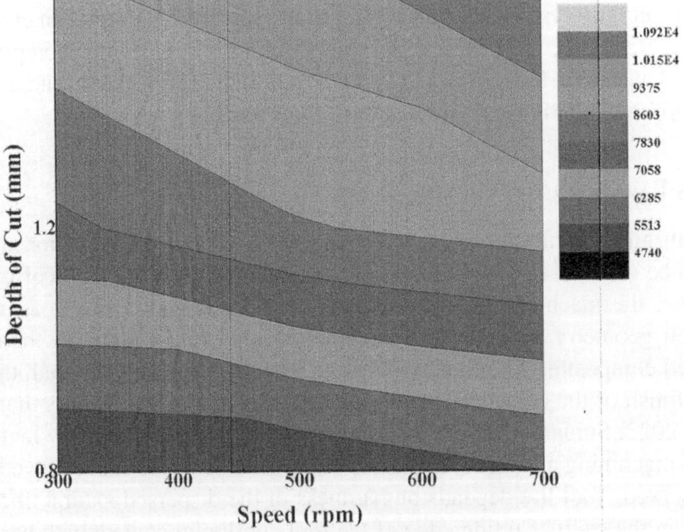

FIGURE 8.3 Material removal rate (mm^3/min) with respect to speed and feed rate.

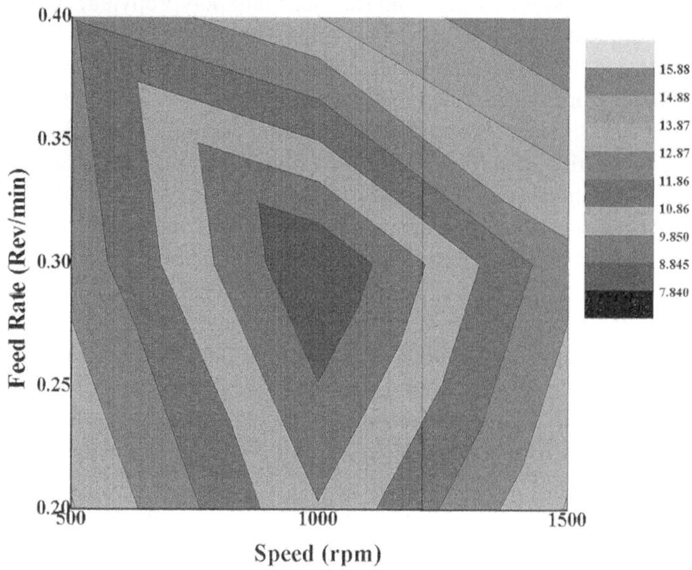

FIGURE 8.4 Surface roughness (μm) with respect to speed and feed rate.

8.3 MILLING OF NATURAL FIBER COMPOSITES

Milling operation is considered as one of the prominent machining processes to attain superior quality component geometry, including precision slots, finished edges, and contours. But while machining a natural fiber composite, proper selection of machining parameters has to be considered; otherwise, it will lead to improper surface integrity (Yahya Hışman Çelik and Alp, 2022). So as to attain a better surface quality with possible material removal rate, the mechanism behind the machining of natural fiber composites has to be deliberated with proper machining parameter selection.

Çelik et al. (2019) studied the effect of milling process parameters (cutting speed and feed rate) while machining jute fiber-reinforced polymer composites fabricated via vacuum infusion method. This study mainly focused on the effect of process parameters on the surface roughness, delamination factor, and cutting forces. Twelve-layer jute fiber-reinforced composites fabricated at different orientation angles were milled using cemented carbide end mill cutters at varying cutting speed and feed rate. It was observed that with increase in feed rate while milling, cutting forces tend to increase as the end mill cutter doesn't get time to cut the reinforced jute fiber. Instead, it tends to break the fiber and remove it in the form of chips. An increase in cutting speed generates friction at the tool-polymer interface, thereby destabilizing the polymer matrix, which in turn reduces the cutting forces generated during the machining process. It was also observed that with an increase in feed rate, the delamination factor also tends to increase, which can be attributed to the surface damage initiated through friction created between tool and the natural fiber composites due to the generation of high cutting force. With an increase in cutting speed, it was observed that the polymer matrix gets softened due to the high temperature generated, reducing the bonding between the natural fibers and the matrix. This also results in surface damage, henceforth increasing the delamination factor with respect to increase in cutting speed. Results from this research proposed that with increase in cutting speed, the surface roughness of the machined surface tends to decrease, while a good surface finish can be attained for natural fiber composites at minimal feed rate. The role of fiber orientation while fabricating natural fiber composites, along with the number of flutes in a milling cutter, also has a great impact in defining the cutting forces, delamination factor, and surface roughness for natural fiber polymer composites while carrying out milling operations.

John et al. (2021) reinforced jute, kenaf, and rice husk into polypropylene matrix to fabricate different natural fiber composites, and the effect of milling operation at different cutting speeds and feed rates was investigated. Based on the results, it was observed that based on the developed natural fiber composites (jute/polypropylene, rice husk/polypropylene, and Kenaf/ polypropylene), cleaner cuts with better surface finish were observed with minimum delamination factor for Kenaf/ polypropylene composites, while rice husk/polypropylene composites exhibited the worst delamination characteristics. This machined surface characteristic can be attributed to the shear-type cutting mechanism that happens for Kenaf/polypropylene composites, while the other two composites exhibit crushed shear breakages and bending of fibers during machining during milling operation, leading to poor surface quality. Along with this, high hardness and abrasive nature of rice husk when compared with kenaf fibers tend to increase the cutting forces generated during milling operation of rice husk/polypropylene composites, leading to the wearing out of tool. This phenomenon leads to the plowing out of material during machining processing damaging the composite surface further. This research also proposed that with increase in the spindle speed, the surface finish of developed polymer composites tends to increase, while increasing feed rate and depth of cut will adversely affect the surface finish of the machined surface.

Slamani et al. (2022) reinforced a 41% volume fraction of flax fiber into epoxy resin via resin transfer molding process to study the effect of milling process parameters on the surface quality of the developed composites. The study provided that the cutting feed rate has a major impact on the cutting force generated during the machining process when compared with spindle speed. With an increase in feed rate during the milling operation, the thickness of sheared uncut chips increases, making the polymer composites to resist further rupture. This resisting phenomenon leads to increase in cutting force for chip removal. It was also observed that the orientation of the fibers while fabricating has a major role in defining the surface characteristics of a machined surface, along with feed rate and cutting speed.

8.3.1 Process Parameters of Milling Operation

The cutting parameters that have to be considered while machining natural fiber composites are cutting speed, feed rate, and depth of cut. Selection of optimized processing parameters for milling natural fiber composites as inapt processing conditions leads to material damage through microcracks, delamination, fiber pull-out, etc., which in turn reduces the mechanical characteristics of the material (Harder et al., 2023; Chegdani and El Mansori, 2021). Identifying the process parameters of natural fiber-based composites mainly depends on the matric and fiber characteristics and processing techniques.

8.3.2 Effect of Process Parameters on Machinability

While milling natural fiber composites, damage occurs, and the most commonly found damage includes delamination, fiber pull-out, fiber breakage, etc., which will lead to diminishing of material properties. Henceforth, the reduction of delamination factor has to be considered during the milling operation of natural fiber composites, and this can be achieved only through proper selection of cutting parameters and tooling. Figure 8.5 displays the effect of cutting speed and feed rate on the delamination factor of a polypropylene/jute composites, which clarifies that with increase in cutting speed and feed rate, delamination factor tends to increase, thereby damaging the surface of the machined area. It is also evident from studies that with increase in depth of cut, the delamination factor due to increase in temperature during the machining process.

With increase in spindle speed, friction generated between end cutter and the workpiece increases, thereby generating high temperatures capable enough to soften the matrix region, thereby leading to de-cohesion of matrix and fiber. This can be considered as the major mechanism behind

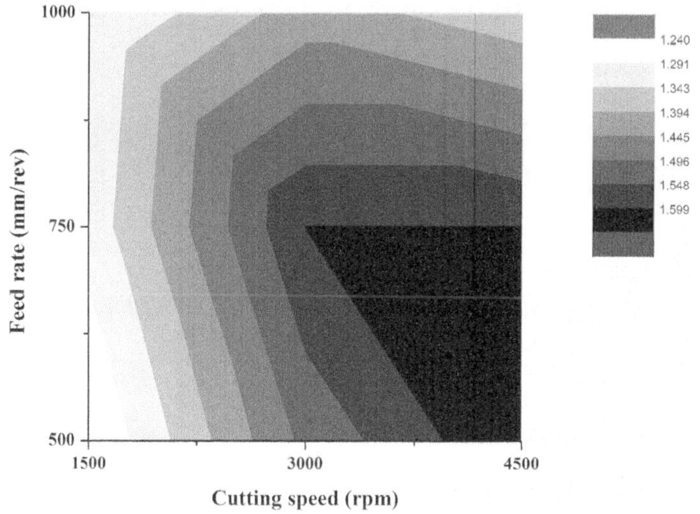

FIGURE 8.5 Delamination factor with respect to speed and feed rate.

the increase in delamination factor with respect to the increase in spindle speed. Similarly, as feed rate is increased, end mill cutters won't be able to carry out complete or proper chipping away of hard fibers, and instead these fibers will be broken and pushed away from the machined surface, thereby increasing the delamination factor. Along with this, the temperature produced at the machined area during the milling operation can also be attributed to an increase in delamination factor with respect to the considered feed rate in this research. It is also evident from Figure 8.5 that after a feed rate level, delamination factor tends to decrease, and this can be attributed to the minimal temperature rise that happens during machining at the machined area. Depth of cut is also a prominent factor leading to delamination during the milling operation. With respect to increase in depth of cut, delamination factor of machined natural fiber composites tends to increase. Studies have stated that as depth of cut increases, cutting force increases, which in turn increases the friction between workpiece and end mill cutter increases, thereby leading to delamination due to the melting of polymer matrix.

Surface roughness is considered to be an important factor when considering a machining process, as a reduction in surface quality, including dimension and tolerance variation, would affect processes, including assembly and joining processes, for which optimized machining parameters have to be identified. From Figure 8.6, it is evident that with an increase in feed rate, surface roughness value tends to increase, thereby reducing the integrity of the machined surface. This reduction in surface integrity can be due to the high compressive load acting on material surface by end mill cutting edges, leading to high machining forces. With increase in cutting force generated, surface roughness value of the machined surface also tends to decrease. Surface finish is dependent on spindle speed such that with increase in spindle speed, surface finish tends to increase. This increase in surface finish with respect to increase in cutting speed, as evident from Figure 8.6, is due to the high stress created on the machining area by mill cutter edges. With increase in spindle speed, temperature generated at machining surface increases, thereby softening the matrix polymer, and the cutting force provided by the cutting edges on the machining area provides a superior surface finish appearance. With increase in depth of cut, surface roughness tends to increase as the cutting force exerted on the machined area by the end cutter edges tends to increase. This increase in cutting force leads to increase in cutting temperature, thereby diminishing the surface roughness of machined natural fiber composites.

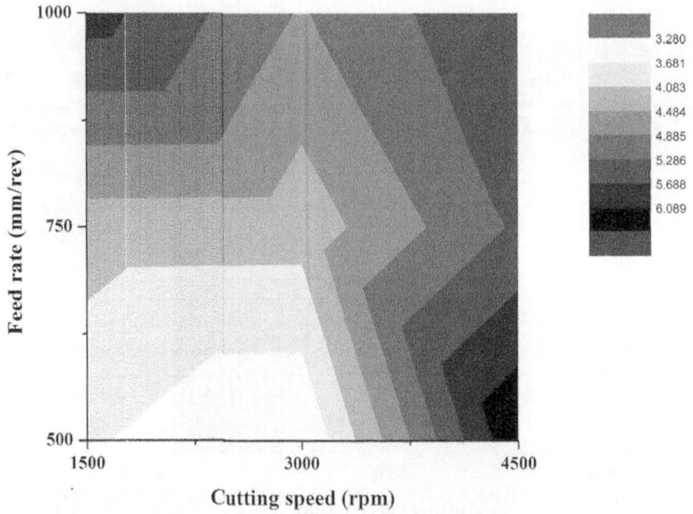

FIGURE 8.6 Surface roughness with respect to speed and feed rate.

8.4 DRILLING OF NATURAL FIBER COMPOSITES

Applications of natural fiber composites include structural applications in automobiles, electronic components, furnishing, etc., wherein assembly is a major and mandatory process. Assembly of natural fibers for the above-mentioned applications includes cutting the material to the required dimensions and joining with the aid of screws and bolts, for which a drilling operation is entailed. However, drilling natural fiber composites of anisotropic nature is a tedious process as machining chances are there for damaging the surface integrity of the material due to various defects, including delamination, fiber pull-out, etc., which in turn would reduce the mechanical strength of the material. This can be overcome only through the identification of proper drilling parameters, tool geometry, tool material, etc. Main drilling parameters include drilling speed and feed rate.

The major irregularities that happen during the conventional drilling of natural fiber composites include delamination, micro-cracking, fiber pull-out, and burning out, of which delamination and fiber pull-out are the major damages that are induced during drilling operations. Delamination can be considered as one of the major failure phenomena that happens during drilling of fiber composites at the start and end of the drilling operation, in which the separation of layer happens during the rotation of drill. Delamination occurring at top surface of the composite layer while drilling due to the pulling up of upper laminated is called as peel-up lamination and pushing out of the laminates by the rotating drill bit at the end of the drilling operation is called as push out delamination, and the same is considered the major damage that occurs during the drilling of a natural fiber polymer composites (Rampal et al., 2022; Malik et al., 2022; Nassar et al., 2017). Fiber pull-out occurs in the natural fiber composites as the generated mechanical forces pull away the fiber from the matrix material by breaking the bond between the fiber and matrix material. This would pave the path to the formation of fatigue cracks, thereby leading to the failure of the natural fiber composites. The major quality attributes that have to be considered while carrying out the drilling operation include circularity, surface roughness, and heat-affected zone.

In a study to define the role of speed, feed, angle point, and tool size on the output responses of Vetiveria zizanioides root-reinforced polymer composites fabricated via compression molding processes, a central composite design was established based on the process parameters, and the drilling was carried out (Boopathi et al., 2023). Based on the results attained, it was monitored that point angle and feed rate have a prominent influence on the entry deface, while spindle speed, feed rate, and point angle majorly affect the exit deface. It was also concluded that tool diameter has minimal influence on deface and thrust force, while all the other factors, including spindle speed, feed, and point angle, have a major influence on the thrust force generation. An increase in thrust force was observed throughout the research with an increase in feed rate, which can be attributed to the increase in axial force developed during the drilling operation, which in turn increases the thrust force. It was also observed that with an increase in the point angle, the area covered by the cutting tool also increases, thereby increasing the thrust force. In another research, an attempt was made to analyze the behavior of different fibers in polymer composites while carrying out a drilling operation. In this research, natural fiber-reinforced polymer composites with a 20 weight fraction of coir, along with 16 weight fraction of natural fibers such as coir, jute, flax, human hair, cotton, sisal, Calotropis, and Kenaf. It was proposed that the strength of fiber, percentage elongation, and the interfacial bonding between the matrix and the fiber highly influence the delamination factor and ovality of the machined surface. It was observed that the hybrid composites with flax or jute fibers along with coir fibers exhibited better delamination factor and ovality than the other set of composites due to their interfacial bonding of fibers with matrix, along with the softness and flexibility of the fibers.

8.4.1 Process Parameters of Drilling Operation

During drilling operations of natural fiber composites, one of the major defects that has been observed is delamination, which leads to interply failure. Sub-surface deformations, including crack development, fiber pull-out, fiber breakage, and matrix fiber de-cohesion, also affect the machinability and surface integrity of natural fiber composites during the drilling operation. In order to carry out drilling operations in natural fiber composites effectively, processing parameters have to be optimized properly, and this includes both machining parameters and tool parameters. Machining parameters include spindle speed and feed rate, while tool parameters include tool geometry, tool material, and types. Machinability of natural fiber composites can be quantified with the aid of material removal rate, surface quality, and cutting forces.

8.4.2 Effect of Process Parameters on Machinability

The main processing parameters that have to optimize while machining natural fiber composites are spindle speed and feed rate; however, delamination and subsurface cracks also have to be considered while carrying out drilling operations. It was evident from research that, while machining natural fiber composites at high spindle speed and feed rate, material removal rate tends to increase. This increase in material removal rate with respect to increase in spindle speed and feed rate can be attributed to the cutting force produced, leading to high temperature and softening of the matric material. Along with this, with an increase in cutting speed, the primary and secondary cutting edges of the drill bit come in contact with the workpiece, leading to cutting away of the material.

Delamination is a major factor to be considered while drilling natural fiber-based composites as the delamination would lead to reduction in mechanical properties and at the same time would have an adverse effect on surface integrity. Optimum spindle speed and feed rate have to be decided while drilling a natural fiber composite as higher spindle speed leads to reduced delamination in the entry and exit points of the drilled area, while at lower spindle speed the presence of cracks and sub-cracks was observed, making the composites unfit for precise applications. Variation of surface roughness with respect to increase in spindle speed is as shown in Figure 8.7.

Surface roughness values attained after drilling a natural fiber composite are found to be less trustworthy when compared with a conventional metal due to fiber pull-outs. However, based on applications, the surface roughness of natural fiber composites has to be evaluated and maintained. An evaluation of surface roughness values of a natural fiber composite with respect to cutting speed and feed rate is as shown in Figure 8.8.

It is evident from the attained results that with increase in feed rate, surface roughness value of the drilled natural fiber composites tends to increase, which can be attributed to the high cutting edge

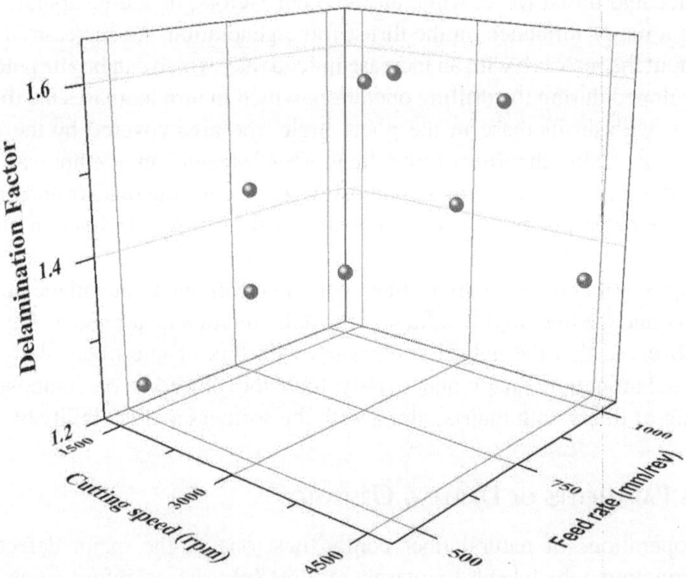

FIGURE 8.7 Delamination factor with respect to speed and feed rate during drilling.

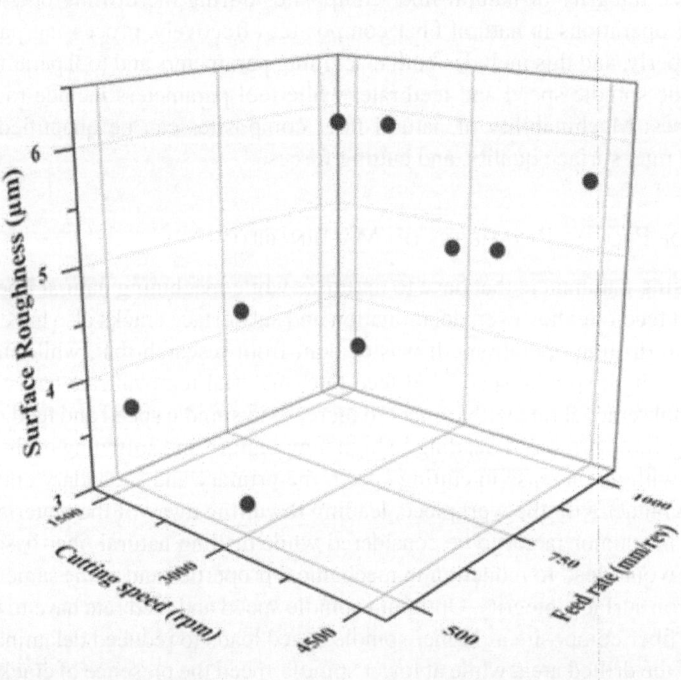

FIGURE 8.8 Surface roughness with respect to speed and feed rate during drilling.

contact pressure exerted on the fiber composites. This leads to a high compressive load acting on the drilled area, thereby reducing the surface integrity of the machined area. Low spindle speed for drilling operation of a natural fiber composite leads to high cutting force at the machining area, leading to high stress and eventually an uneven surface. However, during drilling at high spindle speed, high temperature will be generated in the machining area, leading to softening of the matrix. The cutting edge, when passing above the softened surface, will produce a smeared surface, thereby providing a better surface integrity.

8.5 SUMMARY

The role of conventional machining in shaping the natural fiber composites in accordance with dimensional requirements is briefed in this book chapter. The effectiveness of each machining operation, including turning, milling, and drilling, along with the effect of its process parameters in defining the machinability, is detailed with the aid of results. Studies have proved that the major machining parameters, including cutting speed, feed rate, and depth of cut, have to be optimized and selected properly to carry out natural fiber composites as improper selection results in defects, including delamination, cracks, de-cohesion, and fiber pull-out.

REFERENCES

Alarifi, Ibrahim M. 2023. "A Review on Factors Affecting Machinability and Properties of Fiber-Reinforced Polymer Composites." *Journal of Natural Fibers* 20 (1). https://doi.org/10.1080/15440478.2022.2154304

Boopathi, Sampathi, V. Balasubramani, and R. Sanjeev Kumar. 2023. "Influences of Various Natural Fibers on the Mechanical and Drilling Characteristics of Coir-Fiber-Based Hybrid Epoxy Composites." *Engineering Research Express* 5 (1). https://doi.org/10.1088/2631-8695/acb132

Çelik, Yahya Hışman, and Mehmet Suat Alp 2022. "Determination of Milling Performance of Jute and Flax Fiber Reinforced Composites." *Journal of Natural Fibers* 19 (2): 782–796. https://doi.org/10.1080/15440478.2020.1764435

Çelik, Yahya H., Erol Kilickap, and Ali İmran Kilickap 2019. "An Experimental Study on Milling of Natural Fiber (Jute)- Reinforced Polymer Composites." *Journal of Composite Materials* 53 (22): 3127–3137. https://doi.org/10.1177/0021998319826373

Chegdani, Faissal, and Mohamed El Mansori 2021. Machining Behavior of Natural Fiber Composites. *Encyclopedia of Materials: Composites*. Vol. 3. Elsevier Ltd. https://doi.org/10.1016/B978-0-12-819724-0.00013-6

Delahaigue, J., J. F. Chatelain, and G. Lebrun. 2017. "Machining Analysis of Unidirectional and Bi-Directional Flax-Epoxy Composite Laminates." *Proceedings of the Institution of Mechanical Engineers, Part L: Journal of Materials: Design and Applications* 231 (1–2): 196–209. https://doi.org/10.1177/1464420716671970

Eichhorn, S. J., C. A. Baillie, N. Zafeiropoulos, L. Y. Mwaikambo, M. P. Ansell, A. Dufresne, K. M. Entwistle, et al. 2001. "Current International Research into Cellulosic Fibres and Composites." *Journal of Materials Science* 36 (9): 2107–2131. https://doi.org/10.1023/A:1017512029696

Equbal, Azhar, Mohammad Shamim, Irfan Anjum Badruddin, Md Israr Equbal, Anoop Kumar Sood, Nik Nazri Nik Ghazali, and Zahid A. Khan 2020. "Application of the Combined ANN and GA for Multi-Response Optimization of Cutting Parameters for the Turning of Glass Fiber-Reinforced Polymer Composites." *Mathematics* 8 (6). https://doi.org/10.3390/MATH8060947

Gurunathan, T., Smita Mohanty, and Sanjay K. Nayak. 2015. "A Review of the Recent Developments in Biocomposites Based on Natural Fibres and Their Application Perspectives." *Composites Part A: Applied Science and Manufacturing* 77: 1–25. https://doi.org/10.1016/j.compositesa.2015.06.007

Harder, Nicole, Arturo Rodriguez-Uribe, Michael R. Snowdon, Manjusri Misra, and Amar K. Mohanty. 2023. "Hop Natural Fiber-Reinforced Poly(Butylene Succinate-Co-Butylene Adipate) (PBSA) Biodegradable Plastics: Effect of Fiber Length on the Performance of Biocomposites." *Materials Advances* 4 (6): 1502–1514. https://doi.org/10.1039/d2ma00831a

Hejjaji, Akshay, Redouane Zitoune, Ameur Mohamed Fayçal, and Bougherara Habiba 2018. "5. Challenges of Machining Natural Fiber-Reinforced Composites: A Review." *Drilling Technology*, no. December: 149–160. https://doi.org/10.1515/9783110481204-005

Jani, S. P., A. Senthil Kumar, M. Adam Khan, and M. Uthaya Kumar 2016. "Machinablity of Hybrid Natural Fiber Composite with and without Filler as Reinforcement." *Materials and Manufacturing Processes* 31 (10): 1393–1399. https://doi.org/10.1080/10426914.2015.1117633

Jawaid, M., and H. P.S. Abdul Khalil 2011. "Cellulosic/synthetic fibre reinforced polymer hybrid composites: A review." *Carbohydrate Polymers* 86 (1): 1–18. https://doi.org/10.1016/j.carbpol.2011.04.043

John, Raveen, Richard Lin, Krishnan Jayaraman, and Debes Bhattacharyya 2021. "Effects of Machining Parameters on Surface Quality of Composites Reinforced with Natural Fibers." *Materials and Manufacturing Processes* 36 (1): 73–83. https://doi.org/10.1080/10426914.2020.1813896

Júnior, Marcos Gonçalves, Thiago Valle França, Carlos Alberto Fortulan, Rodrigo Henriques Lopes da Silva, and Cesar Renato Foschini 2022. "Green Ceramic Machining Benefits through Ultrasonic-Assisted Turning: An Experimental Investigation." *International Journal of Advanced Manufacturing Technology* 118 (9–10): 3091–3104. https://doi.org/10.1007/s00170-021-08174-0

Keya Kamrun N., Nasrin A. Kona, Farjana A. Koly, Kazi Madina Maraz, M. Naimul Islam, and Ruhul A. Khan 2019. “Natural Fiber Reinforced Polymer Composites: History, Types, Advantages, and Applications.” *Materials Engineering Research* 1 (2): 69–87. https://doi.org/10.25082/mer.2019.02.006

Khan, Tabrej, Mohamed Thariq Bin Hameed Sultan, and Ahmad Hamdan Ariffin. 2018. “The Challenges of Natural Fiber in Manufacturing, Material Selection, and Technology Application: A Review.” *Journal of Reinforced Plastics and Composites* 37 (11): 770–779. https://doi.org/10.1177/0731684418756762

Lee, Dong-Woo, Jung-Suck Oh, and Jung-Il Song 2017. “Parametric Study for Hole Machining in Natural Fiber Composites.” *Composites Research* 30 (1): 35–40. https://doi.org/10.7234/composres.2017.30.1.035

Lotfi, Amirhossein, Huaizhong Li, Dzung Viet Dao, and Gangadhara Prusty. 2021. “Natural Fiber–Reinforced Composites: A Review on Material, Manufacturing, and Machinability.” *Journal of Thermoplastic Composite Materials* 34 (2): 238–284. https://doi.org/10.1177/0892705719844546

Lu, Na, Robert H. Swan, and Ian Ferguson 2012. “Composition, Structure, and Mechanical Properties of Hemp Fiber Reinforced Composite with Recycled High-Density Polyethylene Matrix.” *Journal of Composite Materials* 46 (16): 1915–1924. https://doi.org/10.1177/0021998311427778

Mahadevappa, Nagamadhu, Vijay K. Shankar, Samuel Sehgal, and Rajath Upadhya. 2020. “Study the impact of drilling process parameters on natural fiber reinforced herringbone epoxy composites.” *Annales de Chimie: Science Des Materiaux* 44 (5): 339–345. https://doi.org/10.18280/acsm.440506

Malik, Khurshid, Faiz Ahmad, and Ebru Gunister. 2022. “Drilling Performance of Natural Fiber Reinforced Polymer Composites: A Review.” *Journal of Natural Fibers* 19 (12): 4761–4779. https://doi.org/10.1080/15440478.2020.1870624

Mansour, Gabriel, Panagiotis Kyratsis, Apostolos Korlos, and Dimitrios Tzetzis. 2021. “Investigation into the Effect of Cutting Conditions in Turning on the Surface Properties of Filament Winding Gfrp Pipe Rings.” *Machines* 9 (1): 1–15. https://doi.org/10.3390/machines9010016

Maya, M. G., Soney C. George, Thomasukutty Jose, M. S. Sreekala, and Sabu Thomas 2017. “Mechanical Properties of Short Sisal Fibre Reinforced Phenol Formaldehyde Eco-Friendly Composites.” *Polymers from Renewable Resources* 8 (1): 27–42. https://doi.org/10.1177/204124791700800103

Naik, Venkatesh, Mohan Kumar, and Vijayananda Kaup. 2022. “A Review on Natural Fiber Composite Materials in Automotive Applications.” *Engineered Science* 18: 1–10. doi:10.30919/es8d589

Nandhakumar, S., K. Mithresh Kanna, A. Mohammed Riyas, and M. Nathin Bharath 2020. “Experimental Investigations on Natural Fiber Reinforced Composites.” *Materials Today: Proceedings* 37 (Part 2): 2905–2908. https://doi.org/10.1016/j.matpr.2020.08.669

Nassar, Mahmoud M. A., Ramanathan Arunachalam, and Khalid I. Alzebdeh 2017. “Machinability of Natural Fiber Reinforced Composites: A Review.” *International Journal of Advanced Manufacturing Technology* 88 (9–12): 2985–3004. https://doi.org/10.1007/s00170-016-9010-9

Pandya, Vishal J., and Pravin P. Rathod. 2020. “Optimization of Mechanical Properties of Green Composites by Gray Relational Analysis.” *Materials Today: Proceedings* 27: 19–22. https://doi.org/10.1016/j.matpr.2019.08.166

Prakash, R. B., S. Satish, and C. J. Thomas Renald 2014. “Investigation on Tensile and Flexural Properties of Coir Fiber Reinforced Isophthalic Polyester Composites.” *International Journal of Current Engineering and Technology* 2: 2347–5161.

Raj, Sonia S., J. Edwin Raja Dhas, and C. P. Jesuthanam 2021. “Challenges on Machining Characteristics of Natural Fiber-Reinforced Composites – A Review.” *Journal of Reinforced Plastics and Composites* 40 (1–2): 41–69. https://doi.org/10.1177/0731684420940773

Rajmohan, T., R. Vinayagamoorthy, and K. Mohan. 2019. “Review on Effect Machining Parameters on Performance of Natural Fibre–Reinforced Composites (NFRCs).” *Journal of Thermoplastic Composite Materials* 32 (9): 1282–1302. https://doi.org/10.1177/0892705718796541

Ramesh, M., A. Gopinath, and C. Deepa 2016. “Machining Characteristics of Fiber Reinforced Polymer Composites: A Review.” *Indian Journal of Science and Technology* 9 (42). https://doi.org/10.17485/ijst/2016/v9i42/101978

Rampal, Gaurav Kumar, Sanjay Mavinkere Rangappa, Suchart Siengchin, and Sunny Zafar 2022. “A Review of Recent Advancements in Drilling of Fiber-Reinforced Polymer Composites.” *Composites Part C: Open Access* 9 (August): 100312. https://doi.org/10.1016/j.jcomc.2022.100312

Slamani, M., H. Chafai, and J. F. Chatelain 2022. “Effect of Milling Parameters on the Surface Quality of a Flax Fiber-Reinforced Polymer Composite.” *Proceedings of the Institution of Mechanical Engineers, Part E: Journal of Process Mechanical Engineering* 2022: 1–8. https://doi.org/10.1177/09544089221126087

Sujuan, Wang, Zhang Tao, Deng Wenping, Sun Zhanwen, and Sandy To 2022. “Analytical Modeling and Prediction of Cutting Forces in Orthogonal Turning: A Review.” *International Journal of Advanced Manufacturing Technology* 119 (3–4): 1407–1434. https://doi.org/10.1007/s00170-021-08114-y

Tan, Boon Khoon, Yern Chee Ching, Sin Chew Poh, Luqman Chuah Abdullah, and Seng Neon Gan 2015. "A Review of Natural Fiber Reinforced Poly(Vinyl Alcohol) Based Composites: Application and Opportunity." *Polymers* 7 (11): 2205–2222. https://doi.org/10.3390/polym7111509

Vinayagamoorthy, R., I. V. Manoj, G. Narendra Kumar, I. Sai Chand, G. V. Sai Charan Kumar, and K. Suneel Kumar. 2018 "Challenges on the Synthesis, Characterization and Machining of Green Fiber Plastics: A Review." *IOP Conference Series: Materials Science and Engineering* 390 (1): 1–7. https://doi.org/10.1088/1757-899X/390/1/012029

Wang, Zimo, Ruiqi Guo, Qiyang Ma, Faissal Chegdani, Bruce Tai, Mohamed El Mansori, and Satish T. S. Bukkapatnam. 2022. "Characterization of the Physical Origins of Acoustic Emission (AE) from Natural Fiber Reinforced Polymers (NFRPs) Machining Processes." *International Journal of Advanced Manufacturing Technology* 118 (3–4): 865–879. https://doi.org/10.1007/s00170-021-07956-w

9 Machining and Machinability of Nanofiber Composites

Advanced and Hybrid Techniques

Bismark Sarkodie, Benjamin Tawiah, and Bin Fei

9.1 INTRODUCTION

Composites are enthralling materials composed of two or more components having contrasting physical, chemical, and mechanical characteristics. Composites are widely used in the aerospace, military, automotive, construction materials, and other engineering fields because of their many advantages, which include good fatigue and corrosion resistance, high specific strength and modulus, ease of fabrication, high design flexibility, and acceptable thermal contraction attributes [1]. Among them, fiber-reinforced composites are significantly functional in the aforementioned areas of application. Traditional fiber-reinforced composites (with diameters in the micrometer range) have excellent structural characteristics like a high specific modulus and strength; notwithstanding, the incredible advancement of nanotechnology presents a further option that uses nanofibers as reinforcement in the fabrication of nanocomposites. Owing to the incredible size effect of nanofibers, nanofiber composites are a novel and adaptable class of materials that provide even better mechanical properties, such as high modulus, high tensile strength, toughness, and large surface area in composites, and thereby improve the structural properties of composites [1–3]. Fiber-reinforced composites are made up largely of matrix and sturdy and rigid fibers (reinforcements). The primary features of nano-reinforcements for such reinforcement effects are that their properties are far higher than those of the reinforcing materials now in use, and their surface area to volume ratio is very high, allowing for stronger interfacial contact with the matrix [1, 4, 5]. Other fibers such as carbon fibers provide electrical and thermal conductivity [6].

The application of composites required the machining of various parts before assembly. Among them, conventional machining has been appropriate under simplified conditions for basic geometries. However, they also require extensive post-machine operations, which are mostly incapable of manufacturing complex geometrical features, making the products more costly [7]. Because of the heterogeneity and anisotropic nature of fiber-reinforced composites, the machinability phases of these materials are distinct from those of metallic materials and their alloys [8]. As a result, during machining, varying forces that result in delamination, fiber pull-out, and many other flaws are encountered. The main global efforts to reduce machining-induced damage have focused on improving process parameters, creating newer tool designs, particularly for cutting fiber-reinforced composites, and creating theoretical models for forecasting critical forces [9]. Advanced machining and hybridization of machining techniques have been identified as an effective approach to properly machining heterogeneous composites with minimal damage to the properties and surface of the composites. Herein, this chapter discusses frequently used nanofibers in composites, advanced and hybrid machining techniques commonly used for fiber-reinforced composites, their effect on the machined composites, and merits and challenges of the technologies in machining nanofiber composites.

DOI: 10.1201/9781032665337-9

9.2 MACHINABILITY OF NANOFIBER COMPOSITES

The machinability of a workpiece is contingent on the elements present in the composites. The individual constituents influence their interaction and their machinability. For instance, owing to poor bonding and reduced compatibility with polymer matrices, higher delamination occurs in natural micrometer fiber composites. Moreover, the lower heat resistance of natural fibers poses significant issues during the machining process [10]. The usage of nanofibers in most composites before machining helps reduce or overcome such challenges. Limited literature is available on natural nanofiber-reinforced composites. Among nanofiber composites, carbon nanofibers find huge applications in advanced composites, machined into the desired form/part, due to their concomitant attributes such as hardness, brittleness, and conductivity. The mechanical behavior of carbon nanofiber composites (CNC) is excellent, including high stiffness-to-weight ratios, strength-to-weight ratios, resistance to corrosion, fracture toughness, and greater dimensional stability. Hence, CNC has gained popularity in several manufacturing industries [11, 12]. The two primary components of CNC composites are resin matrix and reinforcing fibers. These components' different physical and mechanical properties provide the materials with their anisotropic machinability [12]. The tensile strength and modulus of elasticity range from 450–650 ksi and 28–35 $\times 10^6$ psi, respectively [13]. Carbon nanofiber/polymer composites serve as a substitute for typical engineered metallic alloys on account of their better characteristics that perform intriguing functions during and after the machining process [9].

9.3 MACHINING OF NANOFIBER COMPOSITES

In many circumstances, machining composites is unavoidable to ease the assembly of functioning components. For example, holes must be drilled for bolts or rivets [14]. Machining is the process of removing material from a workpiece to fashion a product with the requisite form, size, and surface polish. A hard wedge-shaped tool is used on the workpiece in the traditional method of removing material, also known as conventional machining. The tool makes physical contact with the workpiece, causing stress and causing fracture/damage on the workpiece's surface. Because fiber-reinforced composites (FRCs) are anisotropic materials, drilling increases the complexity of issue, which then influences the strength of the workpiece. Therefore, to achieve excellent mechanical properties without compromising them in machining, it is important to introduce low residual stresses around the hole area during drilling. Other factors that could influence the quality of the drilled hole are fiber/resin pull-out and poor surface hole finish. In FRCs drilling and machining, delamination is the most common issue [15]. Delamination usually occurs at the surface of the drill's entry and, typically, the exit. In advanced engineering composites assembly, it is essential to avoid at least up to about 60% of such defects to ensure the product remains strong and efficient during its lifetime, but delamination can adversely impair the performance of the finished product. Induced thrust force and torque are the origins of delamination in FRC drilling. It has been determined that the delamination factor of drilled FRCs is higher than that of glass fiber-reinforced composites due to the localized buckling of natural fibers compared to glass fibers. Natural FRC delamination decreases with increased cutting speed and increases with increasing feed rate due to an increase in thrust force and drill diameter [16]. As the point angle increases, delamination at the hole entry decreases while delamination at the hole exit increases. Table 9.1 summarizes the major drilling investigations on natural fiber-reinforced composites (NFRCs).

Consequently, traditional machining was shown to be inadequate for extremely hard materials. It is also not suggested for fabricating difficult and smaller forms, as well as achieving high accuracy, precision, and surface integrity [17]. Some damages caused include delamination, fiber pull-outs, and inter-laminar crack propagation [14]. These motivated researchers to develop unique methods of removing the material, known as advanced machining technologies, which are largely used by

TABLE 9.1
Drilling Conditions and Performance Parameters of NFRCs [16]

Fiber Used	Machining Conditions	Performance Parameters	Performance Indicators
Banana, sisal, and roselle	Drill diameter, spindle speed, and feed rate	Thrust force, torque, and delamination	Feed rate and drill diameter are the key performance drivers
Hemp	Spindle speed and feed rate	Deamination factors and tensile strength of hole	Lower feed rate and high spindle speed are required for improved performance
Vetiver and jute	Spindle speed, feed rate, and point angle	Thrust force and torque	Spindle speed has no influence on thrust force and torque
Sisal-glass sandwich	Drill diameter, spindle speed, and feed rate	Thrust force and torque	For improved performance, high spindle speed, low feed rate, and drill diameter are preferred
Sisal-epoxy and sisal polypropylene	Drill geometry, feed rate, and spindle speed	Chip formation characteristics, thrust force, and drill quality	Tooling requisite is substantially different for improved performance
Jute	Feed rate and spindle speed	Delamination factor	Feed rate is a major determinant of drill hole quality
Cenosphere	Spindle speed, feed rate, and drill diameter	Thrust force and surface roughness	Drill hole quality was determined by wt% cenosphere loading
Woven jute	Drill geometry, feed rate, and spindle speed	Thrust force, torque, and delamination	Delamination is caused by the thrust force, with parabolic showing much improved cutting behavior than Jo and twist drilling
Jute, hemp, banana, and glass	Spindle speed and feed rate	Thrust force, torque, and delamination	Natural fiber performs equally to glass fiber – indicating possible replacement of glass fiber with the natural fibers
Basalt and sisal	Spindle speed, feed rate, and drill diameter	Thrust force and delamination	Feed rate influences delamination and also drill diameter
Coire-glass sandwich	Spindle speed, feed rate, and drill diameter	Thrust force, torque, and tool wear	Feed rate influences quality most followed by drill diameter
Human hair and jute	Spindle speed, feed rate, and point angle	Surface roughness	Surface roughness is influenced significantly by drill point angle but has no significant effect on the hole quality

companies. The absence of a wedge-shaped tool in the removal of material is a defining feature of modern machining processes [17]. An extensive description of the mode of materials removal in advanced machining is provided in Section 1.4.

A fiber-reinforced composite's response to machining is dependent on the characteristics of the matrix and fibers, as well as the composite itself. Other factors that impact the choice of a specific process include the type of machining operation, part geometry and size, finish and accuracy requirements, quantity and variety of parts, including the materials used in the parts, availability of cutting tools and the right machine, availability of in-house technology, current machining practices, production schedule, capital requirements, justification for new equipment, and overall costs [13]. In addition to being inhomogeneous, the material behavior of fiber-reinforced

TABLE 9.2
Optimal Requirements for Machining NFRCs [16]

Fibers Used	Major Performance Indicators	Optimal Conditions
Sisal-Jute	Thrust force, torque, and delamination	Spindle speed = 3000r/min Feed rate = 0.04 mm/rev Drill diameter = 6mm
Jute -steel	Thrust force, surface roughness, and damage.	Speed = 1070 r/min Feed = 0.1 mm/rev Point angle = 90^0 Tool size = 8 mm
Jute	Delamination	Speed 500-970 r/min Feed = 0.1 mm/rev Point angle = $90 - 120^0$
Hemp	Delamination	Feed 100 mm/min Spindle speed = 2000 r/m
Sisal-banana	Delamination	Feed rate 0.1mm/rev Spindle speed 1200 r/min Drill diameter = 3 mm
Coire-glass	Thrust force, torque, and tool wear	Feed rate = 0.2 mm/rev Spindle speed = 1500 r/min Drill diameter 8 mm
Medium-density fiberboards	Delamination and exit	Spindle speed = 2950 r/min Feed rate = 1000 mm/min Drill diameter = 4 mm

composites during machining also depends on a variety of fiber and matrix characteristics, such as fiber direction, as well as the volumes of the matrix and fibers.

It is critical to determine the best machining settings for NFRCs in order to obtain superior performance, such as minimal delamination, tool wear, good surface smoothness, and so on. When NFRCs are machined under prime machining settings, labor and operational costs are usually low. Using a response surface model and artificial intelligence, machining parameters such as cutting speed, feed rate, tool geometry, workpiece geometry, fiber orientation, and fiber volume fraction could be adjusted [15]. In addition, an iterative design strategy can be used to optimize the drilling parameters, such as spindle speed and feed rate, based on several performance criteria. Table 9.2 shows a typical ideal setup for machining NFRCs. The majority of the data show that greater spindle speed, lower feed rate, and minimal drill diameter are the optimal parameters for successful machining of NFRCs [16]. However, it is important to state that fiber matrix loading and other processing parameters could influence the supposed ideal condition in Table 9.2.

For cutting nanofiber composites like carbon fiber-reinforced composites, the viability and performance of many advanced machining processes have been assessed, including ultrasonic, abrasive waterjet (AWJM), and laser beam machining (LBM). The advantages include decreased cutting forces, tool wear, and reduced delamination. Yet there are still certain downsides, particularly with laser machining, such as workpiece delamination and the existence of heat-affected zones (HAZ) [19]. Figure 9.1 shows the assessment of HAZ and surface roughness of a machined composite using various tools. The research and development efforts in the field of advanced machining techniques may result in the conception of an ideal procedure that is suited for damage-free machining of a variety of fiber-reinforced composite parts [9].

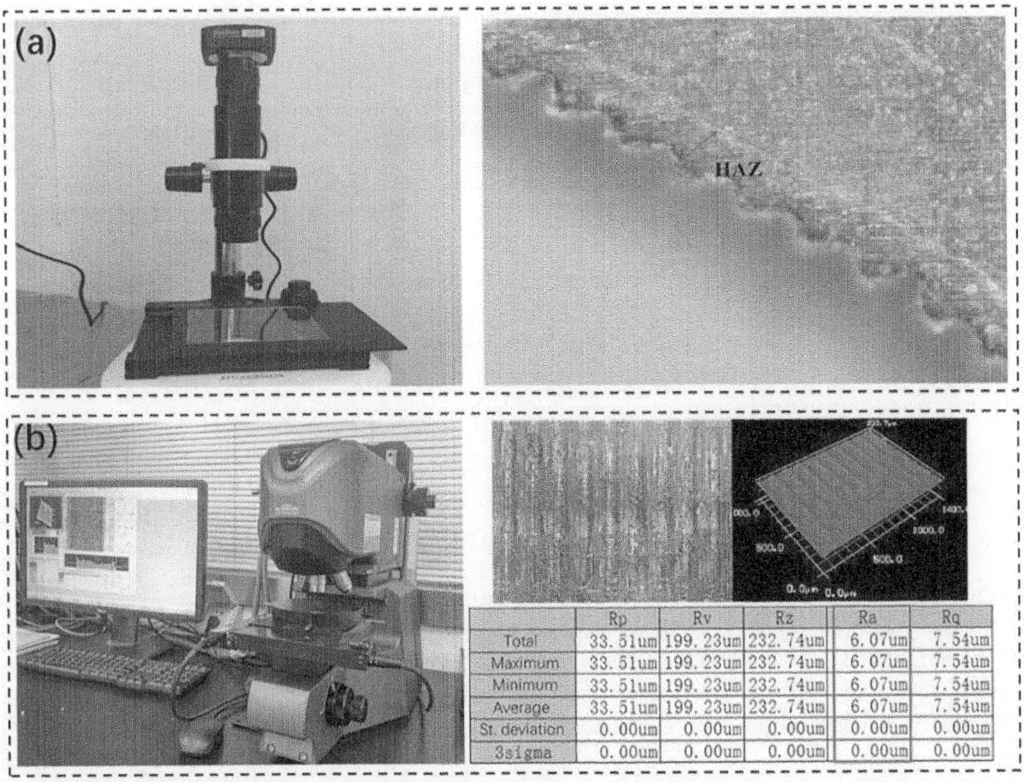

	Rp	Rv	Rz	Ra	Rq
Total	33. 51um	199. 23um	232. 74um	6. 07um	7. 54um
Maximum	33. 51um	199. 23um	232. 74um	6. 07um	7. 54um
Minimum	33. 51um	199. 23um	232. 74um	6. 07um	7. 54um
Average	33. 51um	199. 23um	232. 74um	6. 07um	7. 54um
St. deviation	0. 00um	0. 00um	0. 00um	0. 00um	0. 00um
3sigma	0. 00um	0. 00um	0. 00um	0. 00um	0. 00um

FIGURE 9.1 (a) Examining the heat-affected zone around the cutting hole with the microscope and (b) surface roughness with laser confocal microscope [18].

9.4 ADVANCED MACHINING OF NANOFIBER COMPOSITES

Advanced machining procedures are a specific category of procedures where the substrate and tool do not come into direct contact. Generally, different kinds of energy, including electrical energy, kinetic energy, and chemical energy, are used in advanced machining techniques to cut the product or workpiece. Advanced machining methods for nanofiber composites are diverse and offer the following benefits [20]:

i. High precision and accuracy in component production.
ii. Making complicated and complex forms.
iii. Improved micromachining capabilities.
iv. Faster production rate while processing difficult-to-machine materials than traditional machining procedures.
v. Using appropriate advanced machining technologies, material (such as brittle, ductile, hard, and non-homogeneous) from low to high thickness may be machined.
vi. Precision machining of deep holes with a tiny diameter.
vii. Parts with a high surface polish and tight tolerance are manufactured.

Despite the adaptability of sophisticated machining methods, there are certain drawbacks, such as the following:

i. Quite a hefty initial outlay.
ii. Certain cutting-edge machining techniques have a very low rate of material removal.

iii. Advanced machining techniques demand skilled operators to use them.
iv. Machining techniques such as ECM, Electron Beam Melting (EBM), and electrical discharge machining (EDM); it's not possible to machine electrically non-conductive materials.

A few examples are the use of lasers, EDM, AWJM, ultrasonic machining, electrochemical spark machining, and other advanced machining methods.

A well-guided wedge-shaped tool is not used in advanced machining operations that require creating tensions in the workpiece. The material can be melted or vaporized via a thermal effect, as in the case of laser beam machining, electron beam machining, and EDM. Using chemical and electrochemical principles, machining procedures deliver exceptionally superior surface polish. Based on the mechanism used to remove the material, one may categorize advanced machining methods, as shown in Figure 9.2.

Performing advanced machining processes in non-optimized circumstances may result in poorly shaped and dimensionally inaccurate products, poor surface finishes, removal of less material at a slower pace, and less productivity overall. Consequently, to obtain the optimal input parametric values and obtain the best output response, it is required to utilize proper modeling and optimization approaches in advanced machining processes. Therefore, to produce the best set of machining parameters, it is imperative to select an optimization technique that is capable of solving such single- and multi-objective type complex optimization problems [20]. Advanced machining processes have a large number of input parameters or factors, and the outputs may be multiple.

Industries are using advanced machining techniques because they are affordable and popular [17]. A comprehensive review of abrasive waterjet machining is presented in the literature [17]. The heat flow and the material removal process have been modeled in various ways. Other studies used statistical or numerical approaches to improve the process settings to reduce HAZ, kerf width, and taper (Figure 9.3), increasing energy usage and improving the productivity and quality of the machining. The Taguchi approach and response surface methodology were used in the modeling and optimization processes.

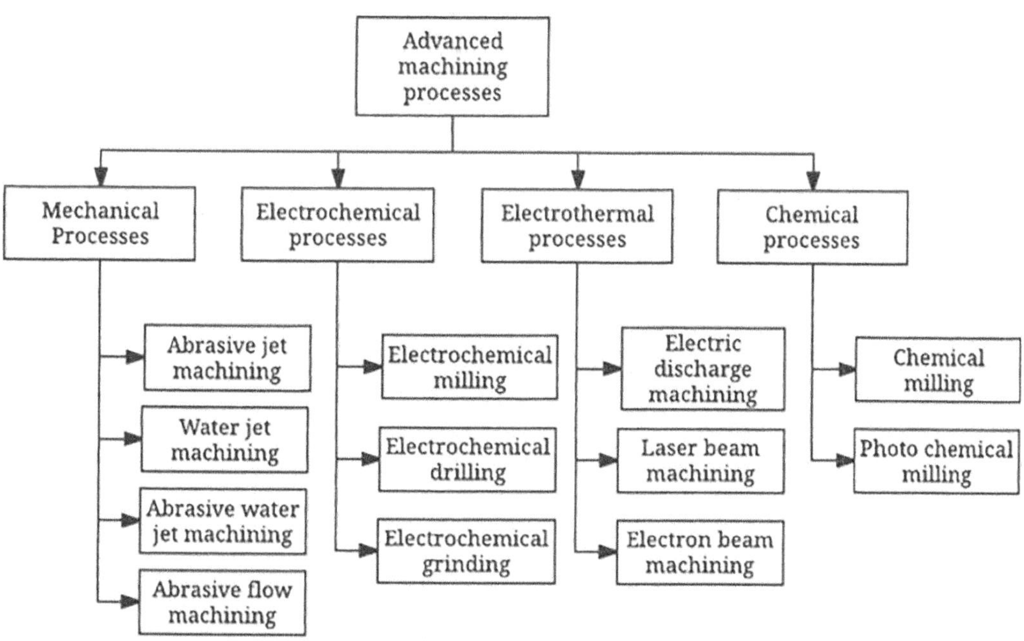

FIGURE 9.2 Classification of advanced machining processes based on the mechanism of material removal [20].

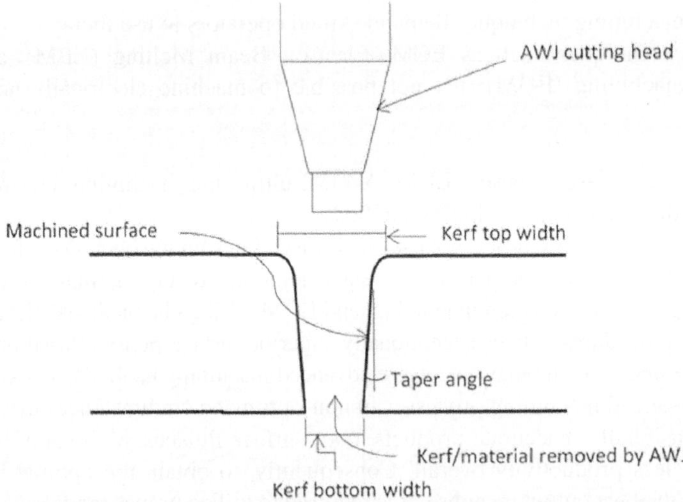

FIGURE 9.3 The kerf taper produced by advanced machining technique: Abrasive waterjet machining [21].

Using the analysis of variance (ANOVA) approach, the influence of process control factors and their interactions on HAZ and cut quality attributes (kerf width, depth, and taper) were examined [22]. Each method consists of specific benefits and drawbacks. The parameters in machining are described below.

$$DF = \frac{D_{\max}}{D_{nom}} \tag{9.1}$$

The delamination factor (DF), as illustrated in Figure 9.4, is obtained by dividing $D_{\mathbf{nom}}$ (the nominal hole diameter) by $D_{\max}$ (the circumference of a circle that contains the maximum delamination damage and is centered on the circle of the drilled hole). Due to the severity of the delamination damage at the hole entry (top surface), DF was only assessed there [19].

Equation (9.2) defines the term "hole taper" as half of the difference between the entrance and exit hole diameters [19].

$$Taper = \frac{\left(D_{entry} - D_{exit}\right)}{2} \tag{9.2}$$

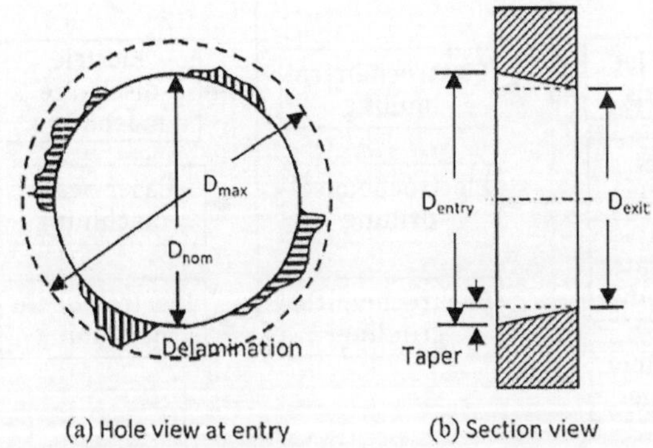

FIGURE 9.4 Evaluation of delamination factor (left) and taper (right) [19].

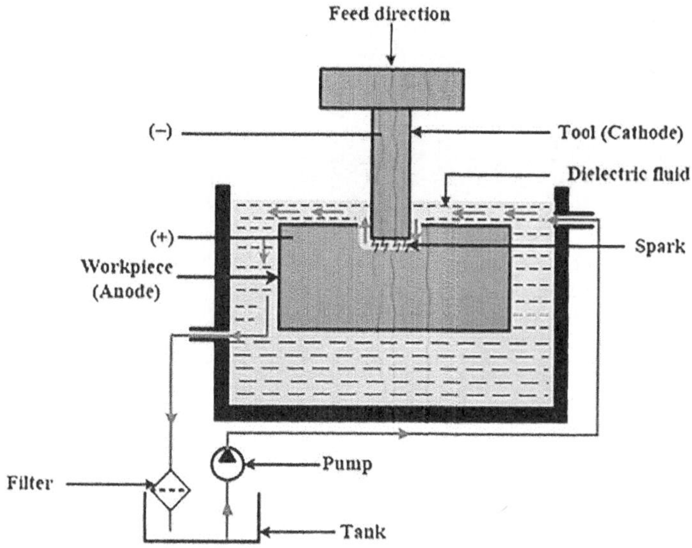

FIGURE 9.5 A representation of the electric discharge machining process [17].

The area of the HAZ is obtained from a cross-sectional view of the hole created using a thin blade abrasive cut-off saw in Scanning Electron Microscope (SEM) images to quantify heat-induced damage [19].

9.4.1 Electrical Discharge Machining Technique

By using electrical sparks that are carefully regulated and produced between an electrode and the workpiece, materials are removed in the EDM process [23]. In order to precisely remove undesirable elements from conductive materials during the course of electric discharge machining (EDM), electrodes are used. Moreover, a dielectric fluid is used in the process, totally immersing the workpiece and electrode to ensure that no major contact forces are generated between them. The capacity to produce electric sparks that remove undesirable areas on the material (workpiece) is the main working principle of EDM. In most cases, there is very little space between the workpiece and the electrode so that they do not come into contact (almost as the thickness of a hair). The primary benefit of electric discharge machining is its ability to trim the workpiece more effectively than other cutting processes. Electric discharge machining also allows users to operate on already hardened work components so far as they are conductive [24]. Images of the outcome of EDM at various phases are provided in Figure 9.6.

For wire EDM of unidirectional carbon fiber-reinforced plastic composite laminates, the effect of cutting direction on workpiece fiber orientation, as well as important operational parameters such as open gap voltage, ignition current, and pulse-on and pulse-off duration, has been studied [23]. Material removal rates were higher when machining parallel to the fiber direction (2.41 mm^3/min) than when machining perpendicular to the fiber direction (2.08 mm^3/min), with ignition current and pulse-off time being statistically significant determinants. This is due to the workpiece's increased electrical conductivity over the fiber length, which results in larger discharge energy. Furthermore, specimens machined perpendicular to the fiber orientation typically show severe delamination on both the top and bottom surfaces, while items machined parallel to the fiber direction are often free of substantial edge flaws.

A study looking into the methods of machining damage and material removal revealed that the primary mechanisms responsible for material removal in the drilling of carbon fiber-epoxy composite by

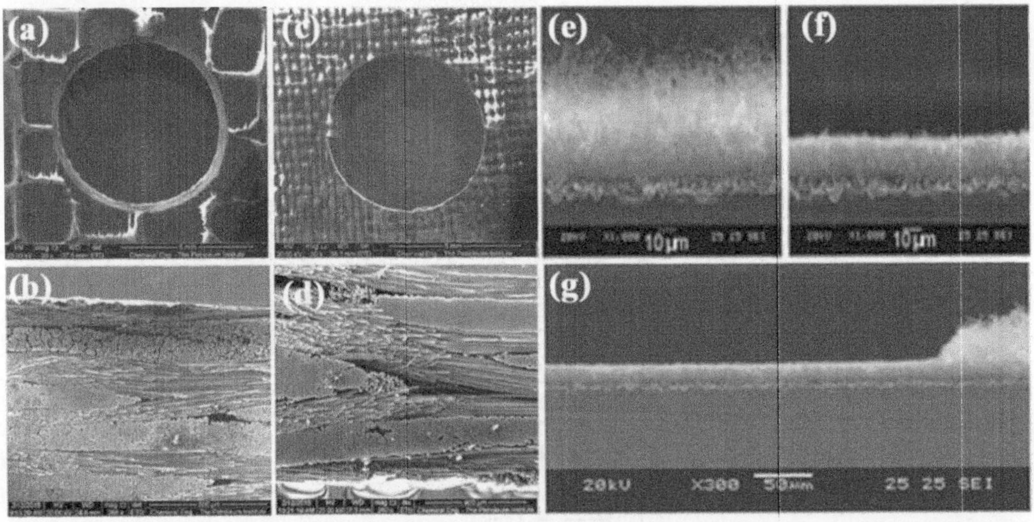

FIGURE 9.6 Images of drill hole at entry (a) and exit (c), morphology of machined composite near the top (b) and bottom (d) [19], and SEM images before (e) and (f) after (g) electric discharge machining carbon nanofiber film [25].

EDM include the dissolution of the polymer matrix and the thermally induced fracture of the carbon fibers [19]. A proposed process was the vaporization of the carbon fibers caused by spark and Joule heating. The diameter of the heat-affected zone in most cases is shown to be largely influenced by pulse-on time, with intermediate pulse-on time having the smallest heat-affected zone. In this case, the width of the heat-affected zone and hole taper achieved by EDM is usually equivalent to or smaller than those attained by laser cutting.

9.4.2 Abrasive Waterjet Machining Technique

The mechanism for the abrasive waterjet is dependent on the erosion effect, which removes the target material by employing water that is flowing at a high flow rate, speed, and pressure via a tiny nozzle (Figure 9.7). To accelerate cutting and enhance cutting quality, hard grains are added to water that is currently flowing. A lean double-acting plunger that is powered by a hydraulic cylinder supplies the abrasive waterjet machine with highly pressured and propelled water. Abrasive waterjet machining equipment and mechanisms are described in detail in the following references [17, 26–29]. Surface roughness, material removal rate, and kerf taper angle are the output characteristics of the AWJM cutting process. These parameters are inextricably linked to the input parameters. The abrasion process influences the strength of the nanofiber-reinforced composite machined by AWJM. As a consequence, the tensile strength of AWJM nanofiber-reinforced composites remains somewhat lower than that of composites cut using standard techniques. [21]. The AWJM process incorporates many process factors and their effects on the target material's machining performance. There are several studies in the literature that predict mathematical models for AWJM quality, and this model is validated by designed experimental conditions [26]. The hydraulic pressure, traverse rate, stand-off distance, and abrasive mass flow rate are among the important tool physiognomies that are taken into account when machining using the AWJM process [30].

Because of its lack of noise, lack of heat effects, minimum stress placed on the workpiece, great machining adaptability, and high flexibility, this technique is less susceptible to material qualities. According to research on the effect of different factors on the width of the kerf taper of carbon-reinforced polymer composite, the stand-off distance and abrasive mass flow have an impact on kerf width, while hydraulic pressure and traverse speed have an impact on the top/bottom kerf width ratio

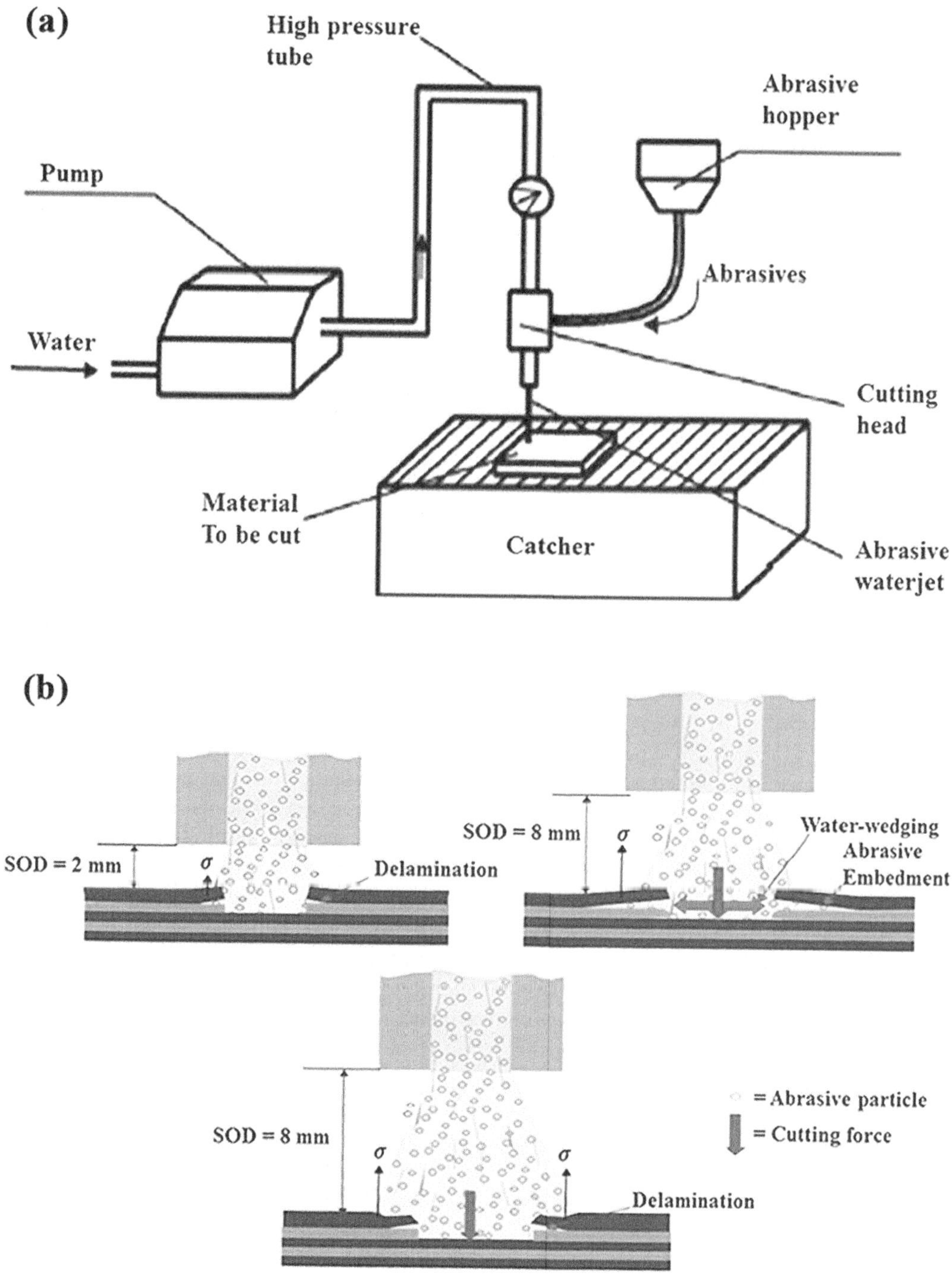

FIGURE 9.7 (a) AWJM system and (b) effect of stand-off distance on delamination [26].

[29]. In an investigation using Taguchi's approach and ANOVA, the kind of abrasive materials, hydraulic pressure, stand-off distance, and traversal rate are additional important control factors, whereas the cutting direction is an inconsequential control factor in determining the surface roughness [31]. Surface roughness was most affected by the types of glass fibers and the thickness of the composite laminate when considering the effect of noise sources. To forecast how the cutting parameters of the AWJM will

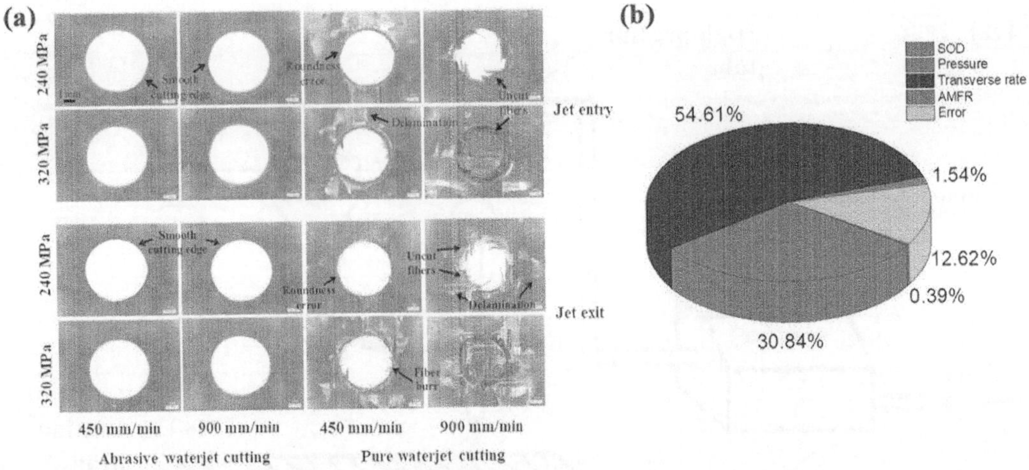

FIGURE 9.8 (a) Entry and exit hole quality with different processing parameters and (b) influence of process parameters on surface roughness [32].

affect the performance of surface roughness, a mathematical model employing piecewise linear regression analysis was created. Via confirmation testing in relation to the selected reference parameter setting, the improvement in the quality attributes has been confirmed. It was verified that the chosen AWJM parameter combination meets the actual requirements for cutting glass fiber-reinforced epoxy composites [31]. It is important to note that the abrasive grains have a significant influence on this machining procedure. The results of AWJM and abrasive-free waterjet in machining support this (Figure 9.8a) [26].

Dhanawade et al. [30] investigated the mechanism of delamination and kerf geometry in carbon epoxy composite abrasive waterjet machining. As a statistical method for designing an experiment, the experiment was done using the response surface method. It has been discovered that delamination lessens with increased pressure and abrasive mass flow rate, along with reduced stand-off distance and traverse rate. When pressure rises, the traverse rate slows, and the stand-off distance shortens, the kerf taper ratio also drops. As the stand-off distance is shortened and the traverse rate is accelerated, the kerf top width shrinks. On the basis of their investigation, mathematical models were developed to predict the maximum delamination length, kerf taper ratio, and kerf top width [30]. According to an ANOVA-based study on the impact of machining parameters on the surface roughness of a carbon fiber/epoxy composite, traverse rate and pressure are extremely important variables in regulating surface roughness (Figure 9.8b) [33]. An optimum level of process parameters (stand-off distance (SOD) – 2.0 mm, pressure – 260 MPa, traverse rate – 50 mm/min, abrasive mass flow rate (AMFR) – 800 g/min) – examined using grey relational analysis, shows the improved surface finish of about 10.75%. In samples that were machined at low pressure and rapid traverse rate, delamination and fiber pull-out were noticeable. Also, work on carbon fiber-reinforced composite was conducted under variable traverse feeds and SOD [34]. Increased traverse feed tends to enhance volumetric removal rate and surface roughness while decreasing mean top and bottom kerf width and mean kerf profile area (Ra). With rising SOD, the volumetric removal rate, average top and bottom kerf widths, average kerf profile area, and surface roughness all rise. The surface roughness rises in the downward cut direction. Low traverse feeds are suggested for better kerf surface quality. When both traverse feed and SOD rise, the kerf taper also rises. The mean top and bottom kerf widths may be generated under the chosen operating circumstances within a tolerance of ±34 μm, and the average surface roughness is equal to 5.1 μm ± 0.2 μm [34]. In addition, Kumar et al. [35] demonstrate how changing the AWJM machining parameters affects the cut quality of an adhesively bonded titanium-carbon fiber-reinforced plastics-titanium hybrid laminate as measured by material removal rate, metal composite interface damage factor, taper ratio, and surface roughness. The best process variables for improving the quality of AWJM cuts were 200 MPa jet pressure, 237.693 mm/min traverse speed, and 1 mm stand-off distance. The computed values for material removal rate, interface damage factor for metal composites, taper ratio, and surface roughness were 5.388 mm^3/s, 1.41, 1.16, and 3.827 m, respectively.

9.4.3 Laser Beam Machining Technique

The basic idea underlying laser beam machining is that when the work material interacts with an intense, highly focused, coherent monochromatic beam of light, the areas earmarked for removal are predominantly detached by melting and/or vaporization. For resin matrix materials, it is shredded by a chemical breakdown. Nanofiber-reinforced composites in general have a high absorption capacity for infrared radiation typical of that generated by a CO_2 laser. As opposed to metals, where the vapor column is enclosed with molten material, the polymers often do not undergo a fusion process. Because of this and the way that plastics conduct heat, vaporization can occur at significantly lower specific powers (10-3-10-5 W/cm^2) [13]. Figure 9.9 displays the process of laser beam machining graphically. Many processes, including melting, evaporation, and chemical degradation, make up the mechanism of material removal during LBM (the disintegration of chemical bonds causes the degradation of materials). A detailed explanation of the laser beam machining process can be found in the literature [38]. The basics of lasers and applications of thermal modeling, notably in laser machining, have also been described in a book chapter by [39]. Also, the speed at which material is removed while utilizing a laser cutter is unaffected by restrictions like maximum tool force, built-up edge development, or tool chatter [37]. The use of fiber optics enables the beams to be focused at certain angles. Parts with intricate forms may be produced using laser milling without the need for costly tooling. One-sided geometry pieces or components that need to be partially machined from just one side are best suited for laser milling [37]. The quantity of energy (E) absorbed per unit area per unit time is affected by the laser power (P), laser scanning speed (V), and laser spot diameter (D), as stated in Equation (9.1), where A is a constant coefficient calculated by the laser absorption coefficient of the composite [36].

$$E = A \times \frac{P}{V \times D^2} \tag{9.3}$$

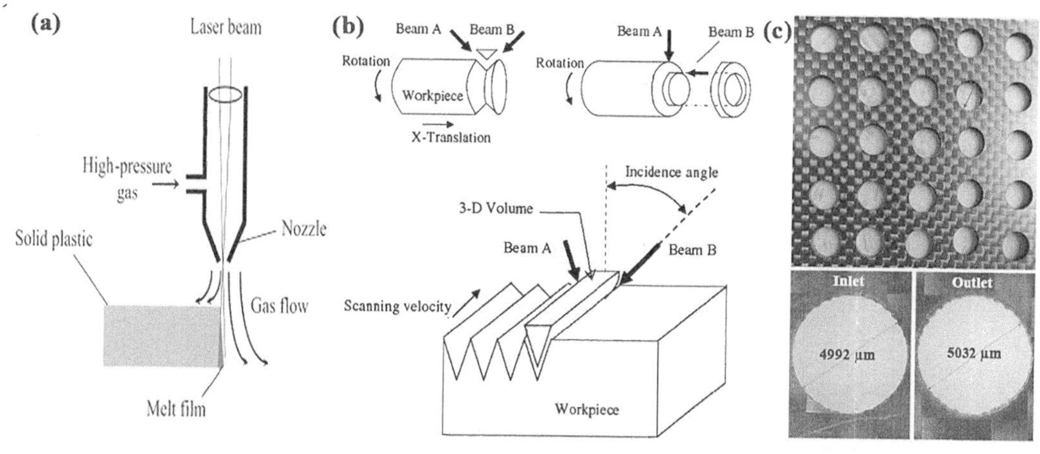

FIGURE 9.9 (a) Schematic diagram of the laser beam machining [36], (b) three-dimensional laser machining [37], and (c) holes of laser-machined carbon nanofiber plate with nanosecond laser [18].

The technology can execute almost any machining operation; however, because of the nonlinear behavior of laser parameters, obtaining a lofty level of quality and performance is quite challenging. Thermal modeling offers the ability to understand how the processing state changes into a readable condition without affecting how it operates. Generally, laser beam machining exhibits a highly precise and controlled process that uses a powerful laser to remove material from a workpiece with great accuracy and efficiency.

9.5 HYBRID MACHINING OF NANOFIBER COMPOSITES

The study of machining complex engineered materials has paid particular attention to machining techniques that are a hybridization of two or more machining techniques. These processes are created to boost the advantages and reduce the shortcomings of each constituent process. Using similar parameter values, hybrid machining systems often surpass their combined efficiency. In most of these processes, the interaction of the components may provide an additional contribution in addition to the performance of the individual component processes. Techniques for hybrid machining are shown in Figure 9.10 [40]. Hybrid machining of fiber nanocomposites typically involves a combination of conventional machining procedures such as milling, drilling, and turning, as well as unconventional methods such as laser machining, ultrasonic machining, and waterjet cutting. These different techniques are often combined to create a more efficient and effective machining process that can handle the unique properties of fiber nanocomposites.

One of the challenges of machining nanofiber composites is that the nanofibers can be easily damaged or pulled out of the matrix material. This can lead to weakened or inconsistent material properties. Hybrid machining techniques can help to minimize this damage by using a combination of cutting and non-cutting actions that can remove material without damaging the fibers. In addition, hybrid machining can be used to create more complex shapes and geometries in nanofiber-reinforced composites. By combining different machining techniques, it is possible to create intricate patterns, pockets, and other features that are challenging to produce with a single machining technique.

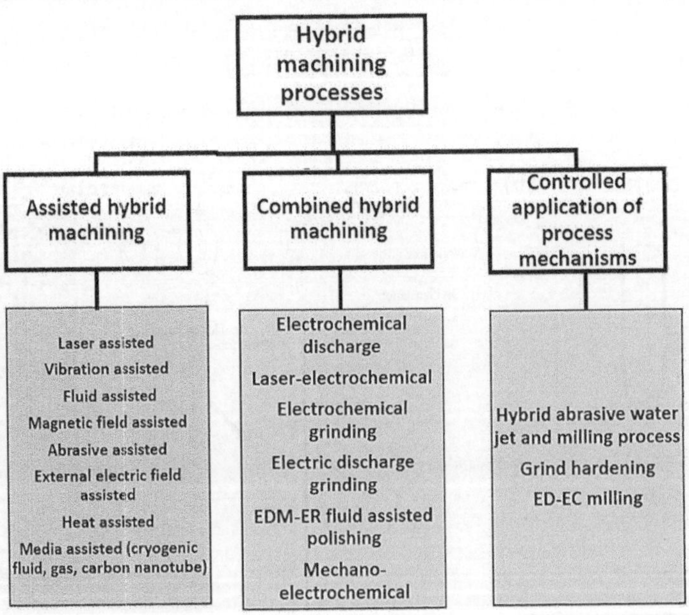

FIGURE 9.10 Broad classification of hybrid machining processes [40].

The majority of hybrid machining techniques were created by merging traditional or sophisticated machining methods with LBM or ultrasonic machining [37]. Some frequently used hybrid machining techniques for machining CNC have been discussed in the subsections below.

9.5.1 Rotary Ultrasonic Machining technique

Rotating ultrasonic machining (RUM), a hybrid machining technology that amalgamates ultrasonic machining and traditional grinding, has been utilized to drill holes in composite materials [42, 43]. The RUM tool vibrates ultrasonically in the direction of the tool axis while rotating at a specific speed, as illustrated in Figure 9.11. The coolant lowers the temperature in the area and removes cutting chips [41]. Experimental research was done by [42] to see if it would be feasible to manufacture carbon fiber-reinforced epoxy using RUM. The thrust force, tool wear, surface roughness, and chips and edge chipping were examined. Thrust force and surface roughness were evaluated as a result of the effects of RUM process factors, thus rotation speed, vibration amplitude, and feed rate. The findings demonstrated that RUM could drill holes in CNC with good efficiency and minimum tool wear. Increased rotational speed and decreased feed rate result in a superior composite surface. For the purpose of RUM edge surface grinding of CNC composites, Wang et al. [41] established a mechanistic model based on both brittle and ductile modes. When numerous process factors are taken into account, the values the model predicted were in consonance with the experimental results, as shown in Figure 9.12.

Also, the impact of several input process parameters (feeding rate, ultrasonic amplitude, tool rotation speed, and depth-of-cut) on the outcomes of experimentation and prediction was evaluated [12] (Figure 9.12). It was discovered that a mechanistic model of feeding-directional cutting forces based on the mode of brittle fracture material removal could be used to forecast how cutting forces would be affected by factors such as *depth of cut, ultrasonic amplitude, feed rate, tool rotation speed, and abrasive properties* [44]. In contrast to traditional drilling, Geng et al. evaluated the diamond core drill's wear mechanism during rotary ultrasonic elliptical machining (r-UEM) of CNC (CD) and the results showed that the drill performance in r-UEM of CNC was much better than that in CD. With less chip adhesion and more grain micro-fracture in r-UEM than in CD, the duration of the stable area increased by 39% and the tool life rose by 28%, respectively. This had a favorable

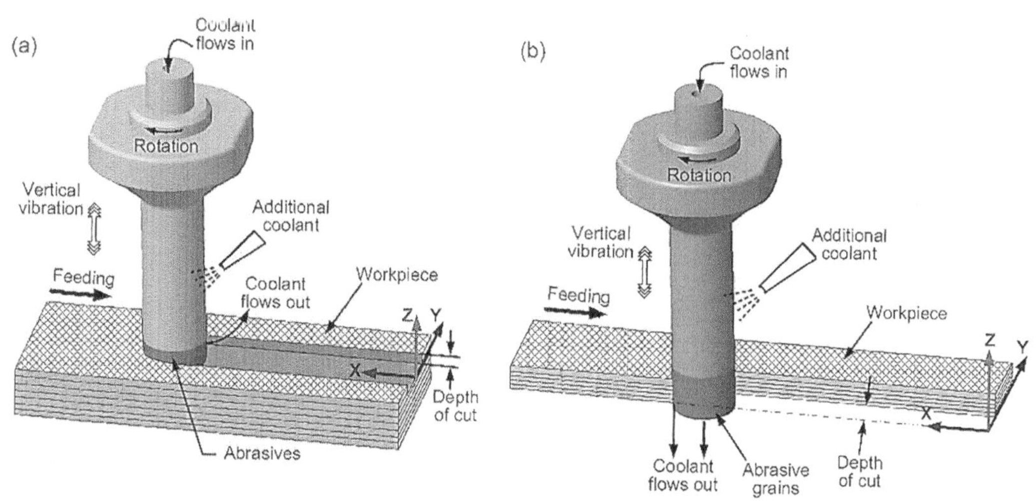

FIGURE 9.11 Illustrations of (a) rotary ultrasonic machining (RUM) end surface grinding process and (b) RUM edge surface grinding process [41].

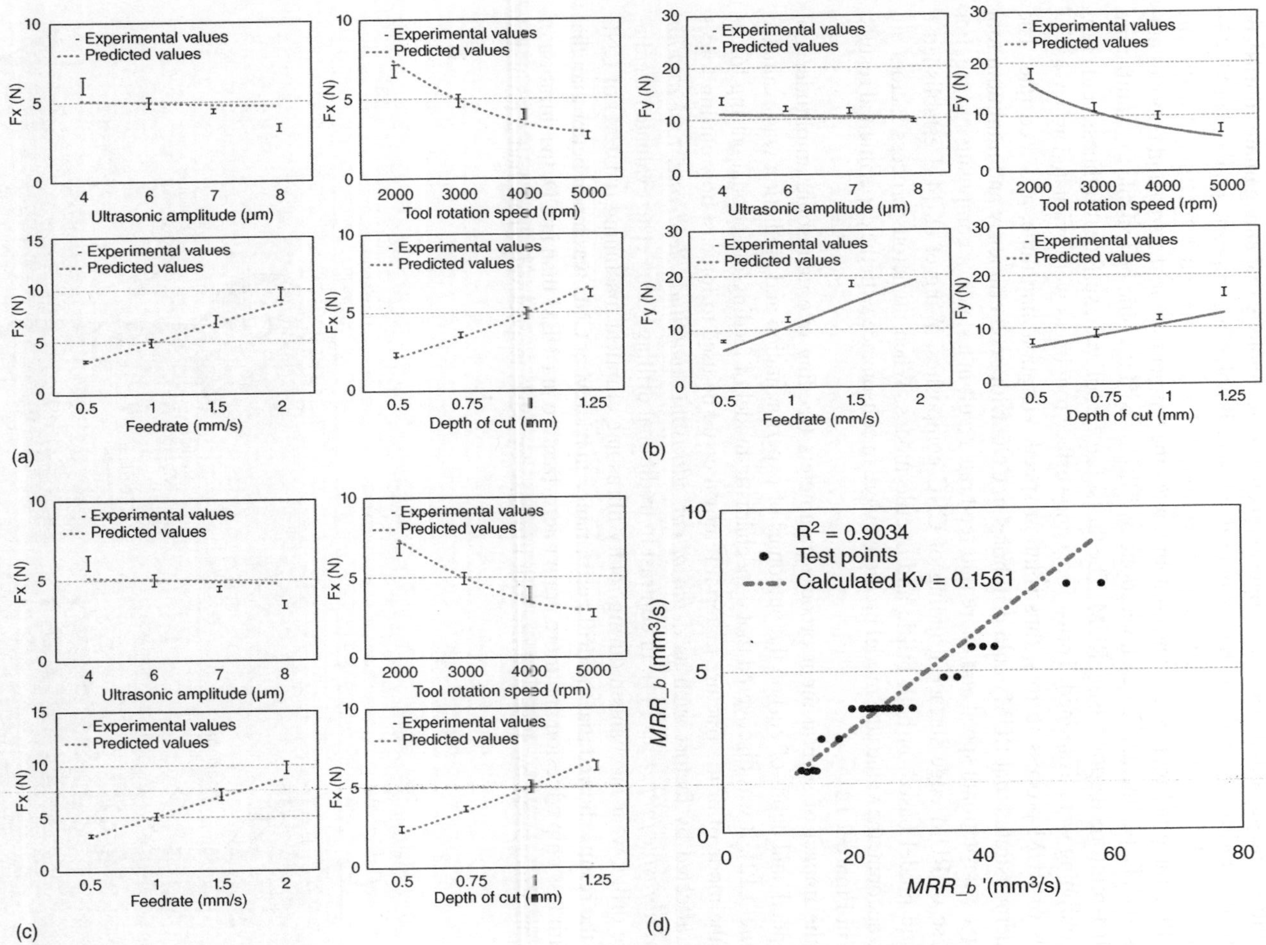

FIGURE 9.12 Appraisal of experimental and predicted results in diverse input considerations [12]: (a) comparisons between a number of feeding-oriented cutting force machining parameters, (b) comparison of various machining parameters for the cutting force and depth-of-cut direction, (c) contrasts between the cutting pressures during feeding and the depth-of-cut directions at different tool variable levels, and (d) the values of the fracture volume factor for different sets of parameters.

impact on the improved drilling performance, including reduced drilling force and a smoother hole surface [45]. A similar result could be observed using rotary ultrasonic helical machining for machining small-diameter CNC holes [11]. Along with a notable increase in the quality of the hole edge and the integrity of the surface, thrust and transverse forces could dramatically decrease. From Figure 9.13c, the hole's inner surface's surface roughness and delamination factor near the hole's exit are significantly lower than they were with CD. This demonstrates the benefits of using CD over ultrasonic machining.

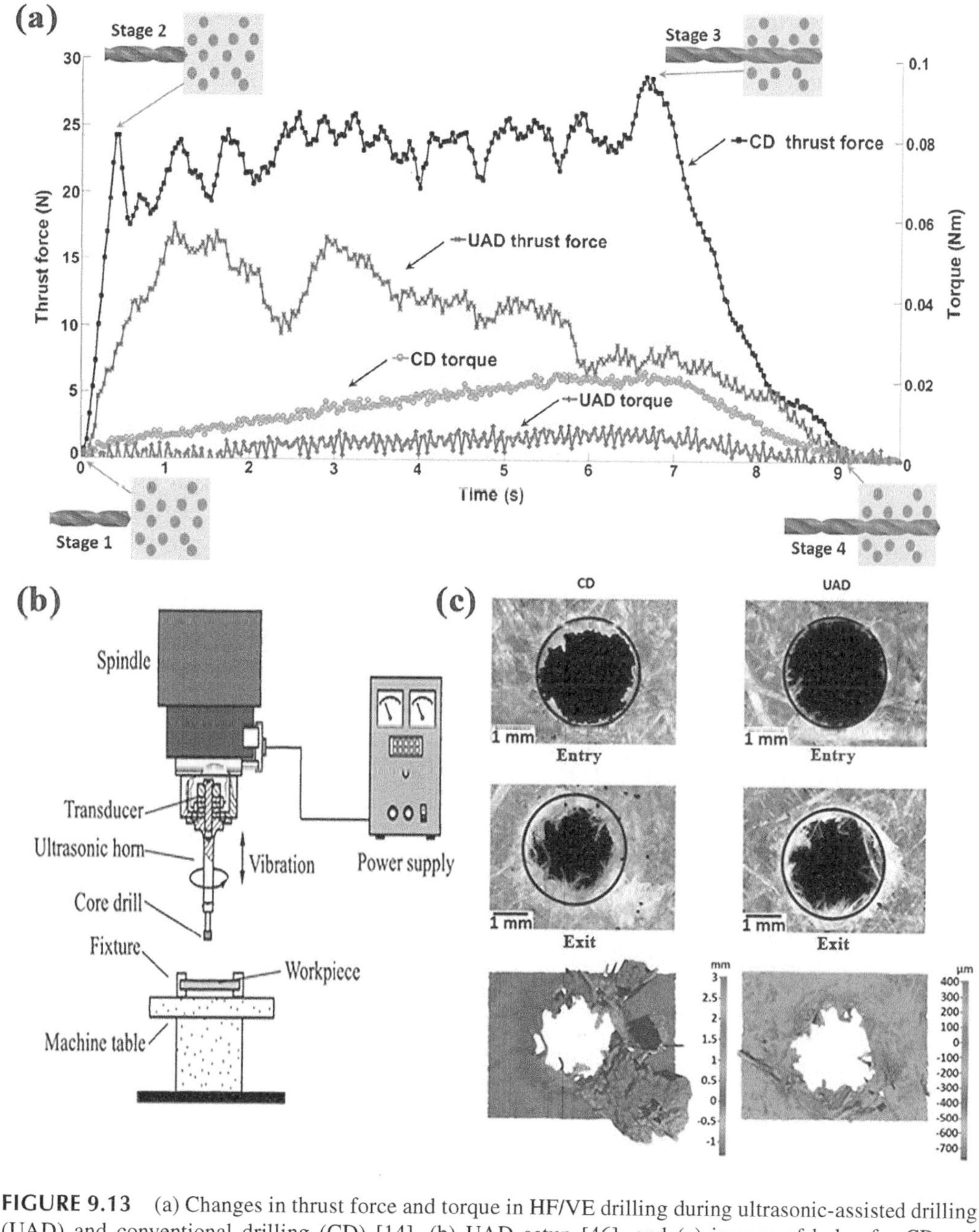

FIGURE 9.13 (a) Changes in thrust force and torque in HF/VE drilling during ultrasonic-assisted drilling (UAD) and conventional drilling (CD) [14], (b) UAD setup [46], and (c) images of holes for CD and UAD [14].

9.5.2 Ultrasonic-Assisted Drilling Technique

One such hybrid machining technique is ultrasonic-assisted drilling (UAD), which applies high-frequency, low-amplitude vibration to drill-bit movement during cutting (Figure 9.14). A significant reduction in thrust force – up to 30% or more – was seen in earlier experimental testing using aerospace-grade carbon/epoxy composites, along with an improvement in hole quality and surface topography [14]. Researchers have shown that ultrasonic machining for CNC drilling works well, minimizes drilling force and tool wear, and enhances hole surface quality [46, 49]. According to Wang et al. [14], drilling results using CD and UAD produced progressive changes in thrust force and torque (Figure 9.13a), as well as images showing how different techniques affect composite surfaces. (Figure 9.13c). Owing to the intrinsic heterogeneity of the composite material, vibrations in the drilling equipment, and high sensitivity of the measuring system, the drill tip fully contacts the workpiece during machining and reaches a plateau with very small oscillations. This is seen in Figure 9.14a. The force decreased to zero after the drill bit had entirely inserted itself into the workpiece [14]. The degree of delamination in the CD is noticed to be greater than that in the UAD, demonstrating that the UAD may successfully lessen the extent of delamination (Figure 9.13c). This is probably due to the lower thrust force in UAD than that in CD.

9.5.3 Electrochemical Discharge Machining Technique

A hybrid form of non-contact machining called electrochemical discharge machining (ECM) combines electric discharge action with electrochemical dissolution. Thermal melting and chemical dissolving are used to remove the substance [40]. In ECM, the workpiece and tool are separated by a tiny distance and function similarly to anodes and cathodes in EDM. EDM uses electrolyte, which is often a concentrated salt solution, to circulate through the machining gap at high speeds to remove reaction products and disperse heat created in the electrolyte cell, as opposed to merely relying on sparks and dielectric fluid to mill a workpiece [50]. It has been determined that ECM is prospective process energy with the capacity to combine with other process energies. Electrical variables (voltage, current, energy), as well as pulse characteristics, may be used to influence material removal rates in ECM (pulse frequency, time, duration, and duty cycle). Process forces and size effects are absent from the equation [51]. In ECM-based integrated processes, material removal is

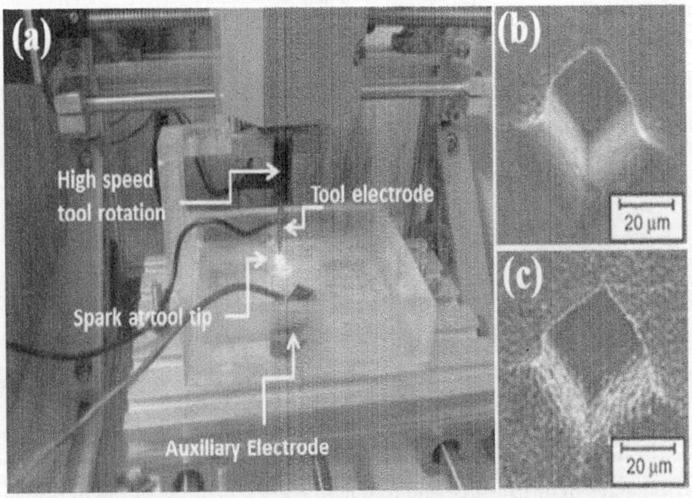

FIGURE 9.14 (a) Experimental setup for ECM [47] and composites machined by (b) ECM and (c) EDM [48].

a joint effort between ECM and the other processes [47]. The resulting synergetic effect significantly exceeds the benefits of individual process energies [51]. Figure 9.14b and c illustrate how EDM and ECM affect composite materials, with ECM enabling better machining. In the EDM process, excellent surface polish decreased HAZ, and dimensional accuracy may be achieved by carefully adjusting the process parameters. EDM has been used to machine carbon fiber-epoxy composite materials [47]. It is noteworthy that increasing tool rotational speed, voltage, and electrolyte concentration results in higher material removal rates and lower overcut hole rates in most natural nanofiber-reinforced polymer composites. After machining the composite, the high-speed electrochemical discharge drilling also correctly seals the fiber ends in the composite's structure, resulting in dimensional precision and superior surface quality. Nevertheless, the look is poor due to the melting and burning of fiber ends.

9.5.4 Laser-Assisted Waterjet Machining Technique

The waterjet laser cutting technique (shown in Figure 9.15) has several merits over traditional lasers since it incorporates the benefits of waterjet and LBM. Most of the energy used for waterjet machining dissipates in the water rather than the material [53]. The material removal by high-speed waterjet is enhanced in laser-waterjet machining, which reduces the thermally induced flaws brought on by laser machining. The laser and waterjet work together to locally heat and soften the workpiece material. The softened material is subsequently ejected by the striking waterjet. The laser-heated area is cooled by the waterjet as well. The method generates surfaces devoid of a heat-affected zone, making it beneficial for cutting carbon fiber composites [40, 53].

The form and temperature distribution of the composite after drilling were found to be considerably influenced by the duty cycle, according to a simulation of the water-guided laser processing of carbon fiber-reinforced polymers. Compared to conventional laser processing, water-guided laser processing can produce superior processing outcomes [54]. To study the transient heating process, Sun et al. [55] created a three-dimensional finite element model. To validate the model, experiments using the same process parameters were carried out. The mechanism involved in the removal of material was elucidated using the finite element model and tests. As the epoxy hits the melting point, it is regarded as removed, and the carbon fiber is removed when the temperature approaches sublimation. As there is almost no heat buildup in between pulses as a result of the waterjet's powerful cooling action, the heat-affected zone width remains constant regardless of the scanning speed. The uniform power distribution in the waterjet causes the kerf sidewall to be comparatively vertical. The findings show that carbon fiber-reinforced polymers may be processed using waterjet-guided laser cutting, which has certain benefits over LBM.

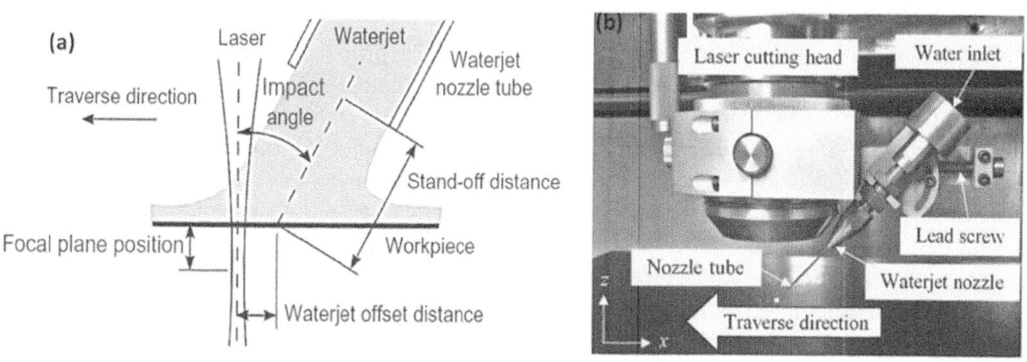

FIGURE 9.15 (a) A line diagram of hybrid laser-waterjet machining and (b) a pictorial image of a hybrid laser-waterjet machine [52].

9.6 ADVANTAGES OF ADVANCED AND HYBRID MACHINING TECHNIQUES

In contrast to traditional production techniques, which require the use of sharp cutting tools, advanced machining technologies do not. They made it possible to create complicated shapes and improved composite micromachining [56].

The waterjet also acts as a means of cooling the area that has been heated by the laser during laser-assisted waterjet machining. The method generates surfaces that are devoid of HAZ, making it effective for cutting a variety of substrates [40, 53].

The best method for machining nanofiber composites that are susceptible to high temperatures is AWJM machining [5] because when the workpiece's surface is washed by the waterjet during the AWJM process, the likelihood of environmental pollution caused by fibrous materials is greatly minimized or avoided [31].

Due to the following benefits of ECM, several hybrid machining techniques are turning to it as a potential process because ECM provides great surface quality and minimal tool wear since atomic-level dissolving takes place [51]. More so, complex shapes can be machined. Compared to EDM, ECM has been applied successively to enhance the surface integrity of features of composites [48].

9.7 CHALLENGES IN ADVANCED AND HYBRID MACHINING TECHNIQUES

Although advanced and hybrid machining techniques are promising, there are some challenges that arise in their application. For instance, the surface roughness of holes in nanofiber composites created by AWJM is mostly higher in contrast to holes created by CD [57]. The AWJM process also generates loud noise and a messy working environment [31]. In LBM, the energy required to melt the nanofiber or to vaporize it is higher than that required to melt or vaporize a polymeric matrix in a nanofiber/polymer composite. Moreover, the thermal conductivity of fiber is much higher than that of the polymeric matrix. For example, in carbon fiber-reinforced composites, the resin is approximately decomposed at 500°C while the carbon fibers are vaporizing at 3300°C. In glass fiber-reinforced polymer composite, the resin is approximately decomposed at 600°C while the glass fibers are vaporized at 2000°C [58]. Also, the heat generated during LBM chars the edges of the hole. This led to the introduction of laser assisted waterjet machining, which has proven to be an effective technique that reduces heat damage. Nevertheless, there are few studies about CNC processing with waterjet-guided laser to date [55]. EDM frequently has tool wear, which results in a change in tool length and a corresponding change in the cathode's immersed depth. This causes insufficient cathode feed, and the precision of the machining cannot be guaranteed. This influences other variables, such as the average spark discharge during machining, and consequently the machining result [59].

9.8 FUTURE OUTLOOK

From the above-discussed techniques, it is noted that CNC is primarily the sole nanofiber composite that has been exploited with advanced and hybrid machining approaches. However, there are many different composites made with all sorts of nanofibers, which find application in many different areas. With the knowledge that the nature of the nanofiber influences the outcome of machining action, it is anticipated that the exploration of advanced and hybrid machining techniques on other nanofiber composites, such as biodegradable cellulose nanofiber composites, would be substantial in the field of composites. Although fair knowledge about laser-assisted waterjet machining has been provided, it is also necessary to comprehend the mechanism of interaction between water jet-guided laser and CNC [55], factoring in the behavior of the workpiece for different input parameters of the machining process. There is not a lot of literature available yet about RUM's power use. Experimental research and literature descriptions have been done on the effects of process factors on the rotation speed, feed rate, ultrasonic power, and CNC type of all the RUM components. To determine how additional input variables, such as abrasive material, depth of cut, and ultrasonic vibration amplitude, affect power utilization, further study is required [12]. Generally, unlike metal composites, advanced and hybrid machining of nanofiber composites is very limited, hence the need for further exploration to widen the application of nanofiber composites in many areas. Since mathematical models have been used to successfully predict the outcome of some machining processes on composites, they could be a perch for the swift assessment of the effect of various advanced and hybrid machining techniques on nanofiber composites.

9.9 CONCLUSION

Advanced and hybrid machining techniques have shown a positive impact on geometrically shaping nanofiber composites. These effective machining techniques can cut intricate profiles and drill holes in a variety of workpiece materials. Unfortunately, during machining operations, the non-homogeneous characteristics of FRCs led to chip fracture and discontinuity. The efficiency of the various techniques in machining is hugely dependent on the input parameters, which generate the cutting force to interact with the workpiece, by jetting, discharge, heating, rotation, ultrasonication, etc., with consideration to the composite material to be machined. Advanced and hybrid techniques have provided a more efficient avenue to produce well-machined nanofiber composites than conventional machining techniques. Among all, a significant number of carbon nanofiber composites are machined with advanced and hybrid techniques due to their unique properties, such as strength and large surface area. Hybridization of traditional and advanced machining methods or two/more advanced machining techniques is employed to overcome various disadvantages in advanced machining. Outcomes such as surface roughness, delamination, and fiber pull-out are all influenced by the input parameters. Mathematical models have been developed to predict the behavior of particular machining parameters on the nanofiber composite. Generally, the difference in properties of nanofibers used for composites makes it necessary to explore advanced and hybrid machining techniques on other nanofiber composites, including biodegradable nanofibers.

REFERENCES

1. Jiang, S., et al., Electrospun nanofiber reinforced composites: A review. *Polymer Chemistry*, 2018, **9**(20): 2685–2720.
2. Ramalingam, M. and S. Ramakrishna, 1 - Introduction to nanofiber composites, in *Nanofiber Composites for Biomedical Applications*, M. Ramalingam and S. Ramakrishna, Editors. 2017, Woodhead Publishing. 3–29.
3. Mazari, S.A., et al., Chapter 18 - Environmental impact of using nanomaterials in textiles, in *Nanosensors and Nanodevices for Smart Multifunctional Textiles*, A. Ehrmann, T.A. Nguyen, and P. Nguyen Tri, Editors. 2021, Elsevier. 321–342.
4. Toriello, M., et al., Progress on the fabrication and application of electrospun nanofiber composites. *Membranes (Basel)*, 2020. **10**(9).
5. Mouw, J.K., G. Ou, and V.M. Weaver, Extracellular matrix assembly: A multiscale deconstruction. *Nature Reviews Molecular Cell Biology*, 2014. **15**(12): 771–785.
6. Jang, D. and S. Lee, Correlating thermal conductivity of carbon fibers with mechanical and structural properties. *Journal of Industrial and Engineering Chemistry*, 2020. **89**: 115–118.
7. Nayak, R., D.M. Pradhan, and A. Sahoo, *Machining of Nanocomposites*. 2022.
8. Kumar, J. and R.K. Verma, Experimental investigation for machinability aspects of graphene oxide/carbon fiber reinforced polymer nanocomposites and predictive modeling using hybrid approach. *Defence Technology*, 2021. **17**(5): 1671–1686.
9. Singh, I. and K. Debnath, *Advanced Machining Techniques for Fiber-Reinforced Polymer Composites*. 2015. 321–344.
10. Alarifi, I.M., A review on factors affecting machinability and properties of fiber-reinforced polymer composites. *Journal of Natural Fibers*, 2023. **20**(1): 2154304.

11. Geng, D., et al., Experimental study on drilling load and hole quality during rotary ultrasonic helical machining of small-diameter CFRP holes. *Journal of Materials Processing Technology*, 2019. **270**: 195–205.
12. Asmael, M., et al., Ultrasonic machining of carbon fiber–reinforced plastic composites: A review. *The International Journal of Advanced Manufacturing Technology*, 2021. **113**(11): 3079–3120.
13. Komanduri, R., Machining of fiber-reinforced composites. *Machining Science and Technology*, 1997. **1**(1): 113–152.
14. Wang, D., et al., Machinability of natural-fibre-reinforced polymer composites: Conventional vs ultrasonically-assisted machining. *Composites Part A: Applied Science and Manufacturing*, 2019. **119**: 188–195.
15. Balasubramanian, K., et al., Machining analysis of natural fibre reinforced composites using fuzzy logic. *IOP Conference Series: Materials Science and Engineering*, 2016. **152**(1): 012051.
16. Rajmohan, T., R. Vinayagamoorthy, and K. Mohan, Review on effect machining parameters on performance of natural fibre–reinforced composites (NFRCs). *Journal of Thermoplastic Composite Materials*, 2019. **32**(9): 1282–1302.
17. Das, M. and U.S. Dixit, Advanced machining processes, in *Introduction to Mechanical Engineering*, J.P. Davim, Editor. 2018, Springer International Publishing: Cham. 269–296.
18. Xu, J., et al., Experimental study on carbon fiber-reinforced composites cutting with nanosecond laser. *Materials*, 2022. **15**(19): 6686.
19. Sheikh-Ahmad, J.Y., Hole quality and damage in drilling carbon/epoxy composites by electrical discharge machining. *Materials and Manufacturing Processes*, 2016. **31**(7): 941–950.
20. Bhowmik, S., Jagadish, and K. Gupta, Introduction, in *Modeling and Optimization of Advanced Manufacturing Processes*, S. Bhowmik, Jagadish, and K. Gupta, Editors. 2019, Springer International Publishing: Cham. 1–14.
21. Masoud, F., et al., Cutting processes of natural fiber-reinforced polymer composites. *Polymers*, 2020. **12**(6): 1332.
22. El-Hofy, M.H. and H. El-Hofy, Laser beam machining of carbon fiber reinforced composites: A review. *The International Journal of Advanced Manufacturing Technology*, 2019. **101**(9): 2965–2975.
23. Abdallah, R., S.L. Soo, and R. Hood, The influence of cut direction and process parameters in wire electrical discharge machining of carbon fibre–reinforced plastic composites. *The International Journal of Advanced Manufacturing Technology*, 2021. **113**(5): 1699–1716.
24. Gous, M.Z., et al., Fabrication and machining of fiber based composite materials using advance machining process, a review. *Materials Today: Proceedings*, 2022. **56**: 3617–3622.
25. Kim, B.H., et al., Electrical discharge machining of carbon nanofiber for uniform field emission. *CIRP Annals*, 2007. **56**(1): 233–236.
26. Thakur, R.K. and K.K. Singh, Abrasive waterjet machining of fiber-reinforced composites: A state-of-the-art review. *Journal of the Brazilian Society of Mechanical Sciences and Engineering*, 2020. **42**(7): 381.
27. Kalla, D., et al., Abrasive waterjet machining of fiber reinforced composites: A review. *AIP Conference Proceedings*, 2012. **1431**: 535–542.
28. Kartal, F., A review of the current state of abrasive water-jet turning machining method. *The International Journal of Advanced Manufacturing Technology*, 2017. **88**(1): 495–505.
29. Bañon, F., et al., A review on the abrasive water-jet machining of metal–carbon fiber hybrid materials. *Metals*, 2021. **11**(1): 164.
30. Dhanawade, A. and S. Kumar, Experimental study of delamination and kerf geometry of carbon epoxy composite machined by abrasive water jet. *Journal of Composite Materials*, 2017. **51**(24): 3373–3390.
31. Azmir, M.A. and A.K. Ahsan, Investigation on glass/epoxy composite surfaces machined by abrasive water jet machining. *Journal of Materials Processing Technology*, 2008. **198**(1): 122–128.
32. Li, M., et al., Experimental study on hole quality and its impact on tensile behavior following pure and abrasive waterjet cutting of plain woven CFRP laminates. *The International Journal of Advanced Manufacturing Technology*, 2018. **99**(9): 2481–2490.
33. Dhanawade, A. and S. Kumar, Study on carbon epoxy composite surfaces machined by abrasive water jet machining. *Journal of Composite Materials*, 2019. **53**(20): 2909–2924.
34. Youssef, H., et al., Accuracy and surface quality of abrasive waterjet machined CFRP composites. *Journal of Composite Materials*, 2021. **55**(12): 1693–1703.
35. Kumar, D. and S. Gururaja, Abrasive waterjet machining of Ti/CFRP/Ti laminate and multi-objective optimization of the process parameters using response surface methodology. *Journal of Composite Materials*, 2019. **54**(13): 1741–1759.

36. Tamrin, K.F., et al., Multiple-objective optimization in precision laser cutting of different thermoplastics. *Optics and Lasers in Engineering*, 2015. **67**: 57–65.
37. Dubey, A.K. and V. Yadava, Laser beam machining—A review. *International Journal of Machine Tools and Manufacture*, 2008. **48**(6): 609–628.
38. Chavoshi, S.Z. and X. Luo, Hybrid micro-machining processes: A review. *Precision Engineering*, 2015. **41**: 1–23.
39. Subramonian, S., 9.09 - Thermal modeling of laser machining process, in *Comprehensive Materials Processing*, S. Hashmi, et al., Editors. 2014, Elsevier: Oxford. 203–217.
40. Saxena, K.K., et al., Chapter 2 - Overview of hybrid machining processes, in *Hybrid Machining*, X. Luo and Y. Qin, Editors. 2018, Academic Press. 21–41.
41. Wang, H., Z.J. Pei, and W. Cong, A mechanistic cutting force model based on ductile and brittle fracture material removal modes for edge surface grinding of CFRP composites using rotary ultrasonic machining. *International Journal of Mechanical Sciences*, 2020. **176**: 105551.
42. Feng, Q., et al., Rotary ultrasonic machining of carbon fiber-reinforced polymer: Feasibility study. *Machining Science and Technology*, 2012. **16**(3): 380–398.
43. Singh, R.P. and S. Singhal, Rotary ultrasonic machining: A review. *Materials and Manufacturing Processes*, 2016. **31**(14): 1795–1824.
44. Wang, H., Z.J. Pei, and W. Cong, A feeding-directional cutting force model for end surface grinding of CFRP composites using rotary ultrasonic machining with elliptical ultrasonic vibration. *International Journal of Machine Tools and Manufacture*, 2020. **152**: 103540.
45. Geng, D., et al., Comparison of drill wear mechanism between rotary ultrasonic elliptical machining and conventional drilling of CFRP. *Journal of Reinforced Plastics and Composites*, 2014. **33**(9): 797–809.
46. Wu, C.Q., et al., Effects of machining conditions on the hole wall delamination in both conventional and ultrasonic-assisted CFRP drilling. *The International Journal of Advanced Manufacturing Technology*, 2019. **104**(5): 2301–2315.
47. Harugade, M., S. Waigaonkar, and N. Mane, Machining of carbon epoxy composite using high speed electrochemical discharge machining. *Materials Today: Proceedings*, 2018. **5**(9, Part 1): 17188–17194.
48. Nguyen, M.D., M. Rahman, and Y.S. Wong, Enhanced surface integrity and dimensional accuracy by simultaneous micro-ED/EC milling. *CIRP Annals*, 2012. **61**(1): 191–194.
49. Huang, C.-R., et al., Modeling evolution of cutting force in ultrasonically assisted drilling of carbon fiber reinforced plastics. *Materials*, 2022. **15**(9): 3392.
50. Bhattacharyya, B., Chapter 1 - Introduction, in *Electrochemical Micromachining for Nanofabrication, MEMS and Nanotechnology*, B. Bhattacharyya, Editor. 2015, William Andrew Publishing. 1–23.
51. Saxena, K.K., et al., Chapter 5 - Electrochemical based hybrid machining, in *Hybrid Machining*, X. Luo and Y. Qin, Editors. 2018, Academic Press. 111–129.
52. Tangwarodomnukun, V., et al., An investigation of hybrid laser–waterjet ablation of silicon substrates. *International Journal of Machine Tools and Manufacture*, 2012. **56**: 39–49.
53. Liu, Y., et al., Overview on the development and critical issues of water jet guided laser machining technology. *Optics & Laser Technology*, 2021. **137**: 106820.
54. Zhang, Y., et al., Numerical simulation of water jet–guided laser micromachining of CFRP. *Materials Today Communications*, 2020. **25**: 101456.
55. Sun, D., F. Han, and W. Ying, Numerical simulation of water jet–guided laser cutting of carbon fiber–reinforced plastics. *Proceedings of the Institution of Mechanical Engineers, Part B: Journal of Engineering Manufacture*, 2019. **233**(10): 2023–2032.
56. Loos, M.R. and I. Manas-Zloczower, *Micromechanical models for carbon nanotube and cellulose nanowhisker reinforced composites. Polymer Engineering & Science*, 2013. **53**(4): 882–887.
57. Montesano, J., H. Bougherara, and Z. Fawaz, Influence of drilling and abrasive water jet induced damage on the performance of carbon fabric/epoxy plates with holes. *Composite Structures*, 2017. **163**: 257–266.
58. Patel, P., P. Gohil, and S. Rajpurohit, Laser machining of polymer matrix composites: scope, limitation and application. *International Journal of Engineering Trends and Technology (IJETT)*, 2013. **4**: 2391–2399.
59. Bian, J., et al., Experimental study of tool wear in electrochemical discharge machining. *Applied Sciences*, 2020. **10**(15): 5039.

10 Joining of Fiber-Reinforced Polymer Composites

Vishal John Mathai and Jerin Jose

10.1 INTRODUCTION

Composite materials, due to their multi-faceted properties, have a wide range of applications in automotive, pipelines for petroleum and natural gas industries, components for energy harvesting systems, marine, and nuclear, to name a few (Sarfraz et al., 2021; Rubino et al., 2020; Lan et al., 2009; Karim et al., 2023; Murray et al., 2021). However, in recent years, the growing awareness of the finite nature of non-renewable resources and a shifting reliance toward sustainable, renewable alternatives have thrust the development of biodegradable and environmentally friendly composite materials into the spotlight. Among these, the interest in natural fiber-reinforced composites (NFRCs) has been on the rise. NFRCs offer a compelling array of advantages, including cost-effectiveness, capacity for biodegradation, lower environmental impact, and better mechanical performance. These characteristics position NFRCs as a promising solution for a multi-faceted application, fueling the ongoing quest for greener and more sustainable materials.

Composite matrix materials come in two primary categories: thermoplastics and thermosetting plastics. Thermoplastics are a class of polymers that exhibit the unique property of softening and melting when exposed to heat. This characteristic enables a range of manufacturing techniques, including thermoforming, extrusion, and injection molding. One of the key advantages of thermoplastics is their ability to be recycled and reused for creating new products, although it's important to note that repeated processing can lead to some degradation in their properties over time.

In comparison to thermosetting plastics, thermoplastics offer several notable benefits. They tend to possess higher levels of ductility and impact resistance, making them suitable for applications where durability is crucial. Another significant advantage of thermoplastics is their shorter processing times in manufacturing. This is because the production of thermosetting plastic components involves a chemical reaction, often referred to as curing or cross-linking, that occurs within the mold. Depending on factors like mold temperature and part thickness, this chemical reaction can take anywhere from several minutes to several hours to complete, contributing to longer manufacturing cycles. In contrast, thermoplastic materials do not require this curing process, resulting in faster and more efficient production.

One of the major challenges that is faced while processing composite structures is the design and manufacturing of those entities as a single part through the process of molding so that subsequent assembly operations can be avoided. Yet, mechanical constraints and other practical considerations frequently compel the joining of polymeric parts, whether to each other or to metallic counterparts, so as to realize the required final assembly. In these instances, the process of joining can still be an efficient production technique, provided that a set of precautions is diligently adhered to and well-established procedures are followed. Consequently, even in cases necessitating component integration, a smooth and effective manufacturing process can be maintained. Nevertheless, joining of composites and research pertaining to joining of composites is becoming increasingly relevant considering its role for load-bearing applications and complex structural assemblies.

DOI: 10.1201/9781032665337-10

10.2 CLASSIFICATION OF JOINING TECHNIQUES FOR FIBER-REINFORCED COMPOSITE MATERIALS

Selecting the right joining process for FRP composites can be challenging due to their complex mechanical properties. The choice depends on factors like the materials used, desired performance, and the specific joining technology. The effectiveness of the joint is closely tied to the characteristics of the chosen process, be it adhesive bonding, mechanical fastening, welding, or others. Proper joint design is essential to ensure compatibility and optimize performance for the intended application, considering factors such as load-bearing capacity, durability, cost-effectiveness, and manufacturability.

Technologies for FRP composite joining can be classified into three primary categories: fastening, bonding, and welding, with each category offering a wide range of joining processes. Fastening and bonding are versatile techniques suitable for facilitating different material joints, including metals, and they do not necessitate the joined parts to be composed of the same material. Within bonding, there are subcategories such as adhesive bonding and solvent bonding. Conversely, welding, a process that requires the interfacing faces of the materials to melt, is primarily applicable for joining thermoplastics. Figure 10.1 provides a broad classification of various joining techniques used for composites.

Subsequent sections will delve into different mechanical fastening and joining techniques commonly employed to connect composite materials to other composites and to metal structures.

10.3 BOLTED JOINTS

Bolted joining stands out as a prevalent method for connecting FRP composites due to its numerous advantages, including straightforward assembly and disassembly, ease of inspection, and simplified maintenance procedures (Fink et al., 2010; Choudhury and Debnath, 2019). Figure 10.2 depicts a generic configuration of a joint realized through bolting. The process of bolted joining involves two fundamental steps: drilling and assembling. The components are first drilled for creating a uniaxial hole, and the entities are then constrained with the help of bolts and nuts (McCarthy et al., 2002). Notably, bolted joining offers the flexibility of connecting specimens with varying thicknesses, making it a versatile choice for FRP composite connections.

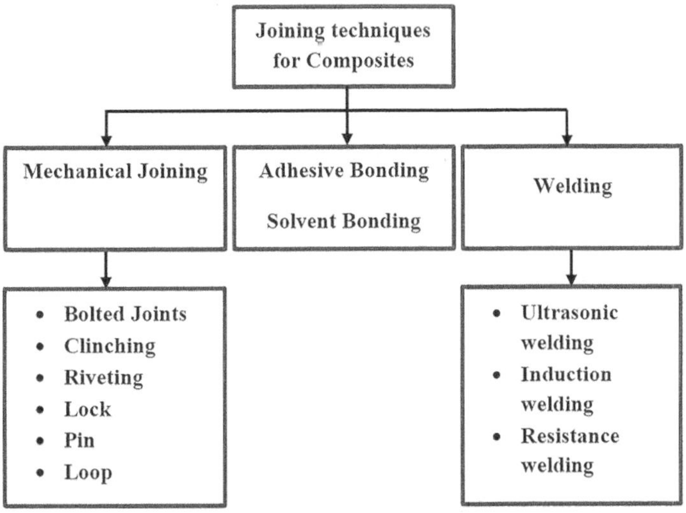

FIGURE 10.1 Classification of various joining techniques for thermoplastics and thermosetting plastics.

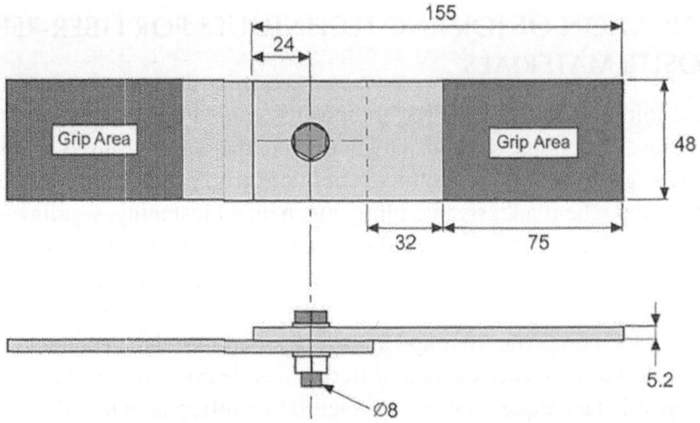

FIGURE 10.2 Schematic representation of bolted joint (McCarthy et al., 2002).

Enhancing the strength of bolted joints in FRP composites often involves the use of multi-bolt configurations (Yun et al., 2014). However, it's important to be aware that an increase in the number of bolt joints will nullify the advantage of the lightweight nature of composite structures.

Additionally, bolted joints in FRP composites can be susceptible to galvanic corrosion, especially when connecting FRP components to metal parts. This corrosion can significantly impact the joint's durability and lifespan. Galvanic corrosion can be reduced by using fasteners or shields made of metal having a lower galvanic potential difference when compared to carbon fiber reinforced polymer (CFRP) composites. Titanium alloys are an example of such materials, but they tend to be costlier and heavier than fasteners made from aluminum alloys (Kashaev et al., 2015).

To address both the weight and corrosion issues, composite material fasteners have been introduced as an innovative solution. These composite fasteners, such as those made from carbon fiber/polyimide, carbon fiber/polyether ether ketone, or carbon fiber/carbon composites (Starikov and Schön, 2001; Guan et al., 2015), offer several advantages. They eliminate the risk of galvanic corrosion, provide radar invisibility, are lightweight, and tend to be cost-effective. These characteristics make composite fasteners a compelling alternative for connecting FRP composites without compromising on key performance attributes.

Although bolted joints may seem straightforward in design and implementation, the selection of appropriate parameters is essential to minimize damage to FRP composites. These key parameters encompass bolt tightness, preload levels, and bolt-hole clearance.

A study conducted by Nassar et al. (2007) investigated the impact of bolt tightness on the mechanical behavior of double-bolted joints. Results suggested an increase in bolt tightness with an increase in the tightness of the joints. Furthermore, a significant reduction in delamination of FRP composites during tensile testing was observed with increased bolt tightness. The application of adequate preload, as demonstrated in the research by Hausmann et al. (2017), has the potential to enhance the load-carrying capacity of bolted joints. However, it has also been reported that the preload doesn't exceed the upper limit of the contact pressure between the head of the bolt and the plates being joined. Additionally, the bolt-hole clearance represents another critical parameter in bolted joints. Research has shown that a larger bolt-hole clearance results in a smaller contact angle, which leads to a pronounced reduction in the interacting area between the join-facilitating entities. Therefore, a comprehensive understanding of these parameters and their careful adjustment is essential for achieving optimal performance and minimizing damage when working with bolted joints in FRP composites.

While a circular cross-section for fasteners is the common choice considering the ease in manufacturing, there have been theoretical investigations suggesting that non-circular cross-sections may offer advantages when joint strength is considered. Wang et al. (1993) and Persson and Madenci (1998) explored the feasibility of changing the bolt cross-section shape from circular to elliptical to enhance joint strength in composite materials. Wang et al. (1993) conducted an analysis that indicated a significant improvement in joint strength. Specifically, joints with a bearing failure mode demonstrated an improvement in strength when elliptical cross-section bolts were used.

However, it's important to note that despite these promising findings, there have been no feasibility studies involving the use of such non-circular bolts in real-world structures. The manufacture of such bolts and the process of creating non-circular holes, that too in composites, present strenuous engineering challenges. Practical considerations and engineering constraints must be carefully evaluated before implementing non-circular fasteners in actual applications.

Another method for increasing the joint efficiency in composite is to bond metallic inserts to the laminates that are coming in contact with the bolt. These bonded metallic inserts alter the way loads are transferred to the laminate by distributing radial tractions from bolt-hole contact across the entire surface of the composite hole, rather than concentrating them on approximately half of the hole surface (Camanho et al., 2005).

However, it's worth noting that the effectiveness of bonded inserts can vary. Researchers have reported negligible improvement in strength of the joint with such modifications. The simulation studies conducted in line with this suggested that the bond between the laminate and insert can fail before the joint reaches its full strength. It has also been demonstrated that aluminum inserts are preferable to steel ones, not only for their lower weight but also for generating lower stresses in the adhesive (Camanho and Matthews, 2000). Rispler et al. (2000) proposed a different approach by placing the metallic insert at the stress-intensive portion of the joint. This insert creates a localized plastic zone, reducing stress concentrations on the composite material and providing stress relief.

Besides improving load distribution and strength, the usage of bonded metallic inserts inside joint holes created in composites offers the advantage of protecting the holes from damage that can happen due to change of the fastener replacement on a regular basis joint failure. Additionally, damaged holes in composite materials can be repaired by bonding such inserts (Camanho and Matthews, 2000).

Moreover, bonding of metallic inserts within holes in CFRPs allows for the use of titanium fasteners, eliminating concerns about galvanic corrosion since the composite part is in contact with a foreign material (adhesive) rather than a metallic component. Disassembly of joints with metal inserts is still possible when a separate insert is bonded to each lap of the joint, although this approach may increase the structure's mass (Camanho et al., 2005). However, bonded inserts may serve as substitutes for washers, which are often used in bolted joints (Choi et al., 2018), potentially mitigating the weight penalty.

One effective method for enhancing the strength of bolted joints involves a technique known as local laminate build-up. However, this approach comes with certain considerations. The increased laminate thickness resulting from this technique introduces additional stresses in the laminate due to eccentricities, specifically when considering single-shear joints, which are commonly used in aerospace structures. Moreover, this method often leads to a significant increase in weight due to the addition of extra composite plies, larger grip lengths, bigger bolt diameters, and heavier metallic fittings (Camanho et al., 2009; Fink and Kolesnikov, 2005).

As an alternative to local build-up, another technique involves the substitution of selected plies of composite with foils of titanium. Titanium boasts excellent specific mechanical properties, corrosion resistance, and a relatively low coefficient of thermal expansion (Kolesnikov et al., 2008).

Its lower levels of thermal expansion coefficient in comparison to metal-based alternatives result in reduced residual thermal stresses during the processes of curing and cooling down to ambient temperature. Stainless steel has also been considered as a material in contention for this application. However, its higher thermal expansion coefficient, greater density, and susceptibility to electrochemical reactions with carbon are some disadvantages (Camanho et al., 2009; Kolesnikov et al., 2008).

The technique employed in the region around the bolt involves the gradual substitution of selected composite plies with foils made of titanium in the bulk to joint transition zone without altering the remaining plies. This local incorporation of metallic foil, replacing the basic plies, results in a joint with improved strength and load-bearing capacities in the hybrid region. The composite joints synthesized in this fashion reduce sensitivity to variations in laminate configuration and environmental effects. Furthermore, this approach takes advantage of the favorable isotropic behavior of metal compared to the complex stress state experienced by bolt-loaded holes in pure FRP composites (Camanho et al., 2009; Fink and Kolesnikov, 2005).

When using bolts in fiber-reinforced polymer (FRP) parts, a significant challenge arises due to the fiber discontinuity at the drilled portion, resulting in a high stress concentration in the hole vicinity. This area has a tendency to become the weakest point in the connected parts because the high strength and stiffness of the fibers are compromised when holes or notches disrupt the load-carrying fibers. To address this issue, the technique of fiber steering has been developed. Fiber steering is employed to counteract this undesirable phenomenon by directing the fibers in a way that aligns with the paths through which the load travels within the structure. The goal is to improve both stiffness and strength. However, it's essential that the steered fiber bunches are accurately aligned with the direction of the load flow to ensure maximum performance (Li et al., 2002; Tosh and Kelly, 2000).

The process of fiber steering involves defining continuous trajectories, often based on principal stress vectors determined from finite element analyses or load paths. Load paths represent regions where the load from the point of application to the point of reaction outside the structure remains constant. The fibers are then placed along these paths either using a robotic system or using a cut-out plate with the paths for manual placement. In the manual placement method, dry fibers are laid onto a tacky surface of prepreg fabric at room temperature (Tosh and Kelly, 2000; Crosky et al., 2006).

Fiber steering is a valuable technique for optimizing the performance of FRP structures, particularly in areas where stress concentrations are a concern due to drilled holes or notches. It helps ensure that the integrity of the composite structure and the capacity of the composite to carry loads are maintained.

Another approach to address the weak points in composite structures resulting from drilled holes involves the use of molded holes or molded-in holes. In this technique, holes are created by aligning reinforcing fibers around the holes before the composite material is cured. This ensures that the fibers remain continuous and can actively participate in load transfer (Sevkat et al., 2011).

The fiber yarns in the vicinity of the molded hole are closely packed, allowing for the continuity of the fibers while also increasing the local fiber volume content. This method helps maintain the structural integrity of the composite material while minimizing the impact of drilled holes, ultimately improving the load-carrying capacity and overall performance of the composite structure (Zitoune et al., 2011; Hufenbach et al., 2011).

Sevkat et al. (2011) conducted a study comparing the bearing strengths of specimens made with glass fiber-based reinforcements using vacuum-assisted resin transfer molding with holes either drilled or molded. In this study, a special metal mold with removable pins was designed and used. The glass fiber fabrics were layered on the mold, and the fibers were aligned around the pins coated with mold release agent for easy removal of the same post-curing. The results showed that the specimens with molded holes exhibited an increase in bearing strength. Observations in similar lines were reported by Zitoune et al. (2011) as well, hence suggesting that the use of molded holes can significantly enhance the bearing strength and fracture behavior of composite specimens compared to drilled holes, especially in certain configurations and materials. However, it is also worth noting that, considering the challenges in manufacturing and metrological requirements, the concept of molded holes is not extensively used.

10.4 HOLE-CLINCHING TECHNIQUES

Hole clinching is a joining technique commonly employed to fasten materials with significant differences in ductility, typically involving a metal upper plate and a composite lower plate. In this process, the upper plate is made of metal due to its greater ductility, while the lower plate is composed of FRP composite material, which is comparatively less ductile. Hole-clinched joining is applicable to both thermoplastic and thermosetting FRP composites, as indicated in previous studies (Lee et al., 2014a, 2014b, 2014c, 2017).

Figure 10.3 depicts the stages of the hole-clinching process. The first step involves pre-drilling a hole in the composite sheet at the desired joining location. Subsequently, a punch is employed to press the upper metal plate into the die cavity along the pre-drilled hole, creating an undercut and securely fastening the two materials together. This approach is particularly useful for joining materials like FRP composites and metals, where differences in ductility would pose challenges for conventional joining methods.

The mechanical performance of hole-clinched joints is primarily influenced by several key factors, including the shape of the clinching tool, the diameter of the hole, and the ply angle of the FRP composite material being used. Notably, the geometrical parameters crucial to the hole-clinching tool include the punch's diameter, corner radius, and the depth of the die, as highlighted in previous research (Lee et al., 2014a, 2014b, 2014c). These factors collectively play a vital role in determining the effectiveness and strength of hole-clinched joints in FRP composite materials.

The formability of the materials being connected plays a pivotal role in achieving successful hole-clinched joints. In prior research, a common approach involved using metal sheets with superior formability as the upper sheet, while composite plates with lower formability were chosen for the lower sheet (Lee et al., 2014a, 2014b, 2014c). Unfortunately, this practice often resulted in inevitable damage to the composite plates during both the drilling and hole-clinching processes. To mitigate such damage and enhance the joinability between aluminum alloy and CFRP, a novel hole-clinched joining technique was introduced by Lee et al. (2017), employing a spring die. The utilization of the spring die led to a significant reduction of approximately 50% in damage to the joint neck. Consequently, this innovative approach using the spring die proved effective in preventing defects around the holes in composite plates.

In contrast to the process of bolting, hole-clinched joining offers a more streamlined mechanical joining process that eliminates the need for additional elements, resulting in cost savings and reduced weight. However, due to an additional drilling stage when compared to bolted joints, the process completion time can be higher. When applying hole-clinched joining to connect composite and metal materials, such as aluminum alloy and steel, it is essential for the metal components to exhibit good ductility. Additionally, it's worth noting that joints produced between composites and metal materials using this technique may be susceptible to galvanic corrosion.

Lambiase and Paoletti (2018) and Lambiase et al. (2017) introduced a hybridized method for executing hole clinching that aimed at enhancing the connection between CFRP and aluminum

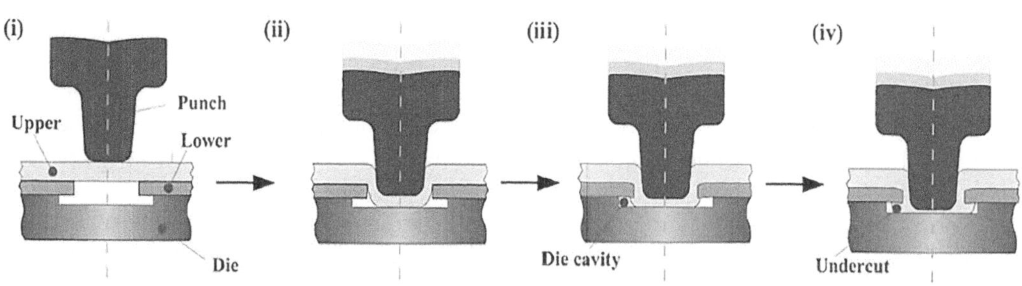

FIGURE 10.3 Concept of joining using hole clinching (Lee et al., 2017).

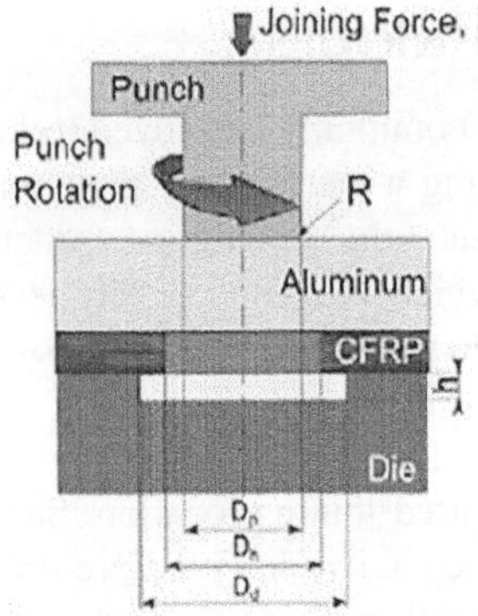

FIGURE 10.4 Hole clinching using a rotating tool (Lambiase et al., 2017).

plates. This innovative approach employed a rotating tool to connect CFRP and aluminum plates, as shown in Figure 10.4. Notably, the ductility of the metal plates was enhanced by applying heat generated through friction during the process, effectively preventing the formation of cracks in the metal sheets.

Another process that comes under this head is mechanical clinching, which is a cold plastic forming technology. The process is reported to have extensive applications in the automobile and aircraft industries. In this process, an extruding action of a punch–die system induces plastic deformation, thereby creating a mechanical interlocking structure, which is not easy to disassemble. It is also being reported that mechanical clinching induces more damage to the composite plates owing to the large level of material flow that happens during the process (Lambiase et al., 2016). Figure 10.5 depicts the schematic diagram of the mechanical clinching process.

The joint characteristics of the mechanically clinched joint are mostly affected by the shape of the punching die and the clinching tool parameters. Lambiase et al. (2016) conducted a study to assess the viability of using glass fiber-reinforced plastic (GFRP) in conjunction with aluminum alloy plates, specifically AA6082-T6 and AA5086, within the context of mechanical clinching. They employed a range of die types, including extensible, grooved, flat, and rectangular shear dies, to investigate their impact on joint quality. Among these die variations, it was observed that joints produced using the extensible die exhibited the highest quality when compared to those formed using other die types. Further, the same author has investigated the effect of punch diameter and punch taper on the joint properties of AA6082–T6 and CFRP sheets and reported that lower taper angle and higher punch diameter are more appropriate for improved mechanically clinched joints; however, this leaves damage on the sheets used (Lambiase and Ko, 2016). The use of a reshaping die at the end of mechanical clinching has also been investigated by the same authors (Lambiase and Ko, 2017) while joining aluminum and CFRP, but they have reported the possibility of several defects like delamination and buckling in case of excessive reshaping.

In order to eliminate the defects of composite plates under high compressive stress during joining, the concept of preheating was also employed prior to mechanical clinching while joining thermoplastic-based composites (Lin et al., 2020). Results suggested that the preheating of composite sheets, owing to the improved malleability, delivered joints with less bucking and delamination. A process in similar lines, which was reported as thermoclinching, was reported by researchers for joining steel sheets with glass fabric reinforced polypropylene, and the process is depicted pictorially in Figure 10.6.

Another variant of clinching process that was performed with the aid of heat is Injection Clinching Joining (ICJ), reported by Abibe et al. (2013, 2011). The step-by-step procedure is pictorially depicted in Figure 10.7. The method initiates by pre-assembling a composite part with a protruding stud to a mating component featuring a corresponding hole, ensuring a snug fit between the stud and

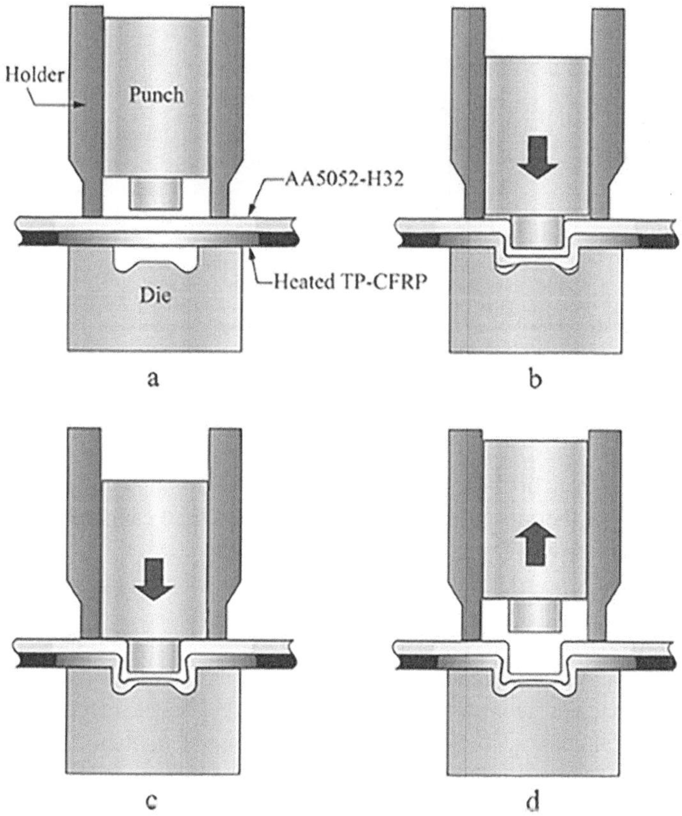

FIGURE 10.5 Mechanical clinching (Lambiase and Ko, 2016; Ouyang and Chen, 2022).

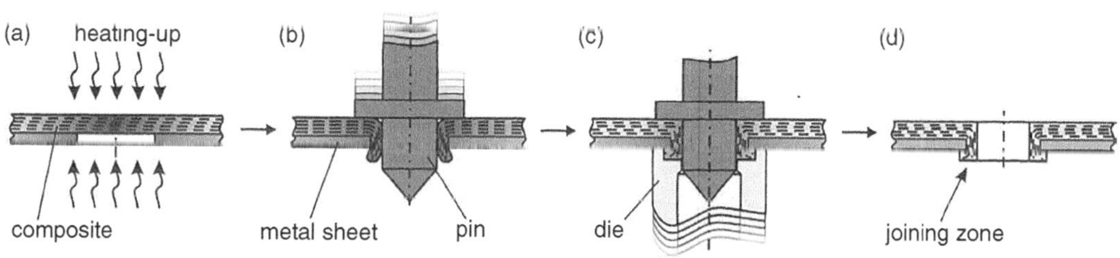

FIGURE 10.6 Thermoclinching process (Gude et al., 2019; Zongkai et al., 2023).

the hole. The process involves the use of a hot case and punch-piston tool system, where the hot case houses the polymeric stud. Subsequently, the stud is subjected to controlled heating at a specific processing temperature for a predetermined time. Following this, the punch-piston mechanism is employed to push the softened polymer material into the hole. To achieve consolidation, the system is then cooled while maintaining pressure, effectively minimizing polymer thermal relaxation and solidifying the joint. Ultimately, this process yields a joint with no additional components apart from the originally joined parts, with the polymeric stud serving as a rivet within the joint structure.

Abibe et al. (2020) introduced a novel joining technique known as friction-based injection clinching joining, which is based on the principles of injection clinching joining. The key distinctions between the two methods lie in the shapes of the tools used and the application of frictional forces. However, there has been relatively limited research conducted on this particular technique.

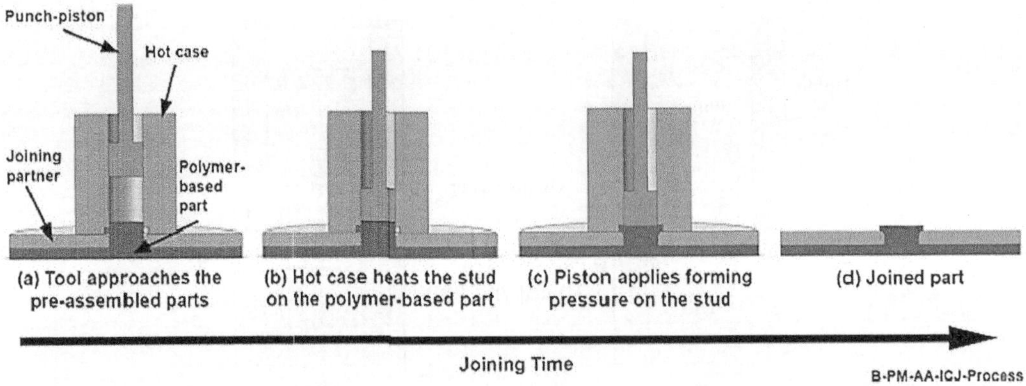

FIGURE 10.7 Injection clinching joining (Abibe et al., 2013).

It is also worth noting that clinching processes or improved clinching processes are applicable for thermoplastic materials as thermoset-based composites once cured cannot be plasticized again and would crack under higher strains imposed during the process.

10.5 RIVETING

Riveting is a semi-permanent joining technique in which a rivet joins the parts. During the process, a hole is drilled in the parts of interest, and a rivet is installed by mechanical means to create a joint. In the case of joining composites, conventional riveting cannot be applied directly for joining as the process will destroy the matrix around the installed rivet, thereby compromising the joint strength.

In order to reduce the demerit, the process has been modified in the name of self-pierce riveting (SPR), which now has a wide range of applications in the automotive industry. The diagrammatic representation of the process is shown in Figure 10.8.

The self-piercing riveting process involves the joining of two stacked sheets, which can be made of the same or different materials, using a rivet that is driven into the stack with the help of a punch. During this process, complete piercing of the top material and partial piercing of the bottom material with the rivet happen. Under the constraining action of the die placed beneath the material and the applied force, the rivet expands, forming an interlock between the stack materials. This process results in plastic deformations occurring in both the rivet and the bottom plate (Abibe et al., 2020; Di Franco et al., 2013).

One of the main advantages, other than the benefits due to reduction in production time and cost, is the non-requirement of prerequisite holes in the plates. Further, the need for surface pretreatment, as in the case of adhesive or solvent bonding, is not required and does not impart any harmful effects

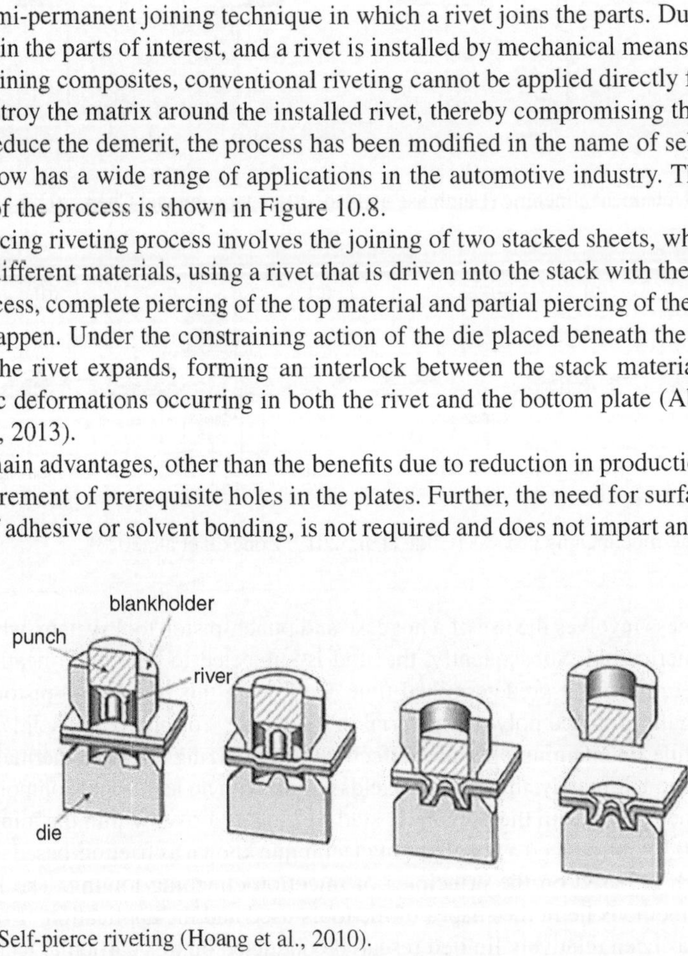

FIGURE 10.8 Self-pierce riveting (Hoang et al., 2010).

on the operator. However, one limitation of the joints realized by this technique is that the disassembly of the parts is not possible without destroying them. The issue of galvanic corrosion can also arise in these joints.

Self-piercing riveting method poses substantial challenges when applied to fiber-reinforced composite materials. A critical and often underestimated drawback is the potential for process-induced damage during the joining process. Unlike metals, composite materials typically lack plastic behavior, rendering them unable to absorb the punch energy through plastic deformation. Consequently, the application of this energy can result in detrimental matrix and fiber damages, such as delaminations and cracking, which may escalate over time, particularly under cyclic stresses, ultimately leading to premature joint failure. Moreover, the rivet's piercing action cuts through the fibers, disrupting crucial load paths within the composite structure, compromising load distribution. Additionally, as the rivet traverses successive plies, the thickness and stiffness of the remaining intact plies decrease. This reduction subjects the final plies to severe bending, which, if it surpasses the peeling stress threshold of the plies, can induce delamination, further undermining the joint's strength and overall integrity. Consequently, the application of short fiber reinforcement (SFR) is limited for brittle and non-plastic composite materials (Meschut et al., 2014; Blaga et al., 2013).

Friction riveting is another branch of riveting techniques that has the potential to be used for joining fiber-reinforced polymer composites (FRPCs). This technique involves embedding a metallic rivet into a base plate made of thermoplastic polymer or a thermoplastic matrix composite (Lambiase et al., 2017). The process consists of three distinct steps: the friction step, the forging step, and consolidation, which is shown in Figure 10.9. A rivet, which is cylindrical in shape, is fixed on a rotating spindle with the thermoplastic polymer or composite plate secured beneath it. The rivet is then moved toward the plate with a rotary motion, and on making contact with the plate, the frictional effect created between the rotating rivet and plate creates localized heating, leading to the melting of the thermoplastic polymer. The rivet is further fed into the material, and when the heat accumulation becomes intense due to the poor thermal properties of the polymer, the rivet starts to soften up. The rotation of the rivet is then reduced, subsequently increasing the axial load, resulting in widening the tip of the rivet, taking on a mushroom-like shape. Finally, on cooling under constant load, the joint is consolidated. During this phase, the rivet firmly embeds itself in the polymeric plate, resulting in the formation of a robust metallic-insert joint (Altmeyer et al., 2014).

Friction-riveting technique also provides the advantages of SFR, like the non-requirement of predrilled holes and surface preparation. The limiting factors of the process are that it can be used only

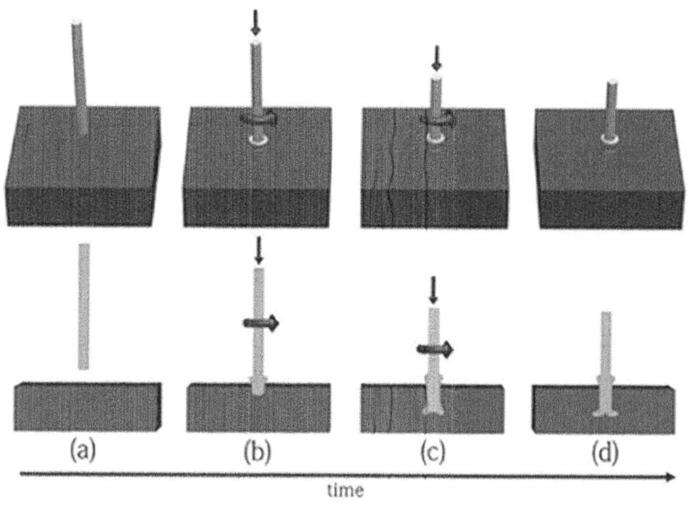

FIGURE 10.9 Friction riveting (Altmeyer et al., 2014).

for thermoplastic materials, as the matrix needs to be melted for joining. Further, the process can be applied for creating spot-like joints. Galvanic corrosion must also be taken into consideration based on the nature of materials being joined.

A variant of the friction-riveting technique that has employed the concept of friction stir welding was proposed by Buffa et al. (2016) for joining AA6082-T6 to self-reinforced polypropylene. A preholed aluminum sheet was used as the top material for establishing the joint. After placing the sheets, a pinless tool was plunged into the aluminum sheet and moved with a constant feed rate along the plate. The frictional heat created due to the interaction between the rotating tool and the aluminum sheet is conducted onto the composite plate below, which softens up. The combined action of heat and the extrusion pressure exerted by the tool causes the softened composite material to flow plastically into the holes in the aluminum. The processed part on subsequent cooling creates a mechanical bond between the sheets.

It is also worth noting that the heat generation during these processes can adversely influence the joint performance. Since the polymer matrix has poor thermal conductivity, parts with such materials have a tendency for heat accumulation when subjected to friction riveting, and this can result in thermal degradation of the polymer-based materials. Hence, considering the mechanical performance of the polymer, it is preferable to minimize the heat accumulation in these materials (Gagliardi et al., 2018; Amancio-Filho et al., 2008).

Electromagnetic riveting is another method of riveting technique that harnesses the power of high-voltage pulsed magnetic fields to swiftly deform and secure rivets. The schematic for the electromagnetic riveting unit is shown in Figure 10.10. This process involves the continuous charging and discharging of a capacitor, creating a dynamic magnetic field that drives a punch to rapidly upset the rivet. The advantages of electromagnetic riveting are evident in its ability to provide a high loading rate, deliver a powerful impact force for secure fastening, maintain stable and reliable riveting quality, and ensure uniform interference for consistent joint strength. Notably, electromagnetic riveting excels when dealing with materials that possess poor cold plasticity, such as titanium alloy, a challenge for conventional riveting techniques. However, it shares the common requirement of predrilled holes with other riveting methods, which can increase the overall time and complexity of the riveting process (Zengqiang and Yangjie, 2020).

The influence of locking type on performance and failure properties of electromagnetic riveted joints was researched by Jiang et al. (2021). It turned out that the locking mode was the main influencing factor compared to the discharge energy, and the selection of an appropriate locking mode is important to improve the quality of joints. Besides, the effect of composite layers' thickness on failure behaviors of electromagnetic riveted joints in the case of high-speed loading was considered

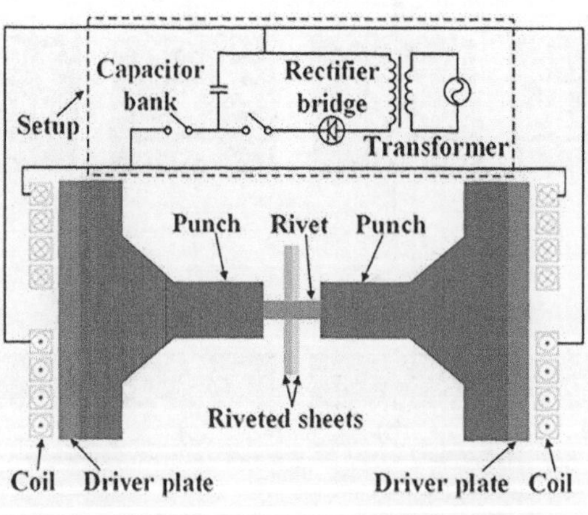

FIGURE 10.10 Electromagnetic riveting (Jiang et al., 2021).

by Cui et al. (2018). It showed that the deformation of aluminum sheets increased with the increase in composite plate thickness. Performance of electromagnetically riveted under adverse conditions was studied by Jiang et al. by comparing the microstructure and properties of the joints (Jiang et al., 2019). The strength of the joints subjected to neutral salt spray conditions was reduced after undergoing three weeks of aging, owing to the destruction of the CFRP sheets. In addition, the fatigue behavior and principle of the joint were also investigated, and results suggested that the shear and fatigue performance of joints decreased linearly with the growth of aging time (Jiang et al., 2017a; Jiang et al., 2017b).

Hybridization of the electromagnetic riveting was also reported by researchers. The viability of combining electromagnetic riveting with hole clinching was investigated by Salamati et al. (2019) with a focus on improvement in material joinability and joint strength. Further, the performance of hybridized joints realized by electromagnetic riveting and self-piercing riveting was researched by Liang et al. (2019), considering the properties and microtopography of the joints. Results suggested that the interlocking value of electromagnetic self-piercing riveting was greater by approximately 150% than conventional self-piercing riveting, thereby showing that the joint strength of electromagnetic self-piercing riveting exhibited higher maximum shear strength than that of conventional self-piercing riveting.

10.6 LOCK JOINTS

In case of the requirement of assembling large elements, the necessity of introducing concentrated forces to the joint may arise. In such cases, a single massive joint that could withstand the force has to be designed. But designing such joints using conventional pre-hole requirement techniques can result in issues of delamination, micro buckling, etc. (Galińska, 2020). Further, such joints may demand bigger holes for joint creation, which will also reduce the joint effectiveness (Li et al., 2002). In this context, a team from Warsaw University of Technology demonstrated the application of a non-adhesive-based form of locked joints (Marjanowski et al., 2017). Such a joint used in a fuselage wing connection of an AOS-71 motoglider is shown in Figure 10.11.

The process begins with cutting a hole in either dry fabric, which is subsequently infused with resin, or in uncured prepreg material. Next, metal rings are precisely placed around the hole, causing the fabric's edge to curl up and become securely locked between these rings. Subsequently, the entire structure undergoes a curing process, which can take place within a vacuum bag or an autoclave. Following the curing phase, a specially designed screw with a hole and a corresponding nut is inserted into the joint's hole to firmly secure it. To complete the assembly of the structure, a bolt is threaded through the hole in the screw, effectively securing the components together.

This solution ensures that the composite fibers are securely integrated with the joint from every angle, effectively gripping them. This stands in contrast to traditional bolted joints, where force is primarily concentrated at the point of contact between the bolt and the composite material. In the case of the non-adhesive form-locked joint, the force is evenly distributed throughout the composite in all directions, enhancing its structural integrity. An additional advantage of this approach is that the hole is cut in the fabric prior to the curing process, eliminating the risk of delamination and other damages typically associated with drilling holes into a cured composite. Furthermore, the non-adhesive form-locked joint allows for easy and convenient assembly and disassembly, providing flexibility in structural modifications as needed (Kelly and Hallström, 2004). Conversely, this joint does come with certain drawbacks. One notable drawback is its inherent complexity, which can make the assembly process more intricate compared to simpler joining methods. Additionally, the presence of extra metal components like the perforated screw and the nut can lead to an increase in the overall mass of the structure, potentially impacting weight-sensitive applications. Furthermore, there is a noteworthy concern regarding the risk of galvanic corrosion, particularly when the components of the joint placed within a carbon fiber-reinforced composite are made of materials such as steel or aluminum alloys. The contact between dissimilar metals in the presence of moisture or other

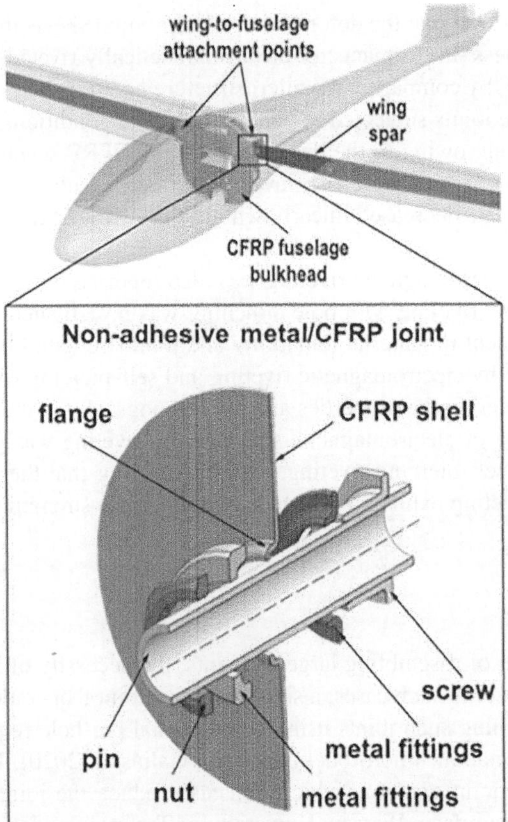

FIGURE 10.11 Non-adhesive form-locked joint (Galińska and Galiński, 2020; Tomasiewicz and Czarnocki, 2016).

corrosive factors can accelerate corrosion, potentially compromising the integrity of the joint over time. Careful material selection and protective measures may be necessary to mitigate this risk effectively.

Research pertaining to the assessment of strength of such joints was carried out by Tomasiewicz and Czarnocki (2016). The authors made joints between stainless steel and carbon fiber epoxy composites, and the results suggested that even though concentrated forces induced strains above and below the joint made, the stress concentration below the joint was less in comparison to an equivalent conventional joint.

10.7 PIN JOINTS

Pin joints are mostly used to establish joints between the metal and a composite part. In this technique, an array of vertical pins is created on the metal part. In the manufacturing process of this joint, the composite plies are positioned atop the metal in such a way that their fibers are arranged around the pins. These pins then penetrate the layers of fabric throughout the thickness of the composite with minimal damage to the fibers. This careful arrangement results in a mechanical interlocking of the laminate with the metal structure, effectively integrating them.

This approach offers the advantage of creating a joint without the need to drill holes into a cured composite, thereby avoiding damage to the fibers and preserving the structural integrity of the composite material. Subsequently, the composite is subjected to the curing process, finalizing the bond between the composite and the metal structure. The curing methods that have been reported to be used are vacuum assisted resin transfer molding (VARTM) and the prepeg technique (Graham et al., 2014; Nguyen et al., 2017; Parkes et al., 2014). The techniques that can be employed for creating the pin arrays can be broadly classified into two categories: surface restructuring and additive manufacturing.

The Comeld method, developed by The Welding Institute, is a surface restructuring technique employing an electron beam (Wang et al., 2016a, 2016b). However, it has several notable drawbacks. First, it offers limited control over the geometry of the pins, which can be a significant limitation when precise pin structures are required for specific applications. Second, the process can result in excessive damage to the surface being restructured, which may compromise the material's integrity and performance. Lastly, the utilization of an electron beam to drive material across the surface can incur substantial costs, making it less economically feasible for certain projects.

Despite these drawbacks, the Comeld method has found applications in the production of pin joints involving titanium and carbon-epoxy composites. This suggests that, in specific cases where the benefits of using this method outweigh its limitations, it can still be a viable option for achieving desired joint configurations and material combinations.

Additive Manufacturing (AM) encompasses a diverse range of techniques, but they all share the fundamental principle of constructing or "building up" objects by adding successive layers of material onto a substrate. In the context of pin manufacturing, AM methods provide advantages such as control over pin geometry and minimal damage to existing surfaces. Two prominent AM techniques for pin production involve metal powder processing: Selective Laser Melting (SLM) and Laser Metal Deposition (LMD). SLM involves the use of a metal powder bed as the base material. A focused laser beam is precisely directed onto this powder bed, selectively melting and fusing layers of the metal powder together. This controlled melting and layer-by-layer build-up enable the creation of intricate and customized pin geometries with precision. LMD, on the other hand, is a process where a high-powered laser beam is used to melt a metal wire or powder as it's continuously deposited onto a substrate. LMD allows for the repair or addition of material to existing components and surfaces, making it a versatile choice for manufacturing pins while minimizing damage to the surrounding area.

Both SLM and LMD offer valuable solutions for pin manufacturing within the realm of AM, providing the flexibility to produce complex shapes and structures with fine control over geometry while preserving the integrity of the base material and surface. Researchers have demonstrated the application of these techniques for pin array fabrication (Graham et al., 2014). However, the technologies are a costly option for industry, even if they are ideal options for research aspects.

Cold Metal Transfer (CMT) is another relatively modern AM technique that facilitates the deposition of molten metal wire onto a metal substrate in a layer-by-layer fashion. Ucsnik et al. (2010) have demonstrated the application of CMT welding for pin generation, in which one end of a wire is arc-welded to the metal substrate, and then the wire is torn apart by applying resistive heating and tensile force. This unique process results in the creation of an array of pins composed of welded wire segments, each possessing the desired shapes and dimensions, and integrally attached to the metal substrate.

Importantly, these surface restructuring and additive manufacturing processes are not limited to a single type of metal. They can be applied to a range of metals, including steel, aluminum, and titanium alloys. This versatility makes them valuable techniques for various applications and allows for the creation of customized components or structures using different metal materials, depending on the specific requirements of the project.

Integration of different additive manufacturing techniques and modification of parent materials that are conventionally used have also been reported by researchers. Chueh et al. (2020) devised an innovative method that combines laser-based powder bed fusion and fused filament fabrication for three-dimensional printing of hybrid metal/polymer objects. In their approach, laser-based powder bed fusion is employed to manufacture an array of metallic interlocking structures with predefined shapes on a metallic base. Subsequently, polymer or reinforced polymer material is extruded and pressed between these metallic structures, resulting in the creation of a mechanical joint. During the polymer extrusion process, the metallic base is heated to facilitate the penetration of the polymer material into the array of metallic structures. The pressing action is performed using a glass window with laser assistance, which enhances the contact between the interlocking structures and the polymer, ensuring a secure and reliable bond. This method represents an exciting advancement in additive manufacturing, enabling the fabrication of hybrid metal/polymer objects with intricate and customizable designs. It leverages the unique capabilities of laser-based powder bed fusion and fused filament fabrication to create robust mechanical joints between metal and polymer components, expanding the possibilities for the production of complex and high-performance composite structures.

Different types of pin geometries have been reported by researchers, like cylindrical, ball-headed pins, grooved, pyramid, helical, and wedged. Some of the feature shapes used to realize the joints are shown in Figure 10.12(a–c).

Ucsnik et al. (2010) researched the Z-pin joining using cylindrical and ball-headed pins during joining of stainless steel 304 and thermosetting CFRP. They have reported that cylindrical pins exhibited pull-out while ball-headed pins were fractured under load. Consequently, the joint tensile strength with ball-head pins has been reported to be higher than the cylinder pin joints. Results for critical parameters like ultimate load and energy absorption too favored joints created using ball-headed pins.

Performance comparison of pyramid, helical, and grooved pins with cylindrical pins was done by Nguyen et al. (2017). Results suggest that the joint strength of joints created using helical and pyramidal pins performed better than cylindrical pins. Further, the failure mechanisms were also different, with cylinder and grooved pins exhibiting pull-out, whereas pyramid and helical pins showed composite failure.

Wang et al. (2016a, 2016b) conducted research to explore the impact of protrusion geometry on joint properties. In their study, they investigated the performance of two distinct bulge shapes: linear and cylindrical patterns. Notably, specimens with cylindrical pins demonstrated a more uniform distribution of stress compared to those with linear bulges under the specified conditions. The observed failure modes in these joints also differed significantly. Linear pin joints tended to fail within the composite layers, while cylinder pin joints tended to splinter at the overlapping regions of the two materials. Interestingly, the failure strength of linear pin joints was approximately double that of cylindrical pin joints. This highlights the remarkable potential for enhancing joint properties through the optimization of bulge shapes (Wang et al., 2016a, 2016b). Tu et al. (2011) also conducted a study on the optimization of protrusion geometry on metal surfaces and reported that the angle and height of the bulge had a significant influence on the concentration of stress near the protrusion.

One of the highlighted advantages of using pin joints in joining composites is the non-requirement of pre-holes in the cured laminate. This eliminates the issues pertaining to delamination and stress concentration in the composite joints. Further, the disruptions caused by the metal pins in the reinforcing fibers will be negligible, too, avoiding the degradation of joint performance. However, the process is not devoid of limitations like the feasibility and cost-effectiveness of manufacturing pin arrays, non-possibility of disassembly, and chances of galvanic corrosion (Galińska and Galiński, 2020). It is also worth noting that while the pin joining technique is generally gentler on composite fibers compared to more intrusive methods like hole drilling, it still introduces localized disruptions to the composite structure. These disruptions manifest as instances of localized fiber waviness, broken fibers, and the formation of resin-rich zones (Nguyen et al., 2016). Other factors that can affect the performance of a composite–metal joint are the poor interface strength, mismatch in thermal expansion, and coefficient of the joining materials. (Nguyen et al., 2016; Mouritz, 2012; Sweeting and Thomson, 2004).

Concept of laser-based material deposition and electroforming has also been reported for establishing polymer–metal joints. It has been reported that electroforming is more viable for this as

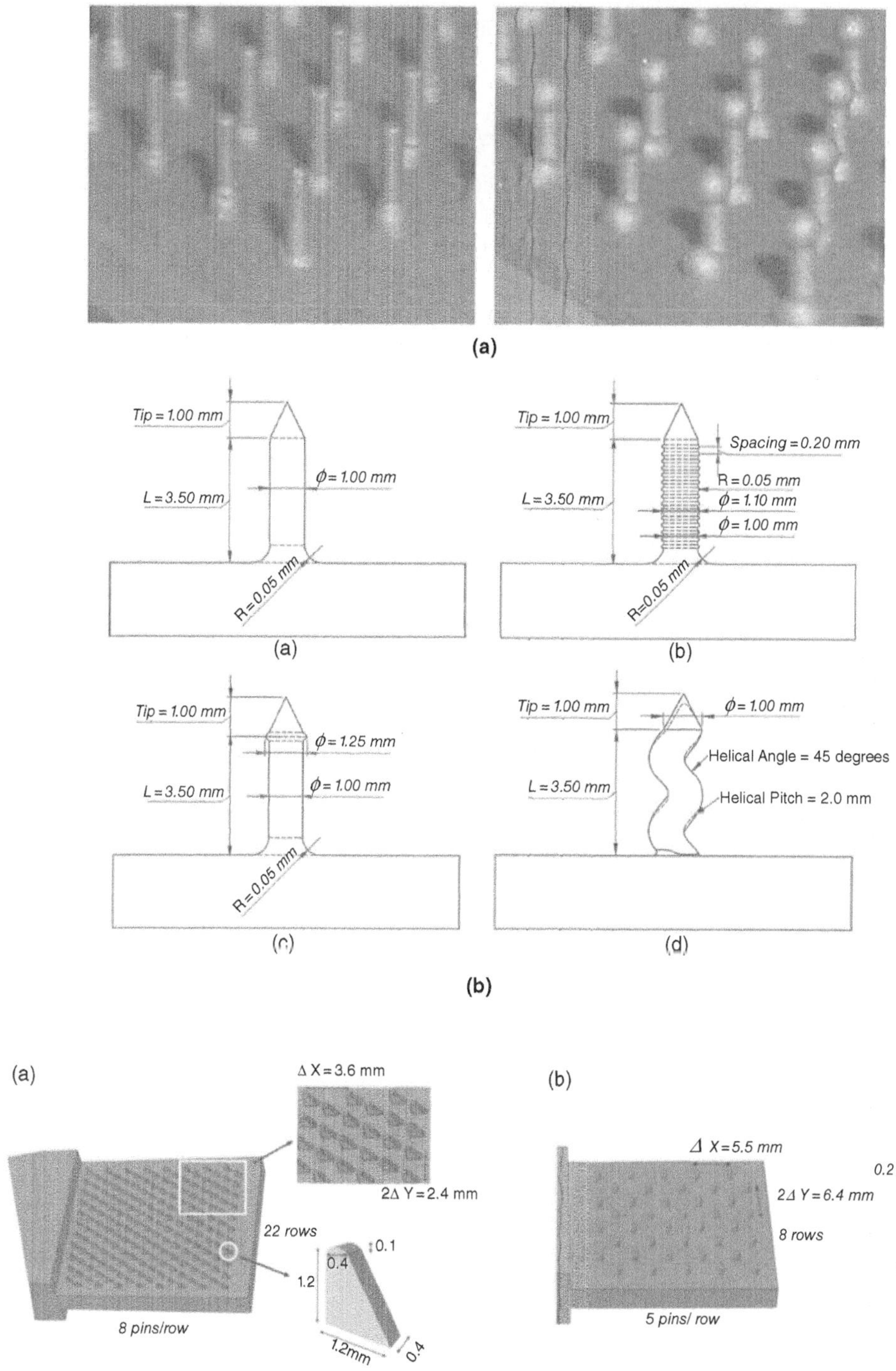

FIGURE 10.12 (a) Cylindrical and ball-headed pins (Ucsnik et al., 2010); (b) cylindrical, grooved, pyramid, and helical pins (Nguyen et al., 2017); and (c) wedged and cylindrical pins (Wang et al., 2016a, 2016b).

electroforming does not require that high temperature for performing the operation compared to a laser (Chueh et al., 2020; Matsuzaki et al., 2019).

10.8 LOOP JOINTS AND PLY INTERWEAVING TECHNIQUES

Loop joints comprise joining carbon fiber-reinforced composites with aluminum alloys using either wire, foil, or fiber as the connecting elements. The concept was demonstrated by a research group named Schwarz-Sibler with the support of the German Research Foundation and is pictorially shown in Figure 10.13.

The wire concept involves the connection of rows of titanium wire loops to an aluminum substrate through welding or casting. These loops serve as guides through which bundles of carbon fibers are threaded. In the foil concept, the loops are substituted with titanium foils, which are also attached to the aluminum substrate. The carbon fibers are inserted between these foils, and the joint is secured by frictional forces. In the fiber concept, the traditional titanium elements are replaced with glass or boron fibers as part of the joining process. Irrespective of the looping type, all the joints are embedded in polymeric resin. These techniques have the benefits of preventing galvanic corrosion due to the presence of the transition structures, non-disruption of the fibers, etc. However, the process is tedious and time-consuming, and the transition points may become a weak point under load. Further, the addition of such transition elements may increase the mass of the joint, which may dent the benefits of its lightweight nature (Lang et al., 2013; Schimanski et al., 2013).

10.9 OTHER JOINING OPTIONS FOR COMPOSITES

Apart from mechanical fastening and joining techniques, two domains with the composites that can be joined are adhesive bonding and fusion bonding.

Adhesive-bonded joints provide notable advantages over mechanical joints, distributing loads more uniformly and minimizing stress concentrations. Nevertheless, it's essential to recognize that stress concentrations persist in both the adhesive and adherends due to inherent material discontinuities at the bonding interface.

The choice of adhesive significantly influences the strength and durability of the bond, a factor particularly pertinent when dealing with thermoplastic polymers. These polymers typically exhibit lower surface energies compared to their thermoset counterparts, posing a challenge for adhesives to effectively wet the adherend surface and establish a strong bond. Therefore, meticulous adhesive selection tailored for bonding with low-energy surfaces becomes imperative.

Adhesives enriched with additives or surface modifiers that enhance wetting on low-energy surfaces can be particularly advantageous. Additionally, rigorous surface preparation is vital for optimizing bond strength. Techniques such as solvent cleaning, abrasion, or plasma treatment are commonly employed to enhance the wettability of adherend surfaces.

While adhesive-bonded joints offer advantages in load distribution and stress alleviation, success hinges on the judicious selection of adhesives and thorough surface preparation. This is particularly

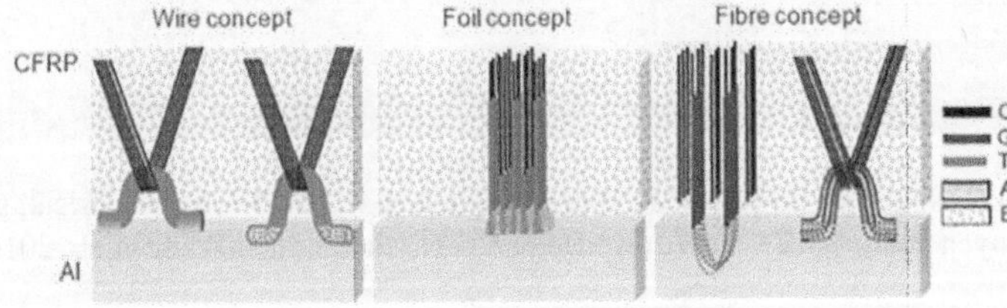

FIGURE 10.13 Concept of composite-to-metal substrate joining using wire, foil, and loop concepts (Schumacher et al., 2013).

crucial when working with materials like thermoplastic polymers, where lower surface energies necessitate a specialized approach to ensure a robust and enduring bond. (Chaves et al., 2014)

Surface treatment plays a pivotal role in adhesive bonding operations, serving multiple purposes. It is employed to eliminate contaminants, enhance wetting of low-energy surfaces, chemically modify the surface (introducing polar groups or coupling agents), and increase surface roughness for improved mechanical interlocking and increased bonding surface area. Various reported methods for surface treatment encompass solvent wiping, mechanical abrasion, acid etching, grit blasting, plasma etching, flame treatment, and corona discharge treatment.

However, it's noteworthy that surface treatments may become unnecessary when utilizing acrylic adhesives. Acrylic adhesives have been specially developed to bond effectively with low surface energy materials, such as thermoplastic polymers. Their inherent formulation enables them to achieve strong bonds without the need for extensive surface preparation.

This characteristic makes acrylic adhesives particularly advantageous in scenarios where simplifying the bonding process is desired, especially when working with low surface energy materials like thermoplastic polymers. While other adhesives may require intricate surface treatments to ensure proper bonding, the unique properties of acrylic adhesives make them a convenient and effective choice for such applications (Pinto et al., 2009).

Certainly, optimizing joint design for high load transfer efficiency is crucial, considering the various factors that influence the failure load of adhesive-bonded joints. While parameters like materials, adherend thickness, and overlap length are often constrained in practical applications, there remains considerable room for effective optimization.

The properties of both the adhesive and adherends, the joint configuration, adhesive and adherend thickness, local geometry, surface preparation, and environmental conditions all play significant roles in determining the joint's failure load. Practical applications often impose restrictions on certain parameters, necessitating a thoughtful and efficient design approach.

In the realm of adhesive bonding for thermoplastics, existing investigations have primarily centered around PEEK-based reinforced composites. However, it is recognized that this focus may not adequately address the diverse needs of current industry applications. In sectors like aerospace and automotive, where materials such as polyetherimide (PEI), polyethersulfone (PES), and polyphenylene sulfide (PPS)-based reinforced composites are prevalent, there is a growing demand to address adhesive bonding capabilities for these specific materials.

To meet the requirements of these industries, further research and development are needed to expand the understanding of adhesive bonding for a broader range of thermoplastic-based composites. This would enable the formulation of optimized joint designs that cater specifically to the unique characteristics and challenges presented by these materials, ensuring efficient load transfer and durability in adhesive-bonded joints within aerospace and automotive applications (Mallick, 2010).

Fusion bonding of thermoplastic polymers involves localized heating above the glass transition temperature for amorphous polymers or the melting temperature for semi-crystalline ones. The process forms a weld between surfaces, cooled under pressure. Care must be taken to keep the temperature below the polymer's degradation point to ensure a strong bond without compromising material integrity.

The heat-affected zone significantly impacts welded joint quality. For optimal fusion bonding, a plastic material should not be welded above 75% of its glass transition point for amorphous polymers or 75% of its melting point for semi-crystalline ones (Suresh et al., 2007). Therefore, studying temperature distribution is crucial for process optimization.

To initiate fusion bonding, surfaces are brought into intimate contact under pressure, and the interface temperature is raised to facilitate molecular diffusion. Minimal surface preparation is needed, reducing execution time. However, caution is necessary to prevent distortions and residual stresses that may lead to deconsolidation in well-consolidated areas due to melting. Constrained setups are employed to maintain the original shape, and mitigating undesirable effects involves reducing thermal gradients among the parts to be joined (Ye et al., 2005; Lionetto et al., 2017).

For semi-crystalline substrates, the dual polymer bonding or thermabond methodology involves using an interlayer of an amorphous polymer for joining (Mcknight et al., 1997; Smiley et al., 1991). In this process, a film of the interlayer polymer is positioned in the bonding areas of the laminate before the consolidation process. During consolidation, both polymers undergo melting, facilitating intermolecular diffusion and creating a bond between them. After cooling, the resulting composite laminate incorporates a thin layer of the interlayer polymer on the bonded surfaces. Subsequently, the interlayer is heated just above its glass transition point for fusion bonding, maintaining a temperature well below the melting point of the semi-crystalline polymer. Successful bonding in this process is attributed to using an interlayer polymer that is molecularly compatible with the reinforced polymer. This compatibility enhances intermolecular diffusion, ensuring a robust and effective bond.

There are several fusion bonding techniques depending on the technology used for heat introduction. The three most promising fusion bonding techniques are ultrasonic welding, induction welding, and resistance welding.

In ultrasonic welding, the components to be joined are pressed together under pressure and exposed to ultrasonic vibrations, typically ranging from 20 to 50 kHz, directed perpendicular to the contact area. Simultaneously, pressure and vibration are applied through a sonotrode connected to a piezoelectric generator, transforming high-frequency alternating current into mechanical vibrations. The parts are secured by a pressure actuator linked to the ultrasonic welder apparatus. The oscillation motion, characterized by low and high amplitude oscillations (sinusoidal strain), generates intermolecular friction, ultimately converting this frictional energy into heat for the welding process (Villegas, 2013).

The strength of ultrasonic welded joints crucially depends on the temperature at the interface, with heat uniformity playing a pivotal role in achieving high-quality results. Tao et al. (2019) observed issues such as cracks and voids in the matrix-fiber interface of ultrasonically welded carbon-reinforced PEEK lap shear coupons, linking these challenges to significant differences in thermal expansion coefficients. Essential process parameters influencing the welding outcome encompass applied pressure, vibration time, solidification time, and the amplitude and frequency of vibration. Further, studies conducted by Sacchetti et al. (2018) on ultrasonically welded samples of carbon-reinforced PEEK demonstrated an improvement in interlaminar fracture toughness with increased amounts of extra resin in the form of a film of PEEK at the bonding interface. These findings emphasize the importance of precise control over welding parameters for optimal joint quality.

Ultrasonic welding, owing to its high processing speed for joining thermoplastic polymer composites, has immense potential for performing spot welding (Mathijsen, 2015).

Induction welding represents a non-contact technique for heating thermoplastic composites with a conducting element. Swift localized heating is achieved using an induction coil generating a time-variable magnetic field. Eddy currents are induced in the electrical conductor, often a steel mesh screen along the bondline, leading to heat generation through Joule losses. This method allows for the rapid attainment of high temperatures. To ensure optimal energy transfer, the coil is positioned as close as possible to the part. Key process parameters governing the welding process include frequency, input power, applied pressure, and welding time (Ahmed et al., 2006).

The generation of heat in induction welding is also influenced by various material properties, including the consolidation style, fiber architecture, and the matrix material used (Yarlagadda et al., 2002). At times, the conductive element remaining in the joint post-welding allows for the potential of reprocessing or repair in instances of incomplete bonding or damage events (Murray et al., 2019). Further, for establishing large area joints, portable induction welding units can be employed, adding flexibility to the welding process.

In composite materials designed for induction welding, the incorporation of conductive elements such as graphite or carbon fibers plays a crucial role in reinforcing thermoplastic resins. These fibers serve as the conductive element, forming closed-loop circuits within the material (Bae et al., 2015). Additionally, micrometer metallic particles embedded in the composite contribute to heat generation, facilitating the welding process. Notably, the utilization of the material itself as a susceptor offers advantages by eliminating the need for contaminating inserts on the bondline, preserving the mechanical performance of the welded joint. Moreover, successful experiments involving ferromagnetic particles, as demonstrated by Bayerl et al. (2012), underscore the efficacy of these elements as heating promoters for melting thermoplastic polymers. As interest in nanostructured polymers grows, the potential application of induction welding becomes apparent, particularly when the polymer matrix incorporates electrically conductive nanofillers, opening avenues for joining nanocomposite parts efficiently (Greco et al., 2016).

In the resistance welding process, a conductive implant is strategically placed between the two components slated for joining. The introduction of electrical current through the insert leads to resistance heating, causing a temperature surge at the bonding interface and subsequently melting the thermoplastic polymer. This method boasts versatility, making it suitable for complex joints and demanding layups, including those with elastic couplings characterized by a high ply count (Samborski, 2017; Rzeczkowski et al., 2020; Samborski et al., 2019). Similar to induction welding, the conductive implant persists within the joint post-welding, offering convenience for potential reprocessing. However, it is important to consider potential drawbacks, as the presence of the implant can influence the mechanical performance of the bond and contribute to both increased assembly weight and a heightened risk of corrosion (Stavrov and Bersee, 2005).

A drawback of resistance welding is the risk of heat flow disorder, leading to local resin boiling in thicker ply packages during autoclaving. To address this issue and ensure quality, it is advisable to join thinner laminates together for the production of test specimens.

In resistance welding for lap shear strength in polymer joints, crucial parameters include welding current, welding duration, and clamping pressure (Panneerselvam et al., 2012). The welding current determines the heat generated, and its control is vital to achieve the desired joint strength without causing polymer degradation. The welding duration, representing the time of current application, influences overall thermal effects on the joint and must be carefully managed to prevent undesirable consequences like fiber motions due to excessive melting. Clamping pressure is essential for maintaining proper contact between materials; excessive pressure can lead to polymer expulsion, while insufficient pressure may result in voids in the joint. Stavrov and Bersee (2005) highlight the resistance of the heating element as a relevant process parameter, contributing to the efficiency of heat generation. Prolonged or overheating can lead to polymer degradation, while too much pressure can reduce bondline thickness, as indicated by the inverse relationship between lap shear strength and bondline thickness observed by Howie et al. (1993). Achieving optimal resistance welding conditions requires a careful balance of these parameters to ensure joint strength without compromising polymer integrity.

10.10 SUMMARY

In this chapter, the advancements in the domain of mechanical joining processes for FRPCs to polymer as well as metal-based counterparts are discussed. The mechanical joining processes, including bolted joining, clinching techniques, riveting techniques, pin, loop, and lock joining, are discussed. Advantages and limitations of each technique have been listed with discussions pertaining to the effect of joining parameters on joint strength and other mechanical properties. On the basis of the survey, it has been inferred that the scope of further research is more in processes like electromagnetic and friction-based riveting techniques. Further, the potentiality of additive manufacturing for creating such joints and parts cost-effectively is indeed a topic to investigate as the current techniques are expensive and limited to certain materials alone. Research pertaining to joining of composites using adhesive bonding is relatively low as the current state may not meet the industry requirements. The selection of fusion bonding techniques is to be done based on the properties of materials of interest, keeping in mind the heat accumulation that may affect the properties of the joint as well as the parent material.

REFERENCES

Abibe, A.B., Amancio-Filho, S.T., Dos Santos, J.F. and Hage Jr, E., 2011. Development and analysis of a new joining method for polymer-metal hybrid structures. *Journal of Thermoplastic Composite Materials*, 24(2), 233–249.

Abibe, A.B., Amancio-Filho, S.T., Dos Santos, J.F. and Hage Jr, E., 2013. Mechanical and failure behaviour of hybrid polymer–metal staked joints. *Materials & Design (1980-2015)*, 46, 338–347.

Abibe, A.B., Sônego, M., Canto, L.B., Dos Santos, J.F. and Amancio-Filho, S.T., 2020. Process-related changes in polyetherimide joined by friction-based injection clinching joining (F-ICJ). *Materials*, 13(5), 1027.

Ahmed, T., Stavrov, D., Bersee, H. and Beukers, A. Induction welding of thermoplastic composites—An overview. *Composites Part A: Applied Science and Manufacturing* 2006, 37, 1638–1651.

Altmeyer, J., Dos Santos, J.F. and Amancio-Filho, S.T., 2014. Effect of the friction riveting process parameters on the joint formation and performance of Ti alloy/short-fibre reinforced polyether ether ketone joints. *Materials & Design*, 60, 164–176.

Amancio-Filho, S.T., Roeder, J., Nunes, S.P., Dos Santos, J.F. and Beckmann, F., 2008. Thermal degradation of polyetherimide joined by friction riveting (FricRiveting). Part I: Influence of rotation speed. *Polymer Degradation and Stability*, 93(8), 1529–1538.

Bae, D., Shin, P., Kwak, S., Moon, M., Shon, M., Oh, S. and Kim, G., 2015. Heating behavior of ferromagnetic Fe particle-embedded thermoplastic polyurethane adhesive film by induction heating. *Journal of Industrial and Engineering Chemistry*, 30, 92.

Bayerl, T., Schledjewski, R. and Mitschang, P., 2012. Induction heating of thermoplastic materials by particulate heating promoters. *Polymers and Polymer Composites*, 20, 333–342.

Blaga, L., Bancilă, R., dos Santos, J.F. and Amancio-Filho, S.T., 2013. Friction Riveting of glass–fibre-reinforced polyetherimide composite and titanium grade 2 hybrid joints. *Materials & Design*, 50, 825–829.

Buffa, G., Baffari, D., Campanella, D. and Fratini, L., 2016. An innovative friction stir welding based technique to produce dissimilar light alloys to thermoplastic matrix composite joints. *Procedia Manufacturing*, 5, 319–331.

Camanho, P.P., Fink, A., Obst, A. and Pimenta, S., 2009. Hybrid titanium–CFRP laminates for high-performance bolted joints. *Composites Part A: Applied Science and Manufacturing*, 40(12), 1826–1837.

Camanho, P.P. and Matthews, F.L., 2000. Bonded metallic inserts for bolted joints in composite laminates. *Proceedings of the Institution of Mechanical Engineers, Part L: Journal of Materials: Design and Applications*, 214(1), 33–40.

Camanho, P.P., Tavares, C.M.L., De Oliveira, R., Marques, A.T. and Ferreira, A.J.M., 2005. Increasing the efficiency of composite single-shear lap joints using bonded inserts. *Composites Part B: Engineering*, 36(5), 372–383.

Chaves, F.J.P., Da Silva, L.F.M., De Moura, M.F.S.F., Dillard, D.A. and Esteves, V.H.C., 2014. Fracture mechanics tests in adhesively bonded joints: A literature review. *Journal of Adhesion*, 90, 955–992.

Choi, J.I., Hasheminia, S.M., Chun, H.J., Park, J.C. and Chang, H.S., 2018. Failure load prediction of composite bolted joint with clamping force. *Composite Structures*, 189, 247–255.

Choudhury, M.R. and Debnath, K., 2019, October. On the analysis of compressive failure load of single-lap bolted joint of green composites. In *IOP Conference Series: Materials Science and Engineering*, 635 1, 012028. IOP Publishing.

Chueh, Y.H., Wei, C., Zhang, X. and Li, L., 2020. Integrated laser-based powder bed fusion and fused filament fabrication for three-dimensional printing of hybrid metal/polymer objects. *Additive Manufacturing*, 31, 100928.

Crosky, A., Kelly, D., Li, R., Legrand, X., Huong, N. and Ujjin, R., 2006. Improvement of bearing strength of laminated composites. *Composite Structures*, 76(3), 260–271.

Cui, J., Dong, D., Zhang, X., Huang, X., Lu, G., Jiang, H. and Li, G., 2018. Influence of thickness of composite layers on failure behaviors of carbon fiber reinforced plastics/aluminum alloy electromagnetic riveted lap joints under high-speed loading. *International Journal of Impact Engineering*, 115, 1–9.

Di Franco, G., Fratini, L., Pasta, A. and Ruisi, V.F., 2013. On the self-piercing riveting of aluminium blanks and carbon fibre composite panels. *International Journal of Material Forming*, 6, 137–144.

Fink, A., Camanho, P.P., Andrés, J.M., Pfeiffer, E. and Obst, A., 2010. Hybrid CFRP/titanium bolted joints: Performance assessment and application to a spacecraft payload adaptor. *Composites Science and Technology*, 70(2), 305–317.

Fink, A. and Kolesnikov, B., 2005. Hybrid titanium composite material improving composite structure coupling. In *Spacecraft Structures, Materials and Mechanical Testing 2005*, 581.

Gagliardi, F., Conte, R., Ciancio, C., Simeoli, G., Pagliarulo, V., Ambrogio, G. and Russo, P., 2018. Joining of thermoplastic structures by friction riveting: A mechanical and a microstructural investigation on pure and glass reinforced polyamide sheets. *Composite Structures*, 204, 268–275.

Galińska, A., 2020. Mechanical joining of fibre reinforced polymer composites to metals—A review. Part I: Bolted joining. *Polymers*, 12(10), 2252.

Galińska, A. and Galiński, C. Mechanical joining of fibre reinforced polymer composites to metals—A review. Part II: Riveting, clinching, non-adhesive form-locked joints, pin and loop joining. *Polymers* 2020, 12, 1681.

Graham, D.P., Rezai, A., Baker, D., Smith, P.A. and Watts, J.F., 2014. The development and scalability of a high strength, damage tolerant, hybrid joining scheme for composite–Metal structures. *Composites Part A: Applied Science and Manufacturing*, 64, 11–24.

Greco, A., Lionetto, F. and Maffezzoli, A., 2016. Orientation of graphene nanoplatelets in thermosetting matrices. *IEEE Transactions on Nanotechnology*, 15, 877–883.

Guan, Z., Mu, J., Su, F., Bian, T., Huang, Y. and Li, Z., 2015. Pull-through mechanical behavior of composite fastener threads. *Applied Composite Materials*, 22, 251–267.

Gude, M., Vogel, C. and Gröger, B., 2019. Simulation-aided development of a robust thermoclinching joining process for hybrid structures with textile reinforced thermoplastic composites and metallic components. *Materialwissenschaft und Werkstofftechnik*, 50, 1027.

Hausmann, J.M., Helfrich, B. and Motsch, N., 2017. Aspects of preloaded bolted joints of fiber reinforced polymers and metals. *Key Engineering Materials*, 742, 401–407.

Hoang, N.-H., Porcaro, R., Langseth, M. and Hanssen, A.-G., 2010. Self-piercing riveting connections using aluminium rivets, *International Journal of Solids and Structures*, 47(3–4), 427–439.

Howie, I., Gillespie, J.W.J., Smiley, A.J. and Smiley, A.J., 1993. Resistance welding of graphite-polyarylsulfone/polysulfone dual-polymer composites. *Journal of Thermoplastic Composite Materials*, 6, 205–225.

Hufenbach, W., Gottwald, R. and Kupfer, R., 2011, August. Bolted joints with moulded holes for textile thermoplastic composites. In *Proc. Int. Conf. on 'Composite materials* (Vol. 18, 1–6).

Jiang, H., Li, G., Zhang, X. and Cui, J., 2017b. Fatigue and failure mechanism in carbon fiber reinforced plastics/aluminum alloy single lap joint produced by electromagnetic riveting technique. *Composites Science and Technology*, 152, 1–10.

Jiang, H., Luo, T., Li, G., Zhang, X. and Cui, J., 2017a. Fatigue life assessment of electromagnetic riveted carbon fiber reinforce plastic/aluminum alloy lap joints using Weibull distribution. *International Journal of Fatigue*, 105, 180–189.

Jiang, H., Sun, L., Dong, D., Li, G. and Cui, J., 2019. Microstructure and mechanical property evolution of CFRP/Al electromagnetic riveted lap joint in a severe condition. *Engineering Structures*, 180, 181–191.

Jiang, H., Zeng, C., Li, G. and Cui, J., 2021. Effect of locking mode on mechanical properties and failure behavior of CFRP/Al electromagnetic riveted joint. *Composite Structures*, 257, 113162.

Karim, M.A., Abdullah, M.Z., Deifalla, A.F., Azab, M. and Waqar, A., 2023. An assessment of the processing parameters and application of fibre-reinforced polymers (FRPs) in the petroleum and natural gas industries: A review. *Results in Engineering*, 18, 101091.

Kashaev, N., Ventzke, V., Riekehr, S., Dorn, F. and Horstmann, M., 2015. Assessment of alternative joining techniques for Ti–6Al–4V/CFRP hybrid joints regarding tensile and fatigue strength. *Materials & Design*, 81, 73–81.

Kelly, G. and Hallström, S., 2004. Bearing strength of carbon fibre/epoxy laminates: Effects of bolt-hole clearance. *Composites Part B: Engineering*, 35(4), 331–343.

Kolesnikov, B., Herbeck, L. and Fink, A., 2008. CFRP/titanium hybrid material for improving composite bolted joints. *Composite Structures*, 83(4), 368–380.

Lambiase, F., Durante, M. and Di Ilio, A., 2016. Fast joining of aluminum sheets with glass fiber reinforced polymer (GFRP) by mechanical clinching. *Journal of Materials Processing Technology*, 236, 241–251.

Lambiase, F. and Ko, D.C., 2016. Feasibility of mechanical clinching for joining aluminum AA6082-T6 and carbon fiber reinforced polymer sheets. *Materials & Design*, 107, 341–352.

Lambiase, F. and Ko, D.C., 2017. Two-steps clinching of aluminum and carbon fiber reinforced polymer sheets. *Composite Structures*, 164, 180–188.

Lambiase, F. and Paoletti, A., 2018. Friction-assisted clinching of aluminum and CFRP sheets. *Journal of Manufacturing Processes*, 31, 812–822.

Lambiase, F., Paoletti, A. and Di Ilio, A., 2017. Advances in mechanical clinching: employment of a rotating tool. *Procedia Engineering*, 183, 200–205.

Lan, F.T., Li, K.Z., Li, H.J., He, Y.G., Shen, X.T. and Cao, W.F., 2009. Joining of carbon/carbon composites for nuclear applications. *Journal of Materials Science*, 44, 3747–3750.

Lang, A., Husemann, L. and Herrmann, A.S., 2013. Influence of textile process parameter on joint strength for integral CFRP-aluminum transition structures. *Procedia Materials Science*, 2, 212–219.

Lee, C.J., Kim, B.M., Kang, B.S., Song, W.J. and Ko, D.C., 2017. Improvement of joinability in a hole clinching process with aluminum alloy and carbon fiber reinforced plastic using a spring die. *Composite Structures*, 173, 58–69.

Lee, C.J., Lee, S.H., Lee, J.M., Kim, B.H., Kim, B.M. and Ko, D.C., 2014a. Design of hole-clinching process for joining CFRP and aluminum alloy sheet. *International Journal of Precision Engineering and Manufacturing*, 15, 1151–1157.

Lee, S.H., Lee, C.J., Kim, B.H., Ahn, M.S., Kim, B.M. and Ko, D.C., 2014b. Effect of tool shape on hole clinching for CFRP with steel and aluminum alloy sheet. *Key Engineering Materials*, 622, 476–483.

Lee, S.H., Lee, C.J., Lee, K.H., Lee, J.M., Kim, B.M. and Ko, D.C., 2014c. Influence of tool shape on hole clinching for carbon fiber-reinforced plastic and SPRC440. *Advances in Mechanical Engineering*, 6, 810864.

Li, R., Kelly, D. and Crosky, A., 2002. Strength improvement by fibre steering around a pin loaded hole. *Composite Structures*, 57(1-4), 377–383.

Liang, J., Jiang, H., Zhang, J., Wu, X., Zhang, X., Li, G. and Cui, J., 2019. Investigations on mechanical properties and microtopography of electromagnetic self-piercing riveted joints with carbon fiber reinforced plastics/aluminum alloy 5052. *Archives of Civil and Mechanical Engineering*, 19, 240–250.

Lin, P.C., Fang, J.C., Lin, J.W., Tran, X.V. and Ching, Y.C., 2020. Preheated (heat-assisted) clinching process for Al/CFRP cross-tension specimens. *Materials*, 13(18), 4170.

Lionetto, F., Pappadà, S., Buccoliero, G., Maffezzoli, A. Finite element modeling of continuous induction welding of thermoplastic matrix composites. *Materials & Design* 2017, 120, 212–221.

Mallick, P.K., 2010. 5—Thermoplastics and thermoplastic–matrix composites for lightweight automotive structures. In *Materials, Design and Manufacturing for Lightweight Vehicles*; Mallick, P.K., Ed., Woodhead Publishing: Cambridge, UK, 174–207.

Marjanowski, J., Tomasiewicz, J. and Frączek, W., 2017. The electric-powered motorglider AOS-71–The study of development. *Aircraft Engineering and Aerospace Technology*, 89(4), 579–589.

Mathijsen, D. The black magic of carbon fiber reinforced thermoplastics. *Reinforced Plastics* 2015, 59, 185–189.

Matsuzaki, R., Kanatani, T. and Todoroki, A., 2019. Multi-material additive manufacturing of polymers and metals using fused filament fabrication and electroforming. *Additive Manufacturing*, 29, 100812.

McCarthy, M.A., Lawlor, V.P., Stanley, W.F. and McCarthy, C.T., 2002. Bolt-hole clearance effects and strength criteria in single-bolt, single-lap, composite bolted joints. *Composites Science and Technology*, 62(10-11), 1415–1431.

McKnight, S.H., Holmes, S.T., Gillespie, J.W., Lambing, C.L.T., Marinelli, J.M. Scaling issues in resistance-welded thermoplastic composite joints. *Advances in Polymer Technology* 1997, 16, 279–295.

Meschut, G., Gude, M., Augenthaler, F. and Geske, V., 2014. Evaluation of damage to carbon-fibre composites induced by self-pierce riveting. *Procedia CIRP*, 18, 186–191.

Mouritz, A.P., 2012. Environmental durability of z-pinned carbon fibre–epoxy laminate exposed to water. *Composites Science and Technology*, 72(13), 1568–1574.

Murray, R.E., Beach, R., Barnes, D., Snowberg, D., Berry, D., Rooney, S., Jenks, M., Gage, B., Boro, T., Wallen, S. and Hughes, S., 2021. Structural validation of a thermoplastic composite wind turbine blade with comparison to a thermoset composite blade. *Renewable Energy*, 164, 1100–1107.

Murray, R.E., Roadman, J. and Beach, R., 2019. Fusion joining of thermoplastic composite wind turbine blades: Lap-shear bond characterization. *Renewable Energy*, 140, 501–512.

Nassar, S.A., Virupaksha, V.L. and Ganeshmurthy, S., 2007. *Effect of Bolt Tightness on the Behavior of Composite Joints*.

Nguyen, A.T., Amarasinghe, C.K., Brandt, M., Feih, S. and Orifici, A.C., 2017. Loading, support and geometry effects for pin-reinforced hybrid metal-composite joints. *Composites Part A: Applied Science and Manufacturing*, 98, 192–206.

Nguyen, A.T., Brandt, M., Feih, S. and Orifici, A.C., 2016. Pin pull-out behaviour for hybrid metal-composite joints with integrated reinforcements. *Composite Structures*, 155, 160–172.

Ouyang, Y. and Chen, C., 2022. Research advances in the mechanical joining process for fiber reinforced plastic composites. *Composite Structures*, 296, 115906.

Panneerselvam, K., Aravindan, S., Haq, A.N., 2012. Study on resistance welding of glass fiber reinforced thermoplastic composites. *Materials & Design*, 41, 453–459.

Parkes, P.N., Butler, R., Meyer, J. and De Oliveira, A., 2014. Static strength of metal-composite joints w2016bith penetrative reinforcement. *Composite Structures*, 118, 250–256.

Persson, E. and Madenci, E., 1998. Composite laminates with elliptical pin-loaded holes. *Engineering Fracture Mechanics*, 61(2), 279–295.

Pinto, A.M.G., Magalhães, A.G., Campilho, R.D.S.G., De Moura, M.F.S.F. and Baptista, A.P.M., 2009. Single-lap joints of similar and dissimilar adherends bonded with an acrylic adhesive. *Journal of Adhesion*, 85, 351–376.

Rispler, A.R., Steven, G.P. and Tong, L., 2000. Photoelastic evaluation of metallic inserts of optimised shape. *Composites Science and Technology*, 60(1), 95–106.

Rubino, F., Nisticò, A., Tucci, F. and Carlone, P., 2020. Marine application of fiber reinforced composites: A review. *Journal of Marine Science and Engineering*, 8(1), 26.

Rzeczkowski, J., Samborski, S. and Valvo, P.S., 2020. Effect of stiffness matrices terms on delamination front shape in laminates with elastic couplings. *Composite Structures*, 233, 111547.

Sacchetti, F., Grouve, W.J., Warnet, L. and Villegas, I.F., 2018. Effect of resin-rich bond line thickness and fibre migration on the toughness of unidirectional Carbon/PEEK joints. *Composites Part A: Applied Science and Manufacturing*, 109, 197–206.

Salamati, M., Soltanpour, M., Zajkani, A. and Fazli, A., 2019. Improvement in joint strength and material joinability in clinched joints by electromagnetically assisted clinching. *Journal of Manufacturing Processes*, 41, 252–266.

Samborski, S. Analysis of the end-notched flexure test configuration applicability for mechanically coupled fiber reinforced composite laminates. *Composite Structures* 2017, 163, 342–349.

Samborski, S., Gliszczynski, A., Rzeczkowski, J. and Wiacek, N., 2019. Mode I interlaminar fracture of glass/epoxy unidirectional laminates. Part I: Experimental studies. *Materials*, 12, 1607.

Sarfraz, M.S., Hong, H. and Kim, S.S., 2021. Recent developments in the manufacturing technologies of composite components and their cost-effectiveness in the automotive industry: A review study. *Composite Structures*, 266, 113864.

Schimanski, K., von Hehl, A. and Zoch, H.W., 2013. Failure behavior of diffusion bonded transition structures for integral FRP-Aluminum compounds. *Procedia Materials Science*, 2, 189–196.

Schumacher, J., Bomas, H. and Zoch, H.-W., 2013. Failure behaviour of advanced seam structures for CFRP-aluminium connections. *Procedia Materials Science*, 2, 227–233.

Sevkat, E., Brahimi, M. and Berri, S., 2011. A novel manufacturing technique to improve bearing strength of pin loaded woven composites. *Advanced Composites Letters*, 20(3), 096369351102000304.

Smiley, A.J., Halbritter, A., Cogswell, F.N. and Meakin, P.J., 1991. Dual polymer bonding of thermoplastic composite structures. *Polymer Engineering & Science*, 31, 526–532.

Starikov, R. and Schön, J., 2001. Quasi-static behaviour of composite joints with countersunk composite and metal fasteners. *Composites Part B: Engineering*, 32(5), 401–411.

Stavrov, D. and Bersee, H.E.N., 2005. Resistance welding of thermoplastic composites-An overview. *Composites Part A: Applied Science and Manufacturing*, 36, 39–54.

Suresh, K., Rani, M.R., Prakasan, K. and Rudramoorthy, R., 2007. Modeling of temperature distribution in ultrasonic welding of thermoplastics for various joint designs. *Journal of Materials Processing Technology*, 186, 138–146.

Sweeting, R.D. and Thomson, R.S., 2004. The effect of thermal mismatch on Z-pinned laminated composite structures. *Composite Structures*, 66(1-4), 189–195.

Tao, W., Su, X., Wang, H., Zhang, Z., Li, H. and Chen, J., 2019. Influence mechanism of welding time and energy director to the thermoplastic composite joints by ultrasonic welding. *Journal of Manufacturing Processes*, 37, 196–202.

Tomasiewicz, J. and Czarnocki, P., 2016. Wing-to-fuselage attachment fitting for composite airframes: Experimental and finite element analysis. *Composites Theory and Practice*, 16, 61–66.

Tosh, M.W. and Kelly, D.W., 2000. On the design, manufacture and testing of trajectorial fibre steering for carbon fibre composite laminates. *Composites Part A: Applied Science and Manufacturing*, 31(10), 1047–1060.

Tu, W., Wen, P.H., Hogg, P.J. and Guild, F.J., 2011. Optimisation of the protrusion geometry in Comeld™ joints. *Composites Science and Technology*, 71(6), 868–876.

Ucsnik, S., Scheerer, M., Zaremba, S. and Pahr, D.H., 2010. Experimental investigation of a novel hybrid metal–composite joining technology. *Composites Part A: Applied Science and Manufacturing*, 41(3), 369–374.

Villegas, I.F., 2013. In situ monitoring of ultrasonic welding of thermoplastic composites through power and displacement data. *Journal of Thermoplastic Composite Materials*, 28, 66–85.

Wang, J.T., Lotts, C.G. and Davis, D.D., 1993. Analysis of bolt-loaded elliptical holes in laminated composite joints. *Journal of Reinforced Plastics and Composites*, 12.

Wang, X., Ahn, J., Kaboglu, C., Yu, L. and Blackman, B.R., 2016a. Characterisation of composite-titanium alloy hybrid joints using digital image correlation. *Composite Structures*, 140, 702–711.

Wang, X., Ahn, J., Lee, J. and Blackman, B.R., 2016b. Investigation on failure modes and mechanical properties of CFRP-Ti6Al4V hybrid joints with different interface patterns using digital image correlation. *Materials & Design*, 101, 188–196.

Yarlagadda, S., Kim, H.J., Gillespie, J.W., Shevchenko, N.B. and Fink, B.K. A study on the induction heating of conductive fiber reinforced composites. *Journal of Composite Materials* 2002, 36, 401–421.

Ye, L., Chen, Z.-R., Lu, M. and Hou, M. De-consolidation and re-consolidation in CF/PPS thermoplastic matrix composites. *Composites Part A: Applied Science and Manufacturing* 2005, 36, 915–922.

Yun, J.H., Choi, J.H. and Kweon, J.H., 2014. A study on the strength improvement of the multi-bolted joint. *Composite Structures*, 108, 409–416.

Zengqiang, C.A.O. and Yangjie, Z.U.O., 2020. Electromagnetic riveting technique and its applications. *Chinese Journal of Aeronautics*, 33(1), 5–15.

Zitoune, R., Crouzeix, L., Collombet, F., Tamine, T. and Grunevald, Y.H., 2011. Behaviour of composite plates with drilled and moulded hole under tensile load. *Composite Structures*, 93(9), 2384–2391.

Zongkai, F., Bai, K. and Chen, C., 2023. The application of ultrasound in joining: Principles, processes and properties, *Journal of Manufacturing Processes*, 101, 269–299.

11 Reuse, Remanufacturing, and Recycling of Natural Fibre Composites

Vishnu Prasad and Amal Alliyankal Vijayakumar

11.1 INTRODUCTION

The construction sector is the main contributor to global carbon emissions. The need for sustainable materials to substitute the existing building materials is an essential requirement to aim for net-zero carbon emissions in the future. Fibre-reinforced polymer (FRP) composites are an excellent material that can attain this goal. FRPs have lower carbon footprints than traditional construction building materials such as steel, concrete masonry, and timber. Similarly, the use of FRPs in other sectors, such as aerospace, construction, automotive, electronics, and marine, also looks forward to the use of sustainable materials [1]. FRPs have excellent mechanical properties and are lightweight, corrosion-resistant, and mouldable. With the increasing use of FRP materials in the construction industry and other engineering applications, the need for reuse and recycling at the end-of-life cycle is a major challenge. Sustainable tools and methods should be used for the effective disposal of FRP waste [2].

Among the categories of composite materials, natural fibre reinforcement is well known for its highlighting advantages of the excellent strength-to-weight ratio, sustainable nature, biodegradability, low cost, and availability compared to synthetic fibres, such as glass and carbon [3]. Natural fibre-reinforced composites (NFRCs) have been used for the last 100 years, and their commercial applications have been noticed for the past 30 to 40 years [4]. Thermoset resins are commonly used polymers for the manufacture of components made of composites in the industry, such as automotive interior parts, sports goods, leisure, and architectural mouldings. [5]. The recycling of thermoset composites requires complicated procedures that involve high temperatures and the presence of a metal catalyst. The use of recyclable thermoplastic resin can be a suitable solution that could use mild recycling conditions to extract the reinforcement and the polymer [6].

FRP composite systems consist of reinforcement and a matrix material. The reinforcement is generally synthetic, which includes carbon, glass, Kevlar, or natural fibres that comprise flax, jute, kenaf, bamboo, etc. There is semi-natural-based reinforcement, which is basalt fibre from basalt rock, which has a growing research interest in composites research. Similarly, natural fibre composites have also gained a lot of research interest; however, they are limited to non-structural applications [5]. Hybridizing them with synthetic fibres is the commonly practised method for this problem. However, this class of fibre composites is well acceptable in the green composites topics as they cause minimal damage to the environment during disposal. The matrix material would be either thermoset or thermoplastic-based, such as epoxy, polyester, and elium. The matrix or resin system binds the fibre reinforcement together and supports the load transfer from fibre to fibre during mechanical loading conditions. Thermoset matrix materials are commonly used for structural engineering applications owing to their high strength. The thermoplastic composites on the other side are recyclable, but the mechanical properties are not as competitive as the thermoset. The cross-linked polymer chains make the thermoset polymer non-recyclable. The recycling of the thermoset destroys

DOI: 10.1201/9781032665337-11

the fibre reinforcement; it is highly difficult to recover it. The weak molecular bond on the thermoplastic composites helps them to be remoulded and reshaped for their next use. As per the reports from the European Composites Industry Association (EuCia), more than 90% of the FRPs are made from glass fibre composites [7]. Among them, about 2/3 are made out of thermoset resin and the remaining from thermoplastic resins [8].

11.2 RECYCLING OF THERMOPLASTICS AND ITS COMPARISON WITH THE VIRGIN MATERIALS

Over the past decade, there have been several research publications in the field of recycling plastics of the fibre-reinforced composites. An early research work by Bertin and Robin [9] reported the characterization of virgin and recycled low-density polyethylene (LDPE)/polypropylene (PP) blends. The use of the twin extrusion method over the single screw extrusion method resulted in better mechanical properties. The regenerated materials showed a 7% improvement in the tensile properties than the virgin materials. A study by Adhikary et al. [10] compared virgin and recycled high-density polyethylene (HDPE). The recycled material delivered an improvement in the tensile strength by 8% over the virgin HDPE. Another interesting study on the recycled polymer blend to use as the matrix in composites was reported by Turku et al. [11]. The construction and household plastic wastes collected from the landfills were recycled. The recycled material was compared with pure PP and PE to understand the quality of recycling by examining the mechanical properties. Characterizations such as tensile properties, scanning electron microscope, and energy-dispersive X-ray spectroscopy were performed for the comparison of the recycled and the pure plastic materials.

A similar study by Taufiq et al. [12] demonstrated the potential of recycled thermoplastics to be used as a replacement for virgin thermoplastics. The blending of two thermoplastics, PP and polyethylene (PE), was considered to form a new polymer system that could then be compared with virgin thermoplastics. The recycled thermoplastics were recovered from rejected-unused disposable diapers. The study characterized the mechanical properties of the polymer on the effect of the manufacturing temperature at 180°C, 190°C, and 200°C. To compare its quality, the virgin materials were also prepared similarly. The tensile properties marked their highest value when the moulding temperature was 180 degrees. Under this condition, the recycled PP/PE blend showed 57% lower tensile strength values when compared to the virgin PP/PE blend. However, the difference in the modulus values was very low by 2%.

11.3 RECYCLING OF THERMOPLASTICS FOR NATURAL FIBRE COMPOSITES APPLICATION

Natural fibres used as reinforcing materials are generally obtained from plants, wood, or animals. Some of the commonly used natural fibres are jute, flax, hemp, kenaf, cotton, etc. The importance of green, sustainable, and environmentally friendly materials has grabbed a lot of attention towards natural fibre composites. The ease of the recycling process of these classes of materials has been an interesting topic of research over the years. Natural fibre composites are extensively used for non-structural applications such as automotive interior components, sports goods, construction, and building materials. [13]. The highlighting advantages of natural fibre composites are their ease of recycling, lower energy consumption during manufacturing, lower health hazard issues, biodegradable with suitable polymer resins such as polylactic acid (PLA), and renewable [14]. The use of recycled polymers, as discussed in the previous session, and their application for the fabrication of composite materials, especially with natural fibres as reinforcement, is an excellent way to reduce the use of non-renewable materials. The following session details some of the applications based on the studies conducted by several researchers.

Polystyrene (PS) is one of the most used plastics for applications ranging from medical to automobile parts. They are considered a material that could be recycled at the end of the life cycle. PS is considered an alternate material to the natural aggregates in concrete manufacturing, which includes applications such as flooring, bricks, plaster, partition walls, and ceiling concrete [15]. The recycled PS is also used for the manufacturing of natural fibre composites. An interesting research work by Agoua et al. [16] presented the fabrication of composites made from the recycling sediments of wood and PS. The thermal conductivity measurement of the recycled material was primarily studied, with the results claiming that the material is suitable for applications as insulating material for the ceiling and dividing walls. Tahar et al. [17] in their results study explore the characterization of the date palm reinforced composites with PS composites. The innovative wood plastic composite material is aimed at developing material that is environmentally friendly and valorization of the vegetal resource. The research work focused on optimizing the fibre size from the range 0.1 to 0.315, 0.315 to 0.5, and 0.5 to 1 mm and the volume fraction at 70, 75, and 80 wt%. The density measurement revealed that the newly developed Leaflets-Polystyrene Composite (LPC) has a density value close to wood, indicating its potential alternative to reduce the use of cutting wood for building material construction requirements. The density values are even close to the medium density fibre board (MDF) and plywood, and the mechanical properties of the LPC materials are within acceptable limits. The values of the flexural modulus and strength are slightly lower than the MDF, which is one of the areas where further research shall be focused on.

Similar to PS, PP is also a commonly used plastic suitable for the fabrication of fibre-reinforced composite materials. The highlighting advantages of being lightweight, corrosion resistance, structural durability, and dielectric vitality are more compared to traditional plastics [18]. The high demand for PP has estimated its waste as 14% compared to the global waste, and the requirement for recycling is high. A research study by Ashori and Nourbakhsh used recycled newspapers as the reinforcement to develop two composite systems made out of recycled PP and recycled polyethylene separately. Studying the fibre loading rate on the tensile properties showed a drop in the tensile strength as the fibre volume fraction increased. The poor interfacial adhesion between hydrophilic and hydrophobic materials was noted towards higher values of fibre volume fraction. The volume fraction of 70% projected the maximum tensile and bending properties. The higher volume fraction also indicated poor resistance to water absorption and an increase in the swelling of the composite samples. The more tubular porous structures, especially at the edges of the samples, are the reason for the increase in the capillary action.

Chun et al. [19] experimented with the fabrication of coconut-reinforced recycled PP composites using a blade mixer setup operation at 180°C and a mixing speed of 50 rpm. The volume fraction of the dried coconut shell powder (CSP) was varied from 10 to 40 php at an interval of 10. The modified CSP with ethanol revealed better compatibility with recycled PP. The differential scanning calorimetry (DSC) analysis revealed that the presence of the CSP content reduced the crystallinity of the composite as the CSP is responsible for hindering the crystallization of PP chains. An investigation by Leu et al. [20] on the wood plastic composites using the wood floor as the reinforcement and recycled PP. The effect and the combination of the use of the coupling agent and the lubricant were investigated by evaluating the mechanical response of the composite. The wood floor with a mesh size of 100–120 with a 47% volume fraction, the coupling agent at 3%, the lubricant at 3%, and the recycled PP at 47% came as the optimized combination. The study inferred the importance of the use of a coupling agent and lubricant. The thermogravimetric analysis shown in Figure 11.1 compares the virgin PP, unmodified, and modified recycled PP composites, showing the improvement in the thermal stability of the recycled composites. The thermal decomposition of the polymer provided the PP resin with a thermal protective layer that improves the thermal stability.

A more substantial approach to the recycling of the LDPE composites with sugarcane fibre reinforcement was studied by Ribeiro et al. [21]. The steam explosion-treated sugarcane fibre projected improved fibre resin interaction and enhanced mechanical performance. The treated fibre surface increased the chances of resin penetration and fibre surface wetting, thereby increasing the surface

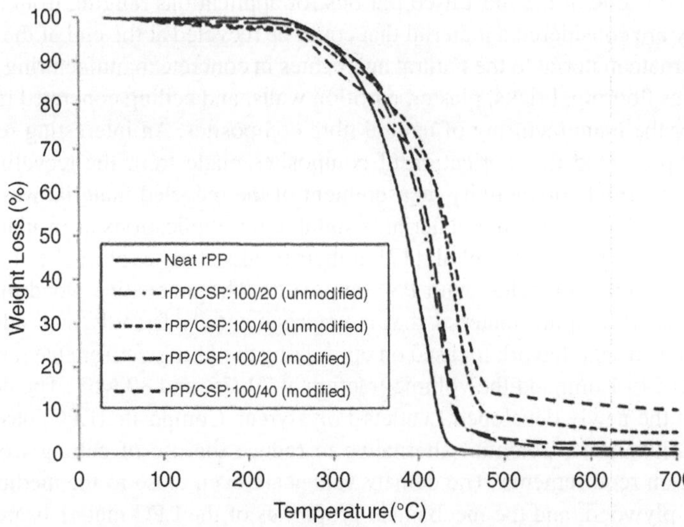

FIGURE 11.1 TGA curves of PP, unmodified, and modified recycled PP composite [19].

area of adhesion with better homogeneity. The mechanical response shows that recycled low-density PE can be considered as a suitable and interesting polymeric resin for the development of green composites. The DSC analysis implies that the presence of the sugarcane fibres improves the crystallinity of the composites with the increase in the fibre volume fraction. The fibre surface treatment enhances the quality of the fibre resin interface interaction, improving the tensile behaviour and the crystallinity of the material. Figure 11.2 shows the surface morphology of the tensile fracture samples of the untreated and surface-treated sugarcane fibre with recycled polymer composites. The scanning electron microscopy (SEM) images signify the importance of the surface treatment that projects the improvement in the fibre matrix adhesion.

11.4 RECYCLING OF THERMOSET RESIN COMPOSITES

As discussed in the previous sections, natural fibre-reinforced composites have interesting demand owing to their lightweight, environmentally friendly, good specific strength, and stiffness properties. The thermoset polymer composites are relatively considered superior in terms of mechanical properties to the thermoplastic polymer composites. The depleting fossil fuel sources and the environmental issues related to them have created a serious need for either a replacement or recycling of these classes of materials. However, when it comes to the recycling of these thermoset composites, they are challenging and complicated. The irreversible and crosslinked polymer chains make them unsuitable for recycling and require additional efforts to make them possible [22]. The prevailing methods for achieving the recycling of the thermoset polymer composites are the inclusion of labile linkages such as Schiff bases, acetal bonds, carbonate bonds, ester bonds, and sulphur-containing structures to the thermoset resin [23].

Hongwei et al. [24] presented the use of vitrimer materials to recycle thermoset polymer composites made out of carbon fibre reinforcement. The composites were thermally reprocessed without damaging the crosslinked network. The epoxy vitrimer contains aromatic diffusible crosslinks, which makes recycling easier. The results claimed the glass transition temperature of the epoxy resin as 147°C, with the tensile strength value at 63 MPa. The carbon fibres were extracted easily as the epoxy resins were dissolved. The recovered carbon fibres registered good mechanical properties when they were reused for composite fabrication. Figure 11.3 shows the stages of recycling epoxy-based composites and the recovery of the carbon fibres. Figure 11.4(a) shows the comparison of the

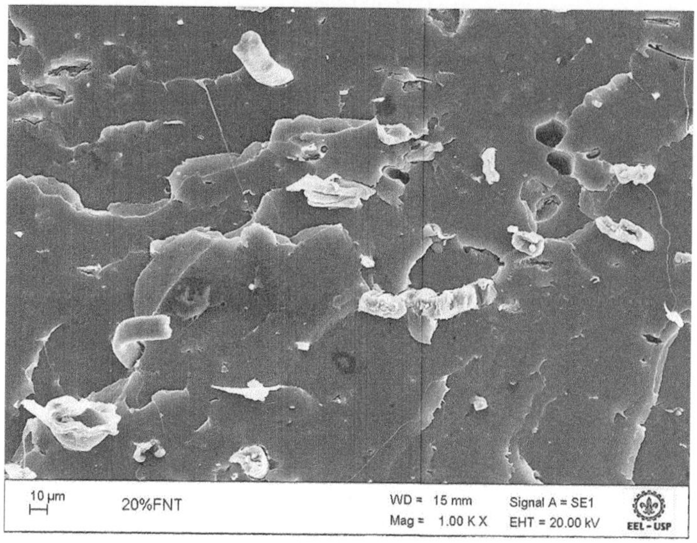

(a)

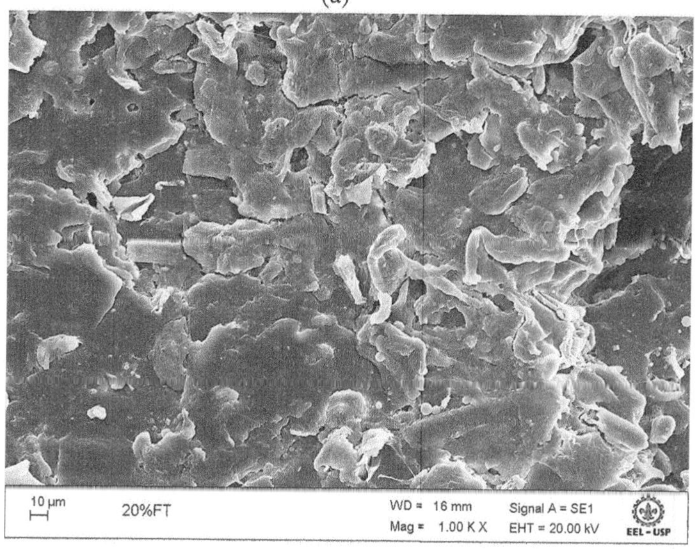

(b)

FIGURE 11.2 SEM images of sugarcane LDPE composites: (a) untreated and (b) steam explosion-treated fibres [21].

dynamic mechanical test results and the tensile test results of the reference and recycled carbon fibre composites. The SEM images of the recycled carbon fibres can be seen in Figure 11.4(b), indicating the quality of the carbon fibres recovered and minimal damages on the fibre surface due to the recycling process.

Hashimoto et al. [25] reported that degradable and chemically recycled epoxy composites have Acetal Linkages. Four types of epoxy resins were compared in this study. Bisphenol A, cresol novolak (CN) with vinyl esters having glycidyl groups, and cyclohexane dimethanol vinyl glycidyl ether were mixed with the amine hardener. Acid-induced degradation method was adopted for the recovery of the carbon fibres. The composite lamina was trimmed to 0.5 grams in weight, soaked in 0.1 mol/L- dilute hydrochloric acid, and mixed with solvent for 12 hours. The generated residue was separated, and the remaining organic material dried.

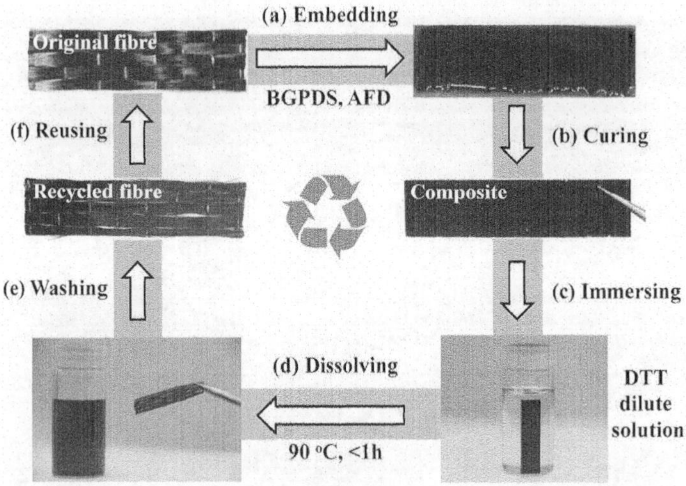

FIGURE 11.3 The recycling of carbon fibre composites: (a) carbon fibre epoxy composite preparation, (b) cured composite, (c) immersing the composites in DTT dilute solution, (d) dissolving the epoxy matrix, (e) washing and drying the extracted carbon fibres, and (f) reuse of the carbon fibres for composite preparation [24].

Research work by Taynton et al. [26] demonstrated the recycling of carbon fibres by the dissolution of polyimide binders. The study claims that by dissolving the polyimide binder, the dynamic covalent bond exchange with an excess of a monomer helps to achieve a closed-loop recycling of carbon fibre reinforced composites (CFRCs). The recovery of full-length fibre, as well as the complete recovery and reuse of the binder material, resulted from the attempt. The recovered samples were further used for the composite preparation, and the flexural test results show the modulus values to be in the range of 87% to 107% for the recycled composites. Another work by Liu et al. [27] discussed the recycling of fully biobased epoxy vitrimer-based carbon fibre composites. The malleability of the epoxy system enables it to be reprocessed through the hot-pressing method. The degradability of the composites was tested, and the recovery of the fibre reinforcement was achieved. The reprocessing of electromyography (EMG) was carried out by hot pressing using a plate vulcanizer. The samples were shredded and placed on steel plates between the hot press at 10 MPa pressure and 190°C for 2 hours. Later, a very large quantity of the sample (50 mg) was added to a beaker containing 10 mL of sodium hydroxide aqueous solution. The degradation was observed by regulating the quantity of glycerin in the resin, the concentration of sodium hydroxide, and the degradation temperature. The recovered carbon fibres were washed and dried in acetone for 24 hours at 50°C. The thermogravimetry analysis (TGA) results of the recycled composites were similar to the reference samples, indicating the quality of the recycling and the recycled composite. The interlaminar shear strength and the tensile test results values of the recycled composites were closer to the virgin composites, claiming the acceptability of the recycling method.

A recent study by Lorena et al. [28] conducted the recycling of a biobased polar epoxy resin reinforced with flax fibres. A chemical recycling method was followed to recover the flax fibres. A small quantity of the flax fibre composite weighing 20 grams was added to a 300 ml solution containing acetic acid with 75% volume. The composite was immersed in the solution for 90 minutes at 80°C. The acid treatment dissolved the thermoset, and the flax fibres were filtered out. Later, the acid solution was roto-evaporated at a pressure of 30 to -0 mbar at 3500 rpm and a temperature of 60°C. The practised process also recovered 225 ml of the acetic acid, which is suitable to be used for further treatment. Figure 11.5 shows the schematic representation of the chemical recycling process. The SEM images of the recycled flax fibres presented in Figure 11.6 indicate the quality of the recovered fibres.

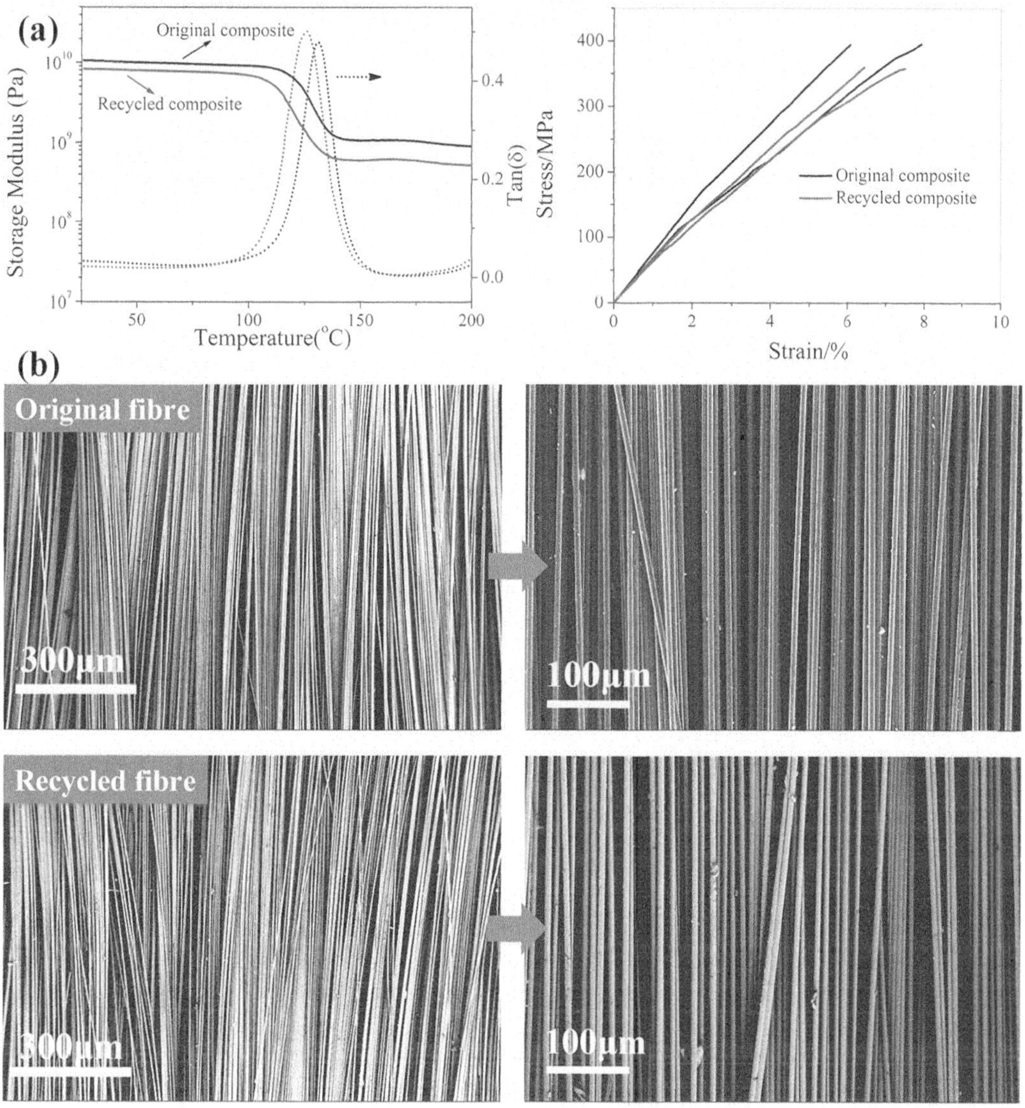

FIGURE 11.4 (a) DMA and tensile test results of the original and recycled composites and (b) SEM images of the original and recycled carbon fibres [24].

11.5 INFLUENCE OF RECYCLING ON MECHANICAL PROPERTIES

The effect of recyclability in polymer composites is analysed by examining one of its significant factors, the mechanical properties. The influence of reprocessing in pure polymers relies on several reasons, such as the processing techniques (thermomechanical, chemical, and ageing), parameters of the processing (UV exposure, temperature, humidity, speed, and time, shear rate), and, finally, the chemical structures of polymers [29]. However, the addition of reinforcements, including particles or fibres, to polymers increases the sophistication between the recycling methods and the resulting material properties. Figure 11.7 depicts the various types of reinforcements used in the polymer composites. In Table 11.1, different types of natural fibre-reinforced polymer composites (NFRPCs) with their properties are listed [35].

FIGURE 11.5 Schematic representation of the chemical recycling process [28].

A brief outline of the mechanical properties after reprocessing several NFRPCs is shown in Table 11.2. The current section discusses the influence of reprocessing/recycling on flexural, tensile, and impact performance of different varieties of NFRPCs researched in recent years. Most research focused on investigating reprocessing by thermo-mechanical recycling methods with the help of injection moulding, extrusion, compression moulding, melt compounding, and shredding/grinding [47, 48]. Whereas, some researchers also studied recycling by using suitable chemical regents, which dissolved the fibres and polymer in necessary solvents.

11.5.1 Effect of Recycling on the Tensile Properties

The tensile strength, elastic modulus, and ultimate stress and strain of a composite material before and after recycling depend on certain factors, such as fibre dispersion and aspect ratio, compatibility between the matrix and the fibre, and properties of the polymer used. Hence, these properties can either be improved or decreased during the recycling cycles. A study conducted by Bakkal et al. [48] processed waste cotton fabrics in LDPE using a customized recycling extruder. The recycling was carried out six times to assess the repeatability of reprocessing on tensile properties. The result showed that the tensile strength and modulus initially increased up to the fourth recycling cycle, and in the latter cycles, they decreased. This decrease in tensile properties has been due to fibre damage. However, the initial increase in tensile properties could be due to an improved aspect ratio and a fine-quality dispersion of cotton fibres upon recycling. In addition, fibre orientation also played a significant role in the performance of the composite. The specimens aligned parallel to the direction of extrusion have the highest tensile strength than samples perpendicular to extrusion. In the meantime,

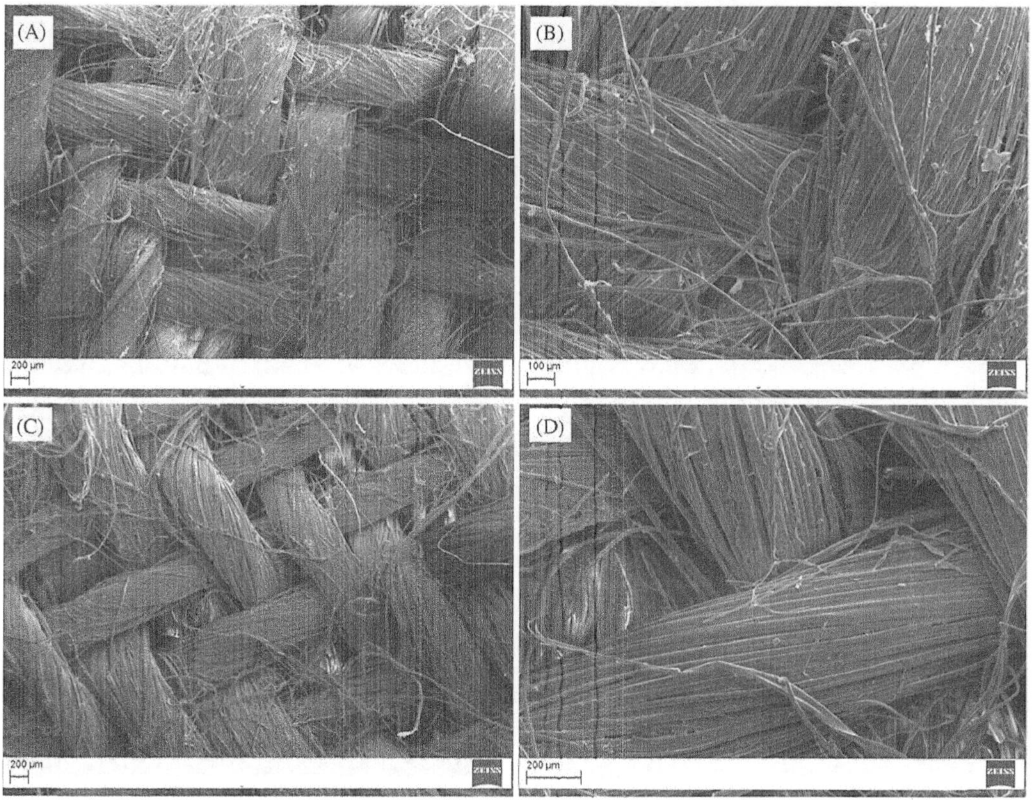

FIGURE 11.6 SEM images of (a and b) flax fibres and (b and c) recycled flax fibres [28].

the tensile strength of neat LDPE decreased as the cycle proceeded, which shows the chain scission and polymer degradation. Similar findings were also observed by Bodur et al. [49]. Likewise, an increase (after reprocessing twice) and successive decrease in tensile properties have been observed in wood fibre PP composite was investigated by Beg et al. [50] and in coriander straw fibre- and wood fibre-reinforced LDPE composites, observed by Uitterhaegen et al. [51]. These patterns were caused by improved fibrillation and dispersion of reinforcements in the matrix, where the decrease in performance is attributed to the degradation of fibre and polymer after continuous recycling.

On the other hand, an initial decrease and following an increase in elastic modulus was observed by Nadali et al. [52] under repetitive recycling of wood flour-Maleic Anhydride polyethylene (HDPE) composites consisting of up to four times grinding and reprocessing under industrial conditions. The decreased fibre size and molecular weight of the polymer with the recycling initially lowered the modulus of elasticity. However, the following improvement in modulus with repeated reprocessing is caused by better polymer crosslinking and increased crystallinity. Lila et al. [53] also noticed a similar trend of decrease and subsequent increase in tensile strength and modulus of bagasse fibre-PP composites upon continuous reprocessing. The increase in tensile strength is due to the increased aspect ratio of fibres because of the crushing of fibre during recycling. Also, the improved degree of crystal order and regularities in molecular arrangements enhanced the elongation under tensile loads.

11.5.2 Effect of Recycling on the Flexural Properties

After reprocessing, the bending performance of NFRPCs shows a similar response to that of the tensile properties. However, the distribution of stresses and modes of failures in bending tests is

FIGURE 11.7 Different types of natural reinforcements: (a) flax [30], (b) bamboo [30], (c) coir [31], (d) Kenaf [32], (e) cellulose [33], and (f) lignin [34].

distinct from tension. In the bending test, the plies/fibres/matrix at the top or bottom layer experiences the highest stress. Whereas in the tensile test, the failure starts propagating from the weak zone under constant axial stress.

For instance, the author Uitterhaegen et al. [51] observed the effect of recycling on flexural and tensile properties of PP and bio-based PE composites reinforced with coriander straw fibre (40% by weight). It is found that the influence of repetitive recycling on the flexural properties was comparatively less than that on the tension. Moreover, a 30% improvement in the flexural strength of PP composites was analysed by ageing them using UV irradiation and thermal cycles, while a small change was observed in tensile properties. This effect may be caused by the action of photo- and thermo-oxidation of the polymer, leading to chain scission; consequently, a lower molecular weight

TABLE 11.1
Details the Comparison of the Mechanical Performance of NFs with Synthetic e-Glass Fibre

Fibre	Density (g/cm³)	Tensile Strength (MPa)	Stiffness/ Young's Modulus (GPa)	Specific Tensile Strength (MPa/g cm⁻³)	Specific Young's Modulus (GPa/g cm⁻³)
E-glass	2.5	2000–3000	70	800–1400	29
Hemp	1.5	550–1110	58–70	370–740	39–47
Ramie	1.5	400–938	44–128	270–620	29–85
Flax	1.5	345–1830	27–80	230–1220	18–53
Sisal	1.3–1.5	507–855	9.4–28	362–610	6.7–20
Jute	1.3–1.5	393–800	10–55	300–610	7.1–39
Silk	1.3	100–1500	5–25	100–1500	4–20
Cotton	1.5–1.6	287–800	5.5–13	190–530	3.7–8.4
Coir	1.2	131–220	4–6	110–180	3.3–5
Wool	1.3	50–315	2.3–5	38–242	1.8–3.8
Feather	0.9	100–203	03-Oct	112–226	3.3–11

TABLE 11.2
Mechanical Performance of Various Recycled NFRPCs after Certain Reprocessing

Natural Fibres & Wt. %	Matrix	Reprocessing Methods & Number of Steps	Flexural Strength & Modulus Improvement (%)	Tensile Strength & Modulus Improvement (%)
Flax, 20	PP [36]	Grinding & injection moulding, 10	Modulus: 10	No improvement
30	PLLA [37]	Extrusion and injection moulding, 5	–	Strength: 70%, modulus: 8%
30	PBS/PLLA [38]	Grinding & injection moulding, 6	–	Decrease in modulus
21	PA11 [39]	Grinding & injection moulding, 7	–	Strength:17%, modulus: 28%
Hemp, 30	PP [40]	Grinding & injection moulding, 6	–	No improvement
Sugarcane Bagasse, 20	PP [41]	Extrusion and injection moulding, 4	Strength:14%, modulus: 19%	–
Sisal, 30	PP [40]	Grinding and injection moulding, 6	–	Strength: decrease, modulus: 10%
30	PLA [42]	Extrusion and injection moulding, 2	–	Strength: 58%, modulus: 5%
30	PHBV [43]	Compounding and compression moulding, 3	No improvement	Strength: decrease
Bamboo, 35	PLA [44]	Granulation, extrusion, pelletization, and injection moulding, 1	Strength: 8%, modulus: 22%	Strength: 7% modulus: 17%
Wood, 10	HDPE [45]	Grinding and injection moulding, 6	Strength: decrease, modulus: no change	Strength: decrease, modulus: no change
40	PVC [46]	Extrusion and injection moulding, 19	Strength: 40%, modulus: no change	–

and carbonyl formation occur. The surface exposed to UV irradiation and oxidation has a significant effect, while the core remains unaffected or unchanged, which governs the tensile properties of the composites. The flexural performance of the composites resulted in an initial increase and followed a decrease as the effect of continuous reprocessing. This similar trend to tensile properties was observed in several studies. As a consequence of recycling, the fibres in the polymer were dispersed more uniformly and broken. Besides, the fibre length was reduced, and the fibre and the polymer began to degrade upon successive reprocessing.

Fazita et al. [44] analysed a variation in the flexural strength of bamboo fabric reinforced in poly(lactic) acid composites upon reprocessing. The recycling process was conducted through the granulation, extrusion, palletization, and injection processes. The improved dispersion of fibres and better crystallinity caused an increase and further decrease in the flexural modulus of composites after continuous recycling. Accordingly, the harmful effect of decreased molecular weight outweighed the positiveness of reprocessing on flexural performance after multiple cycles. Similar findings were also observed by Correa et al. [41] investigating repetitive reprocessing in bagasse fibre/PP composites. Finally, Zhuo et al. [54] also found a beginning increase and, afterwards, a decrease in flexural strength and modulus with successive grinding and injection moulding reprocessing of poplar fibre-reinforced HDPE wood-plastic composites. At the initial stage of recycling, the flexural strength improved by 13% and in the latter cycles, it decreased to 11%. The first increase was due to the better fibre arrangements in the matrix, although this effect was gradually reduced due to the degradation and failure of polymer chains and fibres.

11.5.3 Effect of Recycling on the Impact Strength

Reprocessing also has significant effects on the impact properties of natural fibre-reinforced composites. Soccalingame et al. [55] found that the spruce wood flour-PP composites after repetitive recycling have an increased impact strength. This improvement in impact strength is attributed to the reduction in fibre length, which leads to an increase in the ductility of composites. A similar influence of repetitive recycling in NFRPCs was observed in the findings of [36, 39, 51].

In contrast to the above observations, some investigators have also noted a decrease in the impact strength of NFRPCs after recycling. For example, Fazita et al. [44], Wang et al. [56], and Graupner et al. [57] all found that the impact strength of bamboo fibre/PLA, wood fibre/PP, and cellulose fibre/PLA composites decreased after reprocessing. This deterioration in impact strength is due to the fibre shortening, degradation, and damage of the fibres and the matrix during recycling.

11.6 CONCLUSION

The recycling, reuse, and remanufacturing of NFRPCs offer several advantages in terms of environmental sustainability, resource conservation, and economic growth. These materials are widely used in various applications due to their high strength and durability, but their disposal can pose significant challenges in terms of waste management and environmental impact. By implementing sustainable practices such as recycling, reusing, and remanufacturing, we can reduce the amount of waste generated by these materials and minimize their negative impact on the environment. Moreover, these practices can help conserve natural resources and energy, as well as create economic opportunities through job creation and cost savings for companies. Therefore, it is imperative to adopt sustainable practices for the management of NFRPCs to achieve a more sustainable future.

REFERENCES

1. V. Prasad, K. Sekar, S. Varghese, M.A. Joseph, Evaluation of interlaminar fracture toughness and dynamic mechanical properties of nano TiO_2 coated flax fibre epoxy composites, *Polymer Testing*. 91 (2020) 106784. https://doi.org/10.1016/j.polymertesting.2020.106784

2. J. Qureshi, A review of recycling methods for fibre reinforced polymer composites, *Sustainability*. 14 (2022) 16855. https://doi.org/10.3390/su142416855
3. A. Alliyankal Vijayakumar, V. Prasad, M. Kailathuvalappil Kochunny, Investigation on the effect of stacking order and hybridization on mechanical and water absorption properties of woven flax/bamboo composites, *Polymer Composites*. (2022) pc.26808. https://doi.org/10.1002/pc.26808
4. D.J. Gardner, Y. Han, L. Wang, Wood–plastic composite technology, *Current Forestry Report* 1 (2015) 139–150. https://doi.org/10.1007/s40725-015-0016-6
5. J. Holbery, D. Houston, Natural-fiber-reinforced polymer composites in automotive applications, *JOM*. 58 (2006) 80–86. https://doi.org/10.1007/s11837-006-0234-2
6. Toughening by Nanodroplets: Polymer–Droplet Biocomposite with Anomalous Toughness|*Macromolecules*, (n.d.). https://doi.org/10.1021/acs.macromol.9b02677 (accessed April 2, 2023).
7. *Final Report Summary - EURECOMP (Recycling Thermoset Composites of the SST)* | FP7 | CORDIS | European Commission, (n.d.). https://cordis.europa.eu/project/id/218609/reporting (accessed March 27, 2023).
8. D.E. Witten, V. Mathes, *The Market for Glass Fibre Reinforced Plastics (GRP) in 2020*, (2020).
9. S. Bertin, J.-J. Robin, Study and characterization of virgin and recycled LDPE/PP blends, *European Polymer Journal* 38 (2002) 2255–2264. https://doi.org/10.1016/S0014-3057(02)00111-8
10. K.B. Adhikary, S. Pang, M.P. Staiger, Dimensional stability and mechanical behaviour of wood–plastic composites based on recycled and virgin high-density polyethylene (HDPE), *Composites Part B: Engineering* 39 (2008) 807–815. https://doi.org/10.1016/j.compositesb.2007.10.005
11. I. Turku, T. Kärki, K. Rinne, A. Puurtinen, Characterization of plastic blends made from mixed plastics waste of different sources, *Waste Management & Research* 35 (2017) 200–206. https://doi.org/10.1177/0734242X16678066
12. M.J. Taufiq, Z. Mustafa, M.R. Mansor, Utilisation of recycled thermoplastics sourced from rejected-unused disposable diapers as polymer blends, *Journal of Mechanical Engineering and Sciences* 11 (2017) 3137–3143. https://doi.org/10.15282/jmes.11.4.2017.16.0282
13. V. Prasad, M.A. Joseph, K. Sekar, Investigation of mechanical, thermal and water absorption properties of flax fibre reinforced epoxy composite with nano TiO_2 addition, *Composites Part A: Applied Science and Manufacturing*. 115 (2018) 360–370. https://doi.org/10.1016/j.compositesa.2018.09.031
14. V. Prasad, D. Suresh, M.A. Joseph, K. Sekar, M. Ali, Development of flax fibre reinforced epoxy composite with nano TiO_2 addition into matrix to enhance mechanical properties, *Materials Today: Proceedings*. 5 (2018) 11569–11575. https://doi.org/10.1016/j.matpr.2018.02.125
15. A. Kaya, F. Kar, Properties of concrete containing waste expanded polystyrene and natural resin, *Construction and Building Materials* 105 (2016) 572–578. https://doi.org/10.1016/j.conbuildmat.2015.12.177
16. E. Agoua, E. Allognon-Houessou, E. Adjovi, B. Togbedji, Thermal conductivity of composites made of wastes of wood and expanded polystyrene, *Construction and Building Materials* 41 (2013) 557–562. https://doi.org/10.1016/j.conbuildmat.2012.12.016
17. T. Masri, H. Ounis, L. Sedira, A. Kaci, A. Benchabane, Characterization of new composite material based on date palm leaflets and expanded polystyrene wastes, *Construction and Building Materials*. 164 (2018) 410–418. https://doi.org/10.1016/j.conbuildmat.2017.12.197
18. M.J.H. Khan, M.A. Hussain, I.M. Mujtaba, Polypropylene production optimization in fluidized bed catalytic reactor (FBCR): Statistical modeling and pilot scale experimental validation, *Materials* 7 (2014) 2440–2458. https://doi.org/10.3390/ma7042440
19. K.S. Chun, S. Husseinsyah, F.N. Azizi, Characterization and properties of recycled polypropylene/coconut shell powder composites: Effect of sodium dodecyl sulfate modification, *Polymer-Plastics Technology and Engineering* 52 (2013) 287–294. https://doi.org/10.1080/03602559.2012.749282
20. S.-Y. Leu, T.-H. Yang, S.-F. Lo, T.-H. Yang, Optimized material composition to improve the physical and mechanical properties of extruded wood–plastic composites (WPCs), *Construction and Building Materials* 29 (2012) 120–127. https://doi.org/10.1016/j.conbuildmat.2011.09.013
21. G.L. Ribeiro, M. Gandara, D.D.P. Moreno, C. Saron, Low-density polyethylene/sugarcane fiber composites from recycled polymer and treated fiber by steam explosion, *Journal of Natural Fibers* 16 (2019) 13–24. https://doi.org/10.1080/15440478.2017.1379044
22. S. Utekar, V.K. Suriya, N. More, A. Rao, Comprehensive study of recycling of thermosetting polymer composites – Driving force, challenges and methods, *Composites Part B: Engineering* 207 (2021) 108596. https://doi.org/10.1016/j.compositesb.2020.108596

23. Y. Liu, B. Wang, S. Ma, T. Yu, X. Xu, Q. Li, S. Wang, Y. Han, Z. Yu, J. Zhu, Catalyst-free malleable, degradable, bio-based epoxy thermosets and its application in recyclable carbon fiber composites, *Composites Part B: Engineering.* 211 (2021) 108654. https://doi.org/10.1016/j.compositesb.2021.108654
24. H. Si, L. Zhou, Y. Wu, L. Song, M. Kang, X. Zhao, M. Chen, Rapidly reprocessable, degradable epoxy vitrimer and recyclable carbon fiber reinforced thermoset composites relied on high contents of exchangeable aromatic disulfide crosslinks, *Composites Part B: Engineering.* 199 (2020) 108278. https://doi.org/10.1016/j.compositesb.2020.108278
25. T. Hashimoto, H. Meiji, M. Urushisaki, T. Sakaguchi, K. Kawabe, C. Tsuchida, K. Kondo, Degradable and chemically recyclable epoxy resins containing acetal linkages: Synthesis, properties, and application for carbon fiber-reinforced plastics, *Journal of Polymer Science Part A: Polymer Chemistry* 50 (2012) 3674–3681. https://doi.org/10.1002/pola.26160
26. P. Taynton, H. Ni, C. Zhu, K. Yu, S. Loob, Y. Jin, H.J. Qi, W. Zhang, Repairable woven carbon fiber composites with full recyclability enabled by malleable polyimine networks, *Advanced Materials* 28 (2016) 2904–2909. https://doi.org/10.1002/adma.201505245
27. Y. Liu, B. Wang, S. Ma, T. Yu, X. Xu, Q. Li, S. Wang, Y. Han, Z. Yu, J. Zhu, Catalyst-free malleable, degradable, bio-based epoxy thermosets and its application in recyclable carbon fiber composites, *Composites Part B: Engineering* 211 (2021) 108654. https://doi.org/10.1016/j.compositesb.2021.108654
28. L. Saitta, V. Prasad, C. Tosto, N. Murphy, A. Ivankovic, G. Cicala, G. Scarselli, Characterization of biobased epoxy resins to manufacture eco-composites showing recycling properties, *Polymer Composites* 43 (2022) 9179–9192. https://doi.org/10.1002/pc.27095
29. X. Zhao, K. Copenhaver, L. Wang, M. Korey, D.J. Gardner, K. Li, M.E. Lamm, V. Kishore, S. Bhagia, M. Tajvidi, H. Tekinalp, O. Oyedeji, S. Wasti, E. Webb, A.J. Ragauskas, H. Zhu, W.H. Peter, S. Ozcan, Recycling of natural fiber composites: Challenges and opportunities, *Resources, Conservation and Recycling.* 177 (2022) 105962. https://doi.org/10.1016/j.resconrec.2021.105962
30. A.A. Vijayakumar, N. James, V. Prasad, M. Divakaran, M.K. Kochunny, Tribological and thermogravimetric performance of flax/bamboo hybrid composites for gear applications: Influence of the stacking sequence, *Polymer Composites* 44 (2023) 1038–1053. https://doi.org/10.1002/pc.27152
31. T. Hamouda, A.H. Hassanin, A. Kilic, Z. Candan, M. Safa Bodur, Hybrid composites from coir fibers reinforced with woven glass fabrics: Physical and mechanical evaluation, *Polymer Composites* 38 (2017) 2212–2220. https://doi.org/10.1002/pc.23799
32. N.L. Feng, S.D. Malingam, C.W. Ping, N. Razali, Mechanical properties and water absorption of kenaf/pineapple leaf fiber-reinforced polypropylene hybrid composites, *Polymer Composites* 41 (2020) 1255–1264. https://doi.org/10.1002/pc.25451
33. B. Pereira, F.S. Matos, B.F.A. Valente, N. Von Weymarn, T. Kamppuri, C.S.R. Freire, A.J.D. Silvestre, C. Vilela, From regenerated wood pulp fibers to cationic cellulose: Preparation, characterization and dyeing properties, *Polysaccharides* 3 (2022) 609–624. https://doi.org/10.3390/polysaccharides3030036
34. A.K. Mohanty, S. Vivekanandhan, J.-M. Pin, M. Misra, Composites from renewable and sustainable resources: Challenges and innovations, *Science* 362 (2018) 536–542. https://doi.org/10.1126/science.aat9072
35. H. Jariwala, P. Jain, A review on mechanical behavior of natural fiber reinforced polymer composites and its applications, *Journal of Reinforced Plastics and Composites* 38 (2019). https://doi.org/10.1177/0731684419828524
36. T. Evens, G.-J. Bex, M. Yigit, J. De Keyzer, F. Desplentere, A. Van Bael, The Influence of mechanical recycling on properties in injection molding of fiber-reinforced polypropylene, *International Polymer Processing* 34 (2019) 398–407. https://doi.org/10.3139/217.3770
37. Antoine Le Duigou, Isabelle Pillin, Alain Bourmaud, Peter Davies, Christophe Baley, Effect of recycling on mechanical behaviour of biocompostable flax/poly(l-lactide) composites, *Composites Part A: Applied Science and Manufacturing* 39 (2008) 1471–1478. https://doi.org/10.1016/j.compositesa.2008.05.008
38. Alain Bourmaud, Dan Åkesson, Johnny Beaugrand, Antoine Le Duigou, Mikael Skrifvars, Christophe Baley, Recycling of L-poly-(lactide)-poly-(butylene-succinate)-flax biocomposite, *Polymer Degradation and Stability* 128 (2016) 77–88. https://doi.org/10.1016/j.polymdegradstab.2016.03.018

39. C. Gourier, A. Bourmaud, A. Le Duigou, C. Baley, Influence of PA11 and PP thermoplastic polymers on recycling stability of unidirectional flax fibre reinforced biocomposites, *Polymer Degradation and Stability* 136 (2017) 1–9. https://doi.org/10.1016/j.polymdegradstab.2016.12.003
40. A. Bourmaud, C. Baley, Investigations on the recycling of hemp and sisal fibre reinforced polypropylene composites, *Polymer Degradation and Stability* 92 (2007) 1034–1045. https://doi.org/10.1016/j.polymdegradstab.2007.02.018
41. J.P. Correa-Aguirre, F. Luna-Vera, C. Caicedo, B. Vera-Mondragón, M.A. Hidalgo-Salazar, The effects of reprocessing and fiber treatments on the properties of polypropylene-sugarcane bagasse biocomposites, *Polymers* 12 (2020) 1440. https://doi.org/10.3390/polym12071440
42. Chakaphan Ngaowthong, Martin Borůvka, Luboš Běhálek, Petr Lenfeld, Martin Švec, Rapeephun Dangtungee, Suchart Siengchin, Sanjay Mavinkere Rangappa, Jyotishkumar Parameswaranpillai, Recycling of sisal fiber reinforced polypropylene and polylactic acid composites: Thermo-mechanical properties, morphology, and water absorption behavior, *Waste Management* 97 (2019) 71–81. https://doi.org/10.1016/j.wasman.2019.07.038
43. A. Lagazzo, C. Moliner, B. Bosio, R. Botter, E. Arato, Evaluation of the mechanical and thermal properties decay of x`/sisal and PLA/sisal biocomposites at different recycle steps, *Polymers* 11 (2019) 1477. https://doi.org/10.3390/polym11091477
44. M.R. Fazita, K. Jayaraman, D. Bhattacharyya, M. Hossain, M.K. Haafiz, H.P.S. Abdul Disposal options of bamboo fabric-reinforced poly(lactic) acid composites for sustainable packaging: Biodegradability and recyclability, *Polymers* 7 (2015) 1476–1496. https://doi.org/10.3390/polym7081465
45. D. Åkesson, T. Fuchs, M. Stöss, A. Root, E. Stenvall, M. Skrifvars, Recycling of wood fiber-reinforced HDPE by multiple reprocessing, *Journal of Applied Polymer Science*. 133 (2016). https://doi.org/10.1002/app.43877
46. Laurent Augier, Gianluca Sperone, Carlos Vaca-Garcia, Marie-Elisabeth Borredon, Influence of the wood fibre filler on the internal recycling of poly(vinyl chloride)-based composites, *Polymer Degradation and Stability* 92 (2007) 1169–1176. https://doi.org/10.1016/j.polymdegradstab.2007.04.010
47. S. Bhattacharjee, D.S. Bajwa, Degradation in the mechanical and thermo-mechanical properties of natural fiber filled polymer composites due to recycling, *Construction and Building Materials* 172 (2018) 1–9. https://doi.org/10.1016/j.conbuildmat.2018.03.010
48. M. Bakkal, M.S. Bodur, O.B. Berkalp, S. Yilmaz, The effect of reprocessing on the mechanical properties of the waste fabric reinforced composites, *Journal of Materials Processing Technology* 212 (2012) 2541–2548. https://doi.org/10.1016/j.jmatprotec.2012.03.008
49. M.S. Bodur, M. Bakkal, O.B. Berkalp, T.G. Sadikoglu, F. Chinesta, Y. Chastel, M. El Mansori, *The Effect of Reprocessing on the Tensile Properties of Composites*, in: Paris, (France), 2011: pp. 215–220. https://doi.org/10.1063/1.3552443
50. M.D.H. Beg, K.L. Pickering, Reprocessing of wood fibre reinforced polypropylene composites. Part II: Hygrothermal ageing and its effects, *Composites Part A: Applied Science and Manufacturing* 39 (2008) 1565–1571. https://doi.org/10.1016/j.compositesa.2008.06.002
51. E. Uitterhaegen, J. Parinet, L. Labonne, T. Mérian, S. Ballas, T. Véronèse, O. Merah, T. Talou, C.V. Stevens, F. Chabert, P. Evon, Performance, durability and recycling of thermoplastic biocomposites reinforced with coriander straw, *Composites Part A: Applied Science and Manufacturing* 113 (2018) 254–263. https://doi.org/10.1016/j.compositesa.2018.07.038
52. E. Nadali, M. Layeghi, G. Ebrahimi, R. Naghdi, M. Jonoobi, M.M. Khorasani, Y. Mirbagheri, Effects of multiple extrusions on structure-property performance of natural fiber high-density polyethylene biocomposites, *Materials Research* 21 (2018). https://doi.org/10.1590/1980-5373-mr-2017-0301
53. M.K. Lila, A. Singhal, S.S. Banwait, I. Singh, A recyclability study of bagasse fiber reinforced polypropylene composites, *Polymer Degradation and Stability* 152 (2018) 272–279. https://doi.org/10.1016/j.polymdegradstab.2018.05.001
54. G. Zhuo, X. Zhang, Y. Liu, M. Wang, Effect of multiple recycling on properties of poplar fiber reinforced high density polyethylene wood-plastic composites, *Materials Research Express* 6 (2019) 125514. https://doi.org/10.1088/2053-1591/ab5742
55. L. Soccalingame, D. Perrin, J.-C. Bénézet, S. Mani, F. Coiffier, E. Richaud, A. Bergeret, Reprocessing of artificial UV-weathered wood flour reinforced polypropylene composites, *Polymer Degradation and Stability* 120 (2015) 313–327. https://doi.org/10.1016/j.polymdegradstab.2015.07.013

56. F. Wang, W. Zhong, L. Zhou, Y.-B. Sheng, The phase sensitivities for different phase-shift configurations in an SU(1,1) interferometer*, *Communications in Theoretical Physics* 71 (2019) 1435. https://doi.org/10.1088/0253-6102/71/12/1435
57. N. Graupner, K. Albrecht, G. Ziegmann, H. Enzler, J. Muessig, Influence of reprocessing on fibre length distribution, tensile strength and impact strength of injection moulded cellulose fibre-reinforced polylactide (PLA) composites, *Express Polymer Letters* 10 (2016) 647–663. https://doi.org/10.3144/expresspolymlett.2016.59

12 Preparation, Characterization, and Properties of Hybrid Natural Fiber Composites (HNFCs)

Aneesa Padinjakkara, Jomon Joy, Sunil S. Suresh, and Teena Varghese

12.1 INTRODUCTION

Natural fibers (NFs) are naturally occurring materials that are either derived from biobased or non-bio products. In the case of bio-derived fibers, these result from renewable resources such as plant and animal products. Proteins are considered the constructing blocks of animal-sourced fibers. However, when considering plant-based fibers, cellulose is the major constituent along with the lignin matrix. Consequently, plant fibers (PF) are also recognized as lignocellulose fibers. Nevertheless, non-bio fibers include mineral-based materials like asbestos. However, due to the hazardous nature of asbestos, most countries banned its application in various areas [1–6].

NFs have been used by humans for millennia to fulfill a wide range of needs, from clothing and shelter to everyday items and tools. Derived from plants, animals, and minerals, these fibers have played a vital role in shaping human civilization. Today, as societies worldwide face environmental challenges, NFs have resurfaced as sustainable substitutes for synthetic materials.

The history of NFs dates back thousands of years, when early humans used materials like flax, wool, and silk to create clothing, ropes, and shelter. The discovery and utilization of NFs were critical milestones in human development, enabling early civilizations to adapt to various environments and improve their quality of life. From ancient Egyptians weaving linen to Native Americans crafting clothing from animal hides, NFs were essential to cultural traditions and social progress.

12.2 CLASSIFICATION OF NATURAL FIBERS

NFs are broadly categorized into three main types: plant-based, animal-based, and mineral-based fibers. Classification of NFs is attributed in Figure 12.1 [7].

a. **Plant-Based Fibers:** PFs are derived from different components of plants, such as the stem, leaf, fruit, and seed. Common examples include cotton, flax (linen), jute, hemp, sisal, and bamboo. Cotton, being the most widely used NF globally, is a staple in the textile industry due to its softness, breathability, and versatility. Linen, on the other hand, is valued for its strength and smooth texture, while hemp and bamboo fibers are increasingly appreciated for their eco-friendliness and rapid growth.
b. **Animal-Based Fibers:** Animal fibers (Afs) originate from sources like sheep (wool), goats (cashmere and mohair), rabbits (angora), silkworms (silk), and other animals. Wool,

DOI: 10.1201/9781032665337-12

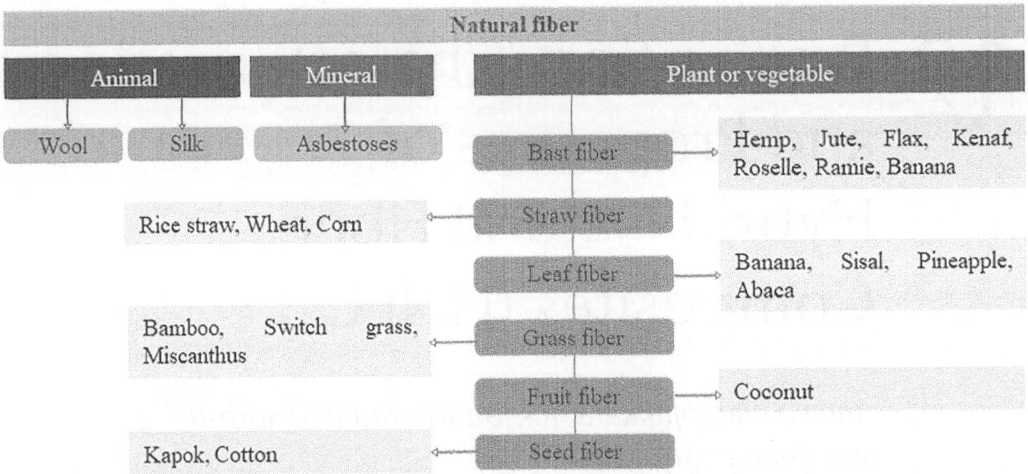

FIGURE 12.1 Classification of NFs based on the sources.

known for its exceptional insulation properties, is popular for warm clothing in colder climates. Cashmere and mohair, with their luxurious softness, are highly sought after for high-quality garments. Silk is renowned for its lustrous appearance and lightweight feel, making it a prized material for luxury clothing and textiles.

c. **Mineral-Based Fibers:** Derived from naturally occurring minerals, mineral fibers include asbestos and fiberglass. Nevertheless, due to health and environmental concerns associated with asbestos, its use has been severely restricted. Meanwhile, fiberglass is still used in certain applications, but its environmental impact has raised concerns.

12.3 PRODUCTION OF NFs

The fabrication of NFs involves various processes, conditional on the specific fiber type.

a. **Plant-Based Fibers:** PFs are cultivated in agricultural settings and undergo harvesting, retting (a process that separates the fibers from the plant's woody components), and subsequent mechanical processing to yield usable fibers. For example, cotton undergoes ginning to remove seeds, while flax requires the process of scutching to remove impurities from the fibers.
b. **Animal-Based Fibers:** AFs are sourced through animal husbandry and shearing. Wool is sheared from sheep, while cashmere and mohair come from goats. Silk is harvested by collecting the silkworm's cocoon and then carefully unwinding the silk thread.
c. **Mineral-Based Fibers:** The production of mineral-based fibers typically involves extraction, processing, and forming of the fibers. Asbestos, in the past, was extensively used for its fire-resistant properties in construction materials. However, due to its carcinogenic nature, its use is now highly regulated or banned in many countries.

NFs are increasingly recognized for their positive impact on the environment and sustainability. Unlike synthetic fibers, which are drawn from nonrenewable petroleum sources, NFs are renewable and biodegradable. This implies that at the end of their lifecycle, they break down naturally, reducing the burden on landfills and oceans. Additionally, cultivating NFs generally requires fewer fossil fuels and energy, thus contributing to lower greenhouse gas emissions.

Moreover, the cultivation of certain NFs like hemp and bamboo can be highly sustainable. These plants grow quickly, requiring minimal water, pesticides, and fertilizers. Furthermore, embracing

NFs can help moderate the dependency on petrochemical-based products and contribute to a more circular economy.

In the case of NF, bio-derived plant- and animal-based fibers are predominantly employed in the polymer composite field as compared with non-bio-derived NFs. This is because biobased fibers are available cheaply and also because of their ease of application. As compared with mineral fibers, biobased fibers can undergo easy surface modification, which also facilitates higher computability of biobased fibers in various polymer matrices [8].

12.4 CHEMICAL COMPOSITION OF NFs

12.4.1 Lignocellulose Fibers or PFs

Lignocellulose fibers commonly originate from plants or vegetables. Based on the applications, PFs have been classified into two major categories. In the primary category, the PFs are those that have been employed due to their higher fiber content, such as sisal, jute, hemp, and kenaf fibers [7, 9]. Later in the second category, PFs are obtained from the by-products of various plants like coir fibers and pineapple fiber, and this category also includes agricultural waste. Furthermore, PFs can be categorized according to the plants' origins. This includes the fibers found in bast, leaves, grass, reed, and other plants [2, 7, 10].

In general, PFs are composed of cellulose and encompass 80% of the crystalline region. Along with cellulose, PFs also contain highly branched polysaccharides named hemicellulose. Chemical constituents like pectin, amorphous lignin, and waxy components are also present in the PFs [11, 12]. Fundamentally, cellulose present in all PFs has a similar chemical structure, even if the degree of polymerization involved in the development of cellulose and its cell geometry varies with various plants. The growth condition of a plant also affects the chemical composition of the fibers as well as its final properties. Some of the selected PFs employed in the polymer composites and their chemical compositions are indicated in Table 12.1 [13]. Various literatures indicate that the properties of PFs are also dependent on the cell dimension of cellulose, micro-fibrillar angle, and defects on fibers [14].

12.4.2 Animal Fibers

The fibers obtained from the animal kingdom are generally called AFs. The major building blocks of AFs are amino acid proteins. Based on their nature, AFs can be classified into two types. AFs based on α-keratin, which is mainly found on hair, leather, wool, etc. The latter category comes with fibroin fibers, found in silk and spider webs, etc. [15–17].

TABLE 12.1
Chemical Composition of Various PFs

Fiber	Cellulose	Hemicellulose	Lignin	Wax
		(wt%)		
Bagasse	55.2	16.8	25.3	–
Bamboo	26–43	30	21–31	–
Flax	71	18.6–20.6	2.2	1.5
Jute	61–71	14–20	12–13	0.5
Abaca	56–63	20–25	7–9	3
Coir	32–43	0.15–0.25	40–45	–
Pineapple	81	–	12.7	–
Rice husk	35–45	19–25	20	14–17

Chemically, proteins are framed by sequences of amino acids with peptide linkages. The hydrogen bond further stabilizes the amino acid and leads to the secondary structure. Fundamentally, three different types of secondary structures are available, such as α-helix, β-sheet, and quaternary shape. The α-helix is formed by a right-hand coiled conformation with prevalent, consistent, and probable sequencing. Similarly, the β-sheet, formed by twisted, pleated sheets, is also stabilized by hydrogen bonds. However, compared with the α-helix, β-sheets are less prevalent. The overall shape of the protein is distinct from the protein fold from the secondary structures. Further, multiple numbers of proteins fold together to form a mono-protein complex, which is defined as the quaternary structure of a protein [2].

12.4.3 Natural Fiber Polymer Composites (NFPCs)

NFs are incorporated into the polymer matrix for the fabrication of different polymer composites. The NFs act as a reinforcement medium in the polymer matrix. The NF-reinforced polymer matrix leads to the development of a polymer composite that imparts higher performance, light weight, biodegradability, and cost-effectiveness.

NFCs are allocated into two categories based on their biodegradation behavior and the polymer matrix employed. NFCs can be framed with renewable and petroleum-based resins. In the case of non-biodegradable NFCs, the resin matrix part is either a non-biodegradable polymer such as polyethylene (PE), bio-PE, PP, bio-PP, nylon, and bio-nylon. Besides, biodegradable polymers such as polylactic acid (PLA), starch, poly(ε-caprolactone) (PCL), and polybutylene adipate terephthalate (PBAT), were employed as the resin matrix to produce the biodegradable composites [18]. Further, completely biobased polymer matrix-based NFCs are known as the green composite. Figure 12.2

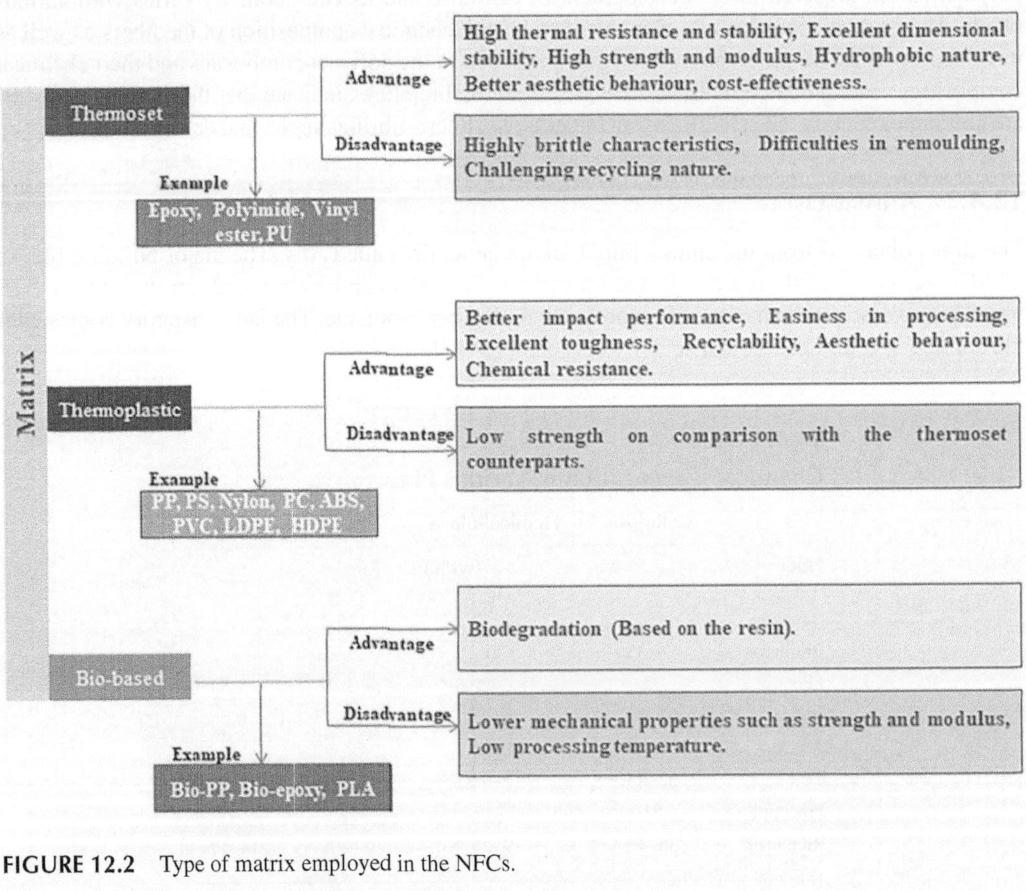

FIGURE 12.2 Type of matrix employed in the NFCs.

TABLE 12.2
Mechanical Properties of Selected PFs

Fiber	Density (g/cm³)	Tensile Strength (MPa)	Young's Modulus (GPa)	Elongation-at-Break (%)
Abaca	1.5	400	10–12	3–10
Coir	1.1	106–175	4–6	17–47
Cotton	1.5–1.6	290–800	5–12	7–8
Banana	1.3	54–754	7–20	10.35
Sisal	1.45	568–640	9–15	3–7
Hemp	1.48	690	60–70	1.6
Jute	1.3	393–770	12–25	7–9

indicates the types of matrices commonly employed in NFC fabrication along with their disadvantage and advantages [19].

The NFCs are constructed of two phases, such as polymer matrix and reinforcement medium. In the NFCs, the major phase is a polymer matrix ether composed of thermoplastic or thermoset resin. These matrices act as a binder for the reinforcements and also perfectly transfer the load to the reinforcement present in the matrix system. In addition to that, the polymer matrix also protects the reinforcement fiber from external damage. In the case of NFCs, reinforcement constituents are fibers that originate from plants, animals, or minerals. The reinforcement medium is embedded in the polymer matrix. Reinforcement acts as a load-bearing medium of the cellulose nanofibrils (CNFs) [20, 21].

The NFCs include the synergistic effects of both polymer matrix and reinforcement medium. The final properties of the fabricated composites depend purely upon the polymer matrix and reinforcement medium. The PFs come with different mechanical properties, which also impart the final characteristics of the composites. Depending upon the applications, the PFs can be selected. Selected PFs and their mechanical properties are indicated in Table 12.2 [22].

12.4.4 Hybrid Natural Fiber-Based Polymer Composites

The incorporation of two or more natural or synthetic fibers in the common polymer matrix is known as the hybridization process. Such developed hybrid composites indicate higher properties than their single fiber-based composites or their parent polymer matrix. Hybridization gives prospects to enhance the function of the composites. The synergistic effect of multiple fibers on the polymer matrix imparts exceptional mechanical, thermal, and rheological properties. Hybrid composites are considered more innovative than single fiber-reinforced composites. Subsequently, a variety of applications may be made using these composites. Due to their peculiar nature, hybrid composites can be employed for higher-end purposes such as automobiles, aerospace, life-protecting products, and marine [22, 23].

In the current situation, a maximum of the hybrid composites are fabricated with a combination of two or more synthetic fibers selected from Kevlar, carbon fiber, glass fiber, etc. Further, researchers are indicating the possibility of a hybrid composite formulation with NFs. In this case, at least one of the fibers employed in the polymer matrix is comprised of NFs. Different types of hybrid composites with NFs are possible, such as synthetic-NF and natural-NF. Various researchers indicated possibilities of hybrid composite formulation with different NFs. Ramasubbu and Madasamy [24] proposed a hybrid NF composite for automobile applications. In this research, the authors employed sisal and kenaf fibers in the epoxy polymer matrix and analyzed their mechanical behavior. The comparative results of sisal/epoxy, kenaf/epoxy composite with sisal/kenaf/epoxy hybrid composite indicate higher mechanical strength for the latter one. The interlocking mechanism of fibers results in better fiber-polymer matrix adhesion, and this further facilitates better property enhancement in the hybrid composites. Kumari et al. [25] fabricated an eco-friendly hybrid composite via the hand lay-up method with banana fiber, sisal fiber, along with sawdust in the epoxy matrix. This research indicated that dielectric permittivity, mechanical properties, and water absorption properties of the composite increased with the sawdust content in the polymer matrix.

Another interesting study by Krishnudu et al. [26] indicated the mutual application of NFs such as coir and silk along with $CaCO_3$ filler on the polymer matrix. The NFs, such as coir and silk, have been introduced in the weight ratios of 10 and 5.7 wt%, respectively. The final blend has been prepared via hand lay-up methodology. The authors further compared the properties of NFs fabricated composite with the composite containing NFs and $CaCo_3$. This study indicates that filler-incorporated NFs composites have higher mechanical and dielectrical properties. Authors believe that the fabricated composite system can be applicable in the automobile and building industries. Balaji et al. [27] considered a polyester matrix for the hybrid natural fiber composite (HNFC) preparation with zea and coir fibers. Zea and coir fibers in the volume ratio of 60:40 were introduced into the polyester matrix and developed as a composite through compression molding (CM) technology. The zea and coir fibers contain approximately 12.6% of hemicellulose in their fiber structure. Superior interfacial adhesion between the fiber and polyester matrix is a result of the fiber's nature. Due to the better interfacial adhesion, the composite developed has shown higher mechanical properties.

The final property of the fabricated composite depends on the fibers and resin employed. Table 12.3 indicates selected composites with their targeted properties.

12.5 FACTORS CONTRIBUTING TO THE PROPERTIES OF THE HNFCs

Various elements are aiding the final mechanical, thermal, and other properties of HNFCs being developed. Generally, the overall performance of the HNFCs is mainly determined by the nature of the resin and fiber. The polymer matrix employed in the composite contributes to the compressive strength and in-plane shear strength. However, other mechanical properties are typically contributed by the fiber present in the polymer matrix.

12.5.1 Fiber Surface

The surface of the fiber plays an important role in the final properties of the composite. In general, NFs possess a highly hydrophilic nature, which leads to incompatibility in the polymer matrix. This leads to a worsening of the mechanical behavior of the final composite. To make the NFs hydrophobic, some chemical treatment should be carried out on the fibers.

The compatibility between fiber and resin is always a challenge to the scientific community. Consequently, researchers analyzed various possibilities for improving the surface behavior of the fibers with different chemical media, such as coupling agents. The pretreatment of the fibers with coupling agents allows for decreasing the water intake capability of the fiber and imparts an improvement in the mechanical properties. The chemical treatment technologies, such as silane treatment, alkylation, benzoylation, acetic acid, and peroxides, are employed over the fibers to improve their surface behavior [40–42]. During the chemical treatment, some of the partial structures delaminated from the fiber and surface become more liable and accountable for the surface interfacial modifications. Different chemical treatment methodologies adopted for fiber modifications are indicated in Figure 12.3 [41].

Alkali treatments on the NFs are also studied by different researchers. This technology is simple and cost-effective, compared with other chemical modification technologies. In general, NaOH concentration from 1% to 10% is widely employed in the treatment of NFs [41]. Saikeng et al. [43] showcased the application of NaOH for the treatment of NFs such as coir and pineapple leaf fiber (PLF). The fibers were chopped into the desired size of 1mm and treated the same with 6% NaOH

TABLE 12.3
Targeted Properties of Different HNFCs with Various Polymer Matrices

Fiber Type					
Natural	Synthetic	Resin Matrix	Targeted Properties	Mentioned Application	Reference
Flax	Carbon fiber, Kevlar	Epoxy	Tensile, flexural, interlaminate shear	Lightweight engineering application	[28]
Jute, sisal	–	Epoxy	Tensile, flexural, hardness	–	[29]
Flax, hemp, jute	–	Epoxy	Tensile modulus, impact performance	Structural application	[30]
Coir, jute	–	Epoxy navolac	Dimensional stability, density	Engineering application	[31]
Flax	Carbon fiber	Epoxy	Tensile strength, tensile modulus	Lightweight application	[32]
Kanaf, sisal		Bioepoxy	Weathering property improvement	Semi-structural application	[33]
Hemp	Glass fiber	PP	Impact performance, stiffness, thermal resistance	Structural application	[34]
Hemp, sisal	–	PLA	Flexural strength, flexural modulus	Automotive, packaging, electronics, agricultural	[35]
Banana, sisal	–	PLA	Tensile strength and modulus, flexural strength and modulus, impact performance	–	[36]
Sugar palm	Glass fiber	PU	Low density, low water absorption, high temperature resistance	Automotive	[37]
Sugar palm	Glass fiber	TPU	Thermal resistance	Automotive	[38]
Banana fiber	Coir	PP/LDPE	Impact performance, flexural strength, modulus	–	
Banana fiber, kenaf		Polyester	Mechanical and thermal properties		[39]
Coir, silk		Polyester	Tensile, flexural, compressive		[40]

for 3h. After the treatment, NFS was washed well with water for the excess removal of salt from NFS and to achieve a neutral pH. The treated fiber was further used for the modification of the PLA matrix. Alkali-treated samples of coir and PLF in the concentration of 3% and 7%, respectively, indicated higher tensile modulus compared with other compositions. However, the treated NFs sample-incorporated polymer matrix has indicated lower thermal stability in comparison with the non-treated NFS included in the resin matrix. The reduction in the thermal stability of treated NFs composites is due to the fact that alkalization tends to remove lignin and hemicellulose from the NF structure.

Apart from the mechanical and thermal properties, the chemical modification of the fibers can also influence properties such as dielectric properties. Agrebi et al. [44] treated PLF and kenaf fiber

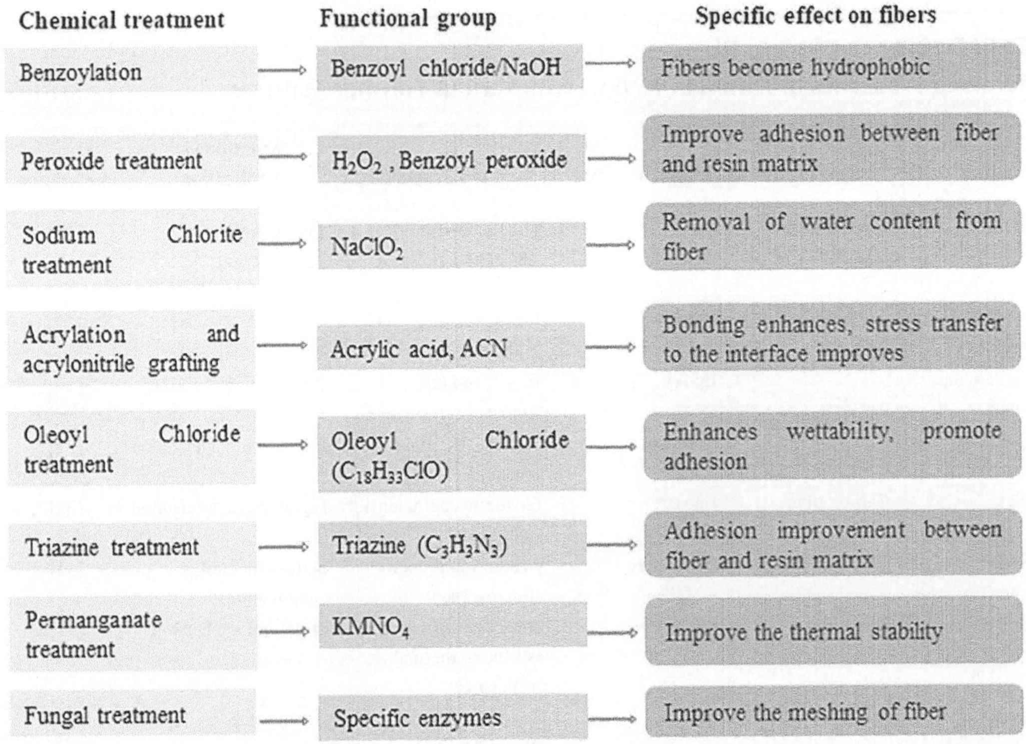

FIGURE 12.3 Types of chemical treatment and their effect on fibers.

with a 2% silane agent for 3h. The treated fibers (50% of total fiber loading) were implemented in novoloc phenolic resin to investigate the dielectric properties at the temperature range of 50–180°C and frequency range of 0.1 Hz to 1 MHz. The silane-treated fibers in the composites result in a lowering of real (ε') value as compared with the untreated fibers. The silane treatment on the fibers results in the conversion of fibers from hydrophilic to hydrophobic. During the silane treatment, the small pores or cavities held by the fibers are closed. Consequently, water molecules cannot easily enter the surface of the fibers as compared with the untreated fibers. This surface modification further results in the lowering of conductivity in the composites.

Similar to the other chemical treatments, benzoylation is also employed to improve the surface behavior of the NFs. Benzoylation decreases the hydrophobicity of the fibers via surface modifications. In this technique, a combination of benzoyl chloride and NaOH is used to treat the fiber surface. The hydroxyl groups present on the fiber surface undergo a reaction with benzoyl chloride. The benzoyl chloride adhered to the fibers can be further eliminated via washing with excess ethanol and water. However, before the treatment with the benzoyl chloride, the –OH groups on the fibers were chemically activated with alkali agents such as NaOH. The benzoylation of the NFs can improve the surface properties, which in turn enhances the adhesion with the polymer matrix.

Researchers are indicating that benzoylation can improve the mechanical and thermal properties of the final composites developed. In a study, [45] sugar palm fiber (SPF) has been treated with benzoyl chloride and NaOH. Further, the obtained treated SPF was combined with glass fiber (GF) and introduced into the epoxy matrix. After the evaluation of the flexural and compressive strength of the fabricated composites, the authors concluded that benzoylation technology can improve the interfacial interactions between the polymer matrix and fibers. In another research, Safari et al. [46] developed an epoxy composite with treated SPF and GF. The composite was constructed in a 90:10 (epoxy:fiber) composition to analyze the synergistic effect of treated SPF (TSPF) and GF on the

TABLE 12.4
Comparative Mechanical Strength Analysis of Epoxy Matrix with TSPF/GF and USPF/GF

Sample	Composition	Tensile Strength (MPa)	Tensile Modulus (MPa)	Tmax Load (N)
Epoxy/GF	90:10	24.9	4385.3	2149.8
Epoxy/TSPF	90:10	22.7	3619.6	2063.4
Epoxy/TSPF/GF	90:7:3	27.2	4308.5	2473.1
Epoxy/TSPF/GF	90:5:5	28.4	5199.6	2580.6
Epoxy/TSPF/GF	90:3:7	31.6	5446.9	2872.2
Epoxy/USPF/GF	90:10	20.3	3347.2	1809.1
Epoxy/USPF/GF	90:7:3	22.0	3836.2	1955.6
Epoxy/USPF/GF	90:5:5	24.1	4291.4	2189.5
Epoxy/USPF/GF	90:3:7	24.4	4262.8	2148.7

epoxy matrix. Further, a similar composition has been prepared with untreated SPF (USPF) and GF on the epoxy matrix for the comparative analysis. Interestingly, results indicate that benzoylation has an impact on water absorption behavior, density, mechanical performance, and thermal stability of the final composites. The major mechanical properties are tabulated in Table 12.4 [46].

The results tabulated in Table 12.4 indicate that the mechanical properties are improved with the addition of TSPF as compared with USPF. Benzoylation of the fiber impacted the surface morphology of the TSPF, which in turn enhances the compatibility between the GF and polymer matrix. The synergism between the GF and TSPF also contributed to better improvement in the mechanical and thermal properties.

In addition to the mono-chemical treatment method, researchers successfully implemented mixed chemical treatment techniques on the NFS. Different literature indicates that mixed treatment techniques have a higher impact on the formulated composites. Neto et al. [47] effectively employed mixed treatment methods on various fibers such as jute, ramie, sisal, and curare. In this research, the authors compared the effect of alkalization on NFs with the combined effect of alkalization and silanization. Results of the studies indicated that the mixed treatment enhances the stiffness of the composites compared with the alkalization effect. The scanning electron microscopy (SEM) analysis of the fractured surface reveals that alkalization improves surface morphology by removing the impurities from the surface of the NFs. At the same time, mixed treatment leads to the creation of a coating over the fibers. Such coatings, in turn, improve the thermal perspective of the fibers.

The chemical treatment on NFS creates different effects, such as delamination of the outer layers of the fibers, development of the flexible layers or coating over the fiber, creation of a bridging group between the fibers and resin matrix, and enhancement of surface morphology, which is reflected in the miscibility between fiber and matrix, creation of covalent bonds between resin and fibers. These effects further lead to the improvement in the resin matrix and fiber adhesion.

12.5.2 Fiber Orientation

Fiber orientation of the fibers on the polymer matrix plays an important role in the final properties of the fiber-reinforced composite materials. Different literature indicates that fiber orientation critically affects the mechanical, electrical, tribological, and thermal conductivity of the composites.

The mechanical properties of the composites purely depend upon the orientation of the fibers present in the matrix system. Yahaya et al. [48] primed hybrid composites with kenaf fiber and aramid for spall-liner applications. To analyze the effect of orientation on the polymer matrix, they have prepared different systems, such as woven, unidirectional, and mat. This research further indicates

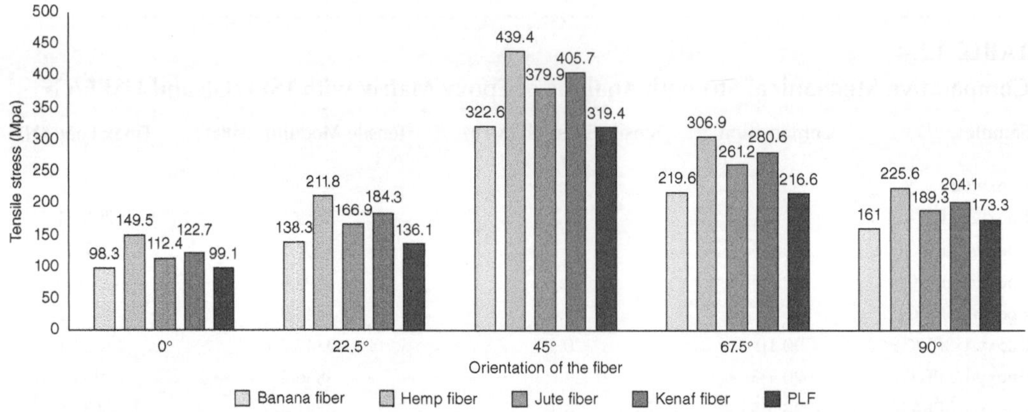

FIGURE 12.4 Comparative tensile stress of the various HNFCs at different orientations of fibers.

that the tensile properties and Charpy impact performance of the woven kenaf system are higher than the unidirectional and mat systems. Similarly, Sathish et al. [49] prepared epoxy composites with banana-kenaf and glass fiber systems. This research indicates that composites with the fiber arrangement of 45° inclination have shown superior results than other orientations.

Many of the researchers indicated that the composites formulated on the fiber direction (0°) have higher strength in assessment with other orientations. At the fiber direction (0°), fiber carries significant strength and stiffness characteristics along with the fiber direction. In an interesting study, Batu and Lemu [50] employed false banana and glass fiber for the fabrication of the composites with epoxy matrix. This study also considered the effect of the orientation of false banana fiber on the mechanical properties of the developed composites. The false banana fibers oriented in 0° (fiber direction) direction have a higher compressive strength of 47.35 MPa, and fibers with 90° orientation (transverse direction) come with a lower value of 13.86 MPa. Additionally, the tensile strength has also shown a similar trend. The fibers oriented in 0° direction have a higher tensile strength of 134.3MPa, and fibers with 90° orientation showed a lower value of 50.31 MPa. Gupta et al. [51] studied the mechanical properties of HNFCs, after formulating composites with epoxy and different NFs at a 30% volume fraction. The tensile stress obtained for the different fibers is indicated in Figure 12.4. The figure pointed out that fiber orientation at 0° has minimum stress as compared with other orientations. Among the various fibers, banana fiber indicated a lower stress value of 98.3 at 0° orientation. Tensile analysis conducted at 1000N hemp fiber indicated a higher maximum stress of 439.4 MPa with the 45° orientation.

12.5.3 Fiber Volume and Length

The load-bearing capacity of the HNFC enormously depends upon the fiber content in the resin matrix. In general, 5–10% NFs are incorporated into the polymer matrix for the improvement of mechanical properties. Several studies have been piloted by various researchers to develop a correlation between the composite properties with fiber volume and length. Prakash et al. [52] studied the effect of fiber volume and length of banana stem, PLF, and glass fiber on the epoxy matrix system. This study indicates that fiber volume and length play an important role in the flexural properties of the generated blend. The flexural strength and modulus values of the composites increased by 40% of the fiber volume. However, above 40% fiber volume, the composites started to decline in their flexural properties. Besides, this research also concluded that fiber length enhances the flexural characteristics of the composites. Aji et al. [53] studied PLF and kenaf fiber-based thermoplastic HNFC composite for the analysis of the effect of fiber addition on the resin matrix. The length of fiber has been kept at 0.25, 0.5, 0.75, and 2 mm under the fiber charging volume from 10 to 70%.

TABLE 12.5
Mechanical Properties of HNFC with Different Weight Percentages of Fiber

Fiber Length (mm)	wt% of Fiber	Tensile Strength (GPa)	Tensile Modulus (GPa)	Flexural Strength (GPa)	Flexural Modulus (GPa)	Impact Strength (J/cm²)	Shear Strength (Gpa)
0	0	0.033	1.00	0.040	1.53	0.40	0.0041
50	31	0.026	1.32	0.039	1.34	1.76	0.0062
	37	0.032	1.31	0.046	1.85	3.58	0.0063
	42	0.033	1.11	0.055	2.18	4.29	0.0076
	47	0.038	1.53	0.056	2.69	5.10	0.0079
	55	0.043	1.52	0.059	3.54	6.02	0.0094
	63	0.033	1.28	–	–	–	–

Note: Fiber length and wt% value zero indicate neat resin. Further, "-" represents the value note given in the literature.

This study indicates approximately 8–12% enhancement in the tensile as well as flexural properties. A study indicated that the length of the PLF fibers positively impacts the mechanical properties such as flexural and tensile properties of the generated thermoplastic composite up to the fiber length of 30% [54]. Velmurugan and Manikandan [55] considered NF and synthetic fibers, such as palmyra fiber and glass fiber, respectively, for the HNFC fabrication. The mechanical behavior of the 50 mm length fiber with different fiber weight percentages is indicated in Table 12.5. It can be concluded from the table that HNFC with 55 wt% of fiber concentration and 50 mm length comes with higher mechanical properties such as tensile strength, flexural strength, impact strength, and shear strength. However, beyond 55 wt% of fiber, the mechanical values of composites are declining in nature.

12.6 TYPE OF HNFC PROCESSING TECHNOLOGIES

Hybrid polymer composites refer to composites consisting of a polymer matrix reinforced with multiple types of reinforcements [56]. These reinforcements can be categorized based on their sources, such as natural or synthetic fibers [57, 58]. Additionally, reinforcements can be categorized as continuous fibers, short fibers, particle reinforcements, mats, or woven fabrics [59, 60]. The arrangement of these reinforcements within the polymer matrix significantly affects the properties of the composites [61]. Frequently used matrices in hybrid polymer composites include polypropylene (PP), polyesters, and epoxies [62–64]. NFs possess numerous advantageous properties that make them viable alternatives to synthetic fibers for environmental concerns [65]. Moreover, synthetic fibers tend to be more expensive, making NF-reinforced polymer composites more cost-effective alternatives. Although they have certain limitations, such as low strength and high water absorption, these challenges can be addressed through the implementation of hybridization techniques [66]. HNFCs, which involve combining NFs with additional reinforcements or matrices, offer a promising solution to improve the mechanical properties and reduce water absorption of NFs, thereby further enhancing their potential and expanding their usage in various industries [67]. Depending on the HNFC to be fabricated, various methodologies are adopted by industries and researchers. This section illustrates some of the major molding techniques employed to fabricate thermoplastic and thermoset HNFCs.

12.6.1 Compression Molding

CM technology is employed for the fabrication of the HNFC by various researchers and industries due to its ease and cost-effectiveness. Both thermoplastic and thermoset polymer-based HNFC can be developed via this method. However, batch-mixed, or twin-screw extruded materials before CM, could offer better mechanical and thermal properties to the HNFC formed with the thermoplastic materials. This is because mixing enhances the compatibility between the polymer matrix and fibers. The sugar palm and kenaf fibers were of interest to Bachtiar and co-workers for the HNFC fabrication with the PP matrix. The sugar palm and kenaf fiber-based HNFC was developed via a melt mixer and followed by CM technology. This study indicates that the fibers play an important role in the final mechanical properties of the composites. The HNFC encompassed with a higher kenaf fiber ratio in the polymer matrix indicated superior tensile strength than the SPF ratio in the polymer matrix [68]. Apart from the PP, the high-density polyethylene (HDPE) matrix is also considered for HNFC development. Compression technology has been employed for the fabrication of HDPE with treated jute fiber (JF)/oil palm-pressed fruit fibers. In this study, researchers employed parameters such as a 130°C temperature and a pressure of 0.2 kPa for 12 minutes [69].

Apart from the fiber size, polymer matrix and CM parameters also reflect the final properties of the fabricated composites. Mann et al. [70] conducted studies on PLA matrix for the fabrication of the HNFC via die molding technologies. In this current study, NFs such as sisal and jute have been employed for the fabrication of the PLA matrix. This study further pointed out that processing parameters such as temperature and pressure also impact the mechanical properties, such as tensile and flexural properties, of the fabricated composites. Velmurugan et al. [71] investigated the effect of CM parameters on the mechanical properties of reinforced PP matrix. To reinforce the PP matrix, the authors employed Calotropis gigantea and hemp fibers after the chemical treatment with a 5% NaOH alkaline solution. The results of the investigation indicated that mold pressure and temperature play an important role in the tensile and flexural strength of the final HNC.

CM can also be useful for the development of HNCs from the thermoset polymer matrix. Epoxy systems are widely studied for the fabrication of thermoset-based composites. Karthi et al. [72] considered a thermoset epoxy matrix system for the fabrication of the composites. In this study, flax and jute fiber were introduced into the epoxy (LY556) system, and HNCs were fabricated via CM techniques. The reinforced system with 30:10% flax and jute fibers, respectively, indicated higher mechanical strength than other systems. The authors believe that the developed composites can be applied in sectors such as automobiles and sheet roofs. Sathish et al. [73] fabricated a thermoset HNFC system with the help of an epoxy (LY556) and catalysis (HY 951). For reinforcement of the polymer matrix, the authors used bamboo and flax after treatment with an alkali solution of 5% NaOH. The final fabrication of the composite has been conducted via CM technology at a temperature of 130°C, and pressure of 35 kg/cm3 for 45 minutes. Yorseng et al. [33] considered a biobased epoxy system SR GreenPoxy 56 along with the hardener SD Surf Clear in the ratio of 27:10 for the thermoset composite development. CM technology has been employed for the final development of the reinforced composite. Apart from epoxy systems, polyester-based HNFCs are also prepared via CM techniques.

In another study, CM technology has been employed by Fei and co-workers [74] for the fabrication of acrylated epoxidized soybean oil (AESO) hybrid composites. AESO mixed with styrene (30%) (denoted as styrene-30). Similarly, AESO is also mixed with glyceryl trimethacrylate (GTMA) as a curing agent in 20, 30, and 40 wt% fractions (named GTMA-20, GTMA-30, and GTMA-40, respectively) to compare the effect during the composite formulation. Both styrene- and GTMA-mixed AESO resins are further reinforced with hybrid fibers such as polyethylene terephthalate, bamboo, polyester, and ethylene-propylene side-by-side fiber mats. The weight ratio of fiber to matrix has been kept at a constant of 1:1. The viscoelastic studies (Figure 12.5) indicate that the composite fabricated with GTMA has higher storage values as compared with styrene-cured AESO composites. This is because GTMA has a higher crosslinking density than styrene with AESO resin. Further, this study indicated that Tg of the composite is influenced by the energy loss of the rPETB-P-ES fibers. The Tan δ peaks of the GTMA-based composites are broader and low intensive as compared with styrene-30. This indicates that GTMA has superior damping capacity as compared with styrene-based composites. The authors further indicate that the hybrid composites can undergo recyclability via chemical recycling methods.

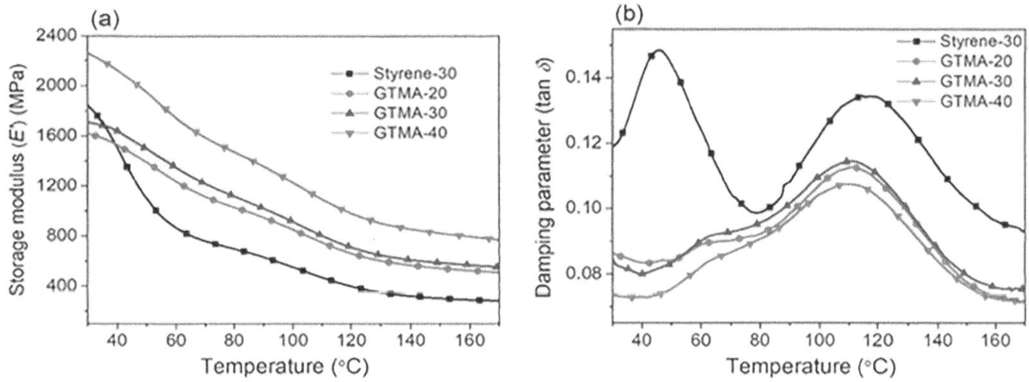

FIGURE 12.5 Viscoelastic properties of hybrid composites: (a) storage modulus and (b) damping parameter

Reprinted (adapted) with permission from Fei, M., Liu, T., Fu, T., Zhang, J., Wu, Y., Qiu, R., & Liu, W. (2019). Styrene-free soybean oil thermoset composites reinforced by hybrid fibers from recycled and natural resources. *ACS Sustainable Chemistry & Engineering*, 7(21), 17808-17816. Copyright (2019) American Chemical Society)

12.6.2 Injection Molding

Injection molding (IM) is considered the simplest way to produce thermoplastic composite products. Many of the automotive industries are using IM technologies for the construction of their products. The advantages of the IM lie in the fact that they can be useful to produce complex parts in any complex geometry without any additional finishing. PP has been considered for the preparation of the HNFC with banana and coconut fibers by Bujjibabu and co-workers. The PP and fibers were mixed in various proportions and prepared as granules via a twin-screw extruder at a constant temperature of 165–170°C. The generated granules are further employed for the composite material via IM technology [75]. Nayak and co-workers considered PET for the preparation of HNFC with areca sheath and Moringa oleifera fruit fibers. The PET and fibers were mixed well in a compounder at a temperature of 260°C with a mixing speed of 100 rpm for 5 minutes. The formulated mixer was later injection molded in a DSM micro-injection molding machine at a temperature of 260°C with an injection speed of 40 mm/s. This developed hybrid composite indicated high-impact properties [76]. Another interesting study by Nyambo et al. [77] explored the possibility of the application of agricultural fibers such as soy, corn, wheat, and their hybrids as the reinforcement medium in the PLA. The fibers/PLA were extruded with the help of a twin-screw compounder at the processing parameters of temperature 180°C, RPM 100, and for a retention time of 180 seconds. The developed polymer was injection molded in a DSM Micro 12 cc molder. In this study, 70 wt% of the PLA and 30 wt% fibers (soy, corn, wheat, and a 10:10:10 % mixture of soy, corn, and wheat) were kept constant in the entire study. The viscoelastic of the specimen indicated that the storage modulus is improved with the content of 30 wt% fibers in the PLA matrix. But, interestingly, hybrid composite (10:10:10 % mixture of soy, corn, and wheat) also has similar storage modulus values. Similarly, tan δ peaks of the PLA and their counterparts are also indicated around 60°C. This further points out that the incorporation of fibers is not involved more in the molecular transitions or Tg of the materials.

12.6.3 Resin Transfer Molding

Resin transfer molding (RTM) is commonly used to produce composites that are appropriate for automotive and aerospace applications. In this technology, fibers are chopped into the desired size, and resin is injected over them. This process is widely accepted for the production of huge-sized composite parts [78]. In the RTM method, post-curing is also required for the proper curing of the

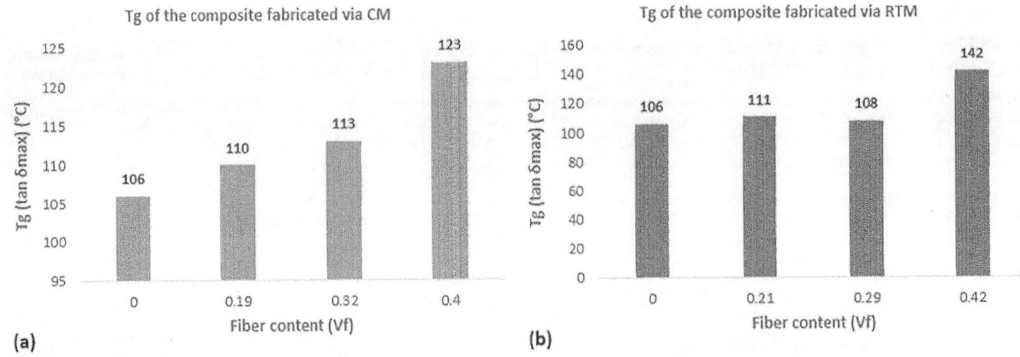

FIGURE 12.6 Tg of the composite fabricated via CM and RTM.

resin part. Idicula et al. [79] formulated a hybrid composite with short banana and sisal fibers with a volume ratio of 1:1, with various fiber loadings on the polyester matrix to compare the effect of CM and RTM technologies on its final properties. After the mechanical studies, the authors established that tensile and flexural properties are affected by the molding technologies. This study further indicates that tensile strength is enhanced with the fiber content in the polyester matrix. The tensile strength of the hybrid composite formulated is proportional to the fiber content in the polyester matrix at higher content tensile strength, and the modulus of the composite is also on the higher side. The higher content of the fibers also influences the Tg of the final fabricated composites. The composite fabricated through the RTM technique indicated a higher Tg (Figure 12.6) as compared to the compression method in a similar fiber volume fraction range. This is associated with the fact that RTM technology leads to higher matrix and fiber interactions. Similarly, Yildirim et al. [80] considered jute fiber and wood particles to produce the hybrid composites via the vacuum-assisted resin transfer molding technique.

12.6.4 Hand Lay-Up Technology

The hand lay-up technology uses open molding treatment to fabricate the polymer composites. Both thermoset and thermoplastic-based composites can be developed via hand lay-up methodology. Besides, the main advantage of this method lies in the fact that the size and shape of the fibers are not a matter while generating the composite material. However, a highly skilled person is required for the fabrication of the composite material. Jawaid et al. [81] indicated the possibility of a hybrid composite production via the hand lay-up method. In this research, oil palm empty fruit bunch (EFB) and jute fibers were initially treated with a chemical agent 2-hydroxy ethyl acrylate to improve the compatibility with the resin. An epoxy resin (ER) matrix is incorporated with the treated fibers via the hand lay-up method. In another study, a thermoset composite has been developed with fibers such as jute and kenaf along with the epoxy. The specimens were prepared with the hand lay-up method in an aluminum dam with the size of 300×300×3 mm^3. After spreading the resin, the air bubbles generated were eliminated with the help of a roller [82]. Besides, abaca- and jute-based ER thermoset hybrid composite has been manufactured via hand lay-up method. The results of the final composite also indicated that abaca and jute hybrid composite has excellent performance than the composite developed alone with abaca fiber [83]. Similarly, jute and flax fibers have been considered by Chaudhary and co-workers for the fabrication of a hybrid composite through hand lay-up technology [84].

Even though a number of processing technologies are available for the fabrication of the HNFC, the type of molding technique is adopted based on the application, targeted properties, etc. Major advantages and disadvantages of the processing techniques are indicated in Figure 12.7 [85].

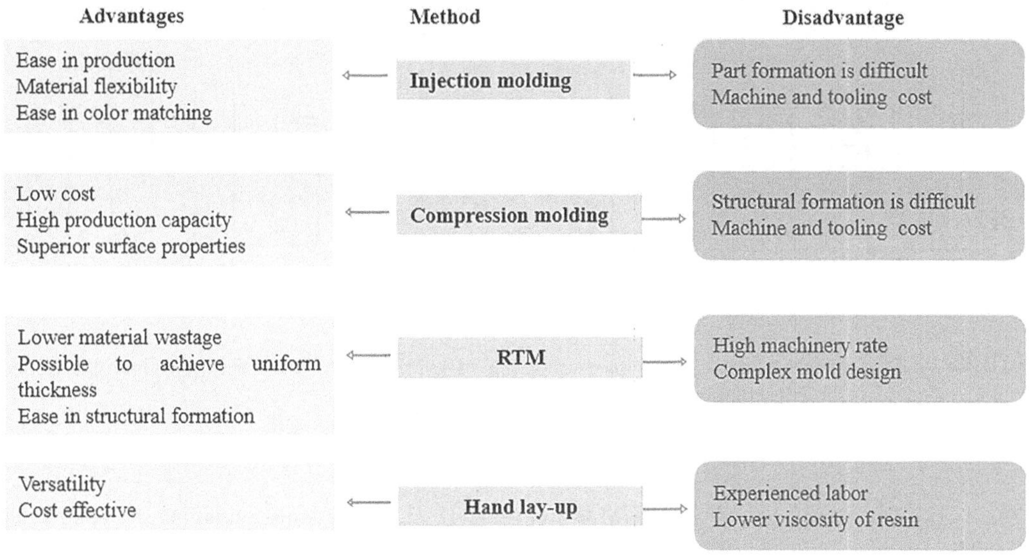

FIGURE 12.7 Advantages and disadvantages of processing techniques.

12.7 CHARACTERIZATION OF HYBRID POLYMER COMPOSITES

The incorporation of a secondary reinforcement or matrix material can significantly enhance the strength of NFs. By combining NFs with other reinforcing materials such as glass fibers or carbon fibers, the resulting hybrid composite can exhibit improved overall strength, allowing for a wider range of applications [86]. Moreover, hybridization techniques can effectively reduce the water absorption properties of NFs. By introducing hydrophobic materials or coatings as secondary matrices, the hybrid composite becomes more resistant to water absorption. This modification enables NFs to retain their mechanical properties even when exposed to moisture or humid environments, expanding their potential applications [87, 88]. The following discussion focuses on the mechanical, thermal, and tribological properties, as well as the water absorption characteristics, of hybrid polymer composites reinforced with NFs.

12.7.1 Mechanical Properties

Numerous studies have explored the potential of enhancing the mechanical properties of NF-reinforced polymer composites through the incorporation of synthetic fibers [89]. Researchers have observed positive outcomes in hybridization, resulting in increased mechanical properties in single NF-reinforced polymer composites [90, 91]. Industrial bio-waste teakwood powder and natural sisal fiber were utilized by Kailainathan S et al. as reinforcements in the fabrication of hybrid polyester matrix composites using the CM technique [92]. The addition of teakwood particulates affects the mechanical properties of the sisal fiber-reinforced composites by improving stress transfer and reducing voids. Similarly, AlMaadeed et al. examined the effects of adding glass fibers to recycled PP/wood flour hybrid composites with varying percentages of reinforcement [93]. Adding wood flour to recycled PP increased its tensile properties, while the tensile strength of the composites significantly increased with the addition of 5 wt% glass fiber. The glass fiber-reinforced composites exhibited higher hardness properties compared to other composites, and the incorporation of glass fiber improved the hardness properties.

Subbiah's group examined the impact of nano-SiO_2 fillers on the mechanical properties of a kenaf fiber epoxy composite using various mass fractions of nano-filler. Hybrid epoxy matrix

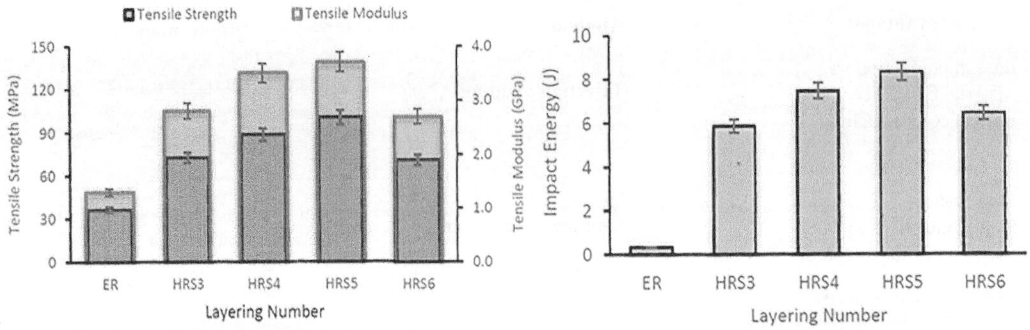

FIGURE 12.8 Tensile properties and impact properties of hybrid rattan strips (RS) and glass fiber (GF)-reinforced epoxy resin (ER) composites

These images were sourced from Irawan et al.'s 2022 Open Access publication.

composites were fabricated using the CM technique [94]. The composite containing 2% nano-filler exhibited the highest tensile strength, which was 20.61% greater than that of the regular composite, 23.71% improvement in compressive strength, and a 22.88% enhancement in impact strength. Hybrid composites were prepared using a hand lay-up technique by reinforcing jute and oil palm fibers with an epoxy matrix [95]. Compared to the oil palm-epoxy composite, the hybrid composites exhibited significantly improved tensile properties as the loading of jute fibers increased, likely due to improved bonding at the fiber/matrix interface.

The mechanical properties of a pure rattan-reinforced epoxy composite improved with an increase in the number of layers [96]. Among different configurations, the composite with three layers of woven rattan exhibited the highest mechanical properties. When combining three layers of rattan and two plies of GF, the hybrid composite showed a significant increase in mechanical strength. Among different composite sequences, the GF/RS/RS/RS/GF hybrid weave exhibited superior mechanical characteristics. Incorporating three layers of RS within the core layer of GF hybrid composites, as opposed to just one layer, notably improved their impact and tensile properties, as shown in Figure 12.8. The tensile fractured surface of bamboo-kenaf/epoxy-OMMT hybrid nano-composites (as depicted in Figure 12.9) exhibited fewer voids, microcracks, and instances of fiber pullout. This reduction can be attributed to the enhanced interfacial adhesion between the fibers and the matrix, resulting in a stronger bond between the components.

Another research group prepared hybrid composites by combining bamboo and epoxy along with the dispersion of 1%, 2%, and 3% carbon nanotubes (CNT) using the hand lay-up technique [97]. The addition of a coupling agent, acetone, enhanced the interaction between the matrix and nanoparticles, leading to a strong interface region. The results showed that the dispersion of CNT in the hybrid composites resulted in improved tensile and flexural strength compared to bamboo and epoxy composites alone. As the concentration of CNT increased, the tensile strength of the hybrid composite also increased. Additionally, the flexural strength showed a significant increase, indicating enhanced resistance to applied loads. The inclusion of high-strain NFs has been observed to enhance the mechanical properties of these composites. The resulting improvements contribute to the overall performance and durability of the materials.

12.7.2 Thermal Properties

The thermal stability of composites can be determined through thermal analysis, which involves studying their thermal properties [98, 99]. In the study conducted by Dhakal et al. [100] reported on the thermal properties of a hybrid composite made of hemp and glass fibers reinforced with polyester. Degradation mechanisms between hemp fiber composites were extremely different from

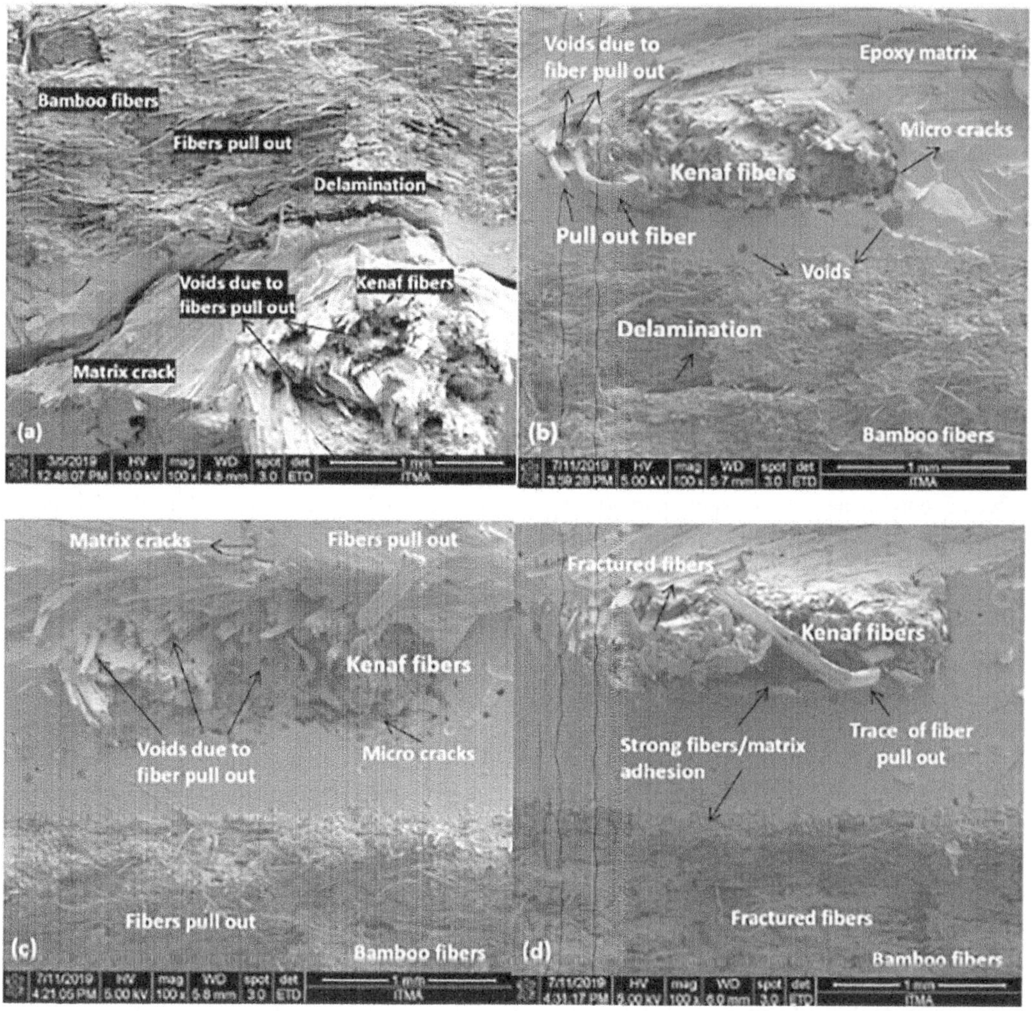

FIGURE 12.9 FESEM images of various composite materials: (a) bamboo-kenaf/epoxy composites, (b) bamboo-kenaf/epoxy hybrid nanocomposites with an MMT component, (c) bamboo-kenaf/epoxy hybrid nanocomposites incorporating HNT, and (d) bamboo-kenaf/epoxy hybrid nanocomposites enriched with OMMT

These images were sourced from Chee et al.'s 2021 Open Access publication.

hemp-glass fiber hybridized polyester composites. The hemp fiber composites undergo a three-step decomposition process, while the hemp-glass fiber hybridized UP composites exhibit only a single-step degradation. Their findings indicated that the hybrid composite exhibited better thermal stability compared to other types of composites. The thermal stability of a woven kenaf/carbon fiber-reinforced epoxy hybrid composite was investigated by Paridah's group [101]. Their studies revealed that the content of kenaf fiber increased in the hybrid composite, specifically with a fabric density of 6 × 6, and the thermal stability of the hybrid composites improved. Notably, utilizing the plain weave design of woven kenaf contributed to enhancing the overall thermal stability in comparison to the satin design. Another study by Boopalan et al. [102] focused on the thermal properties of a hybrid composite comprising jute and banana fibers reinforced with epoxy. The researchers suggested that the thermal stability of the hybrid composite (consisting of 50% jute and 50% sisal at a weight ratio of 30%) was higher compared to other composite variations.

Khalil and his team [103] researched the thermal properties of a hybrid composite made of oil palm and woven jute fibers reinforced with epoxy. The addition of woven jute fibers to the oil palm fiber-reinforced epoxy composite resulted in an improvement in its thermal stability. In a study by Naidu et al. [104], the thermal properties of a composite composed of sisal and glass fibers reinforced with polyester were investigated. The thermal conductivity of the hybrid sisal/glass fiber composite was found to be higher than that of the sisal-reinforced composite but lower than that of the glass fiber-reinforced polyester composite. These results suggested that the hybrid composites exhibited superior thermal stability compared to composites made of single types of fibers.

12.7.3 Water Absorption Properties

Different researchers have reported on the water absorption properties of composites and have observed that hybridization can reduce the water absorption of single fiber-reinforced polymer composites [105, 106]. Prasanna's team studied the water absorption properties of natural fiber-reinforced polyester hybrid composites [107]. They found that the inclusion of bamboo fiber in sisal/unsaturated polyester composites, comprising up to 50% of the total weight, leads to enhanced mechanical properties while simultaneously reducing moisture absorption characteristics. Similarly, Mishra et al. investigated the water absorption behavior of hybrid composites made from palm/sisal and glass fibers reinforced with polyester [108]. They reported that the water absorption properties of the hybrid composites were lower than those of the unhybrid composites. The weight percentage of the fibers influenced the water absorption behavior.

In another study, Thwe et al. examined the environmental behavior of a hybrid composite consisting of bamboo and glass fibers reinforced with PP [109]. They discovered that the moisture absorption level of the bamboo fiber-reinforced PP samples was higher compared to the hybrid bamboo/glass fiber-reinforced polymer composites. As the content of glass fibers increased in the composite, the moisture gain decreased. Recycled PP was reinforced with groundnut shell powder (GSP) in two different particle sizes: 0–250 μm and 250–420 μm [110]. Some of the GSP particles were treated with sodium hydroxide at room temperature. The treated and untreated GSP were then compressed and compounded with recycled PP to create GSP-recycled PP composites. Comparing the two, the treated GSP-recycled PP composites exhibited reduced moisture absorption compared to the untreated GSP-recycled PP composites.

Dhanunjayarao et al. [111] studied the effect of graphite particles on the water uptake along with the diffusivity of hybrid glass\jute composites. By combining jute fibers with E-glass fibers within an epoxy matrix, the water absorption of the resulting composite specimen is decreased in comparison to the jute/epoxy composite. Venkateshwaran et al. studied the water absorption properties of a hybrid composite made from banana and sisal fibers reinforced with epoxy [112]. They found that incorporating sisal fibers in the banana fiber composite, up to a 50% weight ratio, led to a decrease in water absorption properties. The moisture absorption behavior of the composites is represented in Figure 12.10.

In summary, these studies highlight the potential of hybridization to decrease water absorption properties in various fiber-reinforced polymer composites, leading to improved performance in moist environments.

12.7.4 Tribological Properties

In the late 1990s, the extensive use of NFs in polymer tribology emerged as a response to global concerns about environmental changes, pollution, and global warming. Researchers have extensively studied and reviewed the tribological performance and wear mechanisms of polymer composites [113]. The wear of polymer composites is not solely determined by the material itself but also depends on external factors like applied load, sliding speed, temperature, and surface roughness, as

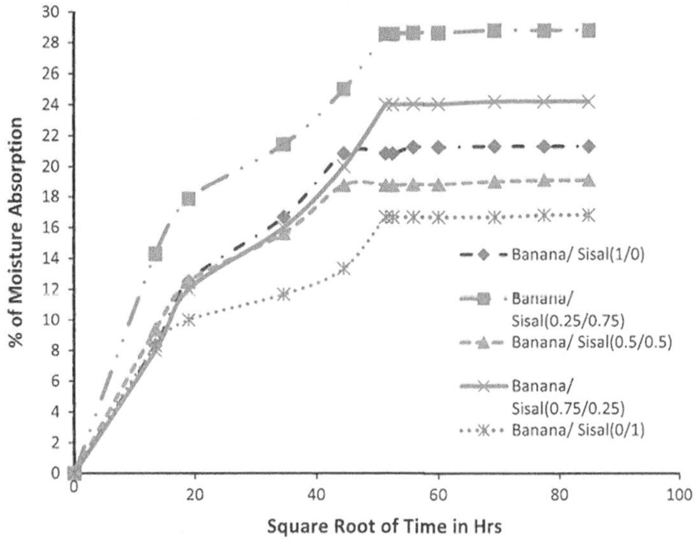

FIGURE 12.10 Moisture absorption characteristics of banana/sisal/epoxy composites

Reproduced with permission from Venkateshwaran et al. Copyright © 2011 Elsevier.

well as the wear mechanism (adhesive or abrasive) [114]. By incorporating fillers into the polymer matrix, the load-carrying capacity and stress transfer in the composite system can be enhanced, resulting in improved wear resistance [115]. Optimal amounts of micro- and nano-particulate reinforcements are key to achieving this enhanced wear resistance. The tribological performance of a material is also influenced by various factors, including fiber loading, surface treatment of fibers, fiber orientation relative to the sliding direction, and fiber length. These parameters play significant roles in determining how well the material performs under sliding conditions [116].

The combination of hybrid NFs with synthetic fibers in polymer composites is gaining significant attention for tribological applications. This approach offers synergistic benefits that enhance the tribological behavior of polymers. By combining the low strength of NFs with the high strength of synthetic fibers, it is possible to create strong and biodegradable hybrid composites [117]. For instance, when jute is used as a reinforcement in epoxy composites, the resulting material may exhibit wear issues under high loads and sliding speeds. However, by incorporating ceramic fillers like Al_2O_3 and SiC into jute-reinforced composites, improved wear resistance can be achieved, particularly under high load and speed sliding conditions. These hybrid composites offer a balance between strength and biodegradability, making them valuable in applications where both performance and environmental considerations are important. In a study by Biswas and Xess, it was found that incorporating a hybrid composition of glass fiber and short bamboo fiber as reinforcement improved the erosive wear characteristics of epoxy [118]. The results indicated that the composite containing 15 wt% bamboo and glass fiber exhibited the lowest erosion rate across different impact velocities, outperforming other composites. This highlights the potential of hybrid natural-synthetic fiber composites for improving the tribological properties of polymers.

In another study conducted by Kumar et al., the frictional coefficient of epoxy composites reinforced with hybrid sisal and glass fibers was investigated [119]. The composites were tested at various sliding speeds, namely 0.2, 2, and 4 mm/s, while maintaining a constant applied load of 10 N. The findings revealed that an increase in fiber length within the composites led to a corresponding increase in the frictional coefficient. Specifically, the greatest improvement was observed in composites reinforced with 2 cm treated fibers compared to those reinforced with untreated fibers. This suggests that optimizing the fiber length and applying appropriate treatment techniques can positively influence the frictional behavior of hybrid sisal/glass fiber-reinforced epoxy composites. Deshpande and colleagues conducted a study on the friction and wear performance of composites made of bidirectional silk fabric reinforced with HDPE and nano clay [120]. Their findings indicated that as the load and sliding distances increased, the wear losses also increased. Additionally, the wear losses were observed to be significant under steady-state conditions when considering different sliding distances. This suggests that the wear behavior of these composites is influenced by both the applied load and the sliding distance, with the steady-state conditions playing a crucial role in determining the overall wear performance. Zhang et al. conducted a study to investigate the impact of TiO_2 nanoparticles on the tribological behavior of polytetrafluoroethylene (PTFE)-cotton fiber composites [121]. Three types of hybrid composites were examined: unfilled composites, composites filled with TiO_2 nanoparticles, and composites filled with modified TiO_2 nanoparticles. A comparison of their tribological performance was made. The study revealed that the hybrid composites filled with grafted TiO_2 nanoparticles exhibited a lower wear rate. This improvement was attributed to the dispersion of agglomerates and the enhancement of interfacial adhesion between the filler and the matrix. These findings suggest that the incorporation of grafted TiO_2 nanoparticles can positively affect the tribological properties of PTFE-cotton fiber composites by reducing wear rates and improving the overall performance of the material.

12.7.5 Morphological Properties

The characteristics of composite materials, especially mechanical properties, are significantly influenced by factors such as the morphology of fillers, filler content, the dispersion and distribution of fillers within the matrix, and the adhesion between fillers and the matrix. Achieving optimal dispersion and distribution of fillers in the polymer matrix is essential, and this can be accomplished through thorough mixing of the components and the implementation of an appropriate mixing method [122]. Tensile testing and SEM were employed in this study to assess the fractured interfaces of composite materials. SEM micrographs revealed features like fiber fracture, pullout, matrix breakage, and voids at the fracture surface, showcasing the significance of factors like wettability and interfacial adhesion in influencing fiber-matrix interactions and fracture mechanisms during tensile testing. The surface characteristics of NFs play a crucial role in determining adhesion with the polymer matrix [123]. A clean, non-oily surface enhances adhesion, while substantial wax or oily content diminishes adhesion due to inadequate wetting. Previous studies have reported improved tensile properties in thermoset polymers reinforced with NFs, and SEM images have been utilized to understand the governing mechanisms of failure. An example of this is the SEM analysis conducted by V. Choudhary et al. on hybrid composites reinforced with jute, hemp, and flax [30]. Their findings align with results from other studies, emphasizing the significance of factors such as wettability and interfacial adhesion in influencing interactions between fibers and the matrix, as well as the mechanisms of fracture observed during tensile testing.

Soundhar and Kandhasamy utilized the CM method to create epoxy hybrid composites reinforced with sisal fiber and crab shell particles [124]. The micrographs clearly demonstrated the high quality of the fiber-matrix interaction, highlighting the effective cohesion achieved during the composite formation process. This observation underscores the success of the interfacial bonding, as revealed by the morphological analysis, in contributing to the overall structural integrity and performance of the composite material.

Another research group utilized SEM to examine the morphological surfaces of jute and coir fibers before incorporating them into composites [31]. Distinctive features of the fibers were revealed, and SEM photomicrographs of cryogenically tensile fracture surfaces indicated significant differences in the interfacial characteristics of pure jute and pure coir composites. The adhesion between fibers and the matrix was assessed through the phenomenon of fiber pull-out in composite

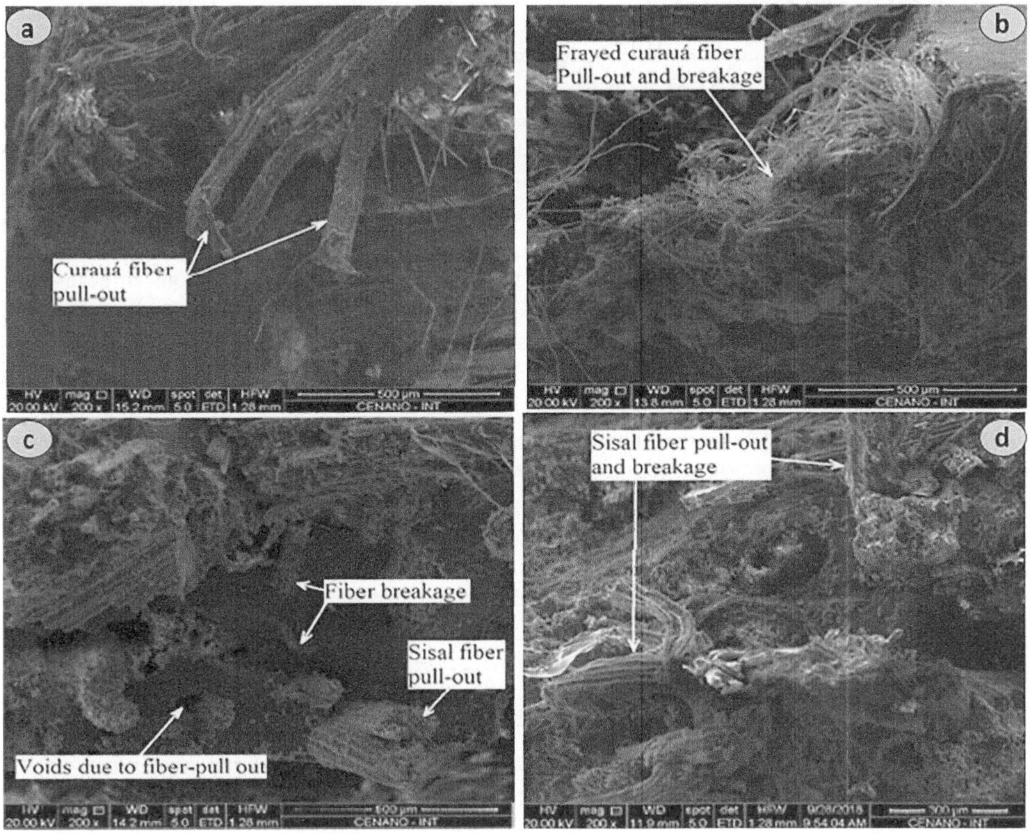

FIGURE 12.11 SEM images depicting the tensile fracture surfaces of hybrid specimens are as follows: (a) alkalinized jute + curauá, (b) mixed-treated jute + curauá, (c) alkalinized jute + sisal, and (d) mixed-treated jute + sisal.

fracture surfaces, serving as an index of adhesiveness. The configuration of hybrid fibers in composites with a layering pattern also influenced adhesion, with well-embedded fibers in the matrix showing fewer pullouts. The literature emphasizes the intricate chemical composition and cell structure of NFs, describing them as essentially composites with rigid cellulose microfibrils embedded in a soft lignin and hemicellulose matrix.

The study conducted on mechanical characterization of intralaminar natural fiber-reinforced hybrid composites by Banea's group highlighted the impact of alkali treatment on NFs, such as sisal, to improve adhesion [125].

SEM images (Figure 12.11) depicted changes in fiber structure and fracture characteristics after treatment, suggesting enhanced adhesion and influencing the overall performance of the composites. The results indicated that alkali treatment enhances adhesion in sisal fibers and influences the fracture characteristics of the composites. The morphological analysis carried out by Jawaid's team demonstrated that the incorporation of coir fibers into oil palm EFB-PP composites results in a reduction of fiber pull-out and protrusion from the surface [126]. This suggests a significant potential for utilizing this hybrid composite in high-end applications, especially when coupled with the use of appropriate coupling agents and additives. The study also reveals that the NFs and the resulting hybrid composites can be effectively injection molded, making them suitable to produce high-demand products. This capability presents a promising solution for utilizing NFs derived from agricultural waste in a sustainable manner.

12.8 APPLICATION OF HNFC

As described in previous sections, HNFCs have superior mechanical and thermal properties as compared to their parent materials. As a result, various industries are trying to replace their existing products with new-generation HNFCs. The proper optimization of the NFs in the resin system can give outstanding properties to the final composites. The HNFCs can be considered whenever the load-bearing capacity of the products is significant. Currently, the biofiber-infused polymer matrix is being used for household and commercial applications such as window frames, doors, tables, and flowerpots.

In addition, these researchers are indicating possibilities of HNFC applications in aerospace. Currently, some of the primary and secondary aircraft products, such as cabinets, blades, and machinery parts, are being replaced with HNFCs. Similarly, automobile segments have also introduced HNFCs in their products due to their lightweight and higher mechanical behavior. HNFCs were found to be applicable in automotive interior and exterior components. The major areas of automotive applications include door panels, bumpers, and interior frames. Both thermoplastics and thermoset resins were employed for the preparation of HNFCs. Thermoplastics such as PP are widely used for the formulation of automobile parts. Consequently, many of the researchers indicated that HNFCs based on PP can have higher properties than PP. Apart from that, thermoset resins are also implemented to produce lightweight automotive products. A study indicates that the hybrid of sisal and kenaf with the epoxy matrix is a suitable candidate for lightweight automobile products [24, 88, 127].

12.9 CONCLUSION AND FUTURE PROSPECTS

In the past several decades, numerous studies have been conducted on the preparation of the HNFC. The selection of NFs and resin systems plays an important role in the final properties of the HNFC. Consequently, this chapter summarized the formulation, characterization technologies, and properties of the HNFCs. Due to the advantages such as renewability, biodegradability, and easy availability, PFs are found to be an attractive candidate in NFCs. The synergetic effects play an important role in the resin matrix system when two or more NFs are present in it. The good distribution of NFs in the resins contributes to the higher mechanical and thermal strength of the final HNFCs. The surface modification technologies on the fibers are also contributing to the balanced properties of the HNFCs. The developed composites can be applied to various sectors such as automotive, aerospace, and construction.

However, further research is essential to improve the moisture resistance and flammability behavior of the HNFCs. In addition, the recyclability of the HNFCs is also a rising question for the scientific community. However, very few studies have been conducted on the HNFCs. So, researchers can concentrate on the various studies on recyclability via mechanical and chemical recycling. Besides, the multiple recyclability behavior of the HNFCs is also studied.

REFERENCES

1. Jagadeesh, P., Puttegowda, M., Boonyasopon, P., Rangappa, S. M., Khan, A., & Siengchin, S. (2022). Recent developments and challenges in natural fiber composites: A review. *Polymer Composites*, *43*(5), 2545–2561.
2. Fuqua, M. A., Huo, S., & Ulven, C. A. (2012). Natural composites. *Polymer Reviews*, *52*(3), 259–320.
3. Onuaguluchi, O., & Banthia, N. (2016). Plant-based natural fibre reinforced cement composites: A review. *Cement and Concrete Composites*, *68*, 96–108.
4. Syduzzaman, M., Al Faruque, M. A., Bilisik, K., & Naebe, M. (2020). Plant-based natural fibre reinforced composites: A review on fabrication, properties and applications. *Coatings*, *10*(10), 973.
5. Kerni, L., Singh, S., Patnaik, A., & Kumar, N. (2020). A review on natural fiber reinforced composites. *Materials Today: Proceedings*, *28*, 1616–1621.

6. Bhattacharyya, D., Subasinghe, A., & Kim, N. K. (2015). Natural fibers: Their composites and flammability characterizations. *Multifunctionality of Polymer Composites*, *1*(1), 102–143.
7. Sathishkumar, T. P., Naveen, J. A., & Satheeshkumar, S. (2014). Hybrid fiber reinforced polymer composites–A review. *Journal of Reinforced Plastics and Composites*, *33*(5), 454–471.
8. Ramachandran, A., Mavinkere Rangappa, S., Kushvaha, V., Khan, A., Seingchin, S., & Dhakal, H. N. (2022). Modification of fibers and matrices in natural fiber reinforced polymer composites: A comprehensive review. *Macromolecular Rapid Communications*, *43*(17), 2100862.
9. Monteiro, S. N., Lopes, F. P. D., Barbosa, A. P., Bevitori, A. B., Silva, I. L. A. D., & Costa, L. L. D. (2011). Natural lignocellulosic fibers as engineering materials—An overview. *Metallurgical and Materials Transactions A*, *42*, 2963–2974.
10. Muniyasamy, S., & Dada, O. E. (2021). Recycling of plastics and composites materials and degradation technologies for bioplastics and biocomposites. In *Waste Management in the Fashion and Textile Industries* (pp. 311–333). Woodhead Publishing.
11. Seki, Y., Selli, F., Erdoğan, Ü. H., Atagür, M., & Seydibeyoğlu, M. Ö. (2022). A review on alternative raw materials for sustainable production: Novel plant fibers. *Cellulose*, *29*(9), 4877–4918.
12. Komuraiah, A., Kumar, N. S., & Prasad, B. D. (2014). Chemical composition of natural fibers and its influence on their mechanical properties. *Mechanics of Composite Materials*, *50*, 359–376.
13. Lee, C. H., Khalina, A., & Lee, S. H. (2021). Importance of interfacial adhesion condition on characterization of plant-fiber-reinforced polymer composites: A review. *Polymers*, *13*(3), 438.
14. Mukherjee, P. S., & Satyanarayana, K. G. (1986). An empirical evaluation of structure-property relationships in natural fibres and their fracture behaviour. *Journal of Materials Science*, *21*, 4162–4168.
15. Ali, M. F., Hossain, M. S., Ahmed, S., & Chowdhury, A. S. (2021). Fabrication and characterization of eco-friendly composite materials from natural animal fibers. *Heliyon*, *7*(5).
16. Kumar, S., Manna, A., & Dang, R. (2022). A review on applications of natural fiber-reinforced composites (NFRCs). *Materials Today: Proceedings*, *50*, 1632–1636.
17. Tonk, R. (2021). Natural fibers for sustainable additive manufacturing: A state of the art review. *Materials Today: Proceedings*, *37*, 3087–3090.
18. Dong, C. (2018). Review of natural fibre-reinforced hybrid composites. *Journal of Reinforced Plastics and Composites*, *37*(5), 331–348.
19. Ismail, S. O., Akpan, E., & Dhakal, H. N. (2022). Review on natural plant fibres and their hybrid composites for structural applications: Recent trends and future perspectives. *Composites Part C: Open Access*, *9*, 100322.
20. Elanchezhian, C., Ramnath, B. V., Ramakrishnan, G., Rajendrakumar, M., Naveenkumar, V., & Saravanakumar, M. K. (2018). Review on mechanical properties of natural fiber composites. *Materials Today: Proceedings*, *5*(1), 1785–1790.
21. Marichelvam, M. K., Kandakodeeswaran, K., & Manimaran, P. (2023). Investigation of mechanical and morphological analysis of natural fiber hybrid composites. *Biomass Conversion and Biorefinery*, *1–14*.
22. Nurazzi, N. M., Asyraf, M. R. M., Fatimah Athiyah, S., Shazleen, S. S., Rafiqah, S. A., Harussani, M. M., … & Khalina, A. (2021). A review on mechanical performance of hybrid natural fiber polymer composites for structural applications. *Polymers*, *13*(13), 2170.
23. Amir, S. M. M., Sultan, M. T. H., Jawaid, M., Ariffin, A. H., Mohd, S., Salleh, K. A. M., … & Shah, A. U. M. (2019). Nondestructive testing method for Kevlar and natural fiber and their hybrid composites. In *Durability and life prediction in biocomposites, fibre-reinforced composites and hybrid composites* (pp. 367–388). Woodhead Publishing.
24. Ramasubbu, R., & Madasamy, S. (2022). Fabrication of automobile component using hybrid natural fiber reinforced polymer composite. *Journal of Natural Fibers*, *19*(2), 736–746.
25. Kumari, N., Paswan, M., & Prasad, K. (2022). Effect of sawdust addition on the mechanical and water absorption properties of banana-sisal/epoxy natural fiber composites. *Materials Today: Proceedings*, *49*, 1719–1722.
26. Mohana Krishnudu, D., Sreeramulu, D., Reddy, P. V., & Rajendra Prasad, P. (2022). Influence of filler on mechanical and di-electric properties of coir and luffa cylindrica fiber reinforced epoxy hybrid composites. *Journal of Natural Fibers*, *19*(1), 339–348.
27. Balaji, N. S., Chockalingam, S., Ashokraj, S., Simson, D., & Jayabal, S. (2020). Study of mechanical and thermal behaviours of zea-coir hybrid polyester composites. *Materials Today: Proceedings*, *27*, 2048–2051.

28. Yashas Gowda, T. G., Madhu, P., Mavinkere Rangappa, S., Siengchin, S., & Jawaid, M. (2022). Mechanical and thermal properties of flax/carbon/kevlar based epoxy hybrid composites. *Polymer Composites*, *43*(8), 5649–5662.
29. Premnath, A. A. (2018). Impact of surface treatment on the mechanical properties of sisal and jute reinforced with epoxy resin natural fiber hybrid composites. *Journal of Natural Fibers*.
30. Chaudhary, V., Bajpai, P. K., & Maheshwari, S. (2018). Studies on mechanical and morphological characterization of developed jute/hemp/flax reinforced hybrid composites for structural applications. *Journal of Natural Fibers*, *15*(1), 80–97. https://doi.org/10.1080/15440478.2017.1320260
31. Saw, S. K., Akhtar, K., Yadav, N., & Singh, A. K. (2014). Hybrid composites made from jute/coir fibers: Water absorption, thickness swelling, density, morphology, and mechanical properties. *Journal of Natural Fibers*, *11*(1), 39–53. https://doi.org/10.1080/15440478.2013.825067
32. Dhakal, H. N., & Sain, M. (2019). Enhancement of mechanical properties of flax-epoxy composite with carbon fibre hybridisation for lightweight applications. *Materials*, *13*(1), 109.
33. Yorseng, K., Rangappa, S. M., Pulikkalparambil, H., Siengchin, S., & Parameswaranpillai, J. (2020). Accelerated weathering studies of kenaf/sisal fiber fabric reinforced fully biobased hybrid bioepoxy composites for semi-structural applications: Morphology, thermo-mechanical, water absorption behavior and surface hydrophobicity. *Construction and Building Materials*, *235*, 117464.
34. Panthapulakkal, S., & Sain, M. (2007). Injection-molded short hemp fiber/glass fiber-reinforced polypropylene hybrid composites—Mechanical, water absorption and thermal properties. *Journal of Applied Polymer Science*, *103*(4), 2432–2441.
35. Pappu, A., Pickering, K. L., & Thakur, V. K. (2019). Manufacturing and characterization of sustainable hybrid composites using sisal and hemp fibres as reinforcement of poly (lactic acid) via injection moulding. *Industrial Crops and Products*, *137*, 260–269.
36. Asaithambi, B., Ganesan, G., & Ananda Kumar, S. (2014). Bio-composites: Development and mechanical characterization of banana/sisal fibre reinforced poly lactic acid (PLA) hybrid composites. *Fibers and Polymers*, *15*, 847–854.
37. Atiqah, A., Jawaid, M., Sapuan, S. M., Ishak, M. R., Ansari, M. N. M., & Ilyas, R. A. (2019). Physical and thermal properties of treated sugar palm/glass fibre reinforced thermoplastic polyurethane hybrid composites. *Journal of Materials Research and Technology*, *8*(5), 3726–3732.
38. Atiqah, A., Jawaid, M., Sapuan, S. M., Ishak, M. R., & Alothman, O. Y. (2018). Thermal properties of sugar palm/glass fiber reinforced thermoplastic polyurethane hybrid composites. *Composite Structures*, *202*, 954–958.
39. Alavudeen, A., Rajini, N., Karthikeyan, S., Thiruchitrambalam, M., & Venkateshwaren, N. (2015). Mechanical properties of banana/kenaf fiber-reinforced hybrid polyester composites: Effect of woven fabric and random orientation. *Materials & Design (1980-2015)*, *66*, 246–257.
40. Li, X., Tabil, L. G., & Panigrahi, S. (2007). Chemical treatments of natural fiber for use in natural fiber-reinforced composites: A review. *Journal of Polymers and the Environment*, *15*, 25–33.
41. Khalid, M. Y., Al Rashid, A., Arif, Z. U., Ahmed, W., Arshad, H., & Zaidi, A. A. (2021). Natural fiber reinforced composites: Sustainable materials for emerging applications. *Results in Engineering*, *11*, 100263.
42. Reddy, R. A., Yoganandam, K., & Mohanavel, V. (2020). Effect of chemical treatment on natural fiber for use in fiber reinforced composites–Review. *Materials Today: Proceedings*, *33*, 2996–2999.
43. Siakeng, R., Jawaid, M., Asim, M., Saba, N., Sanjay, M. R., Siengchin, S., & Fouad, H. (2020). Alkali treated coir/pineapple leaf fibres reinforced PLA hybrid composites: Evaluation of mechanical, morphological, thermal and physical properties. *eXPRESS Polymer Letters*, *14*(8).
44. Agrebi, F., Hammami, H., Asim, M., Jawaid, M., & Kallel, A. (2020). Impact of silane treatment on the dielectric properties of pineapple leaf/kenaf fiber reinforced phenolic composites. *Journal of Composite Materials*, *54*(7), 937–946.
45. Safri, S. N., Sultan, M. T., Saba, N., & Jawaid, M. (2018). Effect of benzoyl treatment on flexural and compressive properties of sugar palm/glass fibres/epoxy hybrid composites. *Polymer Testing*, *71*, 362–369.
46. Safri, S. N. A., Sultan, M. T. H., & Shah, A. U. M. (2020). Characterization of benzoyl treated sugar palm/glass fibre hybrid composites. *Journal of Materials Research and Technology*, *9*(5), 11563–11573.
47. Neto, J. S. S., Lima, R. A. A., Cavalcanti, D. K. K., Souza, J. P. B., Aguiar, R. A. A., & Banea, M. D. (2019). Effect of chemical treatment on the thermal properties of hybrid natural fiber-reinforced composites. *Journal of Applied Polymer Science*, *136*(10), 47154.

48. Yahaya, R., Sapuan, S. M., Jawaid, M., Leman, Z., & Zainudin, E. S. (2016). Effect of fibre orientations on the mechanical properties of kenaf–aramid hybrid composites for spall-liner application. *Defence Technology*, *12*(1), 52–58.
49. Sathish, P., Kesavan, R., Ramnath, B. V., & Vishal, C. (2017). Effect of fiber orientation and stacking sequence on mechanical and thermal characteristics of banana-kenaf hybrid epoxy composite. *Silicon*, *9*, 577–585.
50. Batu, T., & Lemu, H. G. (2020). Investigation of mechanical properties of false banana/glass fiber reinforced hybrid composite materials. *Results in Materials*, *8*, 100152.
51. Gupta, U. S., Dharkar, A., Dhamarikar, M., Choudhary, A., Wasnik, D., Chouhan, P., … & Namdeo, R. (2021). Study on the effects of fiber orientation on the mechanical properties of natural fiber reinforced epoxy composite by finite element method. *Materials Today: Proceedings*, *45*, 7885–7893.
52. Prakash, K. B., Fageehi, Y. A., Saminathan, R., Manoj Kumar, P., Saravanakumar, S., Subbiah, R., … & Rajkumar, S. (2021). Influence of fiber volume and fiber length on thermal and flexural properties of a hybrid natural polymer composite prepared with banana stem, pineapple leaf, and S-glass. *Advances in Materials Science and Engineering*, *2021*, 1–11.
53. Aji, I. S., Zainudin, E. S., Khalina, A., Sapuan, S. M., & Khairul, M. D. (2011). Studying the effect of fiber size and fiber loading on the mechanical properties of hybridized kenaf/PALF-reinforced HDPE composite. *Journal of Reinforced Plastics and Composites*, *30*(6), 546–553.
54. Chollakup, R., Tantatherdtam, R., Ujjin, S., & Sriroth, K. (2011). Pineapple leaf fiber reinforced thermoplastic composites: Effects of fiber length and fiber content on their characteristics. *Journal of Applied Polymer Science*, *119*(4), 1952–1960.
55. Velmurugan, R., & Manikandan, V. (2007). Mechanical properties of palmyra/glass fiber hybrid composites. *Composites Part A: Applied Science and Manufacturing*, *38*(10), 2216–2226.
56. Haris, N. I. N., Hassan, M. Z., Ilyas, R.A., Suhot, M. A., Sapuan, S. M., Dolah, R., Mohammad, R., & Asyraf, M. R. M. (2022). Dynamic mechanical properties of natural fiber reinforced hybrid polymer composites: A review. *Journal of Materials Research and Technology*, *19*, 167–182.
57. Karthik, K., Prakash, J. U., Binoj, J. S., & Mansingh, B. B. (2022). Effect of stacking sequence and silicon carbide nanoparticles on properties of carbon/glass/Kevlar fiber reinforced hybrid polymer composites. *Polymer Composites*, *43*, 6096–6105.
58. Neto, J., Queiroz, H., Aguiar, R., Lima, R., Cavalcanti, D., & Banea, M. D. (2022). A review of recent advances in hybrid natural fiber reinforced polymer composites. *Journal of Renewable Materials*, *10*, 561–589.
59. Joy, J., George, E., Thomas, S. & Anas, S. (2020). Effect of filler loading on polymer chain confinement and thermomechanical properties of epoxy/boron nitride (h-BN) nanocomposites, *New Journal of Chemistry 44*, 4494–4503.
60. Anand Raj, M. K., Muthusamy, S., Panchal, H., Mahmoud Ibrahim, A. M., Alsoufi, M. S., & Elsheikh, A. H. (2022). Investigation of mechanical properties of dual-fiber reinforcement in polymer composite. *Journal of Materials Research and Technology*, *18*, 3908–3915.
61. Nayak, S., Nayak, R. K., & Panigrahi, I. (2022). Effect of nano-fillers on low-velocity impact properties of synthetic and natural fibre reinforced polymer composites–A review. *Advances in Materials and Processing Technologies*, *8*, 2963–2986.
62. Bartos, A., Kócs, J., Anggono, J., Móczó, J., & Pukánszky, B. (2021). Effect of fiber attrition, particle characteristics and interfacial adhesion on the properties of PP/sugarcane bagasse fiber composites. *Polymer Testing*, *98*.
63. Keerthi Gowda, B. S., Easwara Prasad, G. L., & Velmurugan, R. (2020). Prediction of mechanical strength attributes of coir/sisal polyester natural composites by ANN. *Journal of Soft Computing in Civil Engineering*, *4*, 79–105.
64. Alsuwait, R. B., Souiyah, M., Momohjimoh, I., Ganiyu, S. A., & Bakare, A. O. (2023). Recent development in the processing, properties, and applications of epoxy-based natural fiber polymer biocomposites. *Polymers (Basel)*, *15*.
65. Keya, K. N., Kona, N. A., Koly, F. A., Maraz, K. M., Islam, M. N., & Khan, R. A. (2019). Natural fiber reinforced polymer composites: History, types, advantages, and applications. *Materials Engineering Research*, *1*, 69–87.
66. Aisyah, H. A., Paridah, M. T., Sapuan, S. M., Ilyas, R. A., Khalina, A., Nurazzi, N. M., Lee, S. H., & Lee, C. H. (2021). A comprehensive review on advanced sustainable woven natural fibre polymer composites. *Polymers (Basel)*, *13*, 1–45.

67. Gupta, M. K., & Srivastava, R. K. (2016). A review on characterization of hybrid fibre reinforced polymer composite. *American Journal of Polymer Science & Engineering*, *4*, 1–7.
68. Bachtiar, D., Siregar, J. P., Bin Sulaiman, A. S., & Bin Mat Rejab, M. R. (2015). Tensile properties of hybrid sugar palm/kenaf fibre reinforced polypropylene composites. *Applied Mechanics and Materials*, *695*, 155–158.
69. Daramola, O. O., Balogun, O. A., Adediran, A. A., Saka, S. O., Oladele, I. O., & Akinlabi, E. T. (2021). Tensile, flexural, and morphological properties of jute/oil palm pressed fruit fibers reinforced high density polyethylene hybrid composites. *Fibers*, *9*(11), 71.
70. Mann, G. S., Singh, L. P., Kumar, P., Khan, A., Khan, A. A. P., & Asiri, A. M. (2022). Effect of process parameters on the fabrication of hybrid natural fiber composites fabricated via compression moulding process. *Journal of Natural Fibers*, *19*(16), 14803–14812.
71. Velmurugan, G., Siva Shankar, V., Natrayan, L., Sekar, S., Patil, P. P., Kumar, M. S., & Thanappan, S. (2022). Multiresponse optimization of mechanical and physical adsorption properties of activated natural fibers hybrid composites. *Adsorption Science & Technology*, *2022*.
72. Karthi, N., Kumaresan, K., Sathish, S., Prabhu, L., Gokulkumar, S., Balaji, D., ... & Pavithran, S. (2021). Effect of weight fraction on the mechanical properties of flax and jute fibers reinforced epoxy hybrid composites. *Materials Today: Proceedings*, *45*, 8006–8010.
73. Sathish, S., Kumaresan, K., Prabhu, L., & Vigneshkumar, N. (2017). Experimental investigation on volume fraction of mechanical and physical properties of flax and bamboo fibers reinforced hybrid epoxy composites. *Polymers and Polymer Composites*, *25*(3), 229–236.
74. Fei, M., Liu, T., Fu, T., Zhang, J., Wu, Y., Qiu, R., & Liu, W. (2019). Styrene-free soybean oil thermoset composites reinforced by hybrid fibers from recycled and natural resources. *ACS Sustainable Chemistry & Engineering*, *7*(21), 17808–17816.
75. Bujjibabu, G., Das, V. C., Ramakrishna, M., & Nagarjuna, K. (2018). Mechanical and water absorption behavior of natural fibers reinforced polypropylene hybrid composites. *Materials Today: Proceedings*, *5*(5), 12249–12256.
76. Nayak, S., Khuntia, S. K., Mohanty, S. D., Mohapatra, J., & Mall, T. K. (2022). An experimental study of physical, mechanical and morphological properties of alkali treated moringa/areca based natural fiber hybrid composites. *Journal of Natural Fibers*, *19*(2), 630–641.
77. Nyambo, C., Mohanty, A. K., & Misra, M. (2010). Polylactide-based renewable green composites from agricultural residues and their hybrids. *Biomacromolecules*, *11*(6), 1654–1660.
78. Asim, M., Jawaid, M., Saba, N., Nasir, M., & Sultan, M. T. H. (2017). Processing of hybrid polymer composites—A review. *Hybrid Polymer Comwposite Materials*, 1–22.
79. Idicula, M., Sreekumar, P. A., & Joseph, K., Thomas, S. (2009). Natural fiber hybrid composites—A comparison between compression molding and resin transfer molding. *Polymer Composites*, *30*(10), 1417–1425.
80. Yildirim, M., Negawo, T. A., Kilic, A., & Candan, Z. (2020). Development and characterization of hybrid composites from sustainable green materials. *Green Materials*, *9*(4), 182–191.
81. Jawaid, M., Alothman, O. Y., Paridah, M. T., & Khalil, H. S. A. (2014). Effect of oil palm and jute fiber treatment on mechanical performance of epoxy hybrid composites. *International Journal of Polymer Analysis and Characterization*, *19*(1), 62–69.
82. Anand, P., Rajesh, D., Senthil Kumar, M., & Saran Raj, I. (2018). Investigations on the performances of treated jute/Kenaf hybrid natural fiber reinforced epoxy composite. *Journal of Polymer Research*, *25*, 1–9.
83. Ramnath, B. V., Kokan, S. J., Raja, R. N., Sathyanarayanan, R., Elanchezhian, C., Prasad, A. R., & Manickavasagam, V. M. (2013). Evaluation of mechanical properties of abaca–jute–glass fibre reinforced epoxy composite. *Materials & Design*, *51*, 357–366.
84. Chaudhary, V., Sahu, R., Manral, A., & Ahmad, F. (2020). Comparative study of mechanical properties of dry and water aged jute/flax/epoxy hybrid composite. *Materials Today: Proceedings*, *25*, 857–861.
85. Zamora-Mendoza, L., Gushque, F., Yanez, S., Jara, N., Álvarez-Barreto, J. F., Zamora-Ledezma, C., ... & Alexis, F. (2023). Plant fibers as composite reinforcements for biomedical applications. *Bioengineering*, *10*(7), 804.
86. Aravindh, M., Sathish, S., Prabhu, L., Raj, R. R., Bharani, M., Patil, P. P., Karthick, A., & Luque, R. (2022). Effect of various factors on plant fibre-reinforced composites with nanofillers and its industrial applications: A critical review. *Journal of Nanomaterials*, *2022*.
87. Sinha, A. K., Narang, H. K., & Bhattacharya, S. (2020). Mechanical properties of hybrid polymer composites: A review, *Journal of the Brazilian Society of Mechanical Sciences and Engineering*, *42*.

88. Adesina, O. T., Jamiru, T., Sadiku, E. R., Ogunbiyi, O. F., & Beneke, L. W. (2019). Mechanical evaluation of hybrid natural fibre–reinforced polymeric composites for automotive bumper beam: A review. *International Journal of Advanced Manufacturing Technology*, *103*, 1781–1797.
89. Tan, C. S., Chan, M. Y., & Koay, S. C. (2021). Preliminary study of the mechanical properties of hybrid fibres reinforced unsaturated polyester composites. *Journal of Physical Science*, *32*, 45–59.
90. Mulenga, T. K., Ude, A. U., & Vivekanandhan, C. (2020). Concise review on the mechanical characteristics of hybrid natural fibres with filler content. *AIMS Materials Science*, *7*, 650–664.
91. Saravanan, M., Prakash, S., GnanaPrakash, S., Akshaykrishna, M., & Aswin Subhash, S. (2020). Investigation of mechanical properties and characterization of hybrid natural fibre polymer matrix composites. *IOP Conference Series: Materials Science and Engineering*, *993*.
92. Muralikannan, K. S. R., Nijandhan, K., & Venkatachalam, S. (2019). High-strength hybrid particulate-fibre polymer composites: The role of process temperature on the mechanical strength. *Materials Research Express*, *6*, 125313.
93. AlMaadeed, M. A., Kahraman, R., Noorunnisa Khanam, P., & Madi, N. (2012). Date palm wood flour/glass fibre reinforced hybrid composites of recycled polypropylene: Mechanical and thermal properties. *Materials & Design*, *42*, 289–294.
94. Jotiram, G. A., Palai, B. K., Bhattacharya, S., Aravinth, S., Gnanakumar, G., Subbiah, R., & Chandrakasu, M. (2022). Investigating mechanical strength of a natural fibre polymer composite using SiO_2 nano-filler. *Materials Today: Proceedings*, *56*, 1522–1526.
95. Jawaid, M., Abdul Khalil, H. P. S., Hassan, A., Dungani, R., & Hadiyane, A. (2013). Effect of jute fibre loading on tensile and dynamic mechanical properties of oil palm epoxy composites. *Composites Part B: Engineering*, *45*, 619–624.
96. Irawan, A. P., Anggarina, P. T., Utama, D. W., Najid, N., Abdullah, M. Z., Siregar, J. P., Cionita, T., Fitriyana, D. F., Jaafar, J., Hadi, A. E., & Rihayat, T. (2022). An experimental investigation into mechanical and thermal properties of hybrid woven rattan/glass-fiber-reinforced epoxy composites. *Polymers (Basel)*, *14*.
97. Thakur, A., Purohit, R., Rana, R. S., & Bandhu, D. (2018). Characterization and evaluation of mechanical behavior of epoxy-CNT-bamboo matrix hybrid composites. *Materials Today: Proceedings*, *5*, 3971–3980.
98. Rusdi, R. A. A., Halim, N. A., Nurazzi, N. M., Abidin, Z. H. Z., Abdullah, N., Ros, F. C., Ahmad, N., & Azmi, A. F. M. (2022). The effect of layering structures on mechanical and thermal properties of hybrid bacterial cellulose/Kevlar reinforced epoxy composites. *Heliyon*, *8*, e09442.
99. Praveena, B. A., Buradi, A., Santhosh, N., Vasu, V. K., Hatgundi, J., & Huliya, D. (2022). Study on characterization of mechanical, thermal properties, machinability and biodegradability of natural fiber reinforced polymer composites and its applications, recent developments and future potentials: A comprehensive review. *Materials Today: Proceedings*, *52*, 1255–1259.
100. Dhakal, H. N., Zhang, Z. Y., & Bennett, N. (2012). Influence of fibre treatment and glass fibre hybridisation on thermal degradation and surface energy characteristics of hemp/unsaturated polyester composites. *Composites Part B: Engineering*, *43*, 2757–2761.
101. Aisyah, H. A., Paridah, M. T., Sapuan, S. M., Khalina, A., Berkalp, O. B., Lee, S. H., Lee, C. H., Nurazzi, N. M., Ramli, N., Wahab, M. S., & Ilyas, R. A. (2019), Thermal properties of woven kenaf/carbon fibre-reinforced epoxy hybrid composite panels. *International Journal of Polymer Science 2019*.
102. Boopalan, M., Niranjanaa, M., & Umapathy, M. J. (2013). Study on the mechanical properties and thermal properties of jute and banana fiber reinforced epoxy hybrid composites. *Composites Part B: Engineering*, *51*, 54–57.
103. Jawaid, M., Abdul Khalil, H. P. S., & Alattas, O. S. (2012). Woven hybrid biocomposites: Dynamic mechanical and thermal properties. *Composites Part A: Applied Science and Manufacturing 43*, 288–293.
104. Naidu, V. N. P., Reddy, G. R., Kumar, M. A., Reddy, M. M., Khanam, P. N., & Naidu, S. V. (2011). Compressive & impact properties of sisal/glass fiber reinforced hybrid composites. *International Journal of Fibre and Textile Research*, *1*(1), 11–14.
105. Sanjeevi, S., Shanmugam, V., Kumar, S., Ganesan, V., Sas, G., Johnson, D. J., Shanmugam, M., Ayyanar, A., Naresh, K., Neisiany, R. E., & Das, O. (2021). Effects of water absorption on the mechanical properties of hybrid natural fibre/phenol formaldehyde composites. *Scientific Reports*, *11*, 1–11.
106. Bachchan, A. A., Das, P. P., & Chaudhary, V. (2020). Effect of moisture absorption on the properties of natural fiber reinforced polymer composites: A review. *Materials Today: Proceedings*, *49*, 3403–3408.

107. Prasanna Venkatesh, R., Ramanathan, K., & Srinivasa Raman, V. (2016). Tensile, flexual, impact and water absorption properties of natural fibre reinforced polyester hybrid composites. *Fibres and Textiles in Eastern Europe*, *24*, 90–94.
108. Mishra, S., Mohanty, A. K., Drzal, L. T., Misra, M., Parija, S., Nayak, S. K., & Tripathy, S. S. (2003). Studies on mechanical performance of biofibre/glass reinforced polyester hybrid composites. *Composites Science and Technology*, *63*, 1377–1385.
109. Thwe, M. M., & Liao, K. (2003). Durability of bamboo-glass fiber reinforced polymer matrix hybrid composites, *Composites Science and Technology 63*, 375–387.
110. Usman, M. A., Momohjimoh, I., & Usman, A. O. (2020). Mechanical, physical and biodegradability performances of treated and untreated groundnut shell powder recycled polypropylene composites. *Materials Research Express*, *7*.
111. Dhanunjayarao, B. N., Sanivada, U. K., Swamy Naidu, N. V., & Fangueiro, R. (2022). Effect of graphite particulate on mechanical characterization of hybrid polymer composites. *Journal of Industrial Textiles*, *51*, 2594S–2615S.
112. Venkateshwaran, N., ElayaPerumal, A., Alavudeen, A., & Thiruchitrambalam, M. (2011). Mechanical and water absorption behaviour of banana/sisal reinforced hybrid composites, *Materials & Design 32*, 4017–4021.
113. Manjunath, C. B., Srinivasa, C. V., Basavaraju, B., Manjunatha, G. B., & Ashok, R. B. (2020). A review on tribological behaviour of natural fiber reinforced polymer composites. *IOP Conference Series: Materials Science and Engineering 925*.
114. Sarifuddin, N., & Yusuff, M. I. (2021). *Effect of Sliding Conditions on the Wear Behaviour of SAPC2021, 2020*.
115. Mahadevareddy, Adveesh, B., & Hiremath, V. K. (2020). A review on mechanical and tribological properties of graphite, silicon carbide and pineapple leaf fibre reinforced polymer composites. *Journal of Advanced Research in Manufacturing, Material Science & Metallurgical Engineering 7*, 11–22.
116. Jawaid, M., Sapuan, S. M., & Alotman, O. Y. 2017. *Green Biocomposites Manufacturing and Properties*.
117. Talib, A. A. A., Jumahat, A., Jawaid, M., Sapiai, N., & Leao, A. L. (2021). Effect of wear conditions, parameters and sliding motions on tribological characteristics of basalt and glass fibre reinforced epoxy composites, *Materials (Basel)*, *14*, 1–18.
118. Biswas, S., & Xess, P. A. (2012). Erosion wear behaviour of bamboo/glass fiber reinforced epoxy based hybrid composites, *International Journal of Mechanical and Industrial Engineering*, *1*, 78–82.
119. Ashok Kumar, M., Ramachandra Reddy, G., Siva Bharathi, Y., Venkata Naidu, S., & Naga Prasad Naidu, V. (2010). Frictional coefficient, hardness, impact strength, and chemical resistance of reinforced sisal-glass fiber epoxy hybrid composites. *Journal of Composite Materials 44*, 3195–3202.
120. Wani, T. P., Raja, R., Sampathkumaran, P., Seetharamu, S., Sailaja Bhattacharya, R. R. N., & Deshpande, S. (2022). Tribological behaviour of silk fibre reinforced HDPE nano composite. *Tribology - Materials, Surfaces & Interfaces 16*, 226–234.
121. Zhang, H.-J., Zhang, Z.-Z., Guo, F., & Liu, W.-M. (2009). Friction and wear behavior of the hybrid PTFE/cotton fabric composites filled with TiO_2 nanoparticles and modified TiO_2 nanoparticles. *Polymer Engineering & Science 49*, 115–122.
122. Essabir, H., Raji, M., Bouhfid, R., & Qaiss, A. E. K. (2016). Nanoclay and natural fibers based hybrid composites: Mechanical, morphological, thermal and rheological properties, *Engineering Materials*, 29–49. https://doi.org/10.1007/978-981-10-0950-1_2
123. Paiva, M. C., Ammar, I., Campos, A. R., Cheikh, R. B., & Cunha, A. M. (2007). Alfa fibres: Mechanical, morphological and interfacial characterization. *Composites Science and Technology*, *67*(6), 1132–1138.
124. Soundhar, A., & Kandasamy, J. (2021). Mechanical, chemical and morphological analysis of crab shell/sisal natural fiber hybrid composites. *Journal of Natural Fibers 18*, 1518–1532. https://doi.org/10.1080/15440478.2019.1691127
125. Cavalcanti, D. K. K., Banea, M. D., Neto, J. S. S., Lima, R. A. A., da Silva, L. F. M., & Carbas, R. J. C. (2019). Mechanical characterization of intralaminar natural fibre-reinforced hybrid composites. *Composites Part B: Engineering 175*, 107149. https://doi.org/10.1016/j.compositesb.2019.107149
126. Zainudin, E. S., Yan, L. H., Haniffah, W. H., Jawaid, M., & Alothman, O. Y. (2014). Effect of coir fiber loading on mechanical and morphological properties of oil palm fibers reinforced polypropylene composites. *Polymer Composites*, *35*(7), 1418–1425.
127. Chandgude, S., & Salunkhe, S. (2021). In state of art: Mechanical behavior of natural fiber-based hybrid polymeric composites for application of automobile components. *Polymer Composites*, *42*(6), 2678–2703.

13 Surface-Modified Natural Fiber Composites

Rittin Abraham Kurien and Manuel Thomas

13.1 CHEMICAL TREATMENT

The alkali treatment, which is often referred to as mercerization, is the most widely used technique for modifying the surface of natural fibers. According to the findings of the research, the alkali treatment of natural fiber may be used to eliminate waxy residuals, non-cellulosic matter, and hydroxyl ions. This, in turn, leads to an improvement in the interlocking of the polymer matrix with the fiber and the surface roughness [1]. Chemical treatment with alkali can improve the crystallinity of cellulose, which is a factor that determines the mechanical durability of fiber. The elimination of hemicellulose and lignin from fibers is accomplished by treating them with a solution of 10 wt% sodium hydroxide for a period of 2 hours prior to the production of composites. The research was conducted with the purpose of optimizing the concentration of alkali in order to achieve improved bonding between the surfaces of reinforced polymer composites with fiber [2]. The results of a study showed that natural fibers displayed strong interfacial bonding and cellulose crystallinity with epoxy matrix after being treated with NaOH at a concentration of 5 wt% for a period of 2 hours [3]. The occurrence of fibrillation is avoided as a result of this. When NaOH with a concentration that is greater than 5 wt% is employed for the treatment of the fibers, the phenomenon takes place. The flexural strength of fiber-reinforced composites is seen to be superior to that of untreated fiber-reinforced composites when the fiber is treated with 5 wt% sodium hydroxide. When natural fiber is treated with alkali, polymer composites are able to achieve a higher degree of rigidity [4].

The production of NFCs is already underway, and these NFCs might benefit from a thorough and fundamental understanding of the fiber and its surface. When used as a fortification for composites, NFs are associated with low quality, a decrease in interfacial attachment, and inadequate protection from a variety of factors. This is one of the reasons why they are insufficient and restricted. Moisture intake, a restricted most extreme preparation temperature (about 200°C), and a decrease in hardness and dimensional dependability (shrinking and swelling) are all factors that contribute to the condition. Several alternative approaches have been developed in order to modify NFs in order to combat the deficit. Physical methods, chemical methods, organic methods, and nanotechnology (NT) techniques are the four types of approaches that are utilized in the treatment of nanofluorescent substances (NFs). The purpose of these adjustment procedures is to improve the mechanical characteristics of strands, the grip between the framework and the fiber, and to result in the changing of various qualities of the final objects. These techniques are of varying degrees of efficiency.

13.2 FIBER SURFACE-MODIFIED TREATMENT

The use of natural fibers as reinforcement materials with hydrophobic matrices is limited by their affinity for water, which is hydrophilic by nature. Composite materials are limited in their mechanical performance because of weak interfacial bonds [5]. By providing surface treatments to enhance fiber-matrix interface adhesion, natural fibers can display less hydrophilic behavior.

There are a variety of physical and chemical surface treatment techniques that have been identified in previous literature studies. These techniques include plasma, silane, peroxide, alkali,

DOI: 10.1201/9781032665337-13

acrylation, triazine, acetylation, benzoylation, heat treatment, maleated coupling agents, permanganate, sodium chloride, acrylonitrile and acetylation grafting, stearic acid, isocyanate, and fatty acid derivative (oleoyl chloride) that modify the fiber surface [6]. Throughout the textile industry, chemical treatment techniques are a very important component. The improvement of the mechanical qualities of natural fibers, such as jute and hemp, is one of the most important uses of chemical treatment employed in the industry [7].

However, these treatments may affect the strength and durability of fibers as well. Alkali treatment, in particular, has been known to weaken the fibers by breaking down hydrogen bonds and reducing the fiber's tensile strength [8]. Despite this potential negative effect on fiber strength, alkali treatments are still widely used in the industry due to their effectiveness in improving the colorfastness of textiles and their ability to enhance fiber uniformity. Recent research has shown that different types of fibers have varying degrees of sensitivity to alkali treatments [9].

Figures 13.1 and 13.2 show the classification of natural fiber composites and different cell walls present in natural fibers, respectively. Research has been conducted on the application of enzyme treatment in order to enhance both the tensile strength and longevity of fibers. Flexural strength of the abaca/PP composite increased by 45% and 35%, respectively, because of the increased surface area, which facilitates increased interface bonding with matrix material [11]. Coupling agents function as a connection between natural fibers and the polymer matrix, improving interfacial bonding. Maleic anhydride (MA) is the most widely used coupling agent. The utilization of MA-grafted polyethylene or polypropylene as coupling agents has been observed to enhance tensile strength and

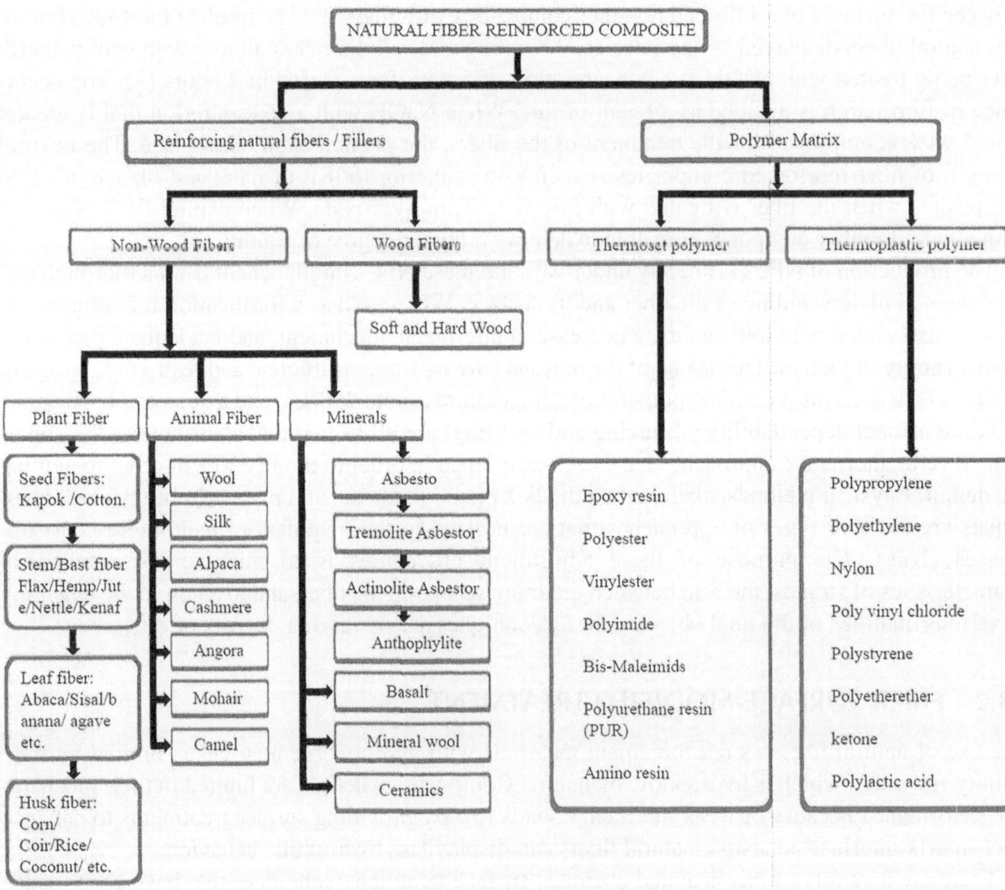

FIGURE 13.1 Classification of natural fiber composite.

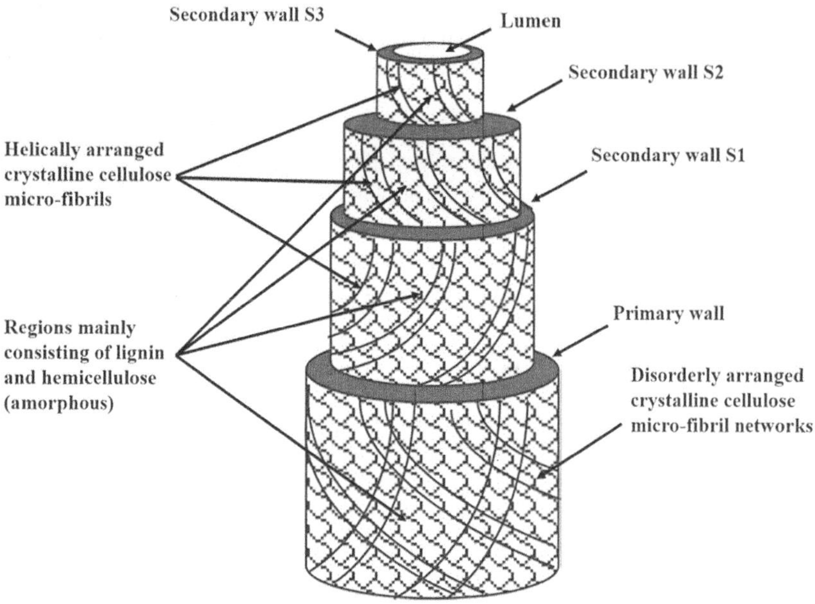

FIGURE 13.2 Natural fiber cell walls [10].

TABLE 13.1
Basic Properties of Natural Fibers [3–6]

Sl No.	Pros	Cons
1	A high level of thermal insulation	Low impact resistance
2	Less weight	Moisture absorption is high
3	Biodegradable	Variations in the qualities of the same fiber that are caused by the use of different processing processes and sources of the fiber
4	Low energy consumption	Poor microbial resistance

elongation at the break point. The primary purpose of implementing surface treatment on natural fibers is to enhance the interfacial adhesion between the fiber and matrix, ultimately leading to improved stress transfer capabilities of the composites [12]. Table 13.1 shows the basic properties of natural fibers [3–6].

13.3 EFFECT OF FIBER SURFACE-MODIFIED TREATMENTS: PHYSICAL, CHEMICAL, AND BIOLOGICAL TREATMENT

The adherence of the fiber to the matrix is negatively impacted when there is a mismatch between hydrophilic natural fibers and a hydrophobic matrix. When the interface between the fiber and the matrix is weak, there is less stress transfer from the matrix to the load-bearing fibers, which results in a reduction in the fiber's reinforced efficiency [13]. The degree of adhesion that exists between the natural fiber and the matrix is a single of the most important factors that determines the mechanical characteristics of a composite that is reinforced with natural fibers [14]. Because of this, a material with a poor fiber-matrix interface is one that is weaker, has less strength, and has a shorter lifespan. Over the course of the past 20 years, a significant number of researchers have concentrated their efforts on enhancing the interfacial adhesion by conducting physical and chemical treatments on the fiber surfaces in order to make them more in accordance with the matrix [15]. In addition, these surface treatments improve the composite's resistance to wear and its durability in the environment (moisture and temperature).

The high moisture absorption rate of natural fibers is a result of their inherent hydrophilic and polar nature. The quality of the fibers can be influenced by environmental factors like rain, soil conditions, sun exposure, and water availability during harvesting seasons, which lead to varying performance and properties. However, their low thermal stability restricts their use in structural applications. Nevertheless, these limitations can be addressed through pretreatment methods that involve physical, chemical, or biological processes [16].

13.4 PHYSICAL MODIFICATIONS

In recent years, natural fibers have garnered a lot of interest for their potential use as reinforcing agents in composite materials. In addition, natural fibers are currently being studied as a possible replacement for synthetic fibers that are not renewable. The unique strengths and modulus of natural plant fibers, as well as their economic viability, low density, decreased tool wear, and better energy recovery, are all positive characteristics of these fibers. The natural fibers have a low thermal stability (the maximum temperature at which they may be used is restricted to 230°C, and they are susceptible to damage in harsh settings.

With the use of biofibers for the reinforcement of polymeric matrices, there is also a significant disadvantage that is linked with this application. The high moisture absorption properties of natural fibers can be attributed to the presence of hydroxyl and other polar groups in their components.

The unfavorable aspects of natural fibers, including inadequate compatibility with matrix and decreased physical properties due to moisture absorption, pose challenges for engineering composites. However, pretreatment methods have been introduced to address these drawbacks. Studies suggest that physical modification can increase the mechanical bond between the fiber and matrix without altering chemical properties. Table 13.2 outlines several examples of physical alterations made to natural fibers.

13.5 CHEMICAL MODIFICATIONS

To a large extent, the mechanical performance and qualities of NFPCs are determined by the interfacial adhesion that exists between two different materials. The mechanical characteristics of the material are poor because there is less adhesion across the phase boundary.

The hydrophilic behavior of natural fibers and the weak bonding between the fiber and the matrix are the primary concerns in the field of non-fiber polymer composites (NFPCs). To a greater degree, these significant obstacles can be handled by the application of chemical modification to natural fibers. In addition to the fact that every other natural fiber is susceptible to absorbing moisture, some resins are also capable of absorbing large amounts of moisture. From this point forward, it is necessary to make modifications to both the fibers and the matrices in order to enhance the interfacial bonding across the boundary. This is done with the intention of improving the functionality and performance of NFPCs.

Natural fibers that are composed of non-cellulose elements, such as hemicelluloses, pectin, and lignin, have a polar behavior and are easily capable of forming bonds of hydrogen with hydroxyl groups (–OH), causing the fiber and matrix to become delaminated, which in turn restricts the mechanical properties of nylon fiber composites (NFPCs). The limitations of natural fibers that were discussed before can be overcome by the application of chemical and physical treatments in advance. During the process of chemically treating fibers, the interfacial adhesion between the fibers and the matrix is strengthened through the operation of chemical reactions. A weak bonding along the interfaces is imposed by two phases of materials that are incompatible with one another. Through the use

TABLE 13.2
Effects of Physical Variations on Natural Fiber

Sl No.	Treatment	Modifications	References
1	Plasma treatment	It is one of the most used techniques, and it makes the surface rougher.	[17]
2	Corona discharge	It alters the fiber's surface energy, makes it more compatible with a hydrophobic matrix, gives jute fibers improved mechanical properties, and improves the tensile strength of hemp fibers.	[18, 19]
3	Steam explosion	This process results in relatively short fibers and lowers fiber strength.	[20]
4	Ultrasound	It can raise the fibers' crystallinity index while being efficient in removing pollutants and other impurities.	[21]
5	UV light	The wavelength of UV utilized ranges from 10 to 400 nm, increasing the polarity and wettability of the fibers as well as the composite's strength due to the reinforcements' and matrix's adherence.	[22, 23]
6	Ozone treatment	It also increases the size and level of crystallinity when lignin is being degraded and the hemicellulose content is being solubilized. The fiber's physicochemical characteristics deteriorated over time.	[23, 24]
7	Laser treatment	The amount of cellulose was unaltered, but evaporation lowered the lignin content of the bagasse fibers by 10%. The chemical surface alterations resembled the thermal breakdown of fibers.	[25]
8	Heat treatment	The temperature used to treat the fibers is nearly the fiber breakdown temperature. There have been reports of an increase in the crystallinity of the fibers, although for bamboo and hemp, the maximum temperature at which the mechanical qualities tend to degrade is 140°C and 160°C, respectively.	[26, 27]

of chemical modification, the hydrophilic behavior of fibers may be significantly reduced, resulting in improved bonding between surfaces across the phase boundary.

The chemical arrangement and makeup of the chemical elements, including cellulose, hemicellulose, lignin, wax, and moisture content, are altered as a result of this pretreatment of fibers, which can be accomplished using any modification process.

This article is a review that discusses several strategies for chemical modification and analyzes the impacts that these approaches have on the mechanical properties of the material. The purpose of this review study is to explore 14 distinct chemical treatments that are available for a wide variety of natural fibers and the influence that these treatments have on composite qualities.

In order to improve the adhesion with reinforcements and the matrix, it has been demonstrated that chemical treatment of fibers is an efficient strategy. There has been a significant amount of study conducted on the effects of chemical processing on natural fiber-reinforced composites (NFCs). The alterations made to natural fibers have been shown to increase adhesion with a variety of matrices [28]. Consequently, this results in stronger interfaces, which in turn contribute to enhanced strength and stiffness, as well as increased efficiency when it comes to the transmission of stress between the fiber and the matrix. The results of an extensive study indicate that chemical treatments are preferable to physical approaches such as corona or plasma treatments. This is due to the fact that chemical treatments have been demonstrated to promote adhesion while simultaneously lowering hydrophilicity [29]. Various chemical methods are utilized for the processing of natural fibers. These processes include alkaline (Mercerization), silane, acetylation, bezoylation, peroxide, maleated coupling, sodium chlorite, acetylation, acrylonitrile grafting, isocyanate, stearic acid, trizane, and oleoyl chloride [10, 30, 31]. The inclusion of a variety of coupling agents has resulted in a considerable improvement in the mechanical characteristics. This is due to the fact that it generates a greater number of anhydride groups with functions and opens up a greater number of sites for chemical linkages. In addition, it has been demonstrated that the incorporation of coupling agents into the matrix results in a reduction in the rate of water absorption.

13.5.1 Alkaline Treatment (Mercerization)

The most affordable and efficient way to change the surface and morphological characteristics of the fiber is alkaline treatment. The initial effect of this treatment is to liberate individual fibers and break up the fiber bundles. The rate of fiber fragmentation and disaggregation increases with additional alkaline treatment. The number of exposed OH groups causes the fiber to change its nature, making it more hydrophobic. Mercerization also enhances the amount of cellulose in the fiber by decreasing the amount of weaker components such as hemicellulose, pectin, and lignin [32]. As a result, the crystallinity of the cellulose also rises. Additionally, mercerization reduces the amount of weaker substances like hemicellulose, pectin, and lignin while increasing the amount of cellulose in the fiber. This increases the crystallinity of the cellulose, which in turn increases the fiber's ability to transfer stress. In order to purify the cellulose content and eliminate organic waste from the fibers, bleaching chemicals like H_2O_2 or $NaOC_{12}$ can be utilized. The fibers should be kept at their optimal level of mercerization to prevent weakening and damage. This also guarantees that the fibers" diameter is significantly reduced, which improves their adhesion to the matrix [10, 33–42].

$$\text{Fiber-OH} + \text{NaOH} = \text{Fiber-O-Na}^{+} + \text{H2O} \tag{13.1}$$

13.5.2 Silane Treatment

Silane, also known as SiH_4, is primarily utilized in composites as a stabilizing agent. Triethoxy octyl silane, pheniltrimethoxysilane, A174 (-methacryl oxypropyl trimethoxy silane), and A151 (vinyltriethoxysilane) are reagents used in silane treatment. The process of silane coupling involves a siloxane bridge, which is produced when alkoxy groups are hydrolyzed in the presence of moisture, to form a chemical link between the matrix and the fiber. The produced silanol has two ends: one is connected to the cellulose OH, while the other undergoes condensation reaction with matrix functional groups. The chemical reaction in Equation (13.2) represents the reaction mechanism of the functional groups of silane. The silane groups" OH attaches to the fiber's surface, increasing the fiber's hydrophobic properties and enhancing its compatibility with the matrix [34].

$$\text{Fiber-OH} + \text{R-Si(OH)}_3 = \text{Fiber-O-Si(OH)}_2\text{-R} \tag{13.2}$$

Prior alkali treatment can boost the efficiency of silane treatment by significantly increasing the amount of reactive sites in the molecule. The silanol bonding between the matrix and the fiber is mostly to blame for this. The molecular continuity created across the interface of the hydrocarbon chain and the composite has decreased the ability of the fiber to inflate into the matrix as a result of silane treatment. The use of silanes (pre-treated with NaOH) to treat oil palm fiber composites has been discussed as having superior heat conductivity to acetylation. According to, pineapple fiber that had been silane-treated had the maximum tensile and impact strength. When compared to other treatments, Kurien et.al discovered that the silane-treated ramie fiber had good tensile and impact strength. Abacca fibers treated with silane displayed improved storage modulus [35].

13.5.3 Acetylation Treatment

Acetylation has a number of advantages, the most important of which are enhanced dimensional stability, increased durability, and increased hydrophobic properties. Reagents that are utilized in this treatment include isopropyl acetate, vinyl acetate, ketene gas, di-ketene liquid, cyclic and non-cyclic anhydrides, acid chlorides, isocyanates, aldehydes, and acetic anhydrides. Among these agents, acetic anhydride is the one that shows the most promise. The cell structure of the fibers is coupled with acetyl groups during the acetylation process [36]. This is accomplished with the assistance of acid catalysts. In order to do this, the fibers are first submerged in acetic acid, and then they are subjected to a subsequent treatment with acetic anhydride. This treatment is carried out at high temperatures for a period of time ranging from 1 to 3 hours, which speeds up the chemical process. The fact that acetic acid and acetic anhydride are not reacting simultaneously during this reaction is something that should be taken into consideration. Acetylation is sometimes referred to as esterification because it is caused by the interaction of the hydroxyl (OH) group and the carboxylic groups with the cellulose constituents of the fiber. Equation (13.3) provides the formula for this reaction. There have been a great number of studies that have demonstrated that fibers that have been pre-treated with an alkaline have better mechanical and physical qualities. The surface morphology of the treated flax fibers was found to be smoother, and wax and cuticles were removed from the surfaces, as demonstrated by scanning electron microscopy (SEM) tests. It is possible to alter the surface shape of the treated fiber by replacing the polar hydroxyl groups within the fiber with less polar acetyl groups. This results in the fiber being more hydrophobic yet compatible with non-polar matrices thanks to the modification. A decrease in the rate of moisture absorption may be achieved by treating natural fibers with acetylation. This results in an improvement in the dimensional stability of the material as well as its mechanical characteristics, including interfacial shear stress, tensile strength, and flexural properties [37].

$$\text{Fiber-OH} + \left(\text{H}_3\text{C-CO}\right)_2 \text{O} = \text{Fiber-O-CO-CH}_3 + \text{CH}_3\text{COOH} \tag{13.3}$$

13.5.4 Benzoylation

This chemical treatment's main goal is to lessen the treated fibers' hydrophilic properties. The fiber must first go through a pretreatment process in which alkali is used to treat the fiber, eliminating lignin, wax, and covering oil. The hydroxyl groups in fiber are activated by pretreatment with alkali; the reagent used for benzoylation is benzoyl chloride, which reacts with the hydroxyl group in cellulosic material to generate C_6H_5CO– (a functional group of benzoyl) in the fiber, improving adherence to the matrix. Equation (13.4) depicts this chemical process. The strength and thermal stability of the reinforcing fibers can be improved by benzoylation treatment. Benzoyl chloride treatment of sugar palm fiber resulted in an increase in tensile strength to 173.99 Mpa, and SEM analysis of the treated fiber verified a significant reduction in fiber diameter from 344.35 um to 480.33 um [38].

$$\text{Fiber} - \text{O - Na} + \text{C}_6\text{H}_5\text{CH}_2\text{Cl} = \text{Fiber} - \text{O-C}_6\text{H}_5\text{CH}_2 + \text{NaCl} \tag{13.4}$$

13.5.5 Peroxide Treatment

The most often used chemicals for peroxide treatment are benzoyl peroxide, methyl ethyl ketone peroxide (MEKP), and dicummyl peroxide. Researchers have noted increased fiber thermal stability, decreased moisture absorption, and improved adhesion with the matrix at the interface. Wax, hemicellulose, and lignin components present in the reinforcement fiber must be removed before the peroxide treatment using an alkali reagent. The mechanism starts when organic peroxides break down into peroxide free radicals. The hydroxyl groups of the reinforcement and matrix are respectively reacted with by the peroxide free radicals that are thus created. Equation (13.5) describes the alleged chemical process [39].

$$\text{R}\dot{\text{O}} + \text{cellulose-H} = \text{R-OH-Cellullose} \tag{13.5}$$

13.5.6 Maleated Coupling Treatment

Maleic anhydride and maleated polyethylene, which change the surface morphology and polymer matrix, are the major reagents utilized in this therapy. A good compatibility between the matrix and fiber reinforcements can be made possible with the aid of maleic coupling agents. Maleic anhydride can be grafted onto the polymer as a result of the cellulose –OH groups' covalent or hydrogen bonds being more reactive. In contrast to other surface treatment techniques, maleated coupling treatment (MCT) improves interfacial bonding while also increasing the strength of the composite [40]. This is done by changing the morphology of the matrix's fibers and polymers. Equation (13.6) illustrates the chemical reaction. According to Equation (13.6), the MCT coats the fiber's surface, which reduces the fiber's propensity to be hydrophilic by excluding OH groups from the surface. Maleic anhydride (MA) can make bonds with the hydroxyl groups of cellulose fiber and the polymer matrix on a C-C basis, creating an effective and efficient bridge interface. The failure examination of the treated fibers revealed breakage via fiber ripping rather than failure at the interface, indicating enhanced fiber resin adherence. Compression molding was used to create a jute fiber/polypropylene composite, and it was claimed that MCT enhanced the dynamic storage modulus and Young's modulus [41].

$$\text{Fiber-OH} + C_4H_2O_3 + \left(C_3H_6\right)_n = \text{Fiber-OH-}C_4H_2O_3 - \left(C_3H_6\right)_n \quad (13.6)$$

13.5.7 Sodium Chlorite Treatment

Natural fiber bleaching can be classified into two categories: (1) reduction bleaching (2) and bleaching via oxidation. Reduction bleaching is often done on protein fibers and involves the use of chemicals like SO_2 and $Na_2S_2O_4$. Cellulosic fibers are bleached via oxidation using chemicals like sodium hypochlorite, H_2O_2, and $NaClO_2$. $NaClO_2$ enhances the flexibility of the fibers while decreasing moisture content and hydrophilicity. By forming lignin compounds, chlorine dioxide lowers the quantity of lignin in the fiber, enhancing the fiber's ability to adhere to the matrix [10].

Its purpose is to lighten the color of the fibers. Under acidic circumstances, it leads to the liberation of chloric acid, which then results in the formation of reactive chlorine dioxide. This, in turn, causes lignin components to react with chlorine dioxide, which ultimately results in the elimination of lignin from the fiber [4]. Additionally, it interacts with the components that make up the fiber. The interaction of these processes contributes to the elimination of polar contaminants from the surface of the fibers and enhances the hydrophobic character of the fibers. Following the elimination of lignocellulosic material, the fibers typically exhibit increased flexibility and are characterized by decreased stiffness qualities.

13.5.8 Isocyanate Treatment

Common reagents used for isocyanate treatment include diphenyl methane diisocyanate, polymeric methane diphenyl isocyanate, and lysine-based isocyanate. Due to the reaction between the functional groups in this treatment ($-N=C=O$) and the OH groups found in the reinforcing fibers, a strong covalent connection is formed, depleting the free OH groups and increasing the fiber's hydrophobicity. For bamboo fiber and PLA composite it showed an improved heat stability with lysine-based diisocyanate. Better tensile strength has been observed for sisal fiber/polyethylene composites made by injection molding after isocyanate treatment [42].

13.5.9 Plasma Treatment

The use of plasma treatment is an efficient technique that may be utilized to modify the surface of natural polymers sans altering the bulk characteristics of these materials. You have the option of generating the plasma discharge by either the corona therapy or the cold plasma treatment. Ionized gas is regarded to be a plasma therapy when it has an equal number of positively and negatively charged molecules that interact with the outer layer of the substance being treated. Both of these approaches are associated with plasma treatment. The rate of the electric discharge is the characteristic that differentiates the two kinds of plasmas with respect to their characteristics. An alternating current discharging at atmospheric pressure can create corona plasma, whereas microwave energy can produce high-frequency cold plasma [43]. Corona plasma is formed by alternating current discharge when the frequency is lower. There was a correlation between the kind of ionized gas and the time of exposure, which had an effect on the alteration of the surfaces of both the wood and the synthetic polymer [44]. In addition, the relationship between the polar component of the surface energy of pine wood and a variety of factors for plasma modification is examined. These characteristics include power, treatment duration, distance of samples to plasma source, type of gas, and the stability of plasma treatment [45]. In the presence of air and nitrogen atmospheres, corona discharge was applied to pulp sheets with moisture levels of up to 85%, and the chemical alteration of the sheets was investigated using dye [46]. On the surface of the treated paper, there was no indication that there was an increase in the number of carboxylic groups.

On the other hand, the quantity of aldehyde groups rose as the duration of the corona treatment increased. In the past, it was demonstrated that the treatment with air corona resulted in a decrease in the molecular weight of cellulose [47]. This was accomplished by measuring the basic viscosity of cellulose solution. When the gas was changed to nitrogen, the corona treatments did not result in a decrease in the inherent viscosity of the cellulose. This demonstrates that the type of gas has an effect on the degree of modification. There was an increase in the amount of water and methanol extractives, which implies that the cellulose and hemicelluloses underwent a transformation.

13.6 BIOLOGICAL MODIFICATIONS

The field of polymer science and technology has documented the existence of environmentally friendly products that are based on polymers. Considering this, natural fiber based biocomposites have garnered a significant amount of attention from researchers due to the fact that they are simple to dispose of once their shelf life has passed without having any negative effects on the environment. Natural fibers can be classified as either plant, animal, or mineral fibers, depending on the type of renewable source that they come from. Fibers that are generated from any portion of the plant are referred to as plant fibers. The majority of the fibers that fall into this category include bast fibers, leaf fibers, seed fibers, fruit fibers, wood fibers, cereal straw fibers, and various grass fibers. The structure of these organisms is composed of cellulose, and they are regarded as a composite structure consisting of cellulose fibrils embedded in lignin matrix. Cellulose, hemicellulose, lignin, pectin, and waxes are the chemical components that make up these fibers. The content of these fibers varies depending on factors such as their origin, age, the retting technique that was utilized, and other factors. Among the most prominent examples of these fibers, which are often referred to as lignocellulosic fibers, are flax, hemp, kenaf, sisal, henequen, banana, oil palm, jute, cotton, coir, and other similar materials. Protein is the most essential structural component of animal fibers, and other key components include hair, silk, and wool. It is the most abundant component.

Even though chemical modifications are effective at improving the mechanical and morphological characteristics of natural fibers, they have the inherent drawback of leaving behind materials that are bad for the environment. To make up for these shortcomings, subsequent research has focused on biological modification techniques. Enzymatic treatment and fungal treatment are two categories for biological alterations based on the mechanism of treatment.

13.7 ENZYMATIC TREATMENT

It is a highly expensive approach for treating fibers, but it has been shown to increase the interface strength of treated fibers in numerous studies. It is used in pilot scales for treating natural fibers. It can selectively eliminate hydrophilic substances, including hemicellulose, pectin, and lignin, which decreases the fibers' capacity to absorb moisture. Numerous studies have suggested that treating reinforcing fibers with enzymes improved the mechanical properties of the finished composite as well as the adhesion of the fibers to the matrix. It uses particular enzymes like hydrolases and oxidoreductases to remove contaminants. The high specificity and fast reaction speed required for effective catalytic activity distinguish chemical catalysis from enzymatic catalysis. A pH of 4 to 8 should be maintained, and the temperature and pressure must both be below 100°C. Tyrosinase, laccase, peroxidase, and other oxidoreductases are categorized as hydrolases, whereas amylase, cellulose, proteases, pectinases, lipases, and esterases are categorized as hydrolases. Common pectinases are commercial blends of several enzymes, including viscozyme, flaxyzye, and pectinex AR, that are used on lignocellulosic fibers. After being treated with enzymes, the fiber gains better wettability, dyeability, and surface texture without losing any fiber quality. It also becomes softer to the touch [48].

According to studies, the oxidoreductase enzyme laccase removes lignin from natural fibers with the least amount of harm to the fibers. When applied with natural phenols like acetosyringora and syringealdehyde has increased antibacterial effects and can reduce the lignin concentration in cellulosic fibers from 35% to 24%. Jute fiber that had undergone enzymatic treatment had decreased toughness and flexural strength. Even though the enzymatic treatment usually removes lignin effectively and without harming the fibers, one major disadvantage is its high operating costs, which can also be decreased by recycling the enzymes after each treatment [49].

13.8 FUNGAL TREATMENT

An effective and creative method to lessen the ecological impact of chemical conduct is the treatment of natural fibers with fungi. The interaction of particular enzymes with non-cellulosic materials and lignin results in the formation of a soluble substrate that is then removed by an aqueous medium, improving the characteristics of the natural fibers. Lack of non-cellulosic components improves the fiber's hydrophobic properties and lowers the rate at which the composites absorb moisture [47]. It has the ability to eliminate undesirable cellulosic elements as lignin, pectin, hemicellulose, wax, etc. By releasing extracellular oxidases onto the surface of natural fibers, Schizophyllum commune, often known as white rot fungi, can degrade lignin and improve the solubility of hemicellulose. By creating hyphae that absorb non-cellulosic materials, fungi can roughen the surface morphology. As a result, the surface of the fibers becomes covered in tiny holes, which improves the interlocking between the fiber and matrix. When kenaf fiber-reinforced composites were subjected to fungus, the amount of pectin decreased by 91.3% [48]. Fungus treatment raised the tensile strength of a hemp fiber/polypropylene composite made by injection molding by 22% and by 32% with additional alkaline treatment. When using a fungus to treat natural fibers, a 22% increase in the stiffness of the composite was observed. Hemp fiber treated with fungal species from the Ophiostoma ulmi genus had improved fiber components compared to untreated fibers [50, 51].

13.9 CONCLUSION

At the moment, the automobile sector is at the forefront of the research and application of composite materials that are derived from fibers like flax and hemp. Applications like structural and nonstructural geotextiles packaging, and molded items are examples of the best potential uses of vast amounts of natural fibers in conjunction with other materials. These applications reveal the greatest possible uses of natural fibers. Specifically, structural beams and paneling have been developed, fabricated, and tested using natural fiber-based composite materials, with a particular emphasis on the resins derived from petroleum and natural fibers. To develop markets in other applications, such as aerospace, maritime, and commercial construction, biocomposites and composites need to achieve durability, dependability standards, serviceability, and high-quality performance. This is necessary in order to broaden markets in other applications. Wood pallets have seen widespread application in a variety of industries, including the automotive, transportation, pharmaceutical, and beverage sectors. Moreover, wood pallets are capable of withstanding heavy load-bearings and can be recycled. Despite this, the lifespan of these pallets is rather limited, regardless of whether they are exposed to a hot or cold climate. They have a limited tolerance to dampness and are sensitive to being attacked by pests.

Recently, there has been an increase in the demand for environmentally friendly items with the objective of lowering the use of fossil fuels, lowering the rate of deforestation, and lowering the rate of global warming. Plastic pallets are often favored over conventional wooden pallets because of their cleanliness, good longevity, resistance to pests and corrosion, resilience to the elements, and minimal rusting. The production of plastic pallets has the potential to safeguard hardwood forests and avert an ecological catastrophe that would otherwise be caused by the excessive cutting down of trees. Their primary benefit is that used or abandoned pallets may be recycled into new pallets, which results in a reduction in the amount of garbage that is sent to landfills. Composite materials made of natural fibers are now being utilized in the process of water and wastewater treatment. For the purpose of capturing and removing pollutants from the air and water, certain composites are developed.

Due to the inherent financial advantages of adopting natural fiber-reinforced composites as opposed to synthetic fiber-reinforced composites, natural fiber composites have emerged as a lucrative option for a number of businesses. Despite the many benefits of natural fibers, one impending problem with NFRPs is their poor compatibility with the matrix. As a result, numerous surface modification techniques are now in use to benefit from the interfacial adhesion between the matrix and the fiber. The many techniques discussed here, including biological treatment, physical treatment, and chemical treatment, are used to improve the composites' general performance.

The widespread and inclusive use of natural fibers in engineering composites is essential for the future of sustainable engineering development. In order to mitigate the impacts of non-biodegradable composites, polymer composites with surface-modified natural fibers are a potential option. Future research suggests that biological and nano treatment of natural fibers may soon be useful in creating composites that are efficient in constructing structures and friendly to the environment.

REFERENCES

1. Raju, A., and Shanmugaraja, M. (2020, December). Recent researches in polylactic acid reinforced with natural fiber composites–A literature review. In *IOP Conference Series: Materials Science and Engineering* (Vol. 988, No. 1, p. 012025). IOP Publishing.
2. Capretti, G. Fundamental and technological aspects of the process of Steam Explosion. *Tinctoria* 1991.
3. Figueiredo, Lintinen, K., Hirvonen, J. T., Kostiainen, M. A., and Santos, H. A. Properties and chemical modifications of lignin: Towards lignin-based nanomaterials for biomedical applications. *Progress in Materials Science* 93, 2018.
4. Faruk, O., Bledzki, A. K., Fink, H. P., and Sain, M. Progress report on natural fiber reinforced composites. *Macromolecular Materials and Engineering* 299, 2014.
5. Verma, and Senal, I. (2019). Natural fiber-reinforced polymer composites. in *Biomass, Biopolymer-Based Materials, and Bioenergy: Construction, Biomedical, and other Industrial Applications*, pp. 103–122.
6. Abdelmouleh, M., Boufi, S., Ben Salah, A., Belgacem, M. N., and Gandini, A. Interaction of silane coupling agents with cellulose. *Langmuir*, 18, 2002.
7. Keener, T. J., Stuart, R. K., and Brown, T. K. Maleated coupling agents for natural fiber composites. *Composites Part A: Applied Science and Manufacturing*, 35, 2004.
8. Ali et al., Hydrophobic treatment of natural fibers and their composites—A review. *Journal of Industrial Textiles*, 2018.
9. Ling, Z., Chen, S., Zhang, X., and Xu, F., Exploring crystalline-structural variations of cellulose during alkaline pretreatment for enhanced enzymatic hydrolysis. *Bioresource Technology*, 224, 611–617, 2017.
10. Kurien, R. A., Selvaraj, D. P., Sekar, M., Koshy, C. P., Paul, C., Palanisamy, S., … & Kumar, P. A comprehensive review on the mechanical, physical, and thermal properties of abaca fibre for their introduction into structural polymer composites. *Cellulose*, 1–22. 2023. https://doi.org/10.1007/s10570-023-05441-z

11. Ferreira, F., Pinheiro, I., de Souza, S., Mei, L., and Lona, L., Polymer composites reinforced with natural fibers and nanocellulose in the automotive industry: A short review. *Journal of Composites Science*, 3, 2, 51, 2019.
12. Akindoyo, O., Beg, M. D. H., Ghazali, S., Heim, H. P., Feldmann, M., and Mariatti, M. Oxidative induction and performance of oil palm fiber reinforced polypropylene composites – Effects of coupling agent and UV stabilizer. *Composites Part A: Applied Science and Manufacturing*, 125, 2019.
13. Kiruthika, V. A review on physico-mechanical properties of bast fiber reinforced polymer composites. *Journal of Building Engineering*, 9, 2017.
14. Xia, X., Liu, W., Zhou, L., Hua, Z., Liu, H., and He, S. Modification of flax fiber surface and its compatibilization in polylactic acid/flax composites. *Iranian Polymer Journal (English Ed.)*, 25, 2016.
15. Annie Paul, S., Boudenne, A., Ibos, L., Candau, Y., Joseph, K., and Thomas, S. Effect of fiber loading and chemical treatments on thermophysical properties of banana fiber/polypropylene commingled composite materials. *Composites Part A: Applied Science and Manufacturing*, 39, 2008.
16. Pothan, A., and Thomas, S. Polarity parameters and dynamic mechanical behaviour of chemically modified banana fiber reinforced polyester composites. *Composites Science and Technology*, 63, 2003.
17. Sanjay, M. R., Siengchin, S., Parameswaranpillai, J., Jawaid, M., Pruncu, C. I., and Khan, A. A comprehensive review of techniques for natural fibers as reinforcement in composites: Preparation, processing and characterization. *Carbohydrate Polymers*, 207, 2019.
18. Marais, S., et al., Unsaturated polyester composites reinforced with flax fibers: Effect of cold plasma and autoclave treatments on mechanical and permeation properties. *Composites Part A: Applied Science and Manufacturing*, 36, 2005.
19. Pizzi et al., High resin content natural matrix-natural fiber biocomposites. *Industrial Crops and Products*, 30, 2009, https://doi.org/10.1016/j.indcrop.2009.03.013
20. Mukhopadhyay, S., and Fangueiro, R. Physical modification of natural fibers and thermoplastic films for composites - A review. *Journal of Thermoplastic Composite Materials*, 22, 2009,
21. Ferreira, P., Cruz, J., and Fangueiro, R. (2018). Surface modification of natural fibers in polymer composites. in *Green Composites for Automotive Applications*.
22. Gassan, and Gutowski, V. S. Effects of corona discharge and UV treatment on the properties of jute-fiber expoxy composites. *Composites Science and Technology*, 60, 2000.
23. Koronis and Silva, A. (2018). *Green Composites for Automotive Applications*.
24. Maqsood, S., Bashir, U., Wiener, J., Puchalski, M., Sztajnowski, S., and Militky, J. Ozone treatment of jute fibers. *Cellulose*, 24, 2017.
25. Botaro, V. R., Dos Santos, C. G., Arantes Júnior, G., and Da Costa, A. R. Chemical modification of lignocellulosic materials by irradiation with Nd-YAG pulsed laser. *Applied Surface Science*, 183, 2001.
26. Ochi, S., Takagi, H., and Niki, R. (2002). Mechanical properties of heat-treated natural fibers. in *High Performance Structures and Materials*.
27. Cao, Y., Sakamoto, S., and Goda, K. (2007). Effects of heat and alkali treatments on mechanical properties of kenaf fibers. in *ICCM International Conferences on Composite Materials*.
28. Perincek, S., Bahtiyari, M. I., Körlü, A. E., and Duran, K., Ozone bleaching of jute fabrics. *AATCC Review*, 7, 3, 34–39, 2007.
29. Zamam, U., Khan, A. H., Hossain, M. A., Khan, M. A., and Khan, R. A. Mechanical and electrical properties of jute fabrics reinforced polyethylene/polypropylene composites: Role of gamma radiation. *Polymer-Plastics Technology and Engineering*, 48, 7, 760–766, 2009.
30. Rahman, G. M. S., Al Mamun, M. A., and Khan, M. A. Effect of γ (Gamma)-radiation on the physico-mechanical properties of grafted jute fabric reinforced polypropylene (PP) composites. *Fibers and Polymers*, 15, 2, 340–346, 2014.
31. Beg, M. D. H., and Pickering, K. L. Mechanical performance of Kraft fiber reinforced polypropylene composites: Influence of fiber length, fiber beating and hygrothermal ageing. *Composites Part A: Applied Science and Manufacturing*, 39, 2008.
32. Kurien, R. A., Selvaraj, D. P., Sekar, M., and Koshy, C. P. (2020, June). Green composite materials for green technology in the automotive industry. In *IOP Conference Series: Materials Science and Engineering* (Vol. 872, No. 1, p. 012064). IOP Publishing. https://doi.org/10.1088/1757-899X/872/1/012064
33. Kurien, R. A., Selvaraj, D. P., and Koshy, C. P. Worn surface morphological characterization of NaOH-treated chopped abaca fiber reinforced epoxy composites. *Journal of Bio-and Tribo-Corrosion*, 7(1), 1–8. 2021. https://doi.org/10.1007/s40735-020-00467-3

34. Kurien, R. A., Biju, A., Raj, K. A., Chacko, A., Joseph, B., and Koshy, C. P. Chicken feather fiber reinforced composites for sustainable applications. *Materials Today: Proceedings*, 2021. https://doi.org/10.1016/j.matpr.2021.10.400
35. Kurien, R. A., Selvaraj, D. P., Sekar, M., Rajasekar, R., and Koshy, C. P. (2021). Experimental investigation on tribological characteristics of NaOH treated chopped abaca fiber reinforced epoxy composites. In *Materials Science Forum* (Vol. 1019, pp. 25–31). Trans Tech Publications Ltd. https://doi.org/10.4028/www.scientific.net/MSF.1019.25
36. Kurien, R. A., Selvaraj, D. P., Sekar, M., Koshy, C. P., and Tijo, D. (2021). Mechanical characterization and evaluation of NaOH treated chopped abaca fiber reinforced epoxy composites. In *Materials Science Forum* (Vol. 1019, pp. 12–18). Trans Tech Publications Ltd. https://doi.org/10.4028/www.scientific.net/MSF.1019.12
37. Kurien, R. A., Selvaraj, D. P., Sekar, M., Koshy, C. P., and Praveen, K. M. Comparative mechanical, tribological and morphological properties of epoxy resin composites reinforced with multi-walled carbon nanotubes. *Arabian Journal for Science and Engineering*, 1–9. 2021. https://doi.org/10.1007/s13369-021-05984-y
38. Kurien, R. A., Santhosh, A., Paul, D., Kurup, G. B., and Reji, G. S. (2020, November). A review on recent developments in kenaf, sisal, pineapple, bamboo and banana fiber-reinforced composites. In *International Conference on Advances in Materials Processing & Manufacturing Applications* (pp. 301–310). Springer, Singapore. https://doi.org/10.1007/978-981-16-0909-1_30
39. Kurien, R. A., Santhosh, A., Paul, D., Kurup, G. B., Reji, G. S., and Selvaraj, D. P. (2020, November). A study on recent developments in jute, cotton, coir, silk and abaca fiber-reinforced composites. In *International Conference on Advances in Materials Processing & Manufacturing Applications* (pp. 375–384). Springer, Singapore. https://doi.org/10.1007/978-981-16-0909-1_38
40. Kurien, R. A., Koshy, C. P., Santhosh, A., Kurup, G. B., Paul, D., and Reji, G. S.. A study on vetiver fiber and lemongrass fiber reinforced composites. *Materials Today: Proceedings*, 68, 2640–2645. 2022. https://doi.org/10.1016/j.matpr.2022.09.563
41. Kurien, R. A., Anil, M. M., Mohan, S. S., and Thomas, J. A. (2023). Natural fiber composites as sustainable resources for emerging applications-a review. *Materials Today: Proceedings*. https://doi.org/10.1016/j.matpr.2023.04.363
42. Kurien, R. A., Biju, A., Raj, A. K., Chacko, A., Joseph, B., Koshy, C. P., and Paul, C.. Comparative mechanical properties of duck feather–jute fiber reinforced hybrid composites. *Transactions of the Indian Institute of Metals*, 1–6. 2023. https://doi.org/10.1007/s12666-023-03015-y
43. Torres, F. G., and Cubillas, M. L. Study of the interfacial properties of natural fiber reinforced polyethylene. *Polymer Testing*, 24. 2005.
44. Kim, S. J., Moon, J. B., Kim, G. H., and Ha, C. S. Mechanical properties of polypropylene/natural fiber composites: Comparison of wood fiber and cotton fiber. *Polymer Testing*, 27. 2008.
45. Nagarajappa, G. B., and Pandey, K. K. UV resistance and dimensional stability of wood modified with isopropenyl acetate. *Journal of Photochemistry and Photobiology B: Biology*, 155. 2016.
46. Jebrane, Pichavant, F., and Sbe, G. A comparative study on the acetylation of wood by reaction with vinyl acetate and acetic anhydride. *Carbohydrate Polymers*, 83. 2011.
47. Azeh, Y., Olatunji, G. A., and Mamza, P. A. Scanning electron microscopy and kinetic studies of ketene-acetylated wood/cellulose high-density polyethylene blends. *International Journal of Carbohydrate Chemistry*, 2012, 1–7. 2012.
48. Marson, G. A., and El Seoud, O. A. Novel, efficient procedure for acylation of cellulose under homogeneous solution conditions. *Journal of Applied Polymer Science*, 74. 1999.
49. Hill, C. A. S. (2006). *Wood Modification: Chemical, Thermal and Other Processes.*
50. Gulati, D., and Sain, M.. Fungal-modification of natural fibers: A novel method of treating natural fibers for composite reinforcement. *Journal of Polymers and the Environment*, 14, 347–352. 2006.
51. Kurien, R. A., Arshad, A., Joseph, A., Sunil, A., Cherian, B. T., Rangappa, S. M., … and Siengchin, S. Agave-jute fiber–reinforced hybrid composite for lightweight applications: Effect of hybridization. *Biomass Conversion and Biorefinery*, 1–14. 2024. https://doi.org/10.1007/s13399-024-05984-6

14 Naturally Derived Materials as Reinforcements to MMCs

Particulates

Merin K. Eapen and Arya Muraleedharan

14.1 INTRODUCTION

Metal matrix composites (MMCs) are being used more frequently for different kinds of applications in engineering and technology, such as sporting equipment, goods, and parts for cars. Unlike typical alloys, MMCs have higher stiffness and wear resistance, lower expansion during heating, lower thermal conductivity, and better damping qualities, and these qualities make them appealing candidates for technological applications.[1] MMCs are made up of two parts. A metal matrix is often a continuous phase of metals and alloys made of aluminium (Al), magnesium (Mg), and other elements that are reinforced with fibres, particles, and whiskers collectively known as reinforcement materials. The matrix is regarded as impure since it is an alloy. While reinforced elements give mechanical properties, the matrix is supported by them, enabling it to adopt the proper form. The novel composite material performs stronger than monolithic material when properly made. In composite materials, reinforcement is provided by increasing the base weight percentage or volume percentage, which significantly alters the material's tribological and mechanical properties.[2] By changing the matrix material with reinforcement materials and manufacturing methods, MMCs may be customized. Due to its versatility, engineers and materials scientists may create MMCs that are tailored to a variety of different industries and applications.[3] Copper, nickel, and steel-based MMCs are particularly used for manufacturing machines, electronic packaging, and industrial purposes. Aluminium, titanium, or magnesium-based MMCs are found to be good for the production of lightweight structures.[4] In earlier days, oxides, nitrides, and carbides were used as reinforcement materials, and over the past 20 years, several processing approaches have been developed by improving the structure and features of particulate-reinforced MMCs.[3] Nowadays, the production characteristics of particulate-enhanced MMCs have gained significant relevance due to their economic viability.[5] However, some of the issues prevent them from being widely used in sectors where cost-effectiveness, ease of manufacture, or easily accessible resources are crucial considerations. A sustainable and ecologically conscious method of material development is the use of green waste products from plants as reinforcement materials in composites made of Al alloy. This strategy is an appealing choice for numerous applications exploring more sustainable material alternatives since it may increase the properties of the material while reducing environmental effects.[6] Naturally formed waste products are found to be abundant sources of silicon and carbon. Hence, they are considered as potential reinforcement materials or fillers in MMCs. They have many benefits over other reinforcements, such as affordability, portability, abundance, and renewability. These biomaterials are ideal for the production of lightweight, high-strength AMCs since they have a low density and comprise ceramic particles like silica (hard ceramic), among others.[7] Aside from the promising uses being investigated in various areas, such as energy production, some agricultural and industrial wastes (fly ash, rice-hull ash, etc.) have exceptional physical and mechanical properties that place them at the forefront for the potential manufacture of smart materials. MMCs have grown beyond

DOI: 10.1201/9781032665337-14

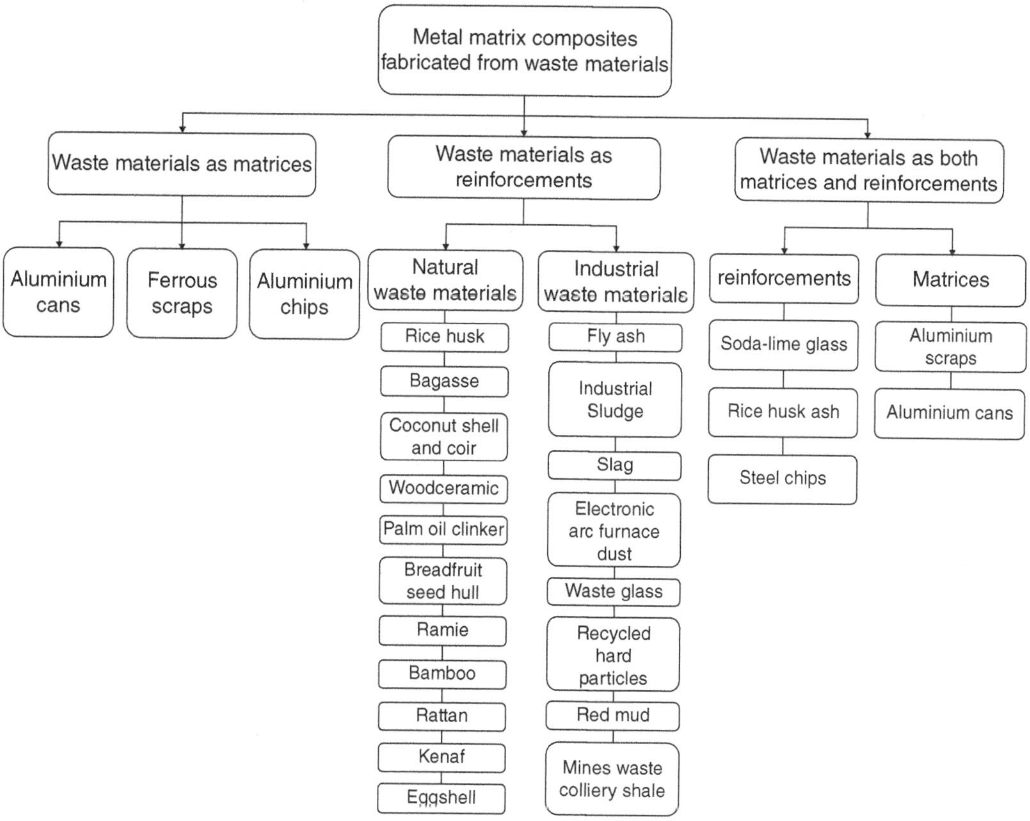

FIGURE 14.1 Classification of metal matrix composites synthesized from waste materials.[8]

their initial use, which focused solely on utilizing their structural or mechanical capabilities, in a variety of ways, giving rise to new ideas and even their definition (Figure 14.1).[9]

14.2 CLASSIFICATION OF MMCs

MMCs can be classified into three types based on reinforcements, such as fibre-reinforced MMCs, multi-layered laminates, and particulate-reinforced MMCs (Figure 14.2).[4]

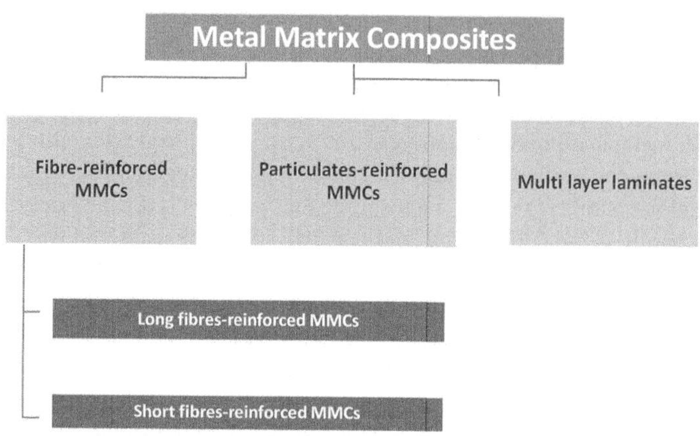

FIGURE 14.2 Classification of metal matrix composites.[10]

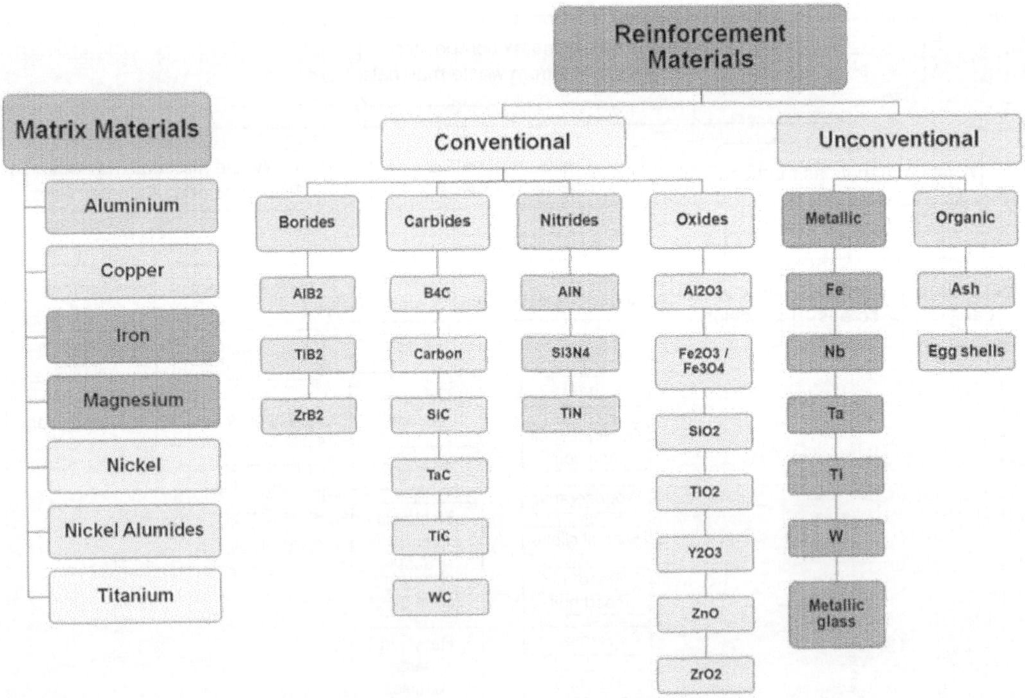

FIGURE 14.3 Usage of matrix and reinforcement materials in MMCs.[10]

14.3 PARTICULATE-REINFORCED MMCs

Particulate-reinforced MMCs are metal matrices reinforced with particulates that are dispersed in the metal matrix randomly or in a particularly defined order. The resultant composite is called a nanocomposite when one of the particle reinforcements' diameter is smaller than 100 nm. The low volume dispersion of hard and strong reinforcing phases at nano-length scale greatly adds to the strengthening of the matrix material, which is why MMCs with nanoscale reinforcements are gaining enormous interest lately. The effective dispersion of nanoscale reinforcement, as opposed to the addition of micron and submicron-length scale reinforcement, increases the strength of the matrix material without negatively influencing its ductility.[4] The importance of particulate-reinforced MMCs lies in their isotropic qualities, low cost, and ease of manufacture (Figure 14.3).[11]

14.4 BAMBOO LEAF ASH

Bamboo is a readily available natural resource that has several advantages for use as reinforcement in composites with metal and polymer matrices. It is simply produced after cutting and lightens the impact on the environment compared to cotton jute, which has excellent properties.[12] Every portion of the bamboo plant has a utility, making it a resource that may be used in a variety of ways. Bamboo culms are used in building and furnishings, and bamboo sticks are used to make sticks, fences, and agarbati. Moreover, the green leaves and roots of bamboo have antioxidant properties that are employed in medicine. However, dry bamboo leaves are often disposed of by burning or landfilling, which harms the environment.[13] BLA particles have low density and low cost and are employed in MMCs like aluminium composites when done with a stir casting process.[14] The chemical constituents/elements contained in the bamboo leaf are silica, aluminium oxide, calcium oxide, potassium oxide, haematite, and titanium oxide.[2] The presence of CaO and high silica concentration is reported in bamboo leaf ash. Silica is renowned for being tough and durable, and CaO adds self-lubricating qualities. Thus, BLA can be used as an innovative, ecologically friendly reinforcement material with outstanding features due to certain characteristics. BLA enriched with silica is being used as a reinforcement material for MMCs, either alone or in combination with other ceramics. It produces cost-effective green MMCs with improved mechanical and tribological qualities. The strong interfacial bonding and uniform reinforcement distribution aid the matrix reinforced with silica-rich BLA in enhancing its mechanical and tribological properties (compared to its base material). The availability of BLA and cheap production costs facilitates the fabrication of inexpensive BLA-reinforced MMCs.[15] However, it has been shown that when the weight percentage of BLA increases, the hardness of the composite diminishes, ultimate tensile strength and percentage elongation too varies.[16] The presence of silica, the primary component of BLA, is the reason for the small loss in hardness. Since silica is a ceramic that is known to be softer than alumina, adding more BLA will inevitably cause the hybrid composite's hardness to decrease. According to data when replacing the expensive alumina with the low-cost agricultural waste BLA, the hybrid composite's hardness decreased by less than 9% while its alumina content was reduced by 40%. Thus, the hybrid composite's hardness qualities will not be significantly affected by the inclusion of BLA. As the BLA content rises, tensile strength falls. In contrast to alumina, which has an elastic modulus of 250GPa, silica is softer and has an elastic modulus that is similar to that of aluminium (60 to 70GPa). As a result, the direct strengthening effect often anticipated from the transfer of stress from the metal matrix to the purportedly tougher and stiffer particles would be slightly less. However, it is reported that there is a loss in specific strength with a rise in BLA content. In the hybrid composite containing 4 wt% BLA, the specific strength variation is less than 9% when the reduction rates are taken into account. In comparison to the maximum 14.3% lower ultimate tensile strength (UTS) value attained for a 40% reduction in the concentration of alumina in the composite, this is a minor improvement. This demonstrates that inexpensive BLA agricultural waste can be used as a complementary material for the creation of alumina-reinforced Al-based MMCs while still achieving equivalent strength-to-weight ratios. It has been reported that the hybrid composites, including BLA, had better fracture toughness than the composites enhanced alone with alumina. This shows that the fracture toughness of the Al-based composite is not compromised by using BLA as a complementary reinforcement to alumina. The decreased number of substantially tougher and more brittle alumina particles in the hybrid composites is probably what accounts for the enhanced fracture toughness. Similar to the majority of hard and brittle ceramic particles, alumina particles are more likely to experience fast fracture propagation.[17] Nevertheless, when BLA is used as a material for reinforcement in an aluminium metal matrix, an increment in hardness, tensile strength, and yield strength of the MMC was found with an increase in BLA content. In terms of elongation, the matrix of the composite was less ductile when compared with the base metal matrix. According to studies, the composite fracture mode shows both macroscopically brittle and microscopically ductile nature.[2]

14.5 RISE HUSK ASH

The rice husk, also known as rice hull, is a component of the rice chaff found in paddy fields. The rice husk often has a high concentration of fibrous components, making it indigestible to the human immune system.[18] When rice is parboiled, the husk is burnt by rice mills to produce steam. The husk then burns down to form ash, which is a more affordable kind of reinforcement than the ceramic particles that are often used. It helps in reducing greenhouse gas emissions by absorbing carbon dioxide for its growth and releasing the same carbon to the environment while burning the rice husk. Its carcinogenic and bio-accumulative effects have led to fatigue, asthma, and loss of appetite. Rice husk in the environment has drawn a lot of criticism.[19] Rice husk is a highly prospective contender among many biomasses with abundant renewable energy sources. It is being used for the production of large-scale goods like SO_2, xylose, and activated carbon. Many industries, including paint, cement pigment, rubber fillers, and MMCs, employ rice husk ash for a variety of purposes.[12] The rice husk is made up of substances such as lignin and opaline silica. These components can be extracted and utilized for various purposes. It has been reported that silica obtained from rice husk has demonstrated tremendous promise in terms of low coefficient of friction, better corrosion and wear resistance, enhanced hardness, high strength, and great machinability.[18] Mechanical characteristics like the hardness of the composite, high tensile strength, and very high fracture toughness, even at lower temperatures, ensure application in aerospace. Notably, nano-sized silicon carbide (SiC) obtained from rice husk shows better properties like flatness, material removal rate, and tool wear rate than micro SiC.[20] The MM composite developed shows homogenous dispersion of rice husk ash particles according to a microstructure investigation. Additionally, it demonstrates that when rice husk ash content rises, toughness as determined by the impact test increases. Incorporating rice husk ash as reinforcement in composites is less dense and more cost-effective and also decreases the carbon footprint.[21] A study that investigated the addition of commercial graphite and nano-silica derived from rice husks to an aluminium matrix found that the composites' hardness, yield strength, and thermal conductivity all significantly increased at varying compositional proportions. Additionally, the composites' linear thermal expansion coefficient decreased. Due to the SiO2 nanoparticles in the hybrid composites forming a strong interface, the Al-matrix hybrid composites demonstrated increased strength. Silica was crucial to the process of densification. The study concludes that Al-matrix hybrid nanocomposites' mechanical characteristics may be improved by adding reinforcement in the form of silica nanoparticles.[18] Hardness is a crucial characteristic that affects how quickly composite materials wear out. A matrix material and a reinforcing material often make up a composite material. Both of these components have varying degrees of hardness, which has a big impact on how quickly the composite wears out. Higher-hardness composite materials, on the other hand, typically show a reduced wear rate. This is because less material is lost since the tougher material can resist the mechanical forces that produce wear more effectively. An increment in the wear resistance of the composite has been reported after the addition of ash particles to the aluminium alloy. The existence of hard phases, specifically the SiO_2 found in rice husk ash, is the cause of increasing the wear rate.[22]

14.6 SUGARCANE BAGASSE ASH

Sugarcane bagasse (SCB), a fibrous byproduct left over from crushing and extracting the juice from sugarcane, is a major waste product of the agricultural industry.[23] After the sugar companies burn the SCB, they produce sugarcane bagasse ash (SCBA). Bagasse is a matted cellulose fibre byproduct of sugarcane processing used in the production of sugar. Both SiC and quartz (SiO_2) are present in significant amounts in the ash.[24] The literature demonstrates the adaptability of sugarcane waste utilization through its conversion to paper, feedstock, and biofuel. According to an examination of SCB, its primary constituents include cellulose, wax, ash, hemicellulose, and lignin. For the objective of developing novel materials with distinct physical and chemical properties, SCB's mix of qualities makes it an excellent component to be used as a reinforcing fibre in composite materials.[23] In literature, it has been found that the majority of SCBA particles have a diameter between 10 and 100 microns and an average size of 34.79 μm. The main crystalline phases present in SCBA are cristobalite and tridymite.[25] The SCBA can be incorporated with other materials like SiC for reinforcement purposes, that is, an increment in tensile strength, hardness, and impact strength of the composite material due to an increase in the presence of SiC, but there is a decrease noted in the elongation of the material because the metal matrix and the reinforcement particles were well bonded together by the tiny grains of SCBA and SiC particles.[26] Regarding the hardness of the composite with SCBA, an increase in the strain in the periphery of the particles in the matrix occurs due to the addition of SCBA particles in the matrix alloy, more specifically, due to the dislocation at the particle boundary by diffraction of light.[25] It has been noted that SCBA shortens the time required for composites to solidify. Al alloys reinforced with ash will increase the strength of the composites because of the improved interfacial bonding. According to the findings, increasing the weight percentage of ash in the composite lowers the rate of wear, while increasing the applied load raises the rate of wear. The rate of wear of the composites rises with a rise in load when sliding at a constant speed. The manufactured composites have superior wear resistance to the master alloy and a slower rate of wear than the master alloy. The particles of ash primarily function as an efficient barrier that lowers the wear rate of the composites as the applied load increases. The uniform distribution of particles of ash in the Al alloy, which denotes good interfacial interaction between the aluminium alloy and the particles of ash, may also enhance the tensile strength of the composites. It was found that the mechanical characteristics of the particulate-reinforced aluminium matrix composites were superior to those of monolithic alloys. It is observed that hardness is also improved. This increase is the result of using ash particles with finer grain sizes and increasing surface area. The uniformly distributed ash particles in a composite offered a higher resistance to plastic deformation, designed to raise the hardness value of the composites.[27]

14.7 BANANA FIBRE

Natural fibres provide significant benefits such as low density, suitable mechanical qualities, high dispersibility, and renewability. They are also biodegradable and recyclable. The ability to adjust the material's mechanical characteristics to a specific application qualifies the fibre-reinforced composite materials to create unique parts. According to the desired outcome, the material may be engineered to display isotropic or highly anisotropic qualities by altering the orientation or location of the fibres.[28] The banana is a plant that is predominantly grown for its fruit, notably in tropical regions of the world. Nearly 60% of the biomass, including the pseudostems, is wasted after harvest. Banana fibre (BF) might be used to make a variety of goods due to its advantageous mechanical and chemical properties when extracted from various portions of the crop. Typically made of helically woven cellulose, micro-fibrils in an amorphous matrix of lignin and BFs are a kind of cellulose. In addition, cellulose content affects mechanical characteristics due to the increased amount of cellulose and small microfibril angle. The significant rise in the mechanical features of composites enhanced with BF is due to the presence of a small amount of cellulose nanofibres from bananas.[29] Similar to traditional materials like glass fibres, BFs have strong specific strength characteristics. Treating the BF helps in eliminating contaminants from the fibre, reducing moisture sorption, permitting mechanical bonding, and consequently increasing matrix reinforcement interaction.[30] In addition to having key qualities like low densities, lightweight, strong tensile strength, and fibre and water resistance, BFs are also environmentally benign. Consequently, these types of trash have a higher likelihood of being used for various activities, including construction and production of building materials.[29]

14.8 LEMON GRASS ASH

Lemon grass is a rich source of lignocellulose, which is made of cellulose, hemicellulose, and lignin with varying percentages. It has also been utilized to make bioenergy and silica and introduced as a reinforcement material in composites.[31] The inclusion of this reinforcement into the alloy of aluminium during a semi-solid stage improved the wettability property, and it was discovered that the reinforced alloy shows thermodynamic stability with the lemon grass ash (LGA) particles producing no intermetallic compounds with the base alloy matrix. The tensile and hardness characteristics of the cast composites were enhanced by adding LGA particles to the Al metal matrix.[32] Both the compressive and tensile strengths of the LGA-reinforced composite were high. The compressive and tensile strengths improved for the composite with LGA content. By acting as load-bearing components, the LGA particles decreased the wear rate and increased content.[33]

14.9 MUSSEL SHELL POWDER

The mussel shell is a waste byproduct of aquaculture. It poses a serious hazard to the ecosystem and may harm the surrounding land and ecology. The mussel shell could be used most effectively when it was processed into powder under controlled circumstances. Increases in temperature, acid concentration, and the effect of mussel shell powder (MSP) are the main factors that affect the corrosion of the composite material.[34] The procedures to achieve MSP with a consistent size include cleaning and washing, followed by drying and grinding. It has been reported that mussel is a reinforcer that is capable of enhancing the mechanical properties and wear rate of the composite. Research reports show that the tensile strength, elongation characteristics, and impact toughness were improved at various MSP compositions.[35] Incorporating MSP into an aluminium-based metal matrix can show improvement in the wear rate behaviour of the composite, along with a rise in applied stress and a drop in percentage weight in MSP. Due to the high wear resistance of the composite, it can be used in automobile applications as rotors, pistons, and brake material.[36] MSP has been considered as a possible reinforcer as it enhanced the mechanical and wear rate behaviours. Due to the greater particle size of MSP, the hardness value reduces as the weight percentage of MSP increases.[35]

14.10 PALM KERNEL SHELL ASH

Palm kernel shell (PKS) contains hemicellulose, cellulose, and lignin. As an alternative to direct combustion, PKS is useful for various applications. PKS provides a good heating value as an energy source but produces smoke during combustion. It can be used as a sorbent material for hydrophobic pollutants or as a filler in hydrophobic materials. PSKs can be used for wastewater treatment and have applications in the food industry.[37] The ash of palm kernel shells (PKSA) is a solid waste from the coconut processing industry, and it is one of the most affordable and low-density reinforcing materials accessible in significant quantities. It has been considered a cost-effective reinforcing material for applications in automobiles and small engine industries. It is anticipated that the addition of PKSA particles to aluminium alloy may encourage the development of another application for this inexpensive waste byproduct.[38] The literature that is now accessible discusses the undergrounded shell particles of PKS, which must be ground into tiny particles before being used in brake lining. PKS has a coefficient of friction on metal surfaces that ranges from 0.37 to 0.52. When utilizing brake lining material, however, a friction coefficient in the range of 0.30 to 0.70 is often preferred. According to research, adding PKS to the mix for making structural lightweight concrete enhanced the mechanical strength, because they are subjected to strong and fluctuating breaking forces. PKS therefore seemed suited for use as the basis material in friction composites. When exposed to an organic solvent, PKS did not change considerably in terms of the weight of the physical structure over a sizable period.[39]

14.11 SEASHELL

Seashells are abundant with calcium carbonates and other macromolecules (proteins, polysaccharides, etc.). It has excellent mechanical properties. These shells mostly consist of crystalline polymorphs of calcite and aragonite. Due to their microstructure, seashells show superior stiffness, strength, hardness, water absorption, crushing strength, and tensile strength. The compressive strength acting on the outer layer of the seashell balances the tensile stress in the inner layer of the seashell. Also, the seashells contain protein up to 5%, which is responsible for the fracture toughness and is comparable to ordinary glass.[40] The microstructural analysis in the periwinkle shell showed particles spread evenly throughout in aluminium alloy. As a result, the strength and hardness of the composite were enhanced because of the capacity of powdered periwinkle shells to refine the alloy grains. The composite may be employed in applications that call for less weight and greater strength.[41] The mechanical characteristics of aluminium alloy composite with seashell reinforcement were greater than those reinforced with alumina. Bio-based seashell powders may be a viable and better replacement material for synthetic ceramics. There was a percentage increase in the values of coefficient of friction and wear rate than those which are unreinforced, compared with the bio-based seashell reinforced composites.[42] The incorporation of seashell particles into aluminium alloy improves its properties such as hardness, density, impact strength, and tensile strength. This composite material may be used in the automotive and aerospace sectors to create lighter, more efficient components. Depending on the application, the weight percentage of the powder varies and it has an impact on the fabricated composite.[43]

14.12 GROUNDNUT SHELL ASH

Commercially groundnut shells are used in feedstock, food, and filler in fertilizers because they contain a variety of bioactive and functional components that are advantageous for humans. However, the majority of abandoned groundnut shells are burnt and buried, which causes environmental pollution. New technologies must be developed to achieve production with no waste and direct these unused resources into meaningful use in biodiesel production, building materials, paper production, and carbon nano-sheet formation.[44] It was found that the microhardness and compression strength of cobalt-chromium composite were increased by reinforcing with groundnut shell ash (GSA), and because of the lower density of GSA, the density of the composite was found to decrease. A high resistance to wear behaviour and improvement in corrosion resistance have also been discovered.[45] In addition to GSA, the hardness of hybrid composites was enhanced, and it decreased with further addition of different percentages of GSA but not less than the hardness of metal alloy. The torsional strength also increases in hybrid material, but when GSA decreases further, the same tendency is observed for impact strength in the hybrid material.[46] According to the findings of experiments performed on aluminium hybrid composites, the addition of GSA causes a slight decrease in the hardness, specific strength, and UTS of the composite. Whereas the fracture toughness improved with the rise of GSA content, the percentage elongation only slightly improved and was largely unchanged. Overall, it can be said that GSA had a favourable effect on the mechanical characteristics of aluminium hybrid composites[47] when compared to bamboo leaf ash and rice husk ash.

14.13 EGGSHELL

Waste composed of chicken eggshells (ESs) is an industrial byproduct, and its disposal poses a serious environmental issue. A brand-new engineering reinforcement, eggshell (ES) has remarkable qualities, including low density, renewable resources, environmental friendliness, and high thermal stability.[48] Previous research has shown that chicken ES, an aviculture waste, is one of the world's biggest environmental concerns, particularly in nations with a thriving egg product business. About 95% of ES is composed of calcium carbonate ($CaCO_3$) in the form of calcite, while the remaining 5% comprises organic substances such as proteins, type X collagen, and sulphated polysaccharides. ES has a considerably lower density than the mineral $CaCO_3$, according to the report, among other things. A biomaterial called an ES is composed of 95% calcium carbonate in the form of calcite and 5% organic compounds such as Al_2O_3, MnO, S, SiO_2, Cl, Cr_2O_3, and P.[49] According to research done on aluminium metal matrix, the presence of ES powder enhances the metal matrix's tensile strength, while also reducing the composite's hardness and raising its compression strength. It is discovered that the composite that contains 7% ES powder has more advantageous characteristics that make it appropriate for enhanced tensile strength and improved compression properties, making it suitable for a variety of applications in various sectors.[50] In comparison to carbonized ES reinforcing material in magnesium matrix, adding 6 wt% of uncarbonized ES particles as reinforcement enhances the matrix hardness, tensile strength, yield strength, and compressive properties of the material. According to the findings, a carbonized ES at 6 wt% improved the hardness and rate of corrosion compared to an ES that strengthened in a magnesium matrix composite that had not been carbonized (Tables 14.1 and 14.2).[56]

TABLE 14.1
Nominal Chemical Composition of Agro-Wastes[51]

Agro-Waste		Chemical Composition (%)									
		Al_2O_3	CaO	Fe_2O_3	K_2O	MgO	Na_2O	SiO_2	SO_3	P_2O_5	Li_2O
Rice husk ash	Before treatment	4.10	0.34	0.64	2.15	0.64	0.15	91.6	–	–	–
	After treatment	3.06	0.56	0.15	2.67	0.73	0.36	91.6	–	–	–
Fly ash	Before treatment	29.60	0.10	0.72	3.53	0.34	–	64.6	–	–	–
	After treatment	25.60	0.10	0.69	3.14	0.56	–	69.5	–	–	–
Palm oil clinkers		3.50	2.30	5.18	4.66	1.24	81.8	0.76	–	–	–
Palm oil fuel ash		5.45	7.50	–	5.30	3.93	–	49.20	1.73	6.41	13.85
Coconut shell ash		5.45	0.57	12.40	0.52	16.2	0.45	45.05	–	–	–

TABLE 14.2
Agricultural and Animal Reinforcement Materials and Corresponding Challenges[52–55]

Sl. No	Reinforcement Material	Chemical Composition (%)	Challenges
1	Rise husk	SiO_2 – 93.4 Al_2O_3 – 0.05 Fe_2O_3 – 0.06 CaO – 0.31 MgO – 0.35 K_2O – 1.4 Na_2O – 0.1 P_2O_5 – 0.8	Tensile strength and hardness increase, but the density decreases due to the presence of porosity.
2	Bamboo ash	SiO_2 – 75.90 Al_2O_3 – 4.13 F_2O_3 – 1.22 CaO – 7.47 MgO – 1.85 K_2O – 5.62 MnO_2 – 0.016 P_2O_5 – 3.22	Decrease in hardness and tensile strength is observed.
3	Eggshell	$CaCO_3$ – 94-97 Mg – 0.2–1.0 $Ca_3(PO_4)2$– 0.2–1.0 Na, Fe, Cu, Mn, K – 0.1 Organic materials – 2–3.3	Hardness and tensile strength improve, but impact toughness reduces.

14.14 CHALLENGES OF AGRICULTURAL REINFORCEMENT MATERIALS

When using liquid-state or semi-liquid-state manufacturing processes, wettability and the formation of solid-liquid interfaces are the major problems. Another significant problem is the dispersion of agricultural waste particulates with aluminium matrix material. The wetting activity between the metal matrix and the agro-based reinforcements determines the strength of the reinforced composites. An increment in the wear rate and reduction in hardness, elongation at break, and tensile strength have also been observed in some studies.[57]

14.15 SUMMARY

Researchers are paying close attention to MMCs because of their prospective applications for them and their excellent corrosion and oxidation resistance. The most popular matrix materials, however, are magnesium and aluminium. This is a result of its excellent mechanical features, superior corrosion resistance, and low density. The development of Al and its alloys has received the majority of attention from researchers and scientists. These days, agro-waste and other bio-based products have been exploited as reinforcement materials for MMCs by numerous researchers. The naturally derived reinforcement materials have various merits, such as cost-effectiveness, biocompatibility, biodegradability, and non-toxicity. These reinforcing materials play a key role in enhancing the mechanical, corrosion, and wear resistance properties of metal matrices. Nevertheless, the individual addition of agro-waste decreases the mechanical and tribological characteristics of a metal. Only a very few and limited research studies have been conducted so far in this area. It is needed to conduct extensive research on this topic to fully exploit the features of bio-based materials. Since the usage of agro-waste and industrial waste reduces environmental pollution, the strengthening of MMCs by using bio-inspired materials may lead to a sustainable future.

REFERENCES

1. Sharma, S. C. Equation for the Density of Particle-Reinforced Metal Matrix Composites: A New Approach. *J. Mater. Eng. Perform.* 2003, *12* (3), 324–330. https://doi.org/10.1361/105994903770343187
2. Chen, X.-G.; da Silva, M.; Gougeon, P.; St-Georges, L. Microstructure and Mechanical Properties of Friction Stir Welded AA6063–B4C Metal Matrix Composites. *Mater. Sci. Eng. A* 2009, *518* (1), 174–184. https://doi.org/10.1016/j.msea.2009.04.052
3. Ibrahim, I. A.; Mohamed, F. A.; Lavernia, E. J. Particulate Reinforced Metal Matrix Composites—A Review. *J. Mater. Sci.* 1991, *26* (5), 1137–1156. https://doi.org/10.1007/BF00544448
4. Seetharaman, S.; Gupta, M. Fundamentals of Metal Matrix Composites. In *Reference Module in Materials Science and Materials Engineering*, 2021. https://doi.org/10.1016/B978-0-12-803581-8.00001-1
5. Saxena, A.; Saxena, K. K.; Jain, V. K.; Rajput, S. K.; Pathak, B. N. A Review of Reinforcements and Process Parameters for Powder Metallurgy-Processed Metal Matrix Composites. *Mater Today Proc.* 2023. https://doi.org/10.1016/j.matpr.2023.02.227
6. Nnakwo, K.; Odo, J.; Ijomah, A. Reinforcing Potential of Irvingia Wombolu Shell Particulates in Al-2Nb-0.5Si Eco-Composite. *Int. J. Novel Res. Civil Struct. Earth Sci.* 2023, *10*, 8–14. https://doi.org/10.5281/zenodo.8344998
7. Kamal, T.; Siddiqui, M. A. Evaluation of Mustard Husk Ash as Reinforcement for Aluminium Matrix Composite. *Mater Res Express* 2018, *5* (8), 086509. https://doi.org/10.1088/2053-1591/aad0e7
8. Bahrami, A.; Soltani, N.; Pech-Canul, M. I.; Gutiérrez, C. Development of metal-matrix composites from industrial/agricultural waste materials and their derivatives. *Crit. Rev. Environ. Sci. Technol.* 2015, *46* (2), 143–208. https://doi.org/10.1080/10643389.2015.1077067
9. Bahrami, A.; Soltani, N.; Pech-Canul, M. I.; Gutiérrez, C. A. Development of Metal-Matrix Composites from Industrial/Agricultural Waste Materials and Their Derivatives. *Crit. Rev. Environ. Sci. Technol.* 2016, *46* (2), 143–208. https://doi.org/10.1080/10643389.2015.1077067
10. Sankaranarayanan, S. K. R. S.; Gupta, M. *Fundamentals of Metal Matrix Composites*. In Elsevier eBooks; 2021; pp 11–29. https://doi.org/10.1016/b978-0-12-819724-0.00001-x
11. Bajakke, P. A.; Malik, V. R.; Deshpande, A. S. Particulate Metal Matrix Composites and Their Fabrication via Friction Stir Processing—A Review. *Mater. Manuf. Process.* 2019, *34* (8), 833–881. https://doi.org/10.1080/10426914.2019.1605181
12. Kolli, M.; Mugada, K. K.; Prakash, N. *Chapter 8 Applications of Green Waste Composite*, 2023, pp 173–184. https://doi.org/10.1515/9783110766523-012
13. Sawarkar, A.; Shrimankar, D.; Kumar, A.; Kumar, A.; Singh, E.; Singh, L.; Kumar, S.; Kumar, R. Commercial Clustering of Sustainable Bamboo Species in India. *Ind. Crop. Prod.* 2020, *154*, 112693. https://doi.org/10.1016/j.indcrop.2020.112693
14. Rao, G.; Bannaravuri, P.; Birru, A. Effect on Corrosion Behaviour of the Surface of Aluminium 4.5 Cu with Bamboo Leaf Ash Composites by Laser Treatment. *Mater. Res. Express* 2020, *7*. https://doi.org/10.1088/2053-1591/ab6c9f

15. Nayak, B.; Jackson Singh, T.; Kumar Nayak, S. A Short Study on the Potential Use and Effect of Bamboo Leaf Ash as Reinforcement in MMCs. *Mater. Today Proc.* 2023. https://doi.org/10.1016/j.matpr.2023.09.066
16. Alaneme, K.; Ademilua, B. O.; Bodunrin, M. Mechanical Properties and Corrosion Behaviour of Aluminium Hybrid Composites Reinforced with Silicon Carbide and Bamboo Leaf Ash. *Tribol. Ind.* 2013, *35*, 25–35.
17. *(PDF) Mechanical Behaviour of Bamboo Leaf Ash Alumina Hybrid Reinforced Aluminium-based Matrix Composites.* https://www.researchgate.net/publication/284673100_Mechanical_behaviour_of_bamboo_leaf_ash_alumina_hybrid_reinforced_aluminium_based_matrix_composites (accessed 2023-10-07).
18. Saini, S.; Gupta, A.; Mehta, A. J.; Pramanik, S. Effect of Graphite and Rice Husk-Derived SiO_2 on Mechanical and Thermal Properties of Re-Sintered Aluminium-Based Metal Matrix Hybrid Nanocomposites. *J. Therm. Anal. Calorim.* 2023, *148* (6), 2335–2344. https://doi.org/10.1007/s10973-022-11897-1
19. Ranjan, A.; Kumar, H.; Nagdeve, L.; Mishra, A.; Gaurav, G. K.; Klemeš, J. J. Mechanical Properties of Rice Husk Ash, an Environmental Pollutant, Based Composites: A Step towards Sustainable Hybrid Composites. *Energy Sources Part Recovery Util. Environ. Eff.* 2022, *44* (4), 9584–9602. https://doi.org/10.1080/15567036.2022.2130476
20. Baghel, M.; Krishna, C.; Sundaramurthy, S. Development of Al-SiC Composite Material from Rice Husk and Its Parametric Assessment. *Mater. Res. Express* 2022, *9*. https://doi.org/10.1088/2053-1591/ac4a2e
21. Udoye, N. E.; Fayomi, O. S. I.; Inegbenebor, A. O. Fractography and Tensile Properties of AA6061 Aluminium Alloy/Rice Husk Ash Silicon Nanocomposite. *Int. J. Chem. Eng.* 2020, *2020*, e8818224. https://doi.org/10.1155/2020/8818224
22. Dwivedi, S. P.; Dwivedi, V. K.; Yadav, R. Tribological and Corrosion Behavior of Eggshell and Agro-Waste RHA Reinforced Al Alloy Composite Material with and without Ball-Milling. *J. Inst. Eng. India Ser. D* 2023. https://doi.org/10.1007/s40033-023-00492-9
23. Lancaster, L.; Lung, M. H.; Sujan, D. Utilization of Agro-Industrial Waste in Metal Matrix Composites: Towards Sustainability. *World Acad. Sci. Eng. Technol.* 2013, *73*.
24. Kandpal, B. C.; Johri, N.; Kumar, N.; Srivastava, A. Effect of Industrial/Agricultural Waste Materials as Reinforcement on Properties of Metal Matrix Composites. *Mater Today Proc* 2021, *46*, 10736–10740. https://doi.org/10.1016/j.matpr.2021.01.575
25. Aluminum Matrix Composite with Sugarcane Bagasse Ash as Reinforcement Material | *Revista Cubana de Física.* http://www.revistacubanadefisica.org/index.php/rcf/article/view/641 (accessed 2023-10-11).
26. Subramanian, S.; Arunachalam, B.; Nallasivam, K.; Pramanik, A. Investigations on Tribo-Mechanical Behaviour of Al-Si10-Mg/Sugarcane Bagasse Ash/SiC Hybrid Composites. *China Foundry* 2019, *16* (4), 277–284. https://doi.org/10.1007/s41230-019-8176-9
27. Shankar, S.; Balaji, A.; Kawin, N. Investigations on Mechanical and Tribological Properties of Al-Si10-Mg Alloy/Sugarcane Bagasse Ash Particulate Composites. *Part. Sci. Technol.* 2018, *36* (6), 762–770. https://doi.org/10.1080/02726351.2017.1301609
28. Chandramohan, D.; Marimuthu, K. Characterization of Natural Fibers and Their Application in Bone Grafting Substitutes. *Acta Bioeng. Biomech.* 2011, *13*.
29. Priyadarshana, R. W. I. B.; Kaliyadasa, P. E.; Ranawana, S. R. W. M. C. J. K.; Senarathna, K. G. C. Biowaste Management: Banana Fiber Utilization for Product Development. *J. Nat. Fibers* 2022, *19* (4), 1461–1471. https://doi.org/10.1080/15440478.2020.1776665
30. Pujari, S. Comparison of Jute and Banana Fiber Composites: A Review. *Int. J. Curr. Eng. Technol.* 2013, 2, 121–126. https://doi.org/10.14741/ijcet/spl.2.2014.22
31. Haque, A. N. M. A.; Remadevi, R.; Naebe, M. Lemongrass (Cymbopogon): A Review on Its Structure, Properties, Applications and Recent Developments. *Cellulose* 2018, *25* (10), 5455–5477. https://doi.org/10.1007/s10570-018-1965-2
32. Jose, J.; Christy, T. V.; Peter, P. E.; Feby, J. A.; George, A. J.; Joseph, J.; Chandra, R. G.; Benjie, N. M. Manufacture and Characterization of a Novel Agro-Waste Based Low Cost Metal Matrix Composite (MMC) by Compocasting. *Mater. Res. Express* 2018, *5* (6), 066530. https://doi.org/10.1088/2053-1591/aac803
33. Murugadoss, P.; Jeyaseelan, C. Utilization of Silicon from Lemongrass Ash Reinforcement with ADC 12 (Al-Si Alloy) Aluminium on Mechanical and Tribological Properties. *Silicon* 2023, *15* (3), 1413–1428. https://doi.org/10.1007/s12633-022-02119-2

34. Characterization and Corrosion Behaviours of Aluminium Alloy/Aquaculture Shell Powder Particulates in Acidic Environment | *SpringerLink*. https://doi.org/10.1007/s40735-020-00427-x (accessed 2023-10-13).
35. Suleiman, I. Y.; Kasim, A.; Sirajo, A.; Idawu, M.; Suleiman, Y.; Kasim, A.; Abdullahi, T.; Munir, Z.; Sirajo. Evaluation of Mechanical, Microstructures and Wear Behaviours of Aluminium Alloy Reinforced with Mussel Shell Powder for Automobile Applications. *J. Mech. Eng.* 2021, *67*. https://doi.org/10.5545/sv-jme.2020.6953
36. Mu'azu, K.; Sirajo, M. Z.; Aliyu, M. M.; Suleiman, I. Y.; Mohammed, A. T. Effect of Particulate Reinforcements at Different Loads on Wear Behaviour of Aluminium Alloy Reinforced with Aquaculture Waste. *J. Mater. Environ. Sci.* 2022, *13*.
37. Okoroigwe, E.; Saffron, C.; Kamdem, P. Characterization of Palm Kernel Shell for Materials Reinforcement and Water Treatment. *J. Chem. Eng. Mater. Sci.* 2014, *5*, 1–6. https://doi.org/10.5897/JCEMS2014.0172
38. Victor, G. A. *Abrasive Wear Behaviour of Al-Cu-Mg/Palm Kernel Shell Ash Particulate Composite.* 2017, No. 31.
39. Elakhame, Z. U.; Alhassan, O.; Samuel, A. Development and Production of Brake Pads from Palm Kernel Shell Composites. *Int. J. Sci. Eng. Res.* 2014, *5*, 10.
40. Vijaya Ramnath, B.; Jeykrishnan, J.; Ramakrishnan, G.; Barath, B.; Ejoelavendhan, E.; Arun Raghav, P. Sea Shells And Natural Fibres Composites: A Review. *Mater Today Proc* 2018, *5* (1, Part 1), 1846–1851. https://doi.org/10.1016/j.matpr.2017.11.284
41. Ibrahim, K. O. Particulate Mollusc Shells as Reinforcements for Aluminium Matrix Composites: A Review. *COVENANT J. Eng. Technol.* 2021, *5*.
42. Rajeshkumar, L.; Saravanakumar, R.; Balaji, D.; Bhuvaneswari, V. Synthesis and Characterization of Bioceramics Reinforced Aluminium Matrix Composites. *Arch. Metall. Mater.* 2022, 67(4), 1217–1226.
43. IJMPE *Influence of Bio-Shells Addition on the Mechanical Properties of Aluminium Alloy*. https://ijmpe.iraj.in/paper_detail.php?paper_id=16718 (accessed 2023-10-13).
44. Duc, P. A.; Dharanipriya, P.; Velmurugan, B. K.; Shanmugavadivu, M. Groundnut Shell-A Beneficial Bio-Waste. *Biocatal. Agric. Biotechnol.* 2019, *20*, 101206. https://doi.org/10.1016/j.bcab.2019.101206
45. *(PDF) Impact of Ground Nut Shell Ash on Cobalt-Chromium Metal Matrix Composites Synthesized using Powder Metallurgy Process*. https://www.researchgate.net/publication/353423465_Impact_of_Ground_Nut_Shell_Ash_on_Cobalt-Chromium_metal_matrix_composites_synthesized_using_Powder_metallurgy_process (accessed 2023-10-13).
46. Ravi Kumar, K.; Kiran, K.; Sreebalaji, V. S. Micro Structural Characteristics and Mechanical Behaviour of Aluminium Matrix Composites Reinforced with Titanium Carbide. *J. Alloys Compd.* 2017, *723*, 795–801. https://doi.org/10.1016/j.jallcom.2017.06.309
47. Alaneme, K. K.; Bodunrin, M. O.; Awe, A. A. Microstructure, Mechanical and Fracture Properties of Groundnut Shell Ash and Silicon Carbide Dispersion Strengthened Aluminium Matrix Composites. *J. King Saud Univ. - Eng. Sci.* 2018, *30* (1), 96–103. https://doi.org/10.1016/j.jksues.2016.01.001
48. Hayajneh, M. T.; Almomani, M. A.; Al-Shrida, M. M. Effects of Waste Eggshells Addition on Microstructures, Mechanical and Tribological Properties of Green Metal Matrix Composite. *Sci. Eng. Compos. Mater.* 2019, *26* (1), 423–434. https://doi.org/10.1515/secm-2019-0027
49. Hassan, S. B.; Aigbodion, V. S. Effects of Eggshell on the Microstructures and Properties of Al–Cu–Mg/Eggshell Particulate Composites. *J. King Saud Univ. - Eng. Sci.* 2015, *27* (1), 49–56. https://doi.org/10.1016/j.jksues.2013.03.001
50. Jannet, S.; Rosari, R.; Ruban, S. R.; Khosla, S.; Sasikumar, U.; Sai, N.; Teja, P. Effect of Egg Shell Powder on the Mechanical and Microstructure Properties of AA 2024 Metal Matrix Composite. *Mater. Today Proc.* 2020, *44*, 135–140. https://doi.org/10.1016/j.matpr.2020.08.546
51. Joseph, O. O.; Babaremu, K. Agricultural Waste as a Reinforcement Particulate for Aluminum Metal Matrix Composite (AMMCs): A Review. *Fibers* 2019, *7* (4), 33. https://doi.org/10.3390/fib7040033
52. Gangisetti, S.; Sahu, R. K.; Parihar, R. S. Chapter 7 Challenges in Green Waste-reinforced Aluminum Composites. In *De Gruyter eBooks*; 2023; pp 163–172. https://doi.org/10.1515/9783110766523-011
53. Ononiwu, N. H.; Akinlabi, E. T.; Ozoegwu, C. G. Sustainability in Production and Selection of Reinforcement particles in Aluminium Alloy Metal Matrix Composites: A Review. *J. Phys. Conf. Ser.* 2019, 1378 (4), 042015. https://doi.org/10.1088/1742-6596/1378/4/042015

54. Korotkova, T.; Ksandopulo, S.; Donenko, A.; Bushumov, S.; Danilchenko, A. Physical Properties and Chemical Composition of the Rice Husk and Dust. *Orient. J. Chem.* 2016, 32 (6), 3213–3219. https://doi.org/10.13005/ojc/320644
55. Olutoge, F. A.; Oladunmoye, O. M. Bamboo Leaf Ash as Supplementary Cementitious Material. *Am. J. Eng. Res. (AJER)* 2017, *6* (6), 1–8.
56. Subash, L.; Natarajan, U. Investigation of Egg Shell and Alumina Reinforced with Magnesium Matrix Composite. *Int. J. Innov. Sci. Res. Technol.* 2020, *5* (4).
57. Parveez, B.; Maleque, M.; Jamal, N. Influence of Agro-Based Reinforcements on the Properties of Aluminum Matrix Composites: A Systematic Review. *J. Mater. Sci.* 2021, *56*. https://doi.org/10.1007/s10853-021-06305-2

15 Automotive Industrial Applications of Natural Fibre Composites (NFCs)

Surej Rajan C., Indira V., and Soney C. George

15.1 INTRODUCTION

Researchers have been inspired to create efficient and unique materials by using natural resources in the automobile industry as a result of sustainability and growing ecological consciousness. Researchers and automakers are increasingly thinking about using natural fibre-reinforced composites (NFRC) since they are lightweight, sustainable, and environmentally benign (Figure 15.1). Natural fibres are starting to replace glass fibres in composite materials as they are less expensive, lighter, and appear to be more ecologically friendly. One of the benefits of natural fibre-based composites in automobile applications is their light weight. It is feasible to reduce vehicle weight by up to 34% by using composite materials made of natural fibres in a few different automotive components. According to a survey from the year 2008, using parts made of biodegradable materials can lower a car's weight by 32 kg. These include bamboo, kenaf, hemp, jute, flax, coir, sisal, banana, and a wide variety of other fibres. To meet the market need for natural fibres rather than synthetic fibres, it is required to employ already accessible materials and look for novel plant fibres that can be extracted using straightforward, affordable procedures without compromising the characteristics of the fibres. The origin of natural fibres in general may be used to classify them, and plant based fibres can be further divided into groups depending on the parts of the plant from which they are derived. Researchers' major focus is on NFRP material properties such as biodegradation, fibre modification, thermal stability, and crystallinity.

It has been demonstrated that natural fibre-reinforced polymers can withstand the structural and durability requirements of interior and exterior car parts. Natural fibre reinforcing has been used successfully in a variety of vehicle parts. To make floor panels, seatback linings, and door cladding, flax, sisal, and hemp are processed. Seat bottoms, back cushions, and head restraints are made of coconut fibre; cotton is used for soundproofing; wood fibre is used for seatback cushions. Acaba is utilized in body panels that are hidden beneath the floor, and other automakers are also using natural substances in their vehicles. At General Motors, package trays and door panels for Saturn L300s and Opel Vectras sold in Europe are made of a kenaf and flax blend, while the seat backs for the Cadillac DeVille and the cargo area floor of the GMC Envoy and Chevrolet TrailBlazer are made of wood fibre. The Ford Freestar's sliding door inserts are constructed of wood fibre. Kenaf has been included in the body of Toyota's i-foot and i-unit concept cars because the company is interested in employing it to manufacture Lexus package shelves. In Europe, Ford puts corn-based Goodyear tyres on its fuel-efficient Fiestas. Because they have less rolling resistance than conventional tyres, Goodyear's corn-infused tyres can give higher fuel efficiency. The growth of the natural fibre composite industry would benefit small business owners and farmers globally, enhance environmental quality by creating a sustainable resource supply chain, and improve the carbon dioxide balance over the course of a vehicle's lifetime with almost zero net greenhouse gas emissions (Naik et al., 2022).

DOI: 10.1201/9781032665337-15

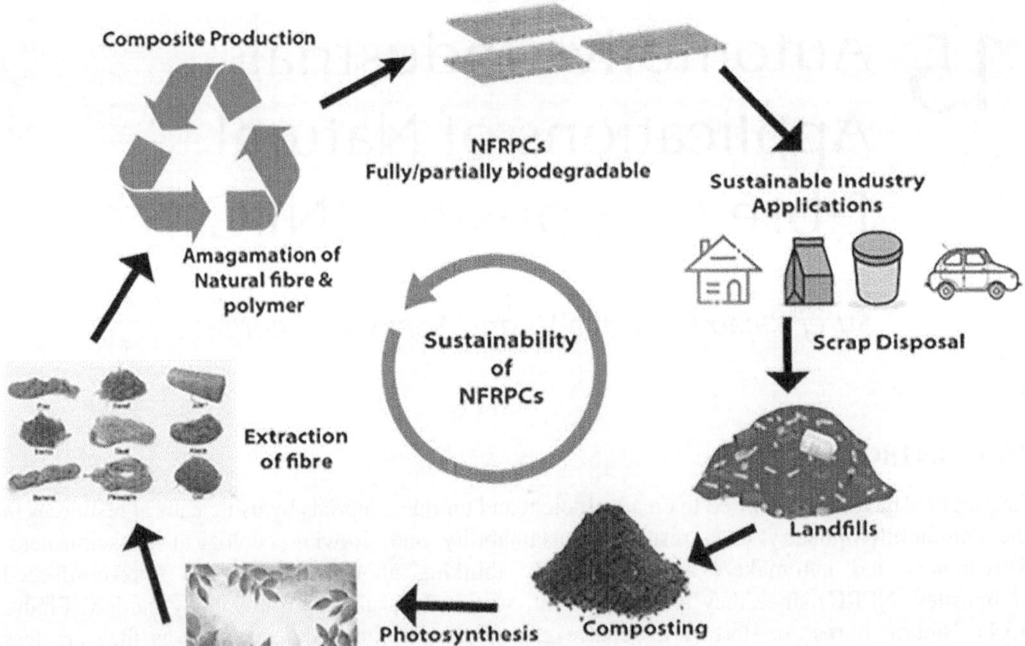

FIGURE 15.1 Sustainability of Natural Fibre reinforced Polymer Composites (NFRPC) (Neto et al., 2021) **Reproduced with permission from MDPI Polymers (2021), as per Creative Common License.**

15.2 NATURAL FIBRE COMPOSITES IN AUTOMOTIVE INDUSTRIES

The cost of the raw materials and components, the weight of the components, and the workability and machinability when using well-established technologies like compression moulding and lamination are major factors in the success of NFRC materials in the automotive industry (Bos, 2004). In order to create eco-friendly composites with better characteristics and reduce the issue of residue accumulation, natural fibres have been added to the polymers (Ferreira et al., 2019). There is growing interest in using natural fibres in composite plastics in many fields, especially the automotive industry, for door panels, seat backs, headliners, package trays, dashboards, trunk liners, and interior parts because of the rising demand for environmentally friendly materials and the excellent qualities of natural fibres (Huda et al., 2008a). The next composite is anticipated to be a hybrid made of two or more fibre types in a polymer matrix. Fibres that may be combined to create hybrid composites include natural and natural, synthetic and synthetic, and synthetic and synthetic. Based on materials, fibres are classified as mineral-based, animal-based, and cellulose-based (Figure 15.2). Due to their low cost, high strength-to-weight ratio, and simplicity of manufacture, hybrid composite materials have several uses in the field of engineering. Although employing NFCs has numerous advantages, there are also drawbacks that limit the technology's ability to be used more widely. Most NFC applications are now limited to automobile interiors because of the natural fibre's susceptibility to moisture and temperature. These natural fibres' surfaces can be altered to decrease the amount of moisture absorption (Witayakran et al., 2017).

15.3 MARKET ESTIMATES

Based on Ken Research Analysis, the market for NFRCs is predicted to grow at a positive compound annual growth rate (CAGR) of over 9% from 2022 to 2030 and will reach about US$12 billion by

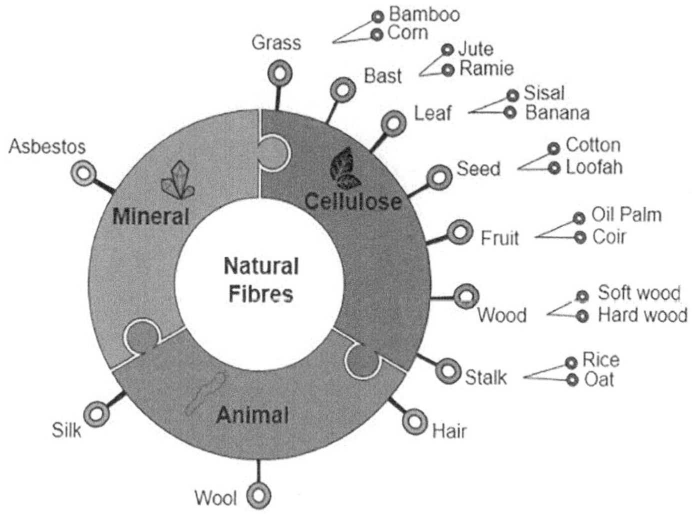

FIGURE 15.2 Schematic of natural fibres' classification (Neto et al., 2021).

Reproduced with permission from MDPI Polymers (2021), as per Creative Common license.

that time. Because natural fibres are more environmentally friendly and contribute to sustainable practices, both industry players and researchers are very interested in using them in polymer composites. The Chinese government expects there to be at least 5,000 fuel cell electric vehicles on the road by 2025, and one million by 2030. The government's support of electric, hybrid, and fuel cell electric cars is predicted to promote the Chinese market for NFRCs.

By 2024, the market for natural fibre composites is anticipated to be worth US$10.89 billion and expand at a CAGR of 11.8%. In the report 'Natural Fibre Composites Market Forecast to 2028 – COVID-19 Impact and Global Analysis – by Raw Material, Technology, End User, and Geography', published by The Insight Partners, the market is predicted to increase from US$4.44 billion in 2022 to US$6.91 billion by 2028, with a CAGR of 7.7% during that time. In 2022, Europe had the biggest market share for natural fibre composites.

15.4 PLANT FIBRES USED IN NATURAL FIBRE COMPOSITES

The natural fibre-based composites have been utilized in furniture, building, packing, shipping pallets, and interior car linings (roof, rear wall, side panel lining). Various plants and animals are used to make natural fibres. Figure 15.3 shows the different kinds of available natural fibres. Components including cellulose, lignin, hemicellulose, pectin, waxes, and water-soluble materials make up plant fibres. Ramie is a perennial herbaceous plant that is widely cultivated in the area and is native to Malaysia, China, and Japan. It has been used there for more than a century to make textiles. Ramie is a fast-growing, non-branching plant that may reach heights of 1–2 m. One of the most significant fibres, kenaf fibres are stiff, robust, and durable with a high resistance to pesticides. Because they are fully biodegradable, kenaf fibres are ecologically beneficial. Cellulosic plants – flax fibres in particular – are more crystalline than fibrous. These fibres have a diameter of 12–16 μm and a length of up to 90 cm. One of the most popular natural fibres, sisal has a wide variety of mechanical benefits and is employed in the automotive sector. The date palm is a palm that is widely cultivated for its fruit. The thermoplastic and thermosetting polymers can be reinforced with these leaf and rachis fibres (Girijappa et al., 2019).

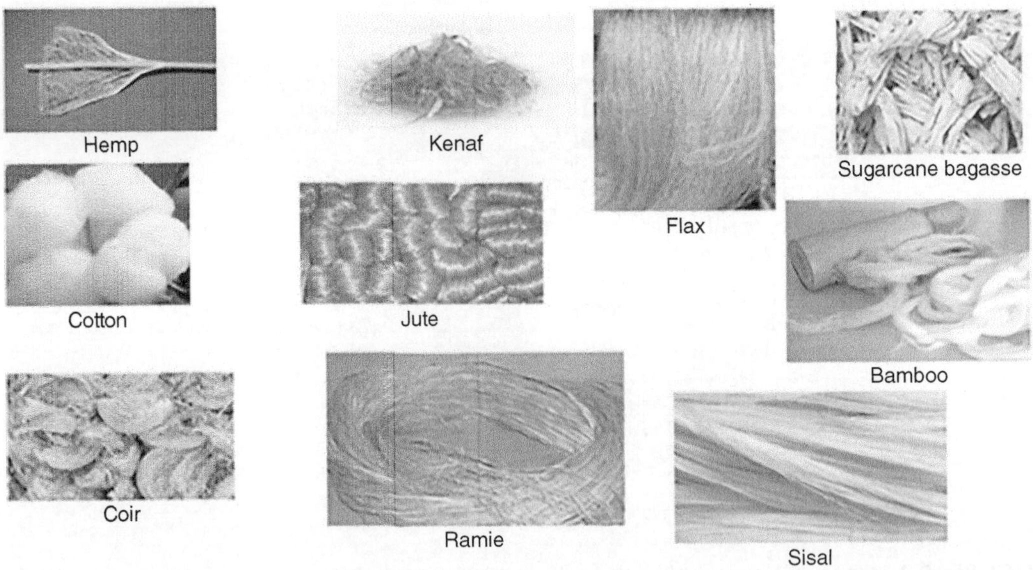

FIGURE 15.3 Different types of natural fibres (Neto et al., 2021).

15.5 MATRIX FOR NATURAL FIBRE-REINFORCED COMPOSITES

15.5.1 Thermoplastics

Thermoplastics retain their plasticity after cooling, which means that by warming them over their processing temperature, they will remelt and can be moulded. Cheaper thermoplastic matrices have a smaller range of application temperatures but lower processing temperatures. These include polyethylene (PE), polyethylene terephthalate (PET), polybutylene terephthalate (PBT), polycarbonate (PC), acrylonitrile butadiene styrene (ABS), polyamide (PA or nylon), and polypropylene (PP). They also use a variety of engineered and commodity plastics. The resin's hardness and moisture resistance make it ideal for use in automotive intake air manifolds and other under-the-hood parts. Polyetheretherketone (PEEK), polyetherketone (PEK), polyamide-imide (PAI), polyarylsulfone (PAS), polyetherimide (PEI), polyethersulfone (PES), polyphenylene sulphide (PPS), and liquid crystal polymer (LCP) are high-performance thermoplastic resins that work well in high-temperature environments and, once hardened, neither absorb water nor degrade when exposed to moisture. These resins, which are reinforced with high-performance fibres, have a long prepreg shelf life without refrigeration and remarkable impact and vibration resistance. The whole liftgate assembly, which is made up of an inner panel made of 30% long glass fibre/PP and an exterior panel made of unreinforced thermoplastic olefin (TPO) compound, is 30% lighter than equivalent stamped steel systems (Ramakrishnan et al., 2022).

15.5.2 Thermosetting Resins

The most popular polymers used in composite materials are thermosets, a class of polymeric resins that become much less fluid and soluble when chemically or thermally cured. When paired with glass fibre reinforcements, polymers like polyester may be used in a range of fabrication processes; however, they are most frequently employed in spray up, compression moulding, injection moulding, and casting. Contrary to polyesters, vinyl ester resins can be utilized to connect superior epoxies with less expensive, faster-curing polyester mixes. For advanced composite matrices, the most common thermosets are epoxies, phenolics, bismaleimides (BMIs), benzoxazines, and polyimides.

Polyester Resins: Due to their adaptability throughout the polymer chain formation process, polyesters are versatile. These resins' key advantages include their superior mechanical, chemical, and electrical compatibility, dimensional stability, low cost, and ease of processing.

Epoxy Resins: An important benefit of epoxy resins over unsaturated polyester resins is that they shrink less. Epoxies can be made with additional chemicals or blended with other epoxies to create performance characteristics. Epoxies are frequently used to produce high-performance composites with superior mechanical properties, resistance to hazardous substances and environments, improved electrical properties, outstanding performance at higher temperatures, and strong substrate adhesion, or a combination of these benefits.

Phenolic Resins: Phenolic composites have excellent fire/smoke toxicity characteristics, high thermal conductivity, and sound-dampening qualities. Phenolic resins are also high-temperature and creep resistant.

Polyurethane: This kind of polymer is used as a coating or an adhesive. When used as a finish for a building's exterior or interior, polyurethanes exhibit strong, flexible, chemical-resistant, and quick-curing qualities. Bumper parts and insulation polyurethanes have increased toughness and abrasion resistance when used as elastomers in things like solid tyres and wheels.

15.5.3 Biodegradable

Polyhydroalkanoates (PHA), polylactides (PLA), polylactide aliphatic copolymer (CPLA), polycaprolactone (PCL), poly(glycolic acid) (PGA), and poly(vinyl alcohol) (PVA) are examples of families of materials that contain biodegradable polymers. Bacterial fermentation of sugar or lipids results in the production of polyhydroalkanoate homopolymers and copolymers, such as poly(betahydroxybutyrate) (PHB) and poly(hydroxybutyrate-co-valerate) (PHBV), which have melting temperatures that range from 40°C to 180°C. By mixing natural/biofibres with biodegradable resins, research is now being used to generate a new class of entirely biodegradable 'green' composites. The main benefits of using green composites are that they are sustainable, entirely biodegradable, and ecologically benign (Huda et al., 2008a).

A biodegradable blend of lactide and aliphatic polyesters is known as polylactide aliphatic copolymer. Depending on how much aliphatic polyester is present in the combination, it can either be a hard plastic like polystyrene (PS) or a soft flexible one like PP. It is stable up to 200°C. By polymerizing caprolactone produced during the chemical production of crude oil, PCL is created. It is simple to produce, has a low melting point (58°C to 60°C), low viscosity, good water, oil, solvent, and chlorine resistance. A synthetic homopolymer of glycolic acid is known as poly(glycolic acid). Given that the monomer is terminated by a hydroxyl group on one end and a carboxyl group on the other, it can be polymerized by a condensation process to produce a polyester. Due to its three hydroxyl groups per monomeric unit, cellulose has a strong hydrophilic nature, whereas biopolymers like PLA and PHB are often hydrophobic. Since the biopolymer (like PLA) and natural fibre that make up the composite are both biodegradable, the composite itself should be as well. There is a need to better understand and explain PLA's qualities as the matrix material for biocomposite materials since they can lead to numerous new prospects in auto industry bioplastic applications. 'Carbon neutral' materials include bioplastics and biofabric. The carbon dioxide the plants collect as they develop balances the carbon dioxide emitted when the plants used to create the components are burnt (Sreekumar and Thomas, 2008).

15.5.4 Composite Manufacturing Techniques

Reinforcing fibres are held in place by a polymer matrix phase in composite materials. Short strands or continuous fibres may make up the fibres. Continuous fibre composites provide higher stiffness and strength. Advanced composites are what they are called, and they frequently have a high fibre-to-resin ratio of 55% to 65% fibre by volume. Advanced composites can also be bonded over another material, usually a polymer foam or an aluminium honeycomb, to create sandwich structures.

Individual reinforcing fibre filaments are bundled into tows (untwisted bundles), skeins (tighter twisted bundles), rovings (slightly twisted bundles), and yarn bundles. In order to create a preform, which is the whole fibre 'skeleton' of a component without the cured resin matrix holding it together, bundles of filaments are frequently woven into textiles. A ply is a unit of fabric that is used to construct a preform. Tackifier, a binder, keeps the plys attached to one another in a preform. This facilitates preform lay-up and lessens fibre wash, which is the movement of fibres during resin infusion.

The volume fraction and fibre-to-resin ratio (or orientation) play a significant role in a composite's structural characteristics. During lay-up, the preform may be crushed numerous times to increase the volume fraction and lessen issues like wrinkling. Debulking is the term for this procedure. The contact between the matrix and the fibres has an impact on the structural characteristics in addition to those of the component materials. A preform that is both simpler to put up and doesn't require additional resin infusion is frequently made using fabric that has already been impregnated with resin (prepreg).

There are several ways to categorize the various fabrication techniques used to create sophisticated composite structures. The procedures are often divided into open-mould and closed-mould methods initially. In contrast to open moulds, which only firmly control the surfaces in contact with the mould, closed moulds provide well-controlled surfaces on all faces of a part. The outer mould line (OML) refers to the controlled surfaces that come into contact with the mould. The inner mould line (IML) is the term for the uncontrolled surfaces. The way resin flows through the preform may also be used to categorize composite production procedures. Deposition of fibres, infusion of fibres with resin, debulking, trimming, and, in the case of composites containing thermoset resin, curing are all general processes in a manufacturing process. Except for curing, each of these process phases will always be present, albeit the sequence in which they occur will differ between procedures.

15.6 AUTOMATED TAPE LAYING

Automated tape laying (ATL) machines are robots with heads that place prepreg tape onto flat or slightly curved surfaces. These robots are typically gantry type. Rollers and heating elements in the head apply pressure and heat to debulk and attach plies together. Although tapes as broad as 300mm are sometimes utilized, a 3–12mm wide tape with continuous fibres embedded in a thermoplastic tape is often employed. The tape is first applied to the tooling, and then layers are created by applying more tape to earlier ones. Each ply is sliced as it is placed by the cutting mechanisms in the head. High material waste occurs as a result of the tape's breadth.

The thermoset resins are used the most frequently. Although hot-head curing is a possibility, autoclave curing is necessary to provide the greatest performance. It is also feasible to install thermoplastic tape, which eliminates the necessity for autoclave processing. In this instance, to guarantee adequate adherence between layers, both the incoming tape and previously deposited material are warmed using laser, gas, or induction heating right before installation. Thermoplastic resins provide better toughness and recyclability as benefits. The ability to generate components in a single automated process without the need for hoover bagging or an autoclave is perhaps the most significant benefit. On flat surfaces, the tape laying method may reach extremely high deposition rates of up to 45kg/h. Its drawbacks include the usage of pricey prepreg components and the restriction to generally flat or gently curved surfaces.

15.7 FILAMENT WINDING

A revolving mandrel serves as a single-sided mould during filament winding. In order to manage fibre orientation, a spool travels along the axis of rotation as the mandrel rotates, drawing fibres from it. Both prepreg tape and fibres impregnated by passing through a resin bath can be utilized with the technique (dry winding and wet winding, respectively). The only types of pieces that can be produced with this method are hollow and constantly convex. These might have axisymmetric shapes or more intricate designs like turbine blades. The mandrel used may be a washable tool made of plaster, salt, sand, ceramic, or a reusable metallic or composite construction with a binder substance that can be washed out of the cured component.

Debulking is accomplished by applying stress to the filaments; therefore, a compaction roller like those used on tape laying machines is not necessary. When the fibres are parallel to the axis of rotation, tension is a very powerful debulking tool. It gets more challenging when fibres are oriented at angles that are closer to the axis of rotation. If more debulking is necessary, an expanding mandrel, a hoover bag, or an autoclave can be used to do the task. Autoclave processing is typically not necessary.

A high fibre content and little material waste may be obtained by the automated and highly reproducible process of filament winding. The cost of the materials is also inexpensive when wet winding is employed. The restriction to axisymmetric or convex sections is the main drawback.

15.8 AUTOMATED FIBRE PLACEMENT

The advantages of both ATL and automated filament winding are combined in automated fibre placement (AFP), also known as automated tow placement (ATP), while avoiding many of their drawbacks. It is now often utilized on state-of-the-art aircraft goods, such as parts of the F-18, V-22, C-17 III, F-22, 787, and A380. A head like that of a tape laying machine concurrently deposits several thin prepreg tapes. This enables the machine to sustain high deposition rates while conforming to significantly tighter curves without experiencing wrinkling. Additionally, it lessens material waste. Onto a spinning mandrel or a stationary mould, material may be placed.

The shape of components is restricted to continuously curved surfaces or hollow structures, depending on how the machine is configured. Bonding a multitude of components together can result in the creation of more intricate pieces. To create skin/stringer structures, for instance, box pieces were initially cut on a revolving mandrel. I-beams are created by cutting them in half and bonding the halves together. The skin is made of fibre and is applied over the I-beams after they have been constructed into slots in an aluminium tool. Similar to how it would happen between the layers of a single component, bonding between previously produced components takes place.

15.9 WET LAYUP AND AUTOCLAVE PROCESSING

Thermoset prepreg on an open mould is commonly used for wet lay-up and autoclave processing. To enable the component removal, a release film is initially placed on the tool. After cutting, the prepreg is put up in the tooling. At this point, ATP and tape lay-up are feasible. On the outside, peel ply, release fabric, and bleeder/breather materials are utilized to add texture, aid in full curing, and enable extra resin to drain. The lay-up is then sealed with a hoover bag. The resin is then heated to cure it and put under pressure in the autoclave to remove any surplus resin. The mould is then taken out of the autoclave, the component is taken out of the mould, and any extra material is cut. The technique's drawbacks include the high cost of prepreg components and the need for an autoclave, which lengthens cycle durations.

15.10 LIQUID COMPOSITE MOULDING

The process of liquid composite moulding (LCM), sometimes referred to as resin infusion, involves inserting a dry fibre preform into a mould tool before infusing thermoset resin into it. Therefore, costly prepreg materials are not necessary.

Vacuum-assisted resin transfer moulding (VARTM), resin transfer infusion (RTI), and Seeman Composites Resin Infusion Moulding Process (SCRIMP) employ open moulds with a vacuum bag to surround the preform and enable resin infusion. Resin transfer moulding (RTM) utilizes a closed mould. RTM typically requires six to 20 minutes for the mould to cure, depending on the size of the item. This can be cut down to one to five minutes by using specific low viscosity resins, such as those used in structural reaction injection moulding (SRIM).

RTM can produce intricate shapes with precise tolerances, but manual labour costs are still significant since it is challenging to automate the lay-up of dry fibre preforms. Automated prototype systems have been created, including a robotic industrial arm with an electrostatic end-effector and a vision-based system that, in 1996, was able to lay out dry fibres at a pace of 15 kg per hour. Despite this success, industry uptake has been gradual, but automated preform manufacturing is now a reality.

Large, inexpensive structures that do not require autoclave processing are frequently built using methods like VARTM, which use vacuum bags over single-sided tooling. Compared to a closed-mould RTM process, these approaches produce longer cycle times but lower tooling costs. Open-mould methods are better suited to big components since tooling costs rise quickly with component size. Autoclave processing is still necessary to produce the components with maximum performance.

15.11 THERMOFORMING

A thermoforming method is used to create thermoplastic sheet or film. Extrusion leads to moulding in the procedure. This procedure involves the use of extruders to create sheets or films that are then drawn into an oven and baked to a certain temperature. Then it is positioned above or inserted into a single-surface, temperature-controlled mould. The most typical mould material is cast or machined aluminium; however, epoxy and wood tooling are also used for low-volume manufacturing. Until it cools, the sheet is kept in place against the mould's surface by vacuum or pressure. After that, the sheet is cut away from the created section. Usually, virgin plastic is added to the trimmed material before being reground and processed into a useful sheet. Thermoforming may be divided into a number of types, such as vacuum forming, pressure forming, twin-sheet forming, and drape forming. The plastic is then taken to a cutting area, where the forms are sliced out of it.

15.12 PULTRUSION

In this procedure, fibres are drawn from a creel and passed through a resin bath and heated dye. The fibre is fully impregnated, the resin content is under control, and the material is curing into its final shape as it travels through the dye. The length of this cured profile is next trimmed. In order to offer fibre direction other than 0, fabrics may also be added to the dye. A procedure known as 'pulforming' allows for some variation to be incorporated into the cross-section, even though pultrusion is a continuous process that results in a profile with a constant cross-section. The materials are pulled through the dye throughout the process for impregnation, and they are then clamped in a mould for curing. As a result, the process is non-continuous yet may accommodate slight cross-sectional variations. Epoxy, polyester, vinyl ester, and phenolic resins are the most frequently used resins.

15.13 BULK MOULDING COMPOUNDS

Thermosets with a polyester foundation are bulk moulding compounds (BMC). Additionally known as dough moulding compounds (DMC) too. In a dough mixer, typically one with a Z-blade, they are made by combining resin (mainly polyester resin), mineral fillers, reinforcing fibres, additives, and peroxide catalysts. The catalysts are organic peroxides such as tertiary butyl perbenzoate or benzoyl peroxide. The catalyst decomposes as a result of the heat and pressure produced during the moulding process. This causes the resin to crosslink and solidify into a network. The molecules are bound together by chemical bonds in the stiff structure created by this cross-linking. Modern composite materials like BMC or DMC are used in a variety of applications in the building and construction business, as well as other industries, including the automotive, public transit, electrical, sanitary, and domestic industries. Excellent environmental and recycling qualities may be found in BMC/DMC. Prepreg manufacture differs significantly from glass sheet moulding compound (SMC) in a significant way. Glass strands are often cut into pieces and dropped onto a resin-filler compound film to create prepregs. Since the cutting is so challenging, this preparation does not work for natural fibres. Different methods are being created. The layer of fibres must have an anisotropic orientation and be sufficiently slack to allow for adequate fibre flow throughout the moulding process. The transportation sector makes substantial use of BMCs. The technical requirements for headlamp reflectors, automotive body panels, and under-bonnet components are very high. The advantages of BMC over other materials and commonly used methods like steel stamping, thermoplastic injection, and aluminium dye casting enable automakers to build more efficient processes and optimize their designs. Headlamp reflectors constructed of BMC are used by almost all auto manufacturers.

15.14 SHEET MOULDING COMPOUND

SMC is a fibre-reinforced composite material made up of fibres, often polyester resin, pigments, fillers, and other additives that have been mixed and processed into sheets to enable compression moulding. It is appropriate for employing automated equipment with large pan volumes. SMC is suited for automotive applications due to its high mechanical qualities and outstanding surface appearance; in the US car sector, a significant amount of SMC is employed in automated equipment for side panels and trim. For this moulding process, natural fibre works well, and fibre dust or short fibres might be utilized to fill and thicken the SMC. Anywhere SMC is present, including radiator supports and grill opening reinforcements, natural fibres like hemp or kenaf have the potential to be used, and various commercial uses currently exist. Natural fibre reinforcements are used in the Ford Freestyle's third-row seatback as well as the Expedition and Lincoln Navigator's door-trim panels Interfacial Bonding between the Fibre and the Matrix

Depending on the method and degree of contact, the strength of the interface has a big impact on composite qualities. The presence of chemical functional groups, mechanical interlocking on rough fibre surfaces, chemical bonding, and inter-diffusion bonding can all be causes of interfacial bonding. There is a chance that many bonding processes might operate simultaneously at one interface. The density of bonds has an impact on contact strength as well. Due to polar fibres and non-polar polymer matrices, interfacial bonding at the NFC interface is often only minimal. In consequence, this has an impact on how well NFCs transfer stress from the matrix to the fibre, hence restricting the mechanical qualities. Chemical processes and coupling substances are the main components of NFC interfacial bonding. The primary goal of these treatments is to increase the wettability of the fibre surface and its potential for chemical interaction with the matrix, which will provide interfacial strength. The matrix's wettability towards the fibre is crucial for the matrix-fibre adhesion, and this may be determined by comparing the surface energies of the fibre and the matrix. For the occurrence of fibre-matrix adhesion, the surface energy of the reinforcements must be higher than that of the matrix. NFCs absorb atmospheric moisture when exposed to the surrounding environment because of their hydrophilicity. As a result, their surface free energy decreases, maybe to the point where their surfaces have lower surface energy than matrices. Coupling agents, sometimes referred to as compatibilizers, serve as a link and join the matrix and fibre together.

15.15 CHEMICAL METHODS

Alkali, acetyl, silane, benzyl, acryl, stearic acid, maleic anhydride, permanganate, peroxide, isocyanate, titanate, and zirconate are just a few of the chemicals that may be used to treat fibres chemically. Alkali, acetyl, and silane are the most often used treatments, and the majority of these treatments attempt to change the surface chemistry. In comparison to other common treatments, natural plant fibres (NPFs) treated with alkali have shown advantages such as improved separation of fibres from fibre bundles, improved removal of undesirable surface constituents, increased tensile strength and stiffness, better thermal stability, and improved interfacial adhesion.

15.16 ALKALI TREATMENT

One of the most often used therapies is the sodium hydroxide (NaOH) alkali treatment. Hemicellulose, lignin, pectin, wax, and fat are all taken out of the NPFs during this procedure. Hemicellulose, lignin, and pectin (cementing components) removal from NPFs causes fibre separation and increases hydroxyl group exposure on the fibre surfaces, which boosts interfacial bonding, fibre roughness, and thermal stability. The tensile strength of date palm fibres was found to be higher than raw fibres after alkali treatment with different concentrations of NaOH (0 wt%, 2 wt%, 5 wt%, 10 wt%) at 25°C for one hour. Sisal fibres were subjected to an hour-long alkali treatment at 30°C with 5 wt% NaOH, which resulted in a minor increase in the mechanical characteristics of sisal/PP composites. Hemp fibres treated with 5 wt% NaOH and 2 wt% Na2SO3 (sodium sulphate) at 120°C for 60 minutes, compared to 10 wt% NaOH and untreated fibres, showed a substantial improvement in average tensile strength for various high-temperature treatments. Low-temperature treatments are most likely to result in the degradation of the crystalline cellulose chains in the microfibrils or bonding between cellulose microfibrils before sufficient removal of weak components from the fibres, so high-temperature alkali treatments appear to be the best way to produce strong and stiff fibres (Sunny and Pickering, 2022).

15.17 MECHANICAL CHARACTERIZATION OF NATURAL FIBRE COMPOSITES

Zhong et al. focused their experiments on sisal fibre/urea-formaldehyde thermoset resin composites, varying the fibre content to 30, 40, 50, 60, and 70% wt%. Sisal fibres were treated with 2% NaOH in water for 24 hours at a constant temperature of 22.2°C, and then washed in distilled water, left to air dry for 15 hours at 70°C, and finally dried in an oven for the same period of time. The inclusion of melamine ($C_3H_6N_6$) improved the interfacial contact between the fibre and polymer matrix. Flexural strength, flexural modulus, and water absorption were all equal to 58.58 MPa, 7.63 GPa, and 0.98% when 30 wt% of the fibre content was used. The highest impact strength of 9.42 kJ/m2 was attained when sisal fibre content was 50% by weight (Zhong et al., 2007). Chow et al. conducted experiments on sisal/PP composites to verify the effect of fibre concentration on the tensile strength, Young's modulus, and water absorption. It was discovered that the composite's tensile strength and Young's modulus dropped the longer it was submerged in water at 90°C. An increase in fibre content led to an increase in water absorption because natural fibres are hydrophilic (Chow et al., 2007). In their investigation, Joseph et al. used three different thermosets – polyester, epoxy, and phenol formaldehyde – and adjusted their fibre contents while contrasting them with low density polyethylene (LDPE). It was discovered that changing each composite's fibre content improved its mechanical qualities (Joseph et al., 1996). Zhao et al. found that the tensile strength and tensile modulus rose, but the impact strength decreased with increasing fibre content of 10, 20, and 30 wt% when sisal fibre/PE composites were manufactured using the pre-impregnation procedure (Zhao et al., 2014). Water absorption is the main disadvantage of NFRCs, despite the fact that their characteristics are improving. Therefore, additional study on improving the characteristics of these natural fibres via surface modification is needed.

Banana and sisal fibres were combined with a polyester matrix, and it was discovered that the sisal-polyester composites had a greater impact strength (Naik et al., 2019). With more sisal fibre included in the material, the impact strength rises proportionately. These composites' mechanical properties are improved by the hybridization of nano-clay (Chowdary et al., 2022). The composite materials' levels of mechanical strength increase as a result of the sisal and jute fibre combinations (Ramesh et al., 2013). The sisal fibre composites that were chemically treated with sodium hydroxide and maleic anhydride garnered a lot of interest in the automobile industry because they demonstrated high mechanical strength (Sood et al., 2015). Sisal and oil palm fibre reinforcement were treated with a 4% alkaline solution of NaOH, which increased the storage modulus and decreased the damping ratio (Barkoula et al., 2008). Wetting characteristics and an increase in fibre surface area were shown to improve the storage modulus. The physical and mechanical properties of the composites are considerably improved by sisal fibre hybridization with other natural fibres. Sisal composites prepare a broad variety of primary components, including door panel, roof panel, floor lamination, wall insulation, and structural application (Choudhary et al., 2023).

The PLA/bamboo composites in this instance had improved tensile strength of 26 to 41 MPa and flexural strength of 85 MPa. Studies on thermosetting polymer composites with bamboo reinforcement that used vinyl ester composites demonstrated tensile strengths of 80 to 120 MPa and elongations of 3% to 5%. These composites are suitable for door panels in the automobile industry and have a wide range of applications (Radzi et al., 2022). The flexural and compressive strengths of the epoxy banana fibre composites are 34.99 MPa and 122.11 MPa, respectively. However, it was discovered that banana fibre/polyester composites that had not been alkali pre-treated produced flexural strengths of 40.16 MPa and compressive strengths of 123.28 MPa. This kind of composite fibre is employed in things like bumpers, glove boxes, door panels, and seat covers. Sisal/PET composites have demonstrated a significant increase in flexural strength and modulus. The tensile and flexural modulus were significantly improved by the reinforcement with 40% sisal fibre. The interiors with specific door panels, seat structures can be designed based on sisal/PET composites. NFCs are utilized to strengthen and have a significant impact on acoustics in the automobile sector. Different natural fibres such as milkweed, kusha grass, sisal, banana, and hay combined with propylene (PP) as 10:90 (wt%) were the subject of numerous acoustic research studies. Sisal and kusha grass significantly improved the tensile strength of PP, whereas hay in composites demonstrated greater tensile strength. The porous nature of milkweed fibres with their air holes makes them well-suited for absorbing sound waves. In the examination of sound absorption, the specimen's increased thickness is also crucial. At higher frequencies, the increase in sound absorption thickness is ineffective (Hariprasad et al., 2020). If they are used for non-structural and semi-structural applications, recognized formulations of kenaf fibres with matrix and certain additives can satisfy the demands of the car industry for interior and exterior components. By enhancing structural factors, including thickness, beam curvature, and strengthening ribs, twisted kenaf fibre material might be enhanced. They are used as front-end modules and bumpers in vehicle components (Omar et al., 2020).

Hemp fibre's impact on various thermoplastic polymer materials was investigated by Stelea L et al., and the materials' characteristics were examined. The polymer matrices chosen were PP, bicomponents, and recycled polyester. The greatest tensile strengths were found in composites with 30% to 50% hemp fibres. Compared to a sample composed solely of hemp, which has an elongation at break of about 71.5%, the composites containing a 25% polymer matrix had a lower elongation at break of under 10%. The results reported by the scanning electron microscopy (SEM) images are supported by the physical-mechanical tensile strengths, which increased in the first stage for composites containing up to around 50% hemp before tending to decrease (Figure 15.4) (Stelea et al., 2022).

On the mechanical and thermal characteristics of hemp/PP composites, three distinct compatibilizers – maleic anhydride grafted polypropylene (PP-MAH), maleic anhydride grafted styrene-(ethylene-co-butylene)-styrene copolymer (SEBS-MAH), and maleic anhydride grafted

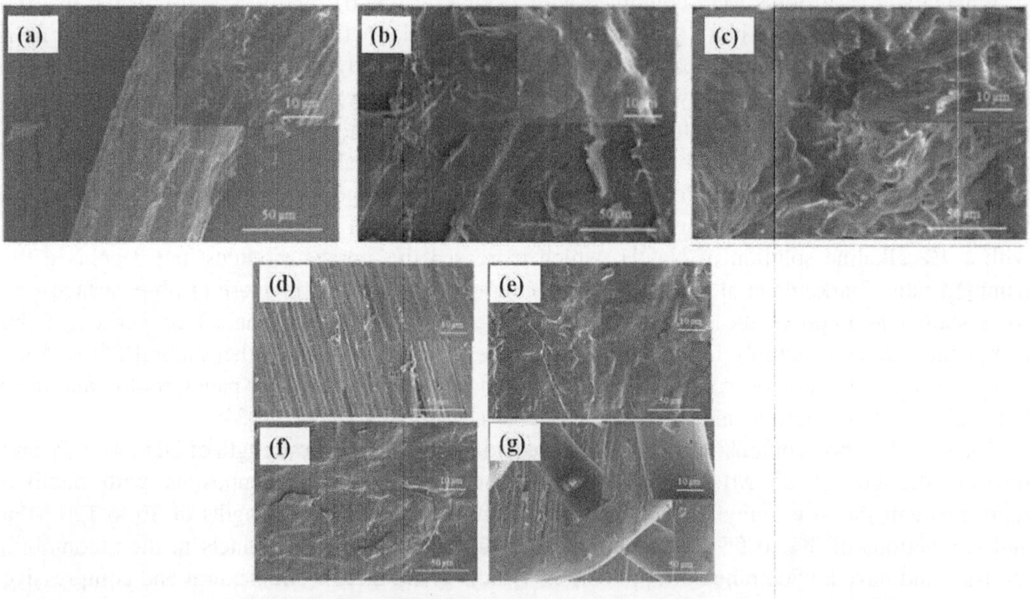

FIGURE 15.4 SEM images of hemp fibres: (a), polypropylene (b), polyester (c), 75H25PP (d), 50H50PP (e), 25H75PP (f), and 50H50PES (g) (Stelea et al., 2022).

poly(ethylene octane) (POE-MAH) – have been studied. The study's findings showed that the interfacial adhesion between fibre and matrix was greatly increased by the application of all the compatibilizers, improving the composites' stiffness, mechanical characteristics, and thermal stability. In comparison to unmodified composites, composites with 5 phr PP-MAH and 6 phr POE-MAH showed increases in tensile strength, flexural strength, notched and unnotched impact strengths of 22%, 8%, 24%, and 82%, respectively (Niu et al., 2011).

When kenaf was used as a reinforcement parameter recycled with PP resin, Neng Sri Suharty et al. noted increases in tensile strength of up to 18%, flexural strength of up to 28%, and impact toughness of up to 27%, which demonstrated a frightening improvement in composite qualities (Suharty et al., 2016). Tensile strength was increased with the usage of Kenaf/Kevlar plain woven hybrid composite, which also showed average stiffness and non-performance to strain characteristics (Salman et al., 2016). The thermal and mechanical characteristics of kenaf composites made from polylactic acid, where the laminate was manufactured by compression moulding and film stacking. Due to better adhesion between the matrix and fibre, chemically treating kenaf fibre with silane and alkali treatment improved the thermal and mechanical properties of the composite (Huda et al., 2008b). In order to replace synthetic fibre in the door panels and package trays of the Saturn L300 and Opel Vectra for the European market, General Motors developed kenaf and flax-based hybrid polymer composites. Now, Malaysian automakers Perodua and Proton use kenaf-based polymer composites in parts of the vehicle's trim, such as the seatbacks, package tray, door panels, and headliners. The potential for commercial production of kenaf into several biocomposites for automobile parts is enormous (Faruk et al., 2012). Brazilian automakers, including General Motors, Ford, Honda, and Volkswagen, currently employ natural fibre biocomposite for their vehicle seats, roofs, dashboard covers, and trunk lids. Examples of these fibres include jute, hemp, and kenaf (Sanjay and Siengchin, 2019). The characteristics of the composite were successfully altered by changing a number of parameters, including the processing technique, temperature, chemical treatment of the fibre, orientation of the fibres, stacking order, volume percentage of fibres, and weaving pattern. As a result, kenaf has been inspiring several researchers to develop kenaf-based hybrid composite materials (Sreenivas et al., 2020).

15.18 THERMO-MECHANICAL CHARACTERIZATION AND THERMAL BEHAVIOUR

Below 150°C, the dehydration of sisal fibre, which is related to the first phase of thermal degradation, may be seen in PLA composites containing sisal fibre. At the initial stage of heat deterioration, weight loss is 9.59%. Sisal fibre's hemicelluloses are thought to have degraded in the second stage of its thermal breakdown, which took place between 205°C and 304°C. At this stage, 20.4% of the fibre's weight has been lost. Between 270°C and 390°C, sisal fibre undergoes its third stage of thermal disintegration, which is ascribed to the breakdown of cellulose. Due to the existence of free radicals in an oxidizing environment, the last stage of the thermal degradation of sisal fibre occurred between 390°C and 475°C (Samouh et al., 2019). The raw PLA's glass transition temperature (Tg) was 58°C, and adding different amounts of sisal fibre (5%, 10%, or 15%) to the PLA matrix had no effect on how easily the PLA's molecular chains moved. It should be noted that adding more sisal fibre to PLA composites enhanced their storage modulus in the glassy state, indicating the fibres' ability to form nuclei (Samouh et al., 2019). In the investigation of sisal-based polyurethane composites, it was discovered that sisal fibres may be added to the bio-based polyurethane matrix in a maximum of 15 wt%, and the resulting materials exhibit good qualities. Bio-PU composites typically decompose thermally in three stages. The thermal stability of sisal fibres and bio-PU composites is marginally increased by chemical modification of sisal fibres, and the rate of thermal degradation is decreased (Głowińska et al., 2017).

Based on the polyaddition reaction of an amine group with an epoxide ring, the composites with various hemp fibre content levels (0, 5, 10, 20, and 30 wt%) were produced. The thermogravimetric analysis (TGA) study showed that the TGA and derivative thermogravimetric (DTG) curves for the examined composites had almost the same trend. All of the composite materials were thermally stable up to around 230°C (the initial thermal decomposition ranged from 213.1°C to 247.7°C). Without respect to the kind of composite, it can be shown from evaluating the tan delta values from the dynamic mechanical analysis (DMA) analysis that the glass transition zone spans a comparable temperature range (Gargol et al., 2021). Na Lu et al. investigated the thermal stability and thermo-mechanical characteristics of the hemp-high-density polyethylene (HDPE) composites in relation to fibre loading and surface functionalization. After fibre chemical treatment, the composites' thermal stability significantly increased. Hemp-HDPE composites with silane and NaOH treatments both have greater heat stability. Between these two types of surface treatment, silane-treated hemp-HDPE composites showed the best thermal stability (Lu and Oza, 2013).

Studies show that even though the addition of kenaf fibre increases the mechanical properties under optimum loading, the thermal properties are negatively affected. El-Shekeil et al. assessed the results of an analysis of composites reinforced with kenaf fibres fabricated at 20%, 30%, 40%, and 50% fibre concentrations. The hardness rises with the addition of 30% fibre content, while the abrasion resistance and thermal stability diminish (El-Shekeil et al., 2012). The thermal properties can be increased by doing some chemical modifications. Faseha Shukor et al. observed that the developed kenaf-reinforced biocomposite's thermal characteristics were improved by the addition of ammonium polyphosphate and an alkaline treatment (Shukor et al., 2014). According to research by El-Shekeil et al., a kenaf-reinforced composite loses thermal stability as its fibre content increases (El-Shekeil et al., 2014). According to Yussuf et al., the thermal stability of kenaf/PLA biocomposite reduces when fibre is added, relative to pure PLA (Yussuf, Massoumi, and Hassan, 2010). According to research by Tawakka et al., when a PLA/kenaf biocomposite is manufactured, adding 5% or 10% (w/w) thymol has no effect on how much warmer the material decomposes than PLA alone. With the addition of 5% and 10% (w/w) thymol to the biocomposites, thermal transitions were shown to be mitigated by differential scanning calorimetry, according to the results (Tawakkal et al., 2014).

15.19 SUMMARY

This chapter provides an extensive update on natural fibre utilisation in the automobile industry. These natural fibres distinctly characterize the automobile industry's dimension at a higher level. The surface-modified natural fibres also push the boundaries of composites' application in the field of electric vehicles. In many automotive applications, glass fibres and mineral fillers are already starting to make way for natural fibres that are more affordable, robust, and eco-friendly. Developed in collaboration with Baden Aniline and Soda Factory (BASF), the unique invention offers weight savings of up to 50% when compared to traditional metal-reinforced sunroof frames. It is composed of 70% sustainable raw material content in the form of natural fibres. International automotive components (IAC) produces the fibre mat internally using a hot-moulding technique and a thermoset binder. Acrodur 950 L binder from BASF holds the fibres together and guarantees the lightweight composite component's heat resistance and loading capability. The natural fibres are strengthened by the low-emission, water-based binder. The mechanical characteristics of natural fibre composites are improved using natural fibres, including sisal, hemp, and kenaf. These natural fibre composites, which perfectly increase the degrees of application, are used in the construction of premium automobiles.

REFERENCES

Barkoula, N. M., B. Alcock, N. O. Cabrera, and T. Peijs 2008. "Flame-Retardancy Properties of Intumescent Ammonium Poly(Phosphate) and Mineral Filler Magnesium Hydroxide in Combination with Graphene." *Polymers and Polymer Composites* 16 (2): 101–113.

Bos, Harriëtte Louise 2004. *The Potential of Flax Fibres as Reinforcement for Composite Materials*. https://doi.org/10.13140/RG.2.1.1540.4643

Choudhary, Sarita, Manoj Kumar Sain, Vikas Kumar, Praveen Saraswat, and Manish Kumar Jindal 2023. "Advantages and Applications of Sisal Fiber Reinforced Hybrid Polymer Composites in Automobiles: A Literature Review." *Materials Today: Proceedings*. https://doi.org/10.1016/j.matpr.2023.02.072

Chow, C. P. L., X. S. Xing, and R. K. Y. Li 2007. "Moisture Absorption Studies of Sisal Fibre Reinforced Polypropylene Composites." *Composites Science and Technology* 67 (2): 306–313. https://doi.org/10.1016/j.compscitech.2006.08.005

Chowdary, Somaiah, Gujjala Raghavendra Mallampati, M. S. R. Niranjan Kumar, Shakuntala Ojha, Manne Anupama Ammulu, and P. Phani Prasanthi 2022. "Evaluation of Mechanical and Tribological Properties of Nano Clay Reinforced Kevlar/Sisal Polyester Composites." *Polymer Composites* 43 (12): 8860–8868. https://doi.org/10.1002/pc.27067

El-Shekeil, Y. A., S. M. Sapuan, K. Abdan, and E. S. Zainudin 2012. "Influence of Fiber Content on the Mechanical and Thermal Properties of Kenaf Fiber Reinforced Thermoplastic Polyurethane Composites." *Materials and Design* 40: 299–303. https://doi.org/10.1016/j.matdes.2012.04.003

El-Shekeil, Y. A., S. M. Sapuan, M. Jawaid, and O. M. Al-Shuja'a 2014. "Influence of Fiber Content on Mechanical, Morphological and Thermal Properties of Kenaf Fibers Reinforced Poly(Vinyl Chloride)/Thermoplastic Polyurethane Poly-Blend Composites." *Materials and Design* 58: 130–135. https://doi.org/10.1016/j.matdes.2014.01.047

Faruk, Omar, Andrzej K. Bledzki, Hans Peter Fink, and Mohini Sain 2012. "Biocomposites Reinforced with Natural Fibers: 2000-2010." *Progress in Polymer Science* 37 (11): 1552–1596. https://doi.org/10.1016/j.progpolymsci.2012.04.003

Ferreira, Filipe V., Liliane M. F. Lona, Ivanei F. Pinheiro, Sivoney F. de Souza, and Lucia H. I. Mei 2019. "Polymer Composites Reinforced with Natural Fibers and Nanocellulose in the Automotive Industry: A Short Review." *Journal of Composites Science* 3 (2). https://doi.org/10.3390/jcs3020051

Gargol, Mateusz, Tomasz Klepka, Łukasz Klapiszewski, and Beata Podkościelna 2021. "Synthesis and Thermo-Mechanical Study of Epoxy Resin-Based Composites with Waste Fibers of Hemp as an Eco-Friendly Filler." *Polymers* 13 (4): 1–17. https://doi.org/10.3390/polym13040503

Girijappa, Thyavihalli, Yashas Gowda, Sanjay Mavinkere Rangappa, Jyotishkumar Parameswaranpillai, and Suchart Siengchin 2019. "Natural Fibers as Sustainable and Renewable Resource for Development of Eco-Friendly Composites: A Comprehensive Review." *Frontiers in Materials* 6 (September): 1–14. https://doi.org/10.3389/fmats.2019.00226

Głowińska, Ewa, Janusz Datta, and Paulina Parcheta 2017. "Effect of Sisal Fiber Filler on Thermal Properties of Bio-Based Polyurethane Composites." *Journal of Thermal Analysis and Calorimetry* 130 (1): 113–122. https://doi.org/10.1007/s10973-017-6293-5

Hariprasad, K., K. Ravichandran, V. Jayaseelan, and T. Muthuramalingam 2020. "Acoustic and Mechanical Characterisation of Polypropylene Composites Reinforced by Natural Fibres for Automotive Applications." *Journal of Materials Research and Technology* 9 (6): 14029–14035. https://doi.org/10.1016/j.jmrt.2020.09.112

Huda, M. S., L. T. Drzal, D. Ray, A. K. Mohanty, and M. Mishra. 2008a "Natural-Fiber Composites in the Automotive Sector." *Properties and Performance of Natural-Fibre Composites*, 221–268. https://doi.org/10.1533/9781845694593.2.221

Huda, Masud S., Lawrence T. Drzal, Amar K. Mohanty, and Manjusri Misra 2008b. "Effect of Fiber Surface-Treatments on the Properties of Laminated Biocomposites from Poly(Lactic Acid) (PLA) and Kenaf Fibers." *Composites Science and Technology* 68 (2): 424–432. https://doi.org/10.1016/j.compscitech.2007.06.022

Joseph, Kuruvilla, Siby Varghese, G. Kalaprasad, Sabu Thomas, L. Prasannakumari, Peter Koshy, and C. Pavithran 1996. "Influence of Interfacial Adhesion on the Mechanical Properties and Fracture Behaviour of Short Sisal Fibre Reinforced Polymer Composites." *European Polymer Journal* 32 (10): 1243–1250. https://doi.org/10.1016/S0014-3057(96)00051-1

Lu, Na, and Shubhashini Oza 2013. "Thermal Stability and Thermo-Mechanical Properties of Hemp-High Density Polyethylene Composites: Effect of Two Different Chemical Modifications." *Composites Part B: Engineering* 44 (1): 484–490. https://doi.org/10.1016/j.compositesb.2012.03.024

Naik, Premkumar, Neelakantha Londe, Laxman Naik, and H. Girish 2019. "Mechanical Properties of Sisal/Banana Hybrid Composite Reinforced Polyester Composites." *International Journal of Mechanical and Production Engineering Research and Development* 9 (6): 161–170.

Naik, Venkatesh, Mohan Kumar, and Vijayananda Kaup 2022. "A Review on Natural Fiber Composite Materials in Automotive Applications." *Engineered Science* 18: 1–10.

Neto, Jorge S. S., Henrique F. M. de Queiroz, Ricardo A. A. Aguiar, and Mariana D. Banea 2021. "A Review on the Thermal Characterisation of Natural and Hybrid Fiber Composites." *Polymers* 13 (24). https://doi.org/10.3390/polym13244425

Niu, Pengfei, Baoying Liu, Xiaoming Wei, Xiaojun Wang, and Jie Yang 2011. "Study on Mechanical Properties and Thermal Stability of Polypropylene/Hemp Fiber Composites." *Journal of Reinforced Plastics and Composites* 30 (1): 36–44. https://doi.org/10.1177/0731684410383067

Omar, Mohd F., Haliza Jaya, and Nik N. Zulkepli 2020. Kenaf Fiber Reinforced Composite in the Automotive Industry. *Encyclopedia of Renewable and Sustainable Materials*. Vol. 9. Elsevier Ltd. https://doi.org/10.1016/b978-0-12-803581-8.11429-8

Radzi, A. M., Sheikh Ahmad Zaki, Mohamad Zaki Hassan, R. A. Ilyas, Khairur Rijal Jamaludin, Mohd Yusuf Md Daud, and Sa'ardin Abd Aziz 2022. "Bamboo-Fiber-Reinforced Thermoset and Thermoplastic Potential Applications." *Polymers* 14.

Ramakrishnan, T., M. D. Mohan Gift, S. Chitradevi, R. Jegan, P. Subha Hency Jose, H. N. Nagaraja, Rajneesh Sharma, P. Selvakumar, and Sintayehu Mekuria Hailegiorgis 2022. "Study of Numerous Resins Used in Polymer Matrix Composite Materials." *Advances in Materials Science and Engineering* 2022. https://doi.org/10.1155/2022/1088926

Ramesh, M., K. Palanikumar, and K. Hemachandra Reddy 2013. "Mechanical Property Evaluation of Sisal-Jute-Glass Fiber Reinforced Polyester Composites." *Composites Part B: Engineering* 48: 1–9. https://doi.org/10.1016/j.compositesb.2012.12.004

Salman, Suhad D., Mohaiman J. Sharba, Zulkiflle Leman, Mohamed T. H. Sultan, Mohamad R. Ishak, and Francisco Cardona 2016. "Tension-Compression Fatigue Behavior of Plain Woven Kenaf/Kevlar Hybrid Composites." *BioResources* 11 (2): 3575–3586. https://doi.org/10.15376/biores.11.2.3575-3586

Samouh, Zineb, Kolos Molnar, François Boussu, Omar Cherkaoui, and Reddad El Moznine 2019. "Mechanical and Thermal Characterization of Sisal Fiber Reinforced Polylactic Acid Composites." *Polymers for Advanced Technologies* 30 (3): 529–537. https://doi.org/10.1002/pat.4488

Sanjay, Mavinkere Rangappa, Suchart Siengchin 2019. Lightweight Natural Fiber Composites *Journal of Applied Agricultural Science and Technology*; 3 (2): 178. https://doi.org/10.32530/jaast.v3i2.108

Shukor, Faseha, Azman Hassan, Md Saiful Islam, Munirah Mokhtar, and Mahbub Hasan 2014. "Effect of Ammonium Polyphosphate on Flame Retardancy, Thermal Stability and Mechanical Properties of Alkali Treated Kenaf Fiber Filled PLA Biocomposites." *Materials and Design* 54: 425–429. https://doi.org/10.1016/j.matdes.2013.07.095

Sood, Mohit, Deepak Dharmpal, and V. K. Gupta 2015. "Effect of Fiber Chemical Treatment on Mechanical Properties of Sisal Fiber/Recycled HDPE Composite." *Materials Today: Proceedings* 2 (4–5): 3149–3155. https://doi.org/10.1016/j.matpr.2015.07.103

Sreekumar, P. A., and S. Thomas 2008. "Matrices for Natural-Fibre Reinforced Composites." *Properties and Performance of Natural-Fibre Composites*, 67–126. https://doi.org/10.1533/9781845694593.1.67

Sreenivas, H. T., N. Krishnamurthy, and G. R. Arpitha 2020. "A Comprehensive Review on Light Weight Kenaf Fiber for Automobiles." *International Journal of Lightweight Materials and Manufacture* 3 (4): 328–337. https://doi.org/10.1016/j.ijlmm.2020.05.003

Stelea, Lucia, Ioan Filip, Gabriela Lisa, Mariana Ichim, Mioara Drobotă, Costică Sava, and Augustin Mureșan 2022. "Characterisation of Hemp Fibres Reinforced Composites Using Thermoplastic Polymers as Matrices." *Polymers* 14 (3). https://doi.org/10.3390/polym14030481

Suharty, Neng Sri, Hanafi Ismail, Kuncoro Diharjo, Desi Suci Handayani, and Maulidan Firdaus. 2016. "Effect of Kenaf Fiber as a Reinforcement on the Tensile, Flexural Strength and Impact Toughness Properties of Recycled Polypropylene/Halloysite Composites." *Procedia Chemistry* 19: 253–258. https://doi.org/10.1016/j.proche.2016.03.102

Sunny, Tom, and Kim L. Pickering 2022. "An Overview of Alkali Treatments of Hemp Fibres and Their Effects on the Performance of Polymer Matrix Composites." *Alkaline Chemistry and Applications* https://doi.org/10.5772/intechopen.100321

Tawakkal, Intan S. M. A., Marlene J. Cran, and Stephen W. Bigger 2014. "Effect of Kenaf Fibre Loading and Thymol Concentration on the Mechanical and Thermal Properties of PLA/Kenaf/Thymol Composites." *Industrial Crops and Products* 61: 74–83. https://doi.org/10.1016/j.indcrop.2014.06.032

Witayakran, S., W. Smitthipong, R. Wangpradid, R. Chollakup, and P.L. Clouston 2017. Natural Fiber Composites: Review of Recent Automotive Trends. *Encyclopedia of Renewable and Sustainable Materials*. Elsevier Ltd. https://doi.org/10.1016/b978-0-12-803581-8.04180-1

Yussuf, A. A., I. Massoumi, and A. Hassan 2010. "Comparison of Polylactic Acid/Kenaf and Polylactic Acid/Rise Husk Composites: The Influence of the Natural Fibers on the Mechanical, Thermal and Biodegradability Properties." *Journal of Polymers and the Environment* 18 (3): 422–429. https://doi.org/10.1007/s10924-010-0185-0

Zhao, Xuefeng, Robert K. Y. Li, and Shu Lin Bai 2014. "Mechanical Properties of Sisal Fiber Reinforced High Density Polyethylene Composites: Effect of Fiber Content, Interfacial Compatibilization, and Manufacturing Process." *Composites Part A: Applied Science and Manufacturing* 65: 169–174. https://doi.org/10.1016/j.compositesa.2014.06.017

Zhong, Jinbiao, Jianping Lv and C. B. Wei "Mechanical properties of sisal fibre reinforced urea- formaldehyde resin composites." *Express Polymer Letters* 1 (2007): 681–687.

16 Environmental Degradation due to Synthetic Fibres

Indira V., Surej Rajan C., Shana Shaji, and Soney C. George

16.1 INTRODUCTION

The global consumption of synthetic fibres is increasing every year to meet the demand. Even though synthetic fibres have many advantages over natural fibres, they are causing different environmental pollution issues. Synthetic fibres are not biodegradable and are not environmentally sustainable. Over several benefits, synthetic fibres are a threat to the environment as they release huge amounts of microplastics, their production needs large quantities of water, their processing causes different effluents to be discharged into water streams, and gaseous pollutants are also released into the atmosphere. Thus, synthetic fibres cause water and air pollution, and they also affect human health and other species on the earth.

In the 2021 report titled ‘From Pollution to Solution’, published by the United Nations Environment Programme (UNEP), a stark warning is issued regarding the repercussions of chemicals found in microplastics, particularly emphasizing their severe impact on women’s health. The report underscores a range of health concerns, such as alterations to human genetics, disruptions in brain development, and changes in respiration rates, among other significant issues. These findings shed light on the urgent need for comprehensive solutions to address the adverse effects associated with microplastic pollution.

Synthetic clothing is mainly composed of polyester and nylon and contributes extensively to the pollution of the environment. Microplastics generated by these synthetic textiles are already poisoning our environment, and the world’s growing manufacturing of synthetic fibres raises concerns that this trend will continue. In numerous environmental samples, including those taken from wastewater treatment plants, fibres are frequently found to constitute the primary component of microplastics. It is estimated that 1.5 million tonnes of the 8 million tonnes of plastic that enter the ocean each year are made up of these microplastics. The International Union for the Conservation of Nature estimates that 35% of the microplastic that ends up in the ocean comes from fibres from synthetic garments (Sophie, 2019).

Over 700,000 individual microfibres can be released into the water by just washing synthetic fabrics; these microfibres eventually find their way into rivers and seas. Little invertebrates that are later eaten by small fish may absorb these microscopic bits of plastic. This unfavourable series of events disturbs the fragile ecosystems of the planet while also introducing plastic into the food chain. A study by the University of Plymouth and the Institute for Polymers, Composites and Biomaterials of the Italian National Research Council (IPCB-CNR) also demonstrated that microfibres can be emitted even when wearing only synthetic clothing. By wearing the clothing, users might theoretically contribute up to 900 million polyester microfibres to the atmosphere each year (Marie, 2021). The popularity of synthetic textile fibres has witnessed a significant increase, with 67% of textile fibres being of synthetic origin in the year 2016 (Table 16.1 and Figure 16.1). Apart from microplastic release and water pollution, these synthetic fibres also cause carbon dioxide (CO_2) emissions, which is a greenhouse gas.

DOI: 10.1201/9781032665337-16

TABLE 16.1
Global Textile Fibre Production in 2016 (Marijnissen and Moskal, 2020)

Type of Man-Made Fibres	Production (Millions of Tonnes:2016)	Type of Natural Fibres	Production (Millions of Tonnes:2016)
Synthetic:			
Polyester staple	16.21	Cotton	22.48
Polyester filament	36.19	Wool	1.10
Nylon	4.79	Silk	0.19
Acrylic	1.79		
Others	0.96		
Total synthetic fibres	59.94		
Cellulosic	5.35		
Total man-made fibres	65.29	Total natural fibres	23.77
Grand Total	89.06		

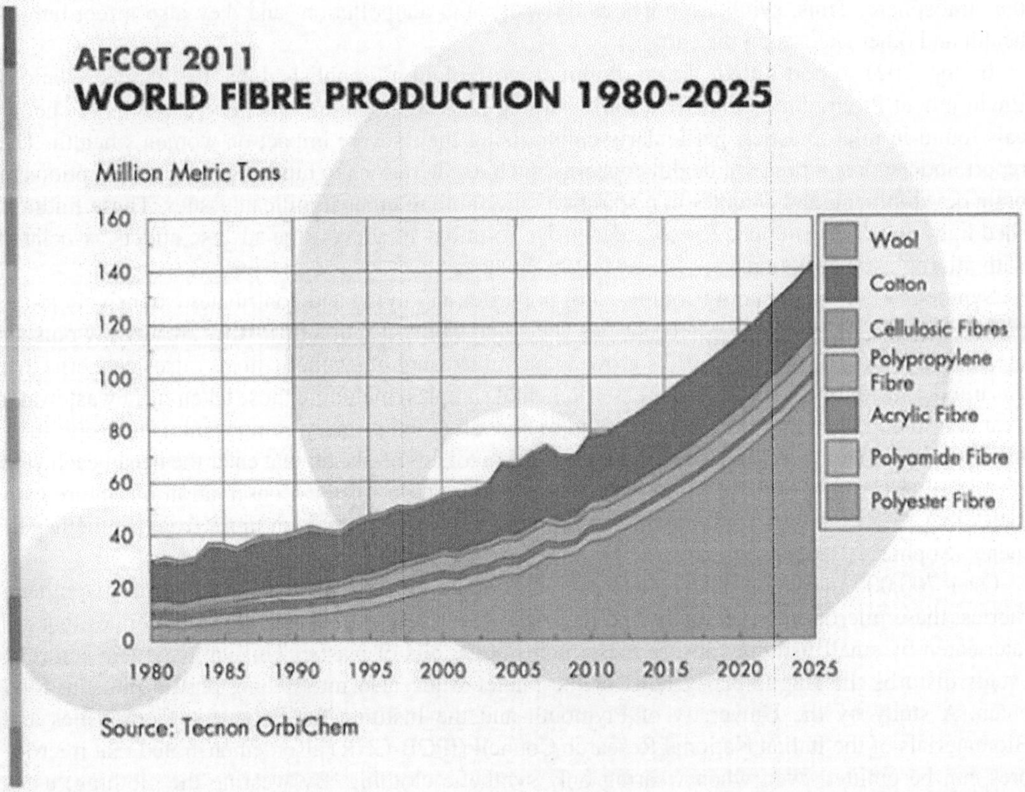

FIGURE 16.1 Historical and anticipated production of textile fibres worldwide (Marijnissen and Moskal, 2020).

16.2 NYLON

Nylon was the first synthetic fibre developed by an American scientist, Wallace H. Carothers, and his research panel, for E.I. Du Pont de Nemours & Company. Chemically, the material was polyhexamethylene adipamide, also recognized as nylon 6,6. Figure 16.2 shows the chemical reaction

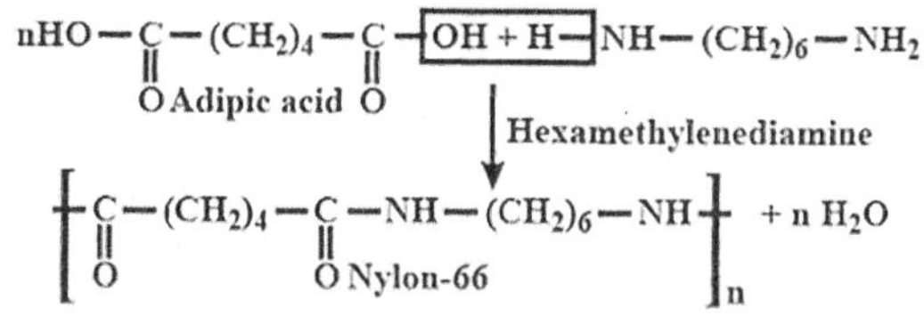

FIGURE 16.2 Chemical reaction resulting in the formation of Nylon 66.

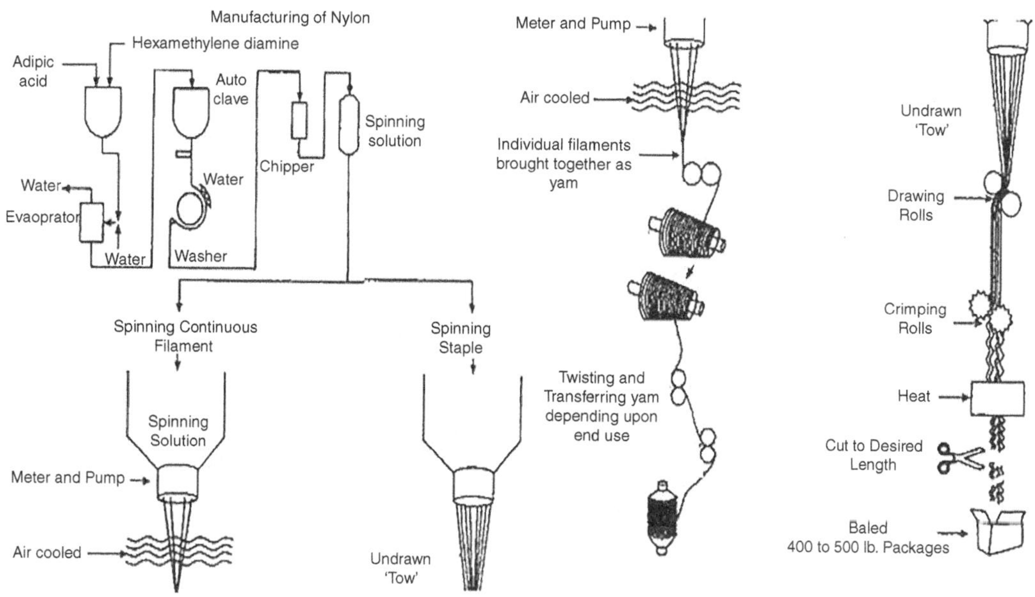

FIGURE 16.3 Nylon manufacturing process.

https://www.brainkart.com/article/Nylon---Manufacturing-Process-and-Properties_1769/#google_vignette.

representing the formation of Nylon 66. Nylon is a common name for a set of synthetic polymers, precisely, semi-aromatic or aliphatic polyamides in which no less than 85% by weight of the amide links (-CO-NH-) are attached directly to cycloaliphatic or aliphatic units. These polyamides are formed primarily by the condensation reaction or by the addition of polymerization between monomers.

Large-scale production of this new fibre began in 1939 at DuPont's factory in Seaford, United States, and is still used extensively to date. This venture was then designated as a momentous landmark by the American Chemical Society in the year 1995. Later, the marketing of DuPont's fibre, nylon 6 (polycaprolactam), was manufactured in Europe using the polymerization of caprolactam. This 'first man-made textile fibre', manufactured 'as strong as steel and as fine as the spider's web', was highly accepted in the market. Nylon was announced as part of 'The World of Tomorrow' at the 1939 New York World's Fair and was highlighted at DuPont's 'Wonder World of Chemistry' at the Golden Gate International Exposition in San Francisco in the year 1939. The manufacturing process of nylon is given in Figure 16.3.

The polyamides are classified and named corresponding to the number of carbon atoms existing in the structural unit of the molecule. For example, nylon created from hexamethylene diamine ($NH_2(CH_2)_6$-NH_2) with six carbon atoms and sebacic acid (COOH-$(CH_2)_8$-COOH) having ten carbon atoms is named as nylon 6,10. A few other nylon categories and their respective raw materials are mentioned in Table 16.2.

TABLE 16.2
Types of Nylon and Its Raw Materials

Nylon	Raw Materials
Nylon 4,6	1,4 diaminobutane, adipic acid
Nylon 6,6	Hexamethylenediamine, adipic acid
Nylon 6,12	Hexamethylenediamine, dodecanedioic acid
Nylon 3	Acrylamide
Nylon 4	2-pyrrolidone
Nylon 6	Caprolactum
Nylon 7	Lactum of heptanoic acid
Nylon 11	W-amino-cendecanoic acid
Nylon 12	Dodelactum

TABLE 16.3
Applications of Nylon in Various Fields

Field	Specific Applications
Engineering	Fish plates for railway line tubing, bearings, tyre cord
Textiles and cosmetics	Garments, brushes
Sports and recreation	Tents, parachutes, tennis rackets, and guitar strings
Medical	Suture threads, gloves, implants, catheters
Electrical	Seals, circuit insulation boards, switch gears
Others	Fishing nets, carpets, upholstery, seat belts, air bags

Nylon was first used to make toothbrush bristles, ropes, and so on. Later, the properties like lightweight, shiny appearance, and ease in handling and washing made it a choice of textile manufacturers, and it became an inevitable matter that we use daily. Nylon is similar to polyester but stronger and more expensive than polyester. Nylon clothes absorb very little water and are easy to wash and dry. It has excellent abrasion resistance and resistance against ordinary chemicals and moths. The persistent nature of nylon means that it is infinitely recyclable. The common applications of nylon fibre are mentioned in Table 16.3. Other special characteristics of nylon fibres are as follows:

1. Superior mechanical stiffness and strength
2. Better fatigue and wrinkle resistance
3. Excellent mechanical damping capability
4. Satisfactory sliding properties
5. Competent electrical insulating properties
6. Fine forbearance to high-energy radiation (gamma- and X-rays)
7. Decent machinability

16.2.1 Environmental Impacts of Nylon

According to the 'Pulse of Fashion Report 2017', nylon crowns the list of synthetic materials that have the greatest impact on the environment. Compared to other plastic-based fibres, the manufacturing and production of nylon is highly energy intensive and produces greenhouse gases. Nitrous oxide is a by-product formed during the process of nylon manufacture, particularly in the

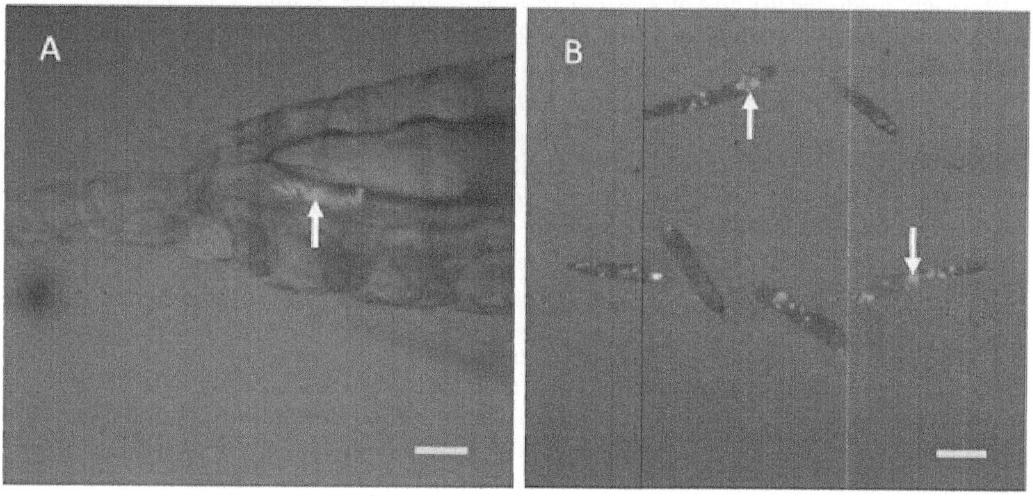

FIGURE 16.4 Ingestion and egestion of microplastics by juvenile Calanus finmarchicus: (a) nylon fibres (10 × 30 μm) in the intestinal tract and (b) nylon granules (10 – 30 μm) in the faecal pellets.

Reproduced with permissions from Cole et al. (2019).

preparation of adipic acid. This nitrous oxide can catalyse stratospheric ozone depletion and thereby global warming. N_2O in the earth's atmosphere is known to be increasing by about 0.2% per year (Weiss 1981; Busenberg, 2014). The atmospheric lifetime of nitrous oxide is approximately 150 years (Ceaon, 1989), and this explains how hazardous its impact is.

Marine contamination, which occurs during the production, synthesis, and usage of synthetic fibres, has become a relevant ecological issue in recent years. The resilience and slow rate of degradation allow these microplastics to survive in the ocean environment for years to decades or longer. These microplastics may be ingested by plankton or other organisms and can enter food chains, finally to human beings (De Falco et al. 2018). They may accumulate within aquatic organisms, resulting in physical impairments like internal abrasions and blockages (Wright et al., 2013). Over 250 marine species are believed to be impacted by plastic ingestion.

A study by Cole et al. explains that nylon granules made significant shifts in prey selectivity, negatively affected lipid accumulation and resulted in premature moulting of a cold water copepod Calanus finmarchicus (Cole et al. 2019). Figure 16.4 shows the nylon fibres and granules in the intestinal tract and faecal pellets of juvenile copepods, which confirm the uptake and egestion of the fibre. This result suggests that microplastic ingestion can affect each and every stage of an organism and can make alterations, thereby intervening in the natural way of life. According to the Ellen MacArthur Foundation, a release of about half a million tonnes of plastic microfibres, equivalent to more than 50 billion plastic bottles, happens into the ocean every year.

Nylon does not blend with natural dyes and low-impact chemical dyes. Therefore, the procedure of colouring nylon creates significant water pollution. Along with this, the importance of water in the production stage of nylon can also create water scarcity. For example, the most popular Nylon 66 requires a very high energy of 138.62 MJ and 663kg of water to create just 1 kilo of fabric (Subramanian, 2014). This shows how intensively this synthetic fibre causes water depletion. Wastewater generated during the production of nylon contains unreacted caprolactum. If this factory wastewater is expelled without undergoing any treatment, it can adversely affect the marine population.

Raw materials in nylon production include fossil fuels. The growing demand for nylon would reduce fossil fuel reserves rapidly. Depletion of these non-renewable resources should be a matter of concern. In addition to this, the extraction of fossil fuels requires drilling and fracking of the earth and is hazardous to the biosphere.

In a study conducted in China, fibrous and granular microplastics as well as nanoplastics of nylon were exposed to rats and observed for the excretion behaviour of consumed nylon particles. This study revealed that large granules were excreted completely, whereas the nanoparticles were not. The missed particles were detected in the blood, indicating the absorption of nylon through the digestive tract (Peng et al. 2022).

The US government conducted a cross-sectional medical survey of current employees of a nylon flock plant to get information regarding a cluster of workers with biopsy-confirmed interstitial lung disease. The study included evaluation surveys, spirometry, chest radiograph and diffusing lung capacity. On the study assessment, workers experienced frequent throat and eye irritation, respiratory symptoms, and generalized aches and fevers. Another important fact is that the majority of the study population experienced better health when away from labour (Washko et al. 2000). An association between nylon inhalation and various ailments can be seen here.

A retrospective study conducted on a cohort from a nylon flocking plant (n = 162) reported a threefold increase in risk of lung cancer (Kern et al. 2011). Additionally, employees from a polyester and polyamide fibre factory in France demonstrated a statistically higher relative risk of mortality from various cancers (1.42 (n = 79, 95% CI 1.06 to 1.89) for high exposure or 1.38 (n = 105, CI 1.05 to 1.81) for former exposure to polymer dust). This group also exhibited a statistically higher risk of lung cancer mortality, regardless of the level or duration of polymer dust exposure (Hours et al. 2007). However, it's important to note that these findings warrant an update with a larger cohort and consideration for confounding factors such as individual smoking histories. Conversely, no association between exposure to synthetic fibres and lung cancer risk was observed in female textile workers in China (Checkoway et al. 2011).

16.3 POLYESTER

Polyester is a kind of polymer derived from petroleum source and is chemically called polyethylene terephthalate (PET). It is mainly used as fibre in clothing (Figure 16.5), and other applications include packaging and engineering applications. The majority of synthetic fibres manufactured globally are made of polyester. The following are some factors that have contributed to PET fibre's overwhelming success: Ease of processing, flexible blending with other natural fibres, recyclability,

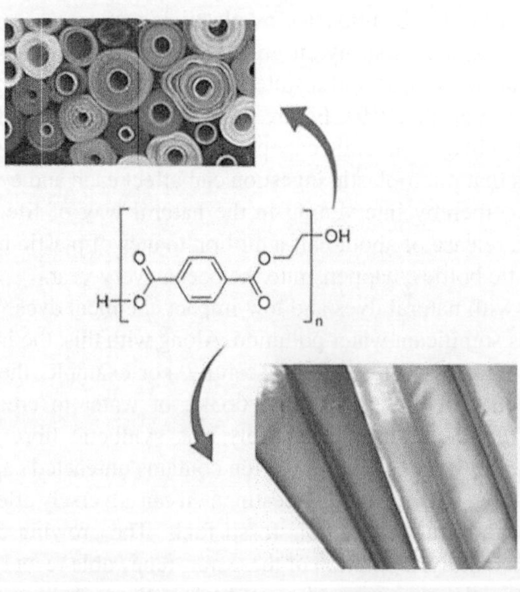

FIGURE 16.5 Polyethylene terephthalate (PET) in the form of fibre and fabric.

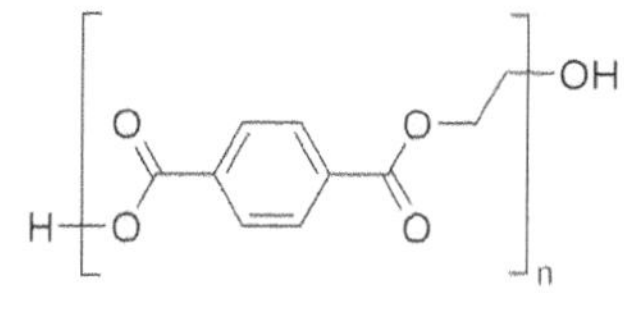

FIGURE 16.6 Chemical structure of polyethylene terephthalate (PET).

good performance, and low production cost. Like other synthetic thermoplastic fibres, PET fibre was first developed as a result of W. H. Carothers' work at DuPont in the 1930s on aliphatic condensation polymers. J. Rex Whinfield and J. T. Dickson, a British research team, concentrated their effort on the aromatic-aliphatic polyester made from terephthalic acid and ethylene glycol, and they were able to significantly increase fibre performance in the early 1940s.

After World War II, commercialization of PET was accelerated by the introduction of Terylene in Great Britain and Dacron by DuPont in the United States. PET fibre also found success in the industrial market as a rubber reinforcement filament yarn, which was primarily used in the sidewalls of radial passenger-car tyres (Jaffe et al., 2020). They have a fair amount of flame resistance, insect and microbial resistance, and are physiologically inert. Polyester is produced by a large number of companies worldwide under many trade names that virtually all have custom-fit qualities.

Polyesters are polycondensation products of dicarboxylic acids with dihydroxy alcohols (diols), or polymers containing ester groups connected to the main chain. The chemical structure of PET is depicted in Figure 16.6. Polycondensation employing diols and a diacid (or an acid derivative) or from a hydroxy acid is the conventional method for making polyesters. To get polymer chains with high enough molecular weights to provide the fibres with relevant mechanical characteristics, a very high conversion rate is preferred. Due to side reactions and the volatilization of monomers, which causes a stoichiometric imbalance of reactants, a high degree of polymerization using this approach is highly challenging to accomplish, despite all precautions. An alternate technique, ring-opening polymerization of lactones, cyclic diesters, and cyclic ketene acetals, has been successfully used to produce large molecular mass polymers under comparatively benign circumstances. Controlling parameters like molecular weight and molecular weight distribution is achievable because of the fact that this polyaddition procedure may be performed with either no or very few side reactions (Brunelle 2008).

Polymer chips are often melted and spun into endless filaments to start the production of synthetic fabrics. These filaments are carded into slivers after being cut into short staple fibres. The next phase is spinning slivers into yarn using various yarn-spinning techniques. Spun yarns are yarns that are created from short staple fibres. In line with this, yarns constructed of infinite filaments are referred to as 'filament yarns'. After that, these yarns can be used to weave or knit further fabrics. A single thread is knitted along a route to create symmetric loops on both sides of the mean path, whereas two threads are woven together perpendicularly to create a knitted cloth. At the finishing stage, a variety of methods can be used to improve the finished product's functionality, appearance, and feel. At this point, surface treatments like fleece can be added to create fabrics with unique textures. Ultimately, the completed fabrics are sewn, fitted, and shipped to clients as clothing. This entire process is schematically represented in Figure 16.7.

16.3.1 Environmental Impacts of Polyester

According to De Falco, F., et al., more than 6,000,000 microfibres are typically discharged from a standard 5 kg wash load of polyester garments, and this amount is dependent on the detergent used. The major factors involved in the microfibre release by washing include fabric and detergent type and wash conditions. Woven polyester fibre can release a larger number of microfibres as compared to knitted polyester fibre. Also, the usage of softeners can reduce the release of microfibres since it

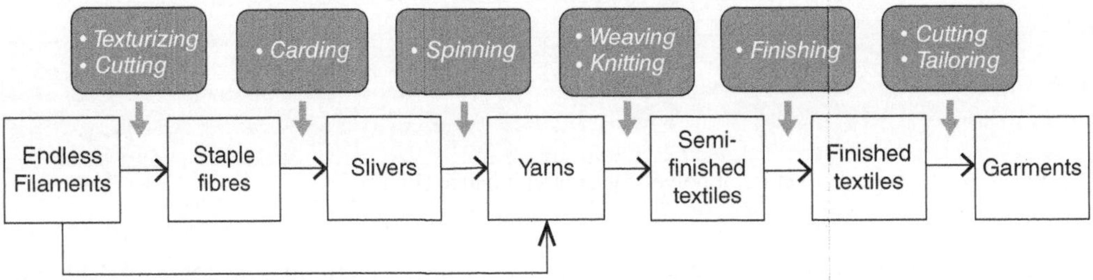

FIGURE 16.7 Manufacturing of polyester textiles (Cai et al., 2020).

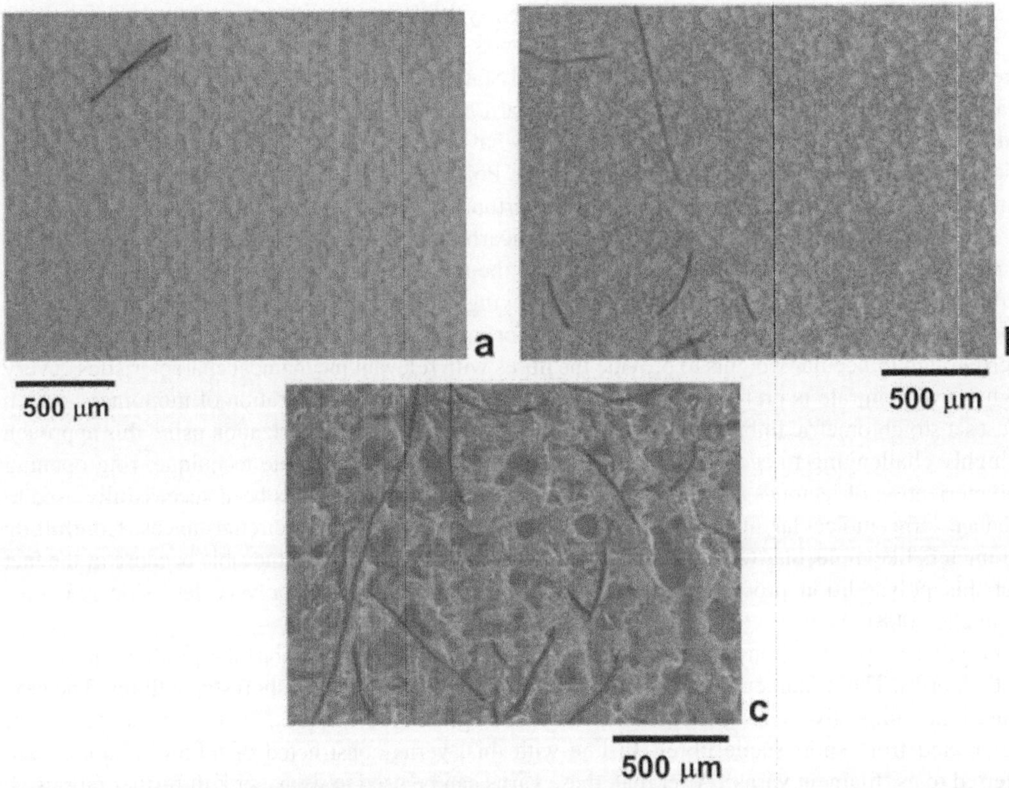

FIGURE 16.8 SEM images of filter surfaces containing microfibres coming from plain weave polyester washed under domestic conditions: (a) water, (b) liquid detergent, and (c) powder detergent.

Reproduced with permissions from De De Falco et al. (2018).

can reduce the friction between the fibres. Figure 16.8 shows that a large number of microfibres are released by washing with powder detergent as compared to water and liquid detergents. This can be due to the presence of inorganic components like zeolites, which are insoluble in water and cause friction with the fabrics. Apart from this, the pH of powder detergents can also increase the microfibre release (De Falco et al. 2018).

Temperature of washing can also influence the number of microfabrics removed upon washing. The number of polyester fibres released while washing polyester fabric at 60°C is higher than the amount released at 30°C. Figure 16.9 shows the release of microfibre from the polyester fabric after washing (Yang et al. 2019). According to a recent modelling research, textile fibres greatly

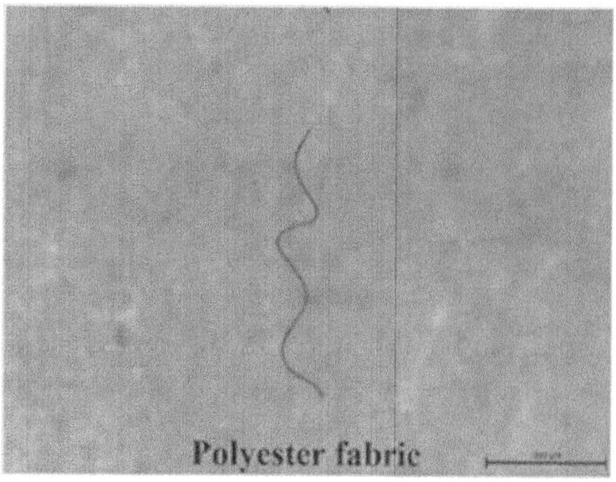

FIGURE 16.9 Microfibre releasing from polyester fabric after washing.

Reproduced with permissions from Yang et al. (2019).

contribute to microplastic emissions into freshwater (Henry et al., 2019; Kawecki and Nowack, 2019). Studies show that microplastic fibre of polyester is generated through the fibre manufacturing process and also during usage. According to Cai Y et al., one of the most important processes leading to microplastic fibre development during the manufacturing of yarn and textiles, respectively, is rotor spinning. Another is surface treatment. Also, the procedures used to cut out textile swatches have a big impact on how many microplastic fibres are extracted. Their study confirms the presence of microplastic fibres in textiles at every stage of production (Cai et al. 2020).

According to the Water Footprint Network, the water footprint of polyester can be as high as '71,000 cubic metres per tonne of fibre'. Although polyester and nylon are not crops like cotton, they generally need water from industrial manufacturing, including water from oil exploration. Moreover, a lot of water is needed during the dyeing process. In order to produce one tonne of coloured fabric, it can require up to 200 tonnes of freshwater, according to Sustain Your Style, a group dedicated to educating consumers about the ethical and environmental problems associated with the fashion and apparel industries. Also, it is possible that the dyes used for these materials may contaminate and pollute the water. The wastewater disposal from these industries is infrequently controlled; dangerous chemicals and dyes can be thrown into rivers nearby. The water in these rivers, if not treated, may be harmful to drink, irritant to the skin, and deplete ecosystems of the freshwater they require to exist. Communities that depend on these rivers suffer disproportionate harm and are frequently obliged to use the contaminated water despite the risks associated with doing so (Marie, 2021). Regarding the water footprint, it was revealed that both the manufacture of virgin polyester fabric and recycled polyester fabric had a significant influence on the eutrophication and shortage of freshwater. The biggest footprint of water eutrophication was generated by chemical oxygen demand, followed by ammonia-nitrogen and five-day biochemical oxygen demand (Qian et al. 2021).

According to the Environmental Protection Agency (EPA), carbon dioxide was responsible for 80% of all US greenhouse gas emissions from human activities. As polyester is made from petroleum, its manufacturing is connected to the effects of the fossil fuel sector on the environment. The World Wildlife Fund points to the fact that petroleum is a fossil fuel, which produces harmful pollutants into the environment when it is used. These emissions include carbon dioxide, which accelerates global warming (Marie, 2021). The manufacture of virgin polyester textiles has a carbon footprint of 119.59 kg CO_2/100 kg. The technique used to produce terephthalic acid accounted for the greatest share, accounting for 45.83%, followed by the processes used to make polyester fabric, ethylene, paraxylene, ethylene glycol, and polyester fibre. The overall carbon footprint of recycling waste polyester was 1154.15 kg CO_2/100 kg, which is around ten times that of producing virgin

TABLE 16.4
CO_2 Emission of Different Synthetic Fibres Based on Energy Consumption (Subramanian, 2014)

Fibre Type	Energy Consumption kWh/kg fibre	CO_2 Emission in kg/kg fibre
Nylon	69	37
Acrylic	49	26
Polyester	35	19
Viscose	28	15

polyester fabrics (Qian et al. 2021). Table 16.4 shows the amount of CO_2 emission in kg per kg of fibre. It was observed that nylon contributes the maximum to CO_2 emission as compared to other synthetic fibres.

16.4 RAYON

Rayon is a synthetic fibre developed by redeveloping cellulose. Pure cellulose, the material found mostly in the cell walls of such woody plants as trees and cotton, makes up rayon fibre. These fibres are known as regenerated cellulose fibres because they are manufactured from cellulose that has been reformed or regenerated. Among all natural polymers, cellulose is expected to be one of the most useful. It is a renewable polymer that degrades naturally. Wood pulp and cotton lint are typical sources used in industry. Ninety-five to ninety-nine per cent of the cellulose is present in highly purified wood pulp. Man-made fibres are produced using such chemical cellulose or dissolving pulps. The chemical structure of cellulose is given in Figure 16.10. Since rayon is made from naturally occurring polymers, it cannot be classified as a synthetic cellulosic fibre. As a result of its filamentous structure and resemblance to silk material, rayon was first referred to as artificial silk. Rayon fabric feels soft like silk and has a texture resembling cotton fabric. Different types of rayon fibres include viscose rayon, cuprammonium rayon, and saponified cellulose acetate.

Viscose was once known as rayon, and it was developed in 1855. When George Audemars plunged a needle into a viscous mixture of mulberry bark pulp and gooey rubber, he discovered that he could create a thread. The first commercial 'synthetic' fibre was developed by Hilaire de Bernigaud in 1884, and he went on to produce the material known as Chardonnay Silk in 1889. However, the fabric was very combustible, thus it was removed from the market. Charles Cross discovered the viscose method in 1891, and in 1905 Courtaulds created the first secure, commercial viscose rayon, also known as artificial silk.

Although rayon is incredibly soft, cool, pleasant, and has excellent absorption abilities, it cannot adequately insulate body heat and should not be used in humid, hot climates. The most common purpose of rayon is to create different types of clothes and home decor, including blankets, sheets, and curtains. Moreover, medical supplies and tyre cables can both be made from rayon. Similar to natural fibres, rayon fibre has a high level of comfort.

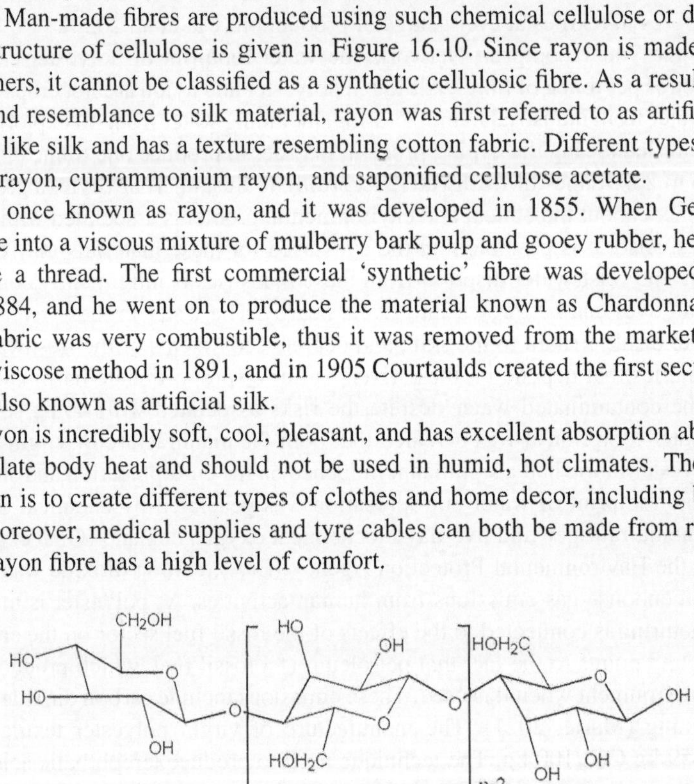

FIGURE 16.10 Chemical structure of cellulose.

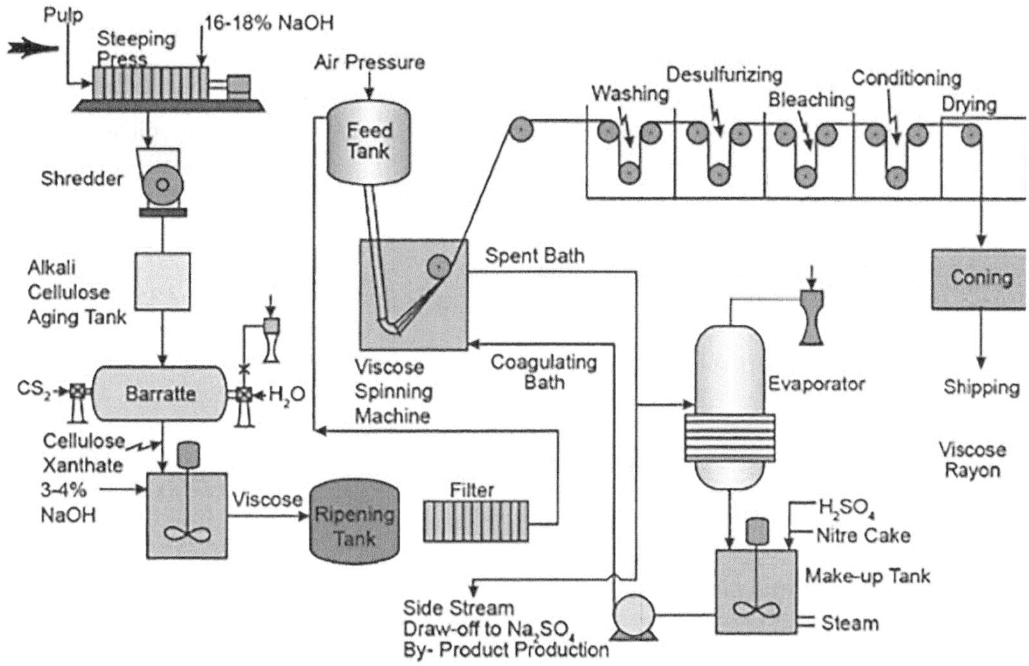

FIGURE 16.11 Viscose rayon manufacturing process

https://www.textileflowchart.com/2015/05/flow-chart-of-viscose-rayon-manufacturing-process.html.

The word 'rayon' (artificial silk) refers to any fibre made of regenerated cellulose, as well as the yarn and fabric that come from it. Scientists tried to create their own synthetic silk, which they called 'artificial silk', because natural silk was more costly and required a great deal of labour to make. In 1855, Frenchman George Audemars succeeded in creating a thread of rayon by plunging a needle into a sticky mixture of mulberry bark pulp and gummy rubber. Although this method was fascinating from a scientific perspective, it was not commercially feasible due to its extreme slowness and high skill and precision requirements. A commission established by the U.S. Department of Commerce and other business groups chose the term 'rayon' for 'Artificial Silk' in 1924. One of two factors contributed to its naming as 'rayon': either its brightness or the structural similarities with cotton (sun = ray, on = cotton). The word 'viscous', which refers to a sticky spinning solution used to make 'rayon', is where the name 'Viscose' originated. As a result, the novel cellulosic derivative is now known as 'Viscose rayon' (Shaikh et al., 2012). Figure 16.11 depicts the manufacturing process diagrammatically. The chemical processes that occur throughout the production process are now briefly summarized.

1. Cellulose is converted to alkali cellulose during steeping.

$$C_6H_9O_4OH + NaOH \rightarrow C_6H_9O_4ONa + H_2O$$

2. Carbon disulphide reacts with alkali cellulose. Sodium cellulose xanthate and sodium tri-thiocarbamate are produced during Xanthation.

$$C_6H_9O_4ONa + CS_2 \rightarrow C_6H_9O_4OCSSNa + Na_2CS_3$$

3. Viscose solution is formed during dissolution.

$$C_6H_9O_4OCSSNa + NaOH \rightarrow \text{Viscose Solution}$$

4. Sodium cellulose xanthate is decomposed to get cellulose during ripening.

$$C_6H_9O_4OCSSNa + H_2O \rightarrow C_6H_9O_4OH + CS_2 + NaOH$$

5. Recovery of cellulose from cellulose xanthate by acid decomposition during spinning.

$$C_6H_9O_4OCSSNa + H_2SO_4 \rightarrow C_6H_9O_4OH + CS_2 + Na_2SO_4$$

16.4.1 Environmental Impacts of Rayon

Rayon has a moderate dry strength, is abrasion-resistant, and feels soft on the skin. It is very pleasant. Although wood pulp, a reasonably priced and renewable resource, is used to make rayon, the manufacturing procedure uses a lot of water and energy, which raises the price. Moreover, it resulted in additional air and water pollution. The production process itself can cause environmental pollution. Major pollution is caused by the emission of carbon disulphide (CS_2). It was the primary industrial solvent to be understood as a neurotoxic agent. There are various reports regarding its central and peripheral neurotoxic influences, especially from viscose rayon factories.

According to the EPA, contact with carbon disulphide in the workplace can have an impact on humans. Short-term exposure can cause light-headedness, headaches, nausea, and visual impairments, whereas long-term exposure may result in neurophysiological as well as behavioural alterations along with reproductive issues such as menstrual disturbances and decreased sperm count (Inga Vallery, 2022). Findings from a study on carbon disulphide exposed-rayon workers pointed out damage in receptiveness of ganglion cells or demyelination of optic nerve fibres (Raitta et al., 1981) and peripheral constriction of the visual field (Kim et al., 1989) in the study sample. The subclinical polyneuropathy caused by CS_2 was suggested to be irreversible, as the cessation of exposure did not amend the reduced conduction velocity of motor nerves (Seppäläinen and Tolonen, 1974). Neurological symptoms such as gait disturbance, extremity weakness, paraesthesia, and speech disturbance affected many of them. Without realizing the severity, labourers get immense health problems from their workspace.

16.5 ACRYLIC FIBRES

Acrylic fibres are a significant class of synthetic fibres formed from polyacrylonitrile and developed by addition of polymerization of acrylonitrile (vinyl cyanide). In real life, the majority of commercial acrylic fibres are copolymers of the vinyl monomers acrylonitrile, vinyl chloride, vinyl acetate, vinyl alcohol, vinylidene chloride, acrylic acid, methacrylic acid, and methacrylate esters. An acrylic fibre must be made of a polymer that contains at least 85% of acrylonitrile by weight (Davis and Shapiro, 1964; Minor, 1982). Dry spinning extrusion or wet spinning extrusion methods are used to produce acrylic fibres. Although the fibres have an adequate degree of chemical stability, at high temperatures they go through a number of changes in their chemical and physical structure. The production of carbon and graphite fibres uses acrylic fibres as a major precursor material. The co-monomer used with acrylonitrile to create the polymer has a significant impact on the physical characteristics of acrylic fibres (Rebenfeld, 2002).

16.5.1 Environmental Impacts of Acrylic Fibres

USEPA (2011) states that air pollutants released during the manufacture of acrylic fibre include acrylonitrile (volatilized residual monomer), solvents, additives, and other organics. The production of 1 kg of acrylic fibre results in 5.4 kg of CO_2 equivalent being released into the atmosphere, according to calculations of the potential for global warming. According to Beton et al., the maximum CO_2 emission was 35.7 kg CO_2 eq/kg fibre, which was the highest value possible (Beton et al., 2014).

According to the findings, the emission of arsenic, cadmium, zinc, and chromium into the air and water during the manufacture of acrylic fibre has the potential to cause cancer. The manufacture of copper wires used in the distribution of energy is linked to arsenic emissions to liquid and air (Beton et al., 2014). The remaining emissions, which include arsenic, are connected to the fabric's dyeing procedure. According to findings from studies of Laing, Barclay, and Buckley, depending on the kind of dye used in the dying process, the loss of dyes to effluent in the textile industry can be estimated to be between 2% and 10% of the total amount of dyes used (Laing, 1991; Barclay and Buckley, 2000). Acrylic fibres are dyed using a variety of dyes that include these emissions, such as acid dyes, basic dyes, and dispersed dyes and phenol compounds used with disperse dyes (Knackmuss, 1996; Shenai, 2001; Kolekar, 2010). According to Ning et al., components of synthetic dyes, polycyclic aromatic hydrocarbons (PAHs), are found as pollutants in textile dyeing sludge, which might endanger the environment when the sludge is disposed of. Before being released into the environment, process effluent must be properly treated (Ning et al., 2014).

16.6 SUMMARY

This chapter gives a detailed update on the environmental issues of different synthetic fibres like nylon, polyester and rayon. The chapter is divided into three main divisions covering the major environmental impacts of all synthetic fibres. Based on the recent literature, the environmental pollution issues like microplastic release, air pollution, water pollution, and human health effects are discussed in each division.

REFERENCES

Barclay S, Buckley C (2000). *Waste Minimisation Guide for the Textile Industry, a Step Towards Cleaner Production (volume 1), by the Pollution Research Group,* University of KwaZulu, Natal Durban-South Africa, January, 92.

Beton A, Debora D, Laura F, Thomas G, Yannick L, Marie D, Anne P, Ines B, Oliver W, Jiannis K, Mauro C, Nicholas D (2014). *JRC Scientific and Technical Reports: Environmental Improvement Potential of Textiles (IMPRO Textiles).* European Commission Joint Research Center Institute for Prospective Technological Studies (IPTS), Seville - Spain. EUR Number: 26316 EN.

Brunelle Daniel J 2008. "Cyclic Oligomer Chemistry." *Journal of Polymer Science, Part A: Polymer Chemistry* 46 (4): 1151 1164. https://doi.org/10.1002/pola.22526

Busenberg Eunyshops. 2014. "*Use of Chlorofluorocarbons (CCL 3 F and CCL 2 F 2) as Hydrologic Tracers and Age-Dating Tools — the Alluvium and Terrace System of Central Oklahoma (CCl3F and CCl2Fz) as Hydrologic Tracers Use of Chlorofluorocarbons and Age-Dating Tools: The Alluvium and Terrace System of Central Oklahoma,*" no. November.

Cai Yaping, Mitrano Denise M, Heuberger Manfred, Hufenus Rudolf, Nowack Bernd 2020. "The Origin of Microplastic Fiber in Polyester Textiles: The Textile Production Process Matters." *Journal of Cleaner Production* 267: 121970. https://doi.org/10.1016/j.jclepro.2020.121970

Ceaon Ralph J C 1989. "Analysis of Sources and Sinks of Atmospheric Nitrous Oxide (NO)." *Journal of Geophysical Research* 94 (D15): 265–283.

Checkoway Harvey, Ray R M, Lundin J I, Astrakianakis G, Seixas N S, Camp J E, Wernli K J, et al. 2011. "Lung Cancer and Occupational Exposures Other than Cotton Dust and Endotoxin among Women Textile Workers in Shanghai, China." *Occupational and Environmental Medicine* 68 (6): 425–429. https://doi.org/10.1136/oem.2010.059519

Cole Matthew, Coppock Rachel, Lindeque Penelope K, Altin Dag, Reed Sarah, Pond David W, Sørensen Lisbet, Galloway Tamara S, Booth Andy M 2019. "*Effects of Nylon Microplastic on Feeding, Lipid Accumulation, and Moulting in a Coldwater Copepod.*" https://doi.org/10.1021/acs.est.9b01853

Davis C W and Shapiro P in *Encyclopedia of Polymer Science and Technology*, Vol 1, pp. 342–373, John Wiley & Sons, Inc., New York, 1964.

De Falco Francesca, Gullo Maria Pia, Gentile Gennaro, Di Pace Emilia, Cocca Mariacristina, Gelabert Laura, Brouta-Agnésa Marolda, et al. 2018. "Evaluation of Microplastic Release Caused by Textile Washing Processes of Synthetic Fabrics." *Environmental Pollution* 236: 916–925. https://doi.org/10.1016/j.envpol.2017.10.057

Henry Beverley, Laitala Kirsi, Klepp Ingun Grimstad 2019. "Microfibres from Apparel and Home Textiles: Prospects for Including Microplastics in Environmental Sustainability Assessment." *Science of the Total Environment* 652: 483–494. https://doi.org/10.1016/j.scitotenv.2018.10.166

Hours, M, Févotte J, Lafont S, Bergeret A. 2007. "Cancer Mortality in a Synthetic Spinning Plant in Besançon, France." *Occupational and Environmental Medicine* 64 (9): 575–581. https://doi.org/10.1136/oem.2006.028282

Inga Vallery Phlegar January 25, 2022. *What Is Rayon- and Is It Sustainable? Consider This Your Go-To Guide*, Accessed on April 01, 2023, https://brightly.eco/blog/what-is-rayon January 2022.

Jaffe Michel, Easts Anthony J, Feng Xianhong 2020. *Polyester Fibers. Thermal Analysis of Textiles and Fibers*. Vol. c. LTD. https://doi.org/10.1016/B978-0-08-100572-9.00008-2

Kawecki Delphine, and Nowack Bernd 2019. "Polymer-Specific Modeling of the Environmental Emissions of Seven Commodity Plastics as Macro- and Microplastics." *Environmental Science and Technology* 53 (16): 9664–9676. https://doi.org/10.1021/acs.est.9b02900

Kern David G, Kern Eli, Crausman Robert S, Clapp Richard W 2011. "A Retrospective Cohort Study of Lung Cancer Incidence in Nylon Flock Workers, 1998–2008." *International Journal of Occupational and Environmental Health* 17 (4): 345–351. https://doi.org/10.1179/107735211799041814

Kim Soon Duck, Song Dong Bin, Yum Yong Tae (1989). Clinical Analysis of 8 Cases of Chronic Carbon Disulfide Poisoning in Workers Engaged in the Viscose Rayon Industry, *Korean Journal of Occupational and Environmental Medicine* 1(2): 186–196.

Knackmuss H J (1996). Basic knowledge and perspectives of bio elimination of xenobiotics compounds. *Journal of Biotechnology* 51(3): 287–295.

Kolekar Y M (2010). *Isolation Characterization and Evaluation of Dye Degradation Potential of the Novel Bacterial Species Alishewanella Soli*. University of Pune, India. http://hdl.handle.net/10603/3221

Laing G I (1991). The Impact of Effluent Regulations on the Dyeing Industry. *Colouration* 12: 56–70.

Marie Wilson October 8, 2021. *The Environmental Cost of Synthetics*, Accessed on April 10, 2023, https://www.trailrunnermag.com/people/environment-people/the-environmental-cost-of-synthetics/

Marijnissen C M, and Moskal Arkadiusz 2020. "Synthetic Nano- and Microfibers." *Synthetic Nano- and Microfibers*. https://doi.org/10.20850/9781716632426

Minor J in *Encyclopedia of Chemical Technology*, Vol. 19, pp. 379–419, John Wiley & Sons, Inc., New York, 1982.

Ning X et al. (2014). Levels, Composition Profiles and Risk Assessment of Polycyclic Aromatic Hydrocarbons (PAHs) in Sludge from Ten Textile Dyeing Plants. *Environmental Research* 132: 112–118.

Peng Chu, He Ning, Wu Yinghong, Lu Yuan, Sun Hongwen, Wang Lei (2022). Excretion Characteristics of Nylon Microplastics and Absorption Risk of Nanoplastics in Rats, *Ecotoxicology and Environmental Safety*, 238: 113586, ISSN 0147-6513.

Qian Weiran, Ji Xiang, Xu Pinghua, Wang Laili. 2021. "Carbon Footprint and Water Footprint Assessment of Virgin and Recycled Polyester Textiles." *Textile Research Journal* 91 (21–22): 2468–2475. https://doi.org/10.1177/00405175211006213

Raitta Christina, Teir Henrik, Tolonen Matti, Nurminen Markku, Helpiö Ermo, Malmström Synnove (1981). "Impaired Color Discrimination Among Viscose Rayon Workers Exposed to Carbon Disulfide." *Journal of Occupational Medicine* 23, 3: 189–192. http://www.jstor.org/stable/45005595

Rebenfeld Ludwig (2002). "*Chapter VI Fibers and Fibrous Materials Tri/Princeton, P. O. Box 625, Princeton, New Jersey 08542, USA*," 199–232.

Seppäläinen Anna Maria, Tolonen Matti (1974). "Neurotoxicity of Long-Term Exposure to Carbon Disulfide in the Viscose Rayon Industry A Neurophysiological Study." *Work, Environment, Health* 11, 3: 145–153. http://www.jstor.org/stable/44378795

Shaikh Tasnim, Chaudhari Satyajeet, Varma Alpa (2012). "Viscose Rayon: A Legendary Development in the Manmade Textile." *Mrs. Alpa Varma/International Journal of Engineering Research and Applications (IJERA)* 2 (5): 675–680. www.ijera.com

Shenai V V (2001). Non-eco Friendly Textile Chemicals and Their Probable Substitutes—An Overview. *Journal of Fibre & Textile Research* 26: 50–54

Sophie Bird May 9, 2019. *Synthetic Fibers in Clothing Contribute to Pollution*, Accessed on April 10, 2023, https://www.indianaenvironmentalreporter.org/posts/synthetic-fibers-in-clothing-contribute-to-pollution

Subramanian Senthilkannan Muthu (2014). Calculating the Water and Energy Footprints of Textile Products, *Assessing the Environmental Impact of Textiles and the Clothing Supply Chain*, Woodhead Publishing, 78–94, ISBN 9781782421047. https://doi.org/10.1533/9781782421122.78

US.EPA, Environmental Protection Agency, Network TT (2011). *Emission Factors AP-42, CH 6: Organic Chemical Process Industry: 6.9 Synthetic Fibers*, Washington, DC - USA. pp: 6.9-2–22.

Washko R M, Day B, Parker J E, Castellan, R M, Kreiss K (2000). Epidemiologic investigation of respiratory morbidity at a nylon flock plant. *American Journal of Industrial Medicine*, 38(6), 628–638. https://doi.org/10.1002/1097-0274(200012)38:6<628::aid-ajim3>3.0.co;2-u

Weiss R F 1981. "The Temporal and Spatial Distribution of Tropospheric Nitrous Oxide." *Journal of Geophysical Research* 86 (8 C): 7185–7195. https://doi.org/10.1029/jc086ic08p07185

Wright Stephanie L, Thompson Richard C, Galloway Tamara S 2013. "The Physical Impacts of Microplastics on Marine Organisms: A Review." *Environmental Pollution* 178: 483–492. https://doi.org/10.1016/j.envpol.2013.02.031

Yang Libiao, Qiao Fei, Lei Kun, Li Huiqin, Kang Yu, Cui Song, An Lihui 2019. "Microfiber Release from Different Fabrics during Washing." *Environmental Pollution* 249: 136–143. https://doi.org/10.1016/j.envpol.2019.03.011

17 Recent Trends and Future Scope of Natural Fiber Composites

Manuel Thomas, Rittin Abraham Kurien, Akash Paul Savio, and Arun Raphel

17.1 INTRODUCTION

Composites are increasingly being used in industry as a substitute for metals by virtue of their lower density and higher specific mechanical properties. The properties of composites can be custom-tailored based on requirements, making them further attractive. The scope of composites is large and is spread over domains ranging from simple material packaging to advanced aerospace applications. The eco-friendly nature, lightweight, cost-effectiveness, and outstanding mechanical properties of natural fiber composites (NFCs) make them a potential alternative to traditional materials in numerous sectors. The composite materials essentially consist of a matrix phase, such as a polymer, which is reinforced by the incorporation of natural fibers. In recent years, there has been a promising inclination toward the use of natural fibers as reinforcement in polymer composites, with the aim of developing cost-effective building materials. This phenomenon is propelled by the escalating worldwide necessity for sustainable and environmentally friendly materials, along with the imperative to identify substitutes for conventional resources such as wood that are progressively depleting. The current chapter provides a comprehensive overview of recent trends, limitations, and future scope of natural fiber composites.

17.2 EVOLUTION OF GREEN COMPOSITES

The use of polymer matrix composites (PMCs) globally is on the rise, increasing its market share among various structural materials. The PMCs can be derived from either thermoplastics or thermosetting polymers, depending on the application. However, a major share of the PMCs is derived from petroleum-based products, which increases the energy cost associated with the composite production. Further, the reuse and recycling of PMCs are still in their infancy, and most of the PMCs after service life cannot be disposed of without significant harm to the environment (1). The environmental setbacks associated with the production and disposal of PMCs have led to the development of green or eco-composites. The conventional composites used for lower-strength applications like packaging panels can be substituted with green composites (2).

A composite material comprises reinforcement and matrix material. For a composite to be termed a green composite, both reinforcement and matrix material need to be biodegradable. Natural fiber reinforcement used with biopolymers matrix generally constitutes green composites. The natural fiber derived from bast (flax, kenaf, hemp, etc.) has superior mechanical properties compared to glass fiber and is biodegradable (3). Unlike the matrix materials used in conventional PMCs, biopolymers serve as the matrix material in green composites. Biopolymers degrade over a period of time and hence can be considered environmentally friendly. Similar to PMCs, green composites can

 DOI: 10.1201/9781032665337-17

be thermosetting or thermoplastic types. Various manufacturing techniques that are employed for the fabrication of thermoplastics can also be used for the fabrication of composites with thermoplastic biopolymer matrix. Thermosetting composites can be fabricated with natural oil-based resins replacing traditional epoxy and vinyl esters (4).

17.3 ADVANCEMENTS IN NATURAL FIBER COMPOSITES

Natural fiber, as the name indicates, is sourced from nature. The use of natural fiber as a substitute for conventional fiber in the manufacturing sector is increasing. Natural fiber can be broadly classified based on its nature of origin into protein-based, cellulose-based, and mineral-based fiber (5). Cellulose-based fibers are plant-based derivatives and include fibers like hemp, jute, and bast fiber. Plant-based fibers are widely accepted and used in large volumes since they have a much larger availability and can be renewed in a short time. Protein-based fibers are derived from animal sources like hair and wool. Asbestos and ceramic fiber fall into the category of mineral fibers. From the industrial perspective, the use of natural fiber as a substitute for conventional fiber possesses various advantages. These materials are cost-effective and are sourced from renewable sources. Also, most of the natural fibers are biodegradable, making them a potential candidate for green composites (6).

For the use of natural fiber composites, much higher refinement is required. Large-scale use of natural fiber-based biocomposites is limited by their lower strength. Composites used for structural applications commonly use longer fibers. However, natural fibers are commonly available as short fibers, limiting their usage for low-strength applications. The aspect ratio of the natural fiber is also important as below certain values of aspect ratios, the natural fiber acts just as a filler material dispersed in polymer matrix (7). Thermosetting biocomposites are used in applications requiring higher strength. Resins derived from natural sources can be used for the fabrication of thermosetting biopolymers (3). Fabrication techniques, including resin transfer molding, vacuum-assisted resin transfer molding, and compression molding, can be used for the fabrication. Thermosetting resins are sourced from natural oils like soybean oil, castor oil, and linseed oil, with suitable chemical treatment (8–10). Polylactic acid (PLA) is a thermoplastic-type biopolymer used as a matrix material (11) and can be derived from natural sources like corn starch, cassava starch, and sugarcane. The ease of availability and reasonably good mechanical properties make them a widely used matrix material for green composites. PLA can be processed similarly to polyethylene terephthalate (PET). Hence, a large number of processing techniques can be used for its manufacturing (12).

A combination of natural fiber and matrix materials derived from biopolymers can be used to create green or eco-composites, which are biodegradable. Further, the raw materials used for the production of biocomposites are replenishable, reducing the overall production cost associated with it. Natural fibers exhibit larger elongation before failure compared to glass fiber, making them suitable for the manufacturing of tougher composites. However, the use of NFCs is typically limited by lower processing temperatures. At higher temperatures above 473K, the natural fiber is susceptible to lignocellulosic degradation and emission of volatile gases (13). This can affect the desired properties of the composites and degrade their functionality. Also, biopolymers, especially thermosetting biopolymers, are formed as a result of chemical reactions. This makes them less susceptible to biodegradation. Green composites are one of the best alternatives to PMCs, which are not biodegradable. The large-scale use of green composites is however limited by higher manufacturing cost, limited manufacturing process, and moisture affinity of the composites manufactured. Commercial usage of green composites is commonly restricted to indoor applications. Further studies are required for large-scale commercial adaptation of green composites in industries so that their performance can be improved and newer efficient techniques for manufacturing can be devised.

17.4 ADVANTAGES AND LIMITATIONS IN NATURAL FIBER COMPOSITES

NFCs provide a variety of beneficial attributes that make them highly sought-after for a wide range of applications. First and foremost, the ability to be renewed is a significant advantage. Natural fibers are intrinsically renewable, as they are obtained from plants, animals, and geological processes. This is in contrast to synthetic fibers, which are often acquired from non-renewable resources such as hydrocarbon derivatives. This characteristic is by sustainability endeavors, since it reduces reliance on limited resources and fosters the application of environmental treatment. NFCs provide significant advantages in terms of cost-competitiveness. Natural fibers are often more affordable compared to synthetic equivalents, making them a financially viable choice for most businesses (9).

Their cost-effectiveness, in addition to their capacity to be renewed, increases their attractiveness as financially feasible and ecologically conscious materials for various applications. However, NFCs have environmental advantages that extend beyond their abilities in terms of renewability and cost-effectiveness. These materials exhibit biodegradability and possess a reduced carbon footprint in comparison to synthetic alternatives, hence aligning effectively with the objectives of enterprises seeking to mitigate their ecological imprint. Also, the inherent characteristics of being lightweight, possessing thermal qualities, and exhibiting mechanical strength render them well-suited for a wide array of applications, spanning from automotive components to aerospace designs. This further emphasizes the adaptability and utility of these materials (11).

Despite these benefits, there are still obstacles that need to be addressed, including the vulnerability to moisture absorption and inadequate wettability. These issues need meticulous evaluation of processing methods and surface treatments to alleviate these limits. In summary, NFCs present themselves as favorable substitutes for conventional materials, providing an array of sustainability, performance, and cost-efficiency in many uses. NFCs have technological, processing, and standardization obstacles, notwithstanding their many advantages. Compared to synthetic materials, these possess restricted mechanical properties, rendering them less appropriate for specific sectors. The complexity and time-consuming nature of processing contribute to an escalation in manufacturing expenses. The absence of standardization in production and testing hinders their adoption in some industries. Subsequent investigations should prioritize the advancement of novel methodologies, enhancement of mechanical features, and establishment of protocols for manufacturing and evalua tion. The use of NFCs has promise in addressing environmental and socio-economic issues, rendering them a crucial element in the manufacturing of inventive and sustainable goods throughout diverse industries. (13).

17.5 POTENTIAL APPLICATION OF NATURAL FIBER COMPOSITES

17.5.1 Aviation Sector

The aerospace industry is one of the main end users of composites. Often, ceramic, metal, and PMCs find application in various aerospace applications. The use of the above classified composites depends on the application process. For high-strength applications at elevated temperatures, metal matrix composites are employed. Ceramic matrix composites are used in high-temperature applications requiring thermal insulation (14). PMCs are versatile materials which can be customized for various strength applications. However, the use of polymer matrix limits the maximum operating temperatures (15). In the aerospace industry, the material choice is mainly governed by two factors, namely safety and weight reduction. Aerospace components are often safety-critical structures and demand a higher degree of engineering. The industry widely uses high-strength, low-density materials like aluminum and titanium alloys for the structures. Composites often have higher strength-to-weight ratios, which makes them a potential candidate suitable for the aerospace industry. Carbon fiber and glass fiber composites are known for their strength and are being largely used as a replacement for conventional materials in aerospace structures (16).

Push toward greener aviation and stringent environmental norms have opened large scope for biocomposites, which are potential alternatives to conventional PMCs used in aerospace applications. Compared to the commercially available PMCs, composites derived from natural fiber possess lower density, increasing the potential for weight savings (17). Natural fiber-based biocomposites are used in components for which strength considerations are not critical. Biocomposites are promising candidate materials for radar housing structures (RADOME) (Figure 17.1) in aircraft. Materials used for RADOME must possess high toughness and lower dielectric constants, which the natural fiber possesses (19). The aircraft structures use damage-tolerant design approach, which mandates specific material data in actual operating conditions (20). However, limited number of works are only reported in the literature, which are not acceptable for damage-tolerant design criteria. Other potential areas of research include wing box structures and components for aircraft interiors (21). Composites made from natural fibers are prone to moisture absorption, which can degrade the overall performance. Natural fibers are subjected to various surface treatments to reduce moisture ingression, improve bonding characteristics, and also to improve specific features like flame retardation and smoke suppressants (22, 23).

In comparison to the exterior aircraft structure, the scope of natural fiber composites is much larger in aircraft interiors (24). Aerospace companies, including Boeing Research and AIRBUS, have kick-started development programs for the utilization of renewable NFCs in aircraft interiors (24, 25). Despite the NFCs having advantages of low density and higher specific stiffness, their extensive use in aerospace applications is limited, and there lies a huge scope for improvement in

FIGURE 17.1 RADOME structure used in aircraft. RADOME houses communication modules and are increasingly fabricated incorporating natural fibers.

Reproduced with permission (18).

many aspects. Some of the limitations include lower strength, durability when subjected to prolonged loading, and increased susceptibility to environmental variables. Further, its use is limited by the higher variation in the reported mechanical behavior.

17.5.2 Application of NFCs in the Automobile and Marine Sector

The need for increased fuel efficiency and reduced carbon footprint has pushed automobile manufacturers toward natural fiber. In automobiles, non-structural components are gradually being replaced by composites containing natural fiber. Overall weight and cost savings are improved by replacing conventional materials with natural fiber-based components (5). The reduced weight improves fuel efficiency and lowers volumes of nonbiodegradable waste. In automobiles, the majority of the natural fiber structures are used in interior components that are not subjected to severe loading. Using NFCs, visually appealing interior parts can be molded at a reduced cost compared to plastic and metallic counterparts. Automobile dashboards, glove boxes, cabin linings, and door panels (Figure 17.2) are broadly fabricated with natural fibers.

The use of natural fiber-based products for exterior applications is limited by their lower strength and hydrophilic nature. However, hybrid NFCs and components prepared from treated NFCs are used in the manufacturing of automobile exterior parts such as car hoods, engine covers, and bumpers (26). Other applications in automobiles include the fabrication of body panels, underfloor panels, boot lids, and spare tire covers (27).

Polymer composites reinforced with glass fiber are widely employed for the fabrication of marine structures. Natural fibers derived from flax, hemp, and jute exhibit lower density than glass fiber, making them a suitable replacement for glass fibers. Natural fibers subjected to appropriate surface treatment for enhancing the mechanical and hydrophobic properties (29) are increasingly used in the fabrication of marine components, including hulls, internal panels, and decks (28). Natural fiber-based sandwich panels can be used for the fabrication of boat hulls, resulting in overall cost reduction (30). Hybridized composites produced from flax and glass fiber reduce moisture ingress and improve the life of composite structures used in marine environments (31). The use of hybrid NFCs in hull fabrication (Figure 17.3) can reduce the glass fiber content by up to 50% without compromising the required mechanical properties (28). Large-scale implementation of natural fiber-reinforced polymers (NFRP) in the marine sector is limited by various factors such as variation in fiber mechanical property, seasonal and geographical influence on fiber availability, and largely by the

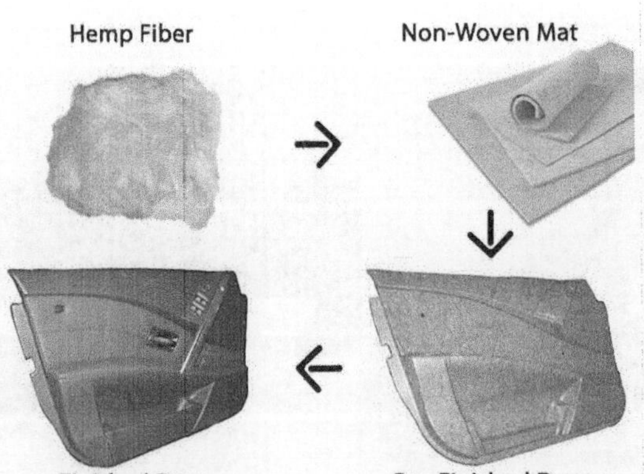

FIGURE 17.2 Various stages associated with the fabrication of automobile door panels from fiber-reinforced polymer composite using natural hemp fiber.

Reproduced with permission (5).

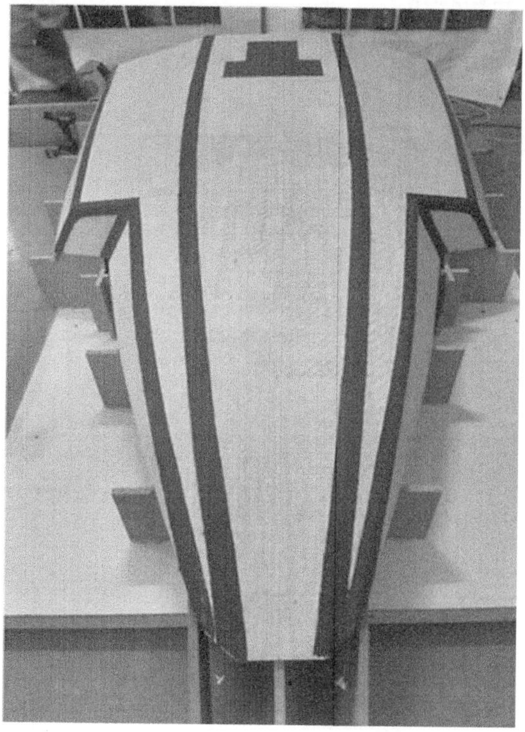

FIGURE 17.3 Hull panels fabricated using natural fiber composites (28).

hydrophilic nature of natural fiber. However, various chemical treatments, fiber processing techniques, and hybridization make natural fibers a potential replacement for glass fibers.

17.5.3 Application of NFCs in the Biomedical Sector

Natural polymers like cellulose, gelatin, chitin, silk, and collagen exhibit good biocompatibility, making them pertinent for bioimplants (32). Natural polymers reinforced with cellulose fibers or whiskers improve the overall mechanical properties of the biocomposites. Biopolymers find extensive application as potential materials suitable for incorporation inside the human body (Figure 17.4).

Chitosan, a biopolymer extracted from the exoskeletons of marine crustaceans is used as a precursor for the fabrication of chitosan fibers. Biopolymers reinforced with chitosan fibers, by their biocompatibility and sustainability, are increasingly used in many biomedical applications such as scaffolds in tissue engineering, wound dressing, and drug delivery (34). In bone tissue regeneration, biodegradable material is used as a substitute to facilitate the regeneration of bone tissue. Chitosan-based biocomposites are used as a substitute material for bones since they can act as sites for bone growth. Chitosan composites are biodegradable and can be formed into porous geometries suitable for bone regeneration (35). Chitosan-based 3-D printed structures (Figure 17.5) are being used as biocompatible scaffolds for supporting cartilage (35). Chitosan-based products are used in wound healing applications. The antibacterial properties of chitosan make it an excellent material for wound healing. Chitosan structures can be formed into numerous shapes, such as sponges, particles, hydrogels, and films (36).

17.6 MECHANICAL BEHAVIOR OF NATURAL FIBERS

The insight into the mechanical behavior of natural fibers is critical for their use as reinforcement in NFCs. Characterization of mechanical properties of NFC is often accomplished with tensile,

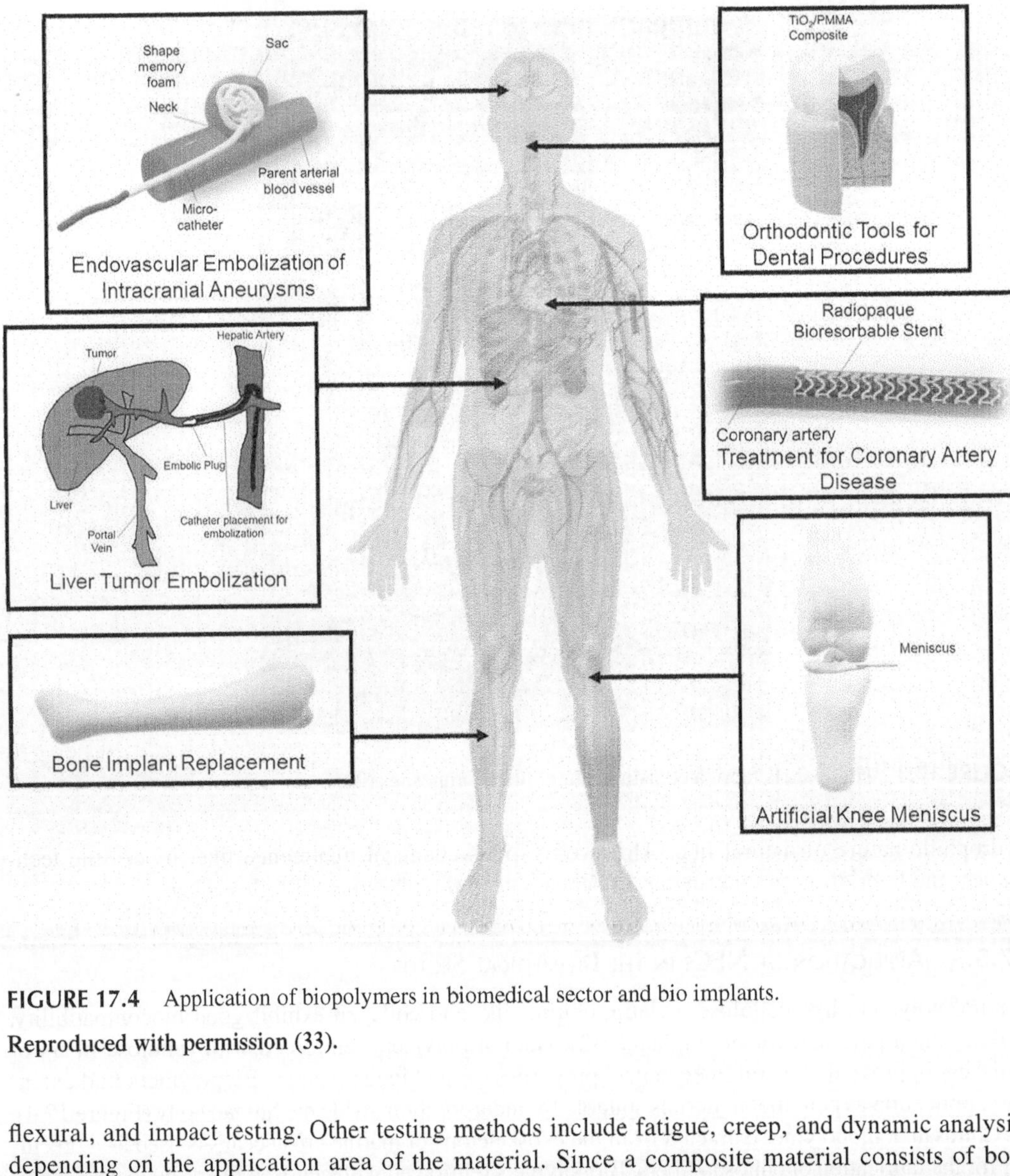

FIGURE 17.4 Application of biopolymers in biomedical sector and bio implants.
Reproduced with permission (33).

flexural, and impact testing. Other testing methods include fatigue, creep, and dynamic analysis, depending on the application area of the material. Since a composite material consists of both matrix and reinforcement, accurate estimation of individual constituent properties is important for the successful prediction of material behavior. Experimental testing of various mechanical and physical aspects of NFC is difficult, and aspects of composite materials are time-consuming and impractical. Similar to the case of metals, the use of computational tools in simulating the mechanical and physical response of composite materials is on the rise. For capturing the accurate mechanical behavior of the materials, finite element (FE) tools require material properties including Young's Modulus (E) Poisson's ratio (ν) and other parameters depending on the type of analysis required to be performed (37). Thus, the estimation of mechanical properties of natural fibers and matrix materials with reasonable accuracy is crucial. The most critical difficulty associated with the NFC is the inherent variability of its associated mechanical properties. Test results reported in the literature exhibit considerable variation in the mechanical properties of natural fibers of different origins.

Figures 17.6 and 17.7 represent the scatter associated with the available data reported in the literature. One of the principal reasons associated with the variation in test results is explained in terms of deviation in fiber diameter (39, 40). Reported mechanical properties are also influenced by the

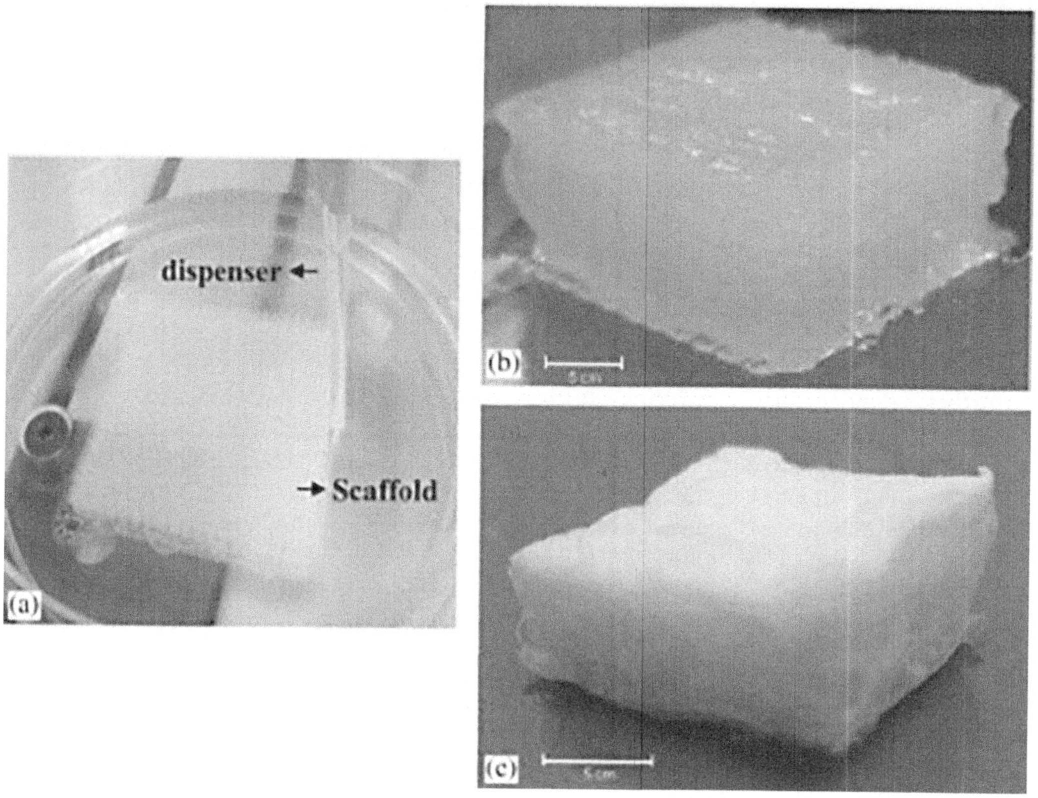

FIGURE 17.5 Printed chitosan scaffolds for tissue engineering.

Reproduced with permission (35).

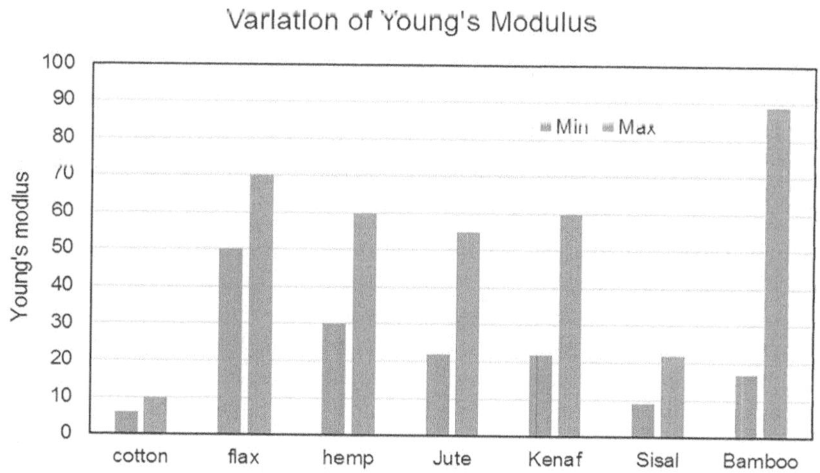

FIGURE 17.6 Maximum and minimum values of Young's Modulus reported from literature for different types of natural fibers, indicating high scatter in material properties (38).

fiber length of the tested fibers. Considerable variation in Young's modulus of natural fibers with variation in fiber diameter is reported in the literature (41, 42). Further, the fiber diameter used for the calculation of Young's modulus is approximated as uniform over the entire fiber length. The fiber diameter is known to vary along the length and is often larger at the center compared to the ends (43).

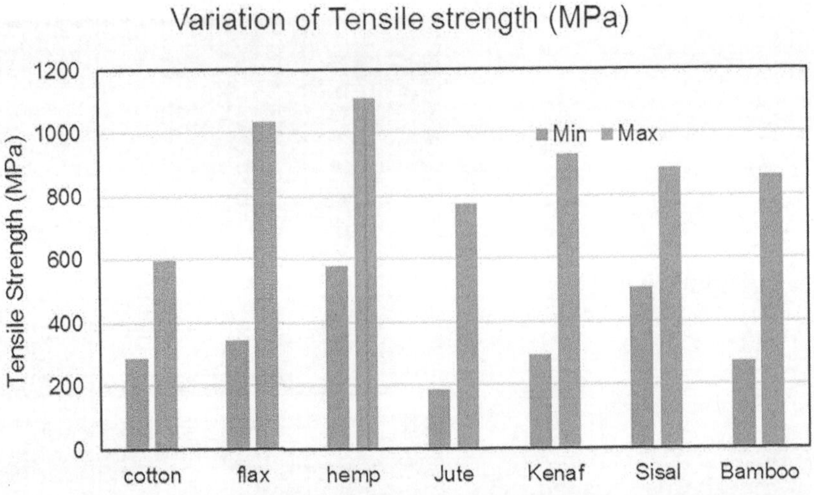

FIGURE 17.7 Maximum and minimum values of tensile strength reported from literature for different types of natural fibers, indicating high scatter in material properties (13).

Tensile properties of natural fibers are often evaluated based on the ASTM D 3822/D 3822M standard (Standard Test Method for Tensile Properties of Single Textile Fibers). According to this standard, the cross-sectional area or diameter of the tested fiber is assumed to be uniform throughout the length, which is never fulfilled for a natural fiber (44). The standard also specifies a minimum gauge length of 10 mm during the testing, which is also seldom satisfied for natural fibers. Further, some of the reported works lack a clear mention of standards used for testing (5). This is one of the major limitations in natural fiber testing, which needs further refinement.

The main constituents of natural fibers are cellulose, lignin, hemicellulose, and pectin (42). Among the key constituents, hemicellulose, being hydrophilic, results in moisture absorption (45). The change in moisture content generally results in varying mechanical properties. The fibers tend to absorb moisture, reducing their Young's modulus and deteriorating the bonding characteristics between fiber and matrix material (46). Another critical problem bottlenecking the commercial-scale usage of NFCs is its durability. The durability effectively determines both the life cycle and the field of application of NFCs. Drastic improvement in reducing moisture absorption is required to improve the durability of NFCs. Natural fiber, being hydrophilic in nature, absorbs water from the operating environment, resulting in the weakening of the fiber matrix interface and swelling (39). The components produced from biodegradable NFCs can sometimes withstand weather and are susceptible to bacterial and fungal attacks in favorable environments. Studies report unfavorable weathering, discoloration, surface degradation, and strength deterioration in NFCs exposed to moisture.

Mathematical models and computational tools are widely used in engineering to improve designs to reduce design time. The use of mathematical models to predict the response of NFCs requires the properties of the fiber and the matrix material. Naturally sourced fibers like jute and kenaf, by default, exist as filaments consisting of individual fibers. The individual fibers themselves can be classified as composites, and the properties of the filament and constituent fiber are a function of a multitude of factors, including age, material source, techniques used for sourcing, and moisture content. This indicates the difficulty associated with the measurement of mechanical properties without significant scatter. Also, various stages associated with fiber processing can lead to fiber degradation and the dissociation of filaments into basic fibers (47). Natural fibers are also subjected to fiber defects like kinks on the surface. When subjected to tensile and flexure loads, these sites act as failure locations (48). Knowledge about the failure of natural fibers is crucial for the development of reliable mathematical models to predict the material behavior under different loading conditions.

In comparison to synthetic fibers, there are only a few mathematical models for predicting the mechanical behavior of natural fibers. The development of accurate material models can significantly improve the application possibilities of natural fibers in numerous engineering fields.

17.7 FUTURE SCOPE OF NATURAL FIBER COMPOSITES

In recent years, there has been a notable increase in interest in natural fiber composites, mostly driven by their distinctive characteristics and potential utility in many industrial sectors. With the increasing need for environmentally friendly and sustainable materials, natural fiber composites are anticipated to have a greater impact in tackling environmental issues and socio-economic difficulties. The construction industry is a significant sector where natural fiber composites find prominent uses. These composite materials have the potential to be utilized in the production of various construction materials, including roofing tiles, wall panels, and flooring, which exhibit performance levels that are either equivalent to or even above those of traditional materials (49). The aforementioned materials exhibit favorable strength-to-weight percentages, exceptional thermal and acoustic insulation characteristics, and heightened fire and moisture resistance. Furthermore, there has been a growing use of natural fiber composites within the automobile sector. Interior components, such as door panels, dashboard trims, and seat backings, are manufactured using them. These composite materials exhibit comparable or even superior mechanical characteristics in comparison to conventional materials, while concurrently providing advantages such as decreased weight and heightened durability.

The automotive industry is anticipated to witness an increased use of natural fiber composites due to their potential to serve as substitutes for conventional glass fiber composites in non-structural applications. These composite materials have several benefits, including decreased weight, enhanced specific stiffness and strength, and prolonged fatigue life. These advantages are of utmost importance for automotive makers aiming to enhance fuel economy and mitigate emissions. Natural fiber composites are anticipated to be utilized in the manufacturing of lightweight and efficient cars in the future, therefore enhancing the sustainability of the transport sector (50).

The use of natural fiber composites in civil engineering, especially in residential buildings, holds significant promise for broadening their range of applications. Although their present usage is primarily confined to outside purposes, like siding and roofing, there is an increasing inclination toward investigating their potential utilization for interior components, including flooring and wall panels. This has the potential to result in a decrease in the use of synthetic substances and promote the adoption of more environmentally friendly construction methods. The growing need for environmentally friendly and energy-efficient structures has led to an increased recognition of the potential of natural fiber composites in the advancement of sustainable housing solutions (51). The aircraft sector is expected to witness the integration of natural fiber composites, which have the potential to be utilized in non-structural applications. These composite materials have several benefits, including decreased weight, enhanced specific stiffness and strength, and prolonged fatigue life. These advantages are crucial for the advancement of aircraft that are both lightweight and efficient. Natural fiber composites are anticipated to play a crucial role in the manufacturing of future airplanes as the aviation industry aims to decrease its carbon emissions and enhance fuel economy (52).

Natural fiber composites are emerging as a promising contender in the electrical and electronics sectors, with the capacity to secure a substantial portion of the market in the coming years. The use of natural fiber composites is anticipated to expand in diverse applications, including insulating materials and electronic components, because of the growing need for environmentally sustainable and lightweight materials. Natural fiber composites have the potential to make major improvements to the development of sustainable and energy-efficient products within the electrical and electronics sector. The use of natural fiber composites in the sports goods sector has attracted considerable attention due to their potential for producing equipment that remains lightweight and durable. These composites provide notable benefits like exceptional strength, low weight, and water resistance, rendering them well-suited for a diverse array of recreational uses. With the growing demand for environmentally friendly and excellent sporting goods, it is expected that natural fiber composites will play a crucial role in manufacturing innovative and sustainable equipment (53).

17.8 CONCLUSION

In summary, the advancements and developments in natural fiber composites provide an attractive foundation for their extensive utilization in diverse sectors. These composites offer multiple advantages, including sustainability, cost-effectiveness, and ecological advantages, which are well-suited to meet the increasing need for sustainable materials. The progression of green composites underscores the transition toward environmentally sustainable alternatives, driven by apprehensions over the ecological consequences associated with normal materials. Although natural fiber composites provide many benefits, they encounter difficulties such as inconsistent mechanical characteristics and sensitivity to moisture absorption. Nevertheless, current research and development endeavors are focused on mitigating these challenges and improving the efficacy and suitability of natural fiber composites.

When considering the future prospects of natural fiber composite materials, it becomes evident that their potential uses extend across several sectors, including buildings, automotive, aerospace, and electronics industries. The use of natural fiber composites in diverse applications within the construction sector has potential advantages, since they exhibit greater strength-to-weight ratios and improved thermal and acoustic insulation characteristics. Within the automobile industry, these composite materials have the capacity to reduce weight and enhance fuel efficiency, therefore aiding in the achievement of sustainability objectives. Moreover, natural fiber composites are positioned to make substantial progress in the fields of civil engineering, aviation production, electrical engineering, electronics, and sports products. They provide a blend of low weight, long-lasting characteristics, and ecological sustainability. The ongoing progress in research and development is anticipated to have a significant impact on the future of sustainable materials in many industries, fostering innovation and promoting sustainability.

REFERENCES

1. Netravali AN, Chabba S. Composites get greener. *Mater Today* 2003; 6(4): 22–29.
2. Sanjay MR, Madhu P, Jawaid M, Senthamaraikannan P, Senthil S, Pradeep S. Characterization and properties of natural fiber polymer composites: A comprehensive review. *J Clean Prod* [Internet]. 2018; 172: 566–581. Available from: https://doi.org/10.1016/j.jclepro.2017.10.101
3. Elisa Zini MS. Green composites: An overview. *Polym Polym Compos*. 2011; 32(12): 1905–1915.
4. Khot SN, Lascala JJ, Can E, Morye SS, Williams GI, Palmese GR, et al. Development and application of triglyceride-based polymers and composites. *J Appl Polym Sci*. 2001; 82(3): 703–723.
5. Pecas, P. Carvalho, H. Salman, H. Leite, M. Natural fibre composites and their applications: A review. *J Compos Sci*. 2018; 2: 66.
6. Shekar HSS, Ramachandra M. Green composites: A review. *Mater Today Proc* [Internet]. 2018; 5(1): 2518–2526. Available from: https://doi.org/10.1016/j.matpr.2017.11.034
7. Takagi H, Ichihara Y. Effect of fiber length on mechanical properties of "green" composites using a starch-based resin and short bamboo fiber. *JSME Int J Ser A Solid Mech Mater Eng*. 2004; 47(4): 551–555.
8. Hong CK, Wool RP. Development of a bio-based composite material from soybean oil and keratin fiber. *J Appl Polym Sci*. 2005; 95(6): 1524–1538.
9. Lee N, Kwon O-J, Chun BC, Cho JW, Park J-S. Characterization of castor oil/polycaprolactone polyurethane biocomposites reinforced with hemp fiber. *Fiber Polym*. 2009; 10: 154–160.
10. Raquez J-M, Deléglise M, Lacrampe M-F, Krawczak P. Thermosetting (bio) materials derived from renewable resources: A critical review. *Prog Polym Sci*. 2010; 35(4): 487–509.
11. Plackett D, Andersen TL, Pedersen WB, Nielsen L. Biodegradable composites based on L-polylactide and jute fiber. *Compos Sci Technol*. 2003; 63(9): 1287–1296.

12. Staiger MP, Tucker N. Natural-fibre composites in structural applications. In: *Properties and performance of natural-fibre composites*. Elsevier; 2008: 269–300.
13. *Thermogravimetric stability of polymer composites reinforced with less common lignocellulosic fiber–An Overview.*
14. Curtin WA. Theory of mechanical properties of ceramic-matrix composites. *J Am Ceram Soc*. 1991; 74(11): 2837–2845.
15. Baker A, Dutton S, Kelly D. *Composite materials for aircraft structures*. 2004.
16. Soutis C. Fibre reinforced composites in aircraft construction. *Prog Aerosp Sci*. 2005; 41(2): 143–151.
17. Arockiam NJ, Jawaid M, Saba N. Sustainable bio composites for aircraft components. In: *Sustainable composites for aerospace applications*. Elsevier; 2018: 109–123.
18. Nurazzi NM, Asyraf MRM, Fatimah Athiyah S, Shazleen SS, Rafiqah SA, Harussani MM, Kamarudin SH, Razman MR, Rahmah M, Zainudin ES, Ilyas RA, Aisyah HA, Norrrahim MNF, Abdullah N, Sapuan SM, Khalina A. (2021). A review on mechanical performance of hybrid natural fiber polymer composites for structural applications. *Polymers* 13(13). MDPI AG. https://doi.org/10.3390/polym13132170
19. Haris MY, Laila D, Zainudin ES, Mustapha F, Zahari R, Halim Z. Preliminary review of biocomposites materials for aircraft radome application. *Key Eng Mater*. 2011; 471: 563–567.
20. Swift T. Damage tolerance capability. *Int J Fatigue*. 1994; 16(1): 75–94.
21. Boegler O, Kling U, Empl D, Isikveren AT. *Potential of sustainable materials in wing structural design*. Deutsche Gesellschaft für Luft-und Raumfahrt-Lilienthal-Oberth eV Bonn, Germany; 2015.
22. Atkinson PA, Haines PJ, Skinner GA. Inorganic tin compounds as flame retardants and smoke suppressants for polyester thermosets. *Thermochim Acta*. 2000; 360(1): 29–40.
23. Cisneros-López EO, González-López ME, Pérez-Fonseca AA, González-Núñez R, Rodrigue D, Robledo-Ortíz JR. Effect of fiber content and surface treatment on the mechanical properties of natural fiber composites produced by rotomolding. *Compos Interfaces*. 2017; 24(1): 35–53.
24. Mansor MR, Nurfaizey AH, Tamaldin N, Nordin MNA. Natural fiber polymer composites: Utilization in aerospace engineering. In: *Biomass, biopolymer-based materials, and bioenergy*. Elsevier; 2019. 203–224.
25. Mascle C, Baptiste P, Sainte Beuve D, Camelot A. Process for advanced management and technologies of aircraft EOL. *Procedia CIRP*. 2015; 26: 299–304.
26. Sangilimuthukumar J, Kumar Senthil Muthu T, Senthilkumar K, Chandrasekar M, Siengchin S. Potential of natural/synthetic hybrid composites for automotive applications. In *Green hybrid composite in engineering and Non-engineering applications*. 2023: pp. 179–195. Springer
27. Ngo T-D. Natural fibers for sustainable bio-composites. *Nat Artif Fiber Reinforced Compos Renew Source*. 2018; 3: 107–126.
28. El Hawary O, Boccarusso L, Ansell MP, Durante M, Pinto F. An overview of Natural fiber composites for marine applications. *J Marine Sci Eng*. 2023; 11(5). MDPI. https://doi.org/10.3390/jmse11051076
29. Valadez-Gonzalez A, Cervantes-Uc JM, Olayo R, Herrera-Franco PJ. Effect of fiber surface treatment on the fiber–matrix bond strength of natural fiber reinforced composites. *Compos B: Eng*. 1999; *30*(3): 309–320
30. Corradi S, Isidori T, Corradi M, Soleri F, Olivari L. Composite boat hulls with bamboo natural fibres. *Int J Mater Prod Technol*. 2009; 36(1/2/3/4): 73. https://doi.org/10.1504/ijmpt.2009.027821
31. Calabrese L, Fiore V, Scalici T, Valenza A. Experimental assessment of the improved properties during aging of flax/glass hybrid composite laminates for marine applications. *J Appl Polym Sci*. 2019; 136(14). https://doi.org/10.1002/app.47203
32. Nair LS, Laurencin CT. Biodegradable polymers as biomaterials. *Prog Polym Sci (Oxford)*. 2007; 32(8–9): 762–798. https://doi.org/10.1016/j.progpolymsci.2007.05.017
33. Toh HW, Toong DWY, Ng JCK, Ow V, Lu S, Tan LP, Wong PEH, Venkatraman S, Huang Y, Ang HY. Polymer blends and polymer composites for cardiovascular implants. *Eur Polym J*. 2021; 146, Elsevier Ltd. https://doi.org/10.1016/j.eurpolymj.2020.110249
34. Kim IY, Seo SJ, Moon HS, Yoo MK, Park IY, Kim BC, Cho CS. Chitosan and its derivatives for tissue engineering applications. *Biotechnol Adv*. 2008; 26(1): 1–21. https://doi.org/10.1016/j.biotechadv.2007.07.009
35. Di Martino A, Sittinger M, Risbud MV. Chitosan: A versatile biopolymer for orthopaedic tissue-engineering. *Biomaterials*. 2005; 26(30): 5983–5990. https://doi.org/10.1016/j.biomaterials.2005.03.016
36. Patrulea V, Ostafe V, Borchard G, Jordan O. Chitosan as a starting material for wound healing applications. *Eur J Pharm Biopharm*. 2015; 97: 417–426. Elsevier. https://doi.org/10.1016/j.ejpb.2015.08.004

37. Alhijazi M, Zeeshan Q, Qin Z, Safaei B, Asmael M. Finite element analysis of natural fiber composites: A review. *Nanotechnol Rev*. 2020; 9(1): 853–875.
38. Gurunathan T, Mohanty S, Nayak SK. A review of the recent developments in biocomposites based on natural fiber and their application perspectives. *Compos Part A Appl Sci Manuf*. 2015; 77: 1–25. Available from: http://dx.doi.org/10.1016/j.compositesa.2015.06.007
39. Satyanarayana KG, Arizaga GGC, Wypych F. Biodegradable composites based on lignocellulosic fiber—An overview. *Prog Polym Sci*. 2009; 34(9): 982–1021.
40. Huda MS, Drzal LT, Mohanty AK, Misra M. Effect of chemical modifications of the pineapple leaf fiber surfaces on the interfacial and mechanical properties of laminated biocomposites. *Compos Interfaces*. 2008; 15 (2–3): 169–191
41. Bledzki AK, Gassan J. Composites reinforced with cellulose based fiber. *Prog Polym Sci*. 1999; 24(2): 221–274.
42. Baley C. Analysis of the flax fiber tensile behaviour and analysis of the tensile stiffness increase. *Compos Part A: Appl Sci Manuf*. 2019; 8: 9–14.
43. Lamy B, Baley C. Stiffness prediction of flax fiber-epoxy composite materials. *J Mater Sci Lett*. 2000 Jun 1; 19: 979–980.
44. ASTM Standard D 3822. Standard test method for tensile properties of single textile fiber. 2020.
45. Davies GC, Bruce DM. Effect of environmental relative humidity and damage on the tensile properties of flax and nettle fiber. *Textile Res J*. 1998; 68: 623–629.
46. Hearle JWS, Susutoglu M. *Interpretation of the mechanical properties of wool fiber*. 1985.
47. Rowell RM, Anand R, Caul DF, Jacobson RE, Gifford O, Drive P. *Utilization of natural fiber in plastic composites: Problems and opportunities*. Department of Forestry, 1630 Linden Drive: 23–52.
48. Bos HL, Donald AM. In situ ESEM study of the deformation of elementary flax fiber. *J Mater Sci*. 1999; 34: 3029–3034.
49. Nguong CW Lee SNB, Sujan D. "A review on natural fibre reinforced. *Polym Compos*. 2013; 7(1).
50. Verma D and Senal I. "6 - Natural fiber-reinforced polymer composites: Feasibiliy study for sustainable automotive industries," in *Biomass, biopolymer-based materials, and bioenergy*, D Verma, E Fortunati, S Jain, X Zhang, Eds., in Woodhead Publishing Series in Composites Science and Engineering. Woodhead Publishing, 2019: 103–122. https://doi.org/10.1016/B978-0-08-102426-3.00006-0
51. Kumar R Ul, Haq MI, Raina A, Anand A. "Industrial applications of natural fibre-reinforced polymer composites – challenges and opportunities." *Int J Sustain Eng*. May 2019; 12(3): 212–220. https://doi.org/10.1080/19397038.2018.1538267
52. Gomez-Campos A et al. "Natural fibre polymer composites - A game changer for the aviation sector?." *J Cleaner Prod*. Mar. 2021; 286: 124986. https://doi.org/10.1016/j.jclepro.2020.124986
53. Peças P, Carvalho H, Salman H, Leite M. "Natural fibre composites and their applications: A review." *J Compos Sci*. Dec. 2018; 2(4): Art. no. 4. https://doi.org/10.3390/jcs2040066

Index

Pages in *italics* refer to figures and pages in **bold** refer to tables.

I

J

K

L

M

N

Q

R

S